PARTICLE DETECTORS

Second Edition

The scope of the detection techniques in particle detectors is very wide, depending on the aim of the measurement. Each physics phenomenon can be used as the basis for a particle detector. Elementary particles have to be identified with various techniques, and relevant quantities like time, energy, and spatial coordinates have to be measured. Particle physics requires extremely high accuracies for these quantities using multi-purpose installations as well as dedicated experimental set-ups. Depending on the aim of the measurement, different effects are used. Detectors cover the measurement of energies from very low energies (micro-electron-volts) to the highest of energies observed in cosmic rays.

Describing the current state-of-the-art instrumentation for experiments in high energy physics and astroparticle physics, this new edition covers track detectors, calorimeters, particle identification, neutrino detectors, momentum measurement, electronics and data analysis. It also discusses up-to-date applications of these detectors in other fields such as nuclear medicine, radiation protection and environmental science. Problem sets have been added to each chapter and additional instructive material has been provided, making this an excellent reference for graduate students and researchers in particle physics. This title, first published in 2008, has been reissued as an Open Access publication on Cambridge Core.

CLAUS GRUPEN is Professor Dr in the Department of Physics at Siegen University. He was awarded the Special High Energy and Particle Physics Prize of the European Physical Society for establishing the existence of the gluon in independent and simultaneous ways, as member of the PLUTO experiment at DESY in 1995.

BORIS SHWARTZ is a Leading Researcher at the Budker Institute of Nuclear Physics. He has worked on the development and construction of the detectors used in several projects, including the KEDR and CMD-2 detectors, and WASA and Belle experiments.

CAMBRIDGE MONOGRAPHS ON PARTICLE PHYSICS, NUCLEAR PHYSICS AND COSMOLOGY 26

General Editors: T. Ericson, P.V. Landshoff

Simulation of a Higgs-boson production in proton–proton interactions in the ATLAS experiment at the Large Hadron Collider (LHC) at CERN. The Higgs decays into a pair of Z bosons, each of which decays in turn into muons pairs. The four muons are indicated by the four straight lines. The hadronic background originates from the interactions of spectator quarks and other interactions in the same beam crossing. (With permission of the CERN photo archive.)

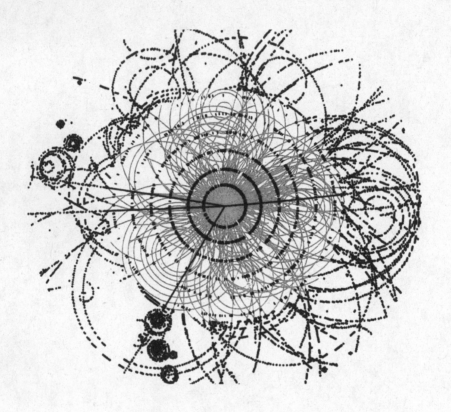

PARTICLE DETECTORS

Second Edition

CLAUS GRUPEN
University of Siegen

BORIS SHWARTZ
Budker Institute of Nuclear Physics, Novosibirsk

with contributions from
HELMUTH SPIELER, *Lawrence Berkeley National Laboratory*
SIMON EIDELMAN, *Budker Institute of Nuclear Physics, Novosibirsk*
TILO STROH, *University of Siegen*

CAMBRIDGE
UNIVERSITY PRESS

CAMBRIDGE
UNIVERSITY PRESS

Shaftesbury Road, Cambridge CB2 8EA, United Kingdom

One Liberty Plaza, 20th Floor, New York, NY 10006, USA

477 Williamstown Road, Port Melbourne, VIC 3207, Australia

314–321, 3rd Floor, Plot 3, Splendor Forum, Jasola District Centre, New Delhi – 110025, India

103 Penang Road, #05-06/07, Visioncrest Commercial, Singapore 238467

Cambridge University Press is part of Cambridge University Press & Assessment,
a department of the University of Cambridge.

We share the University's mission to contribute to society through the pursuit of
education, learning and research at the highest international levels of excellence.

www.cambridge.org
Information on this title: www.cambridge.org/9781009401494

DOI: 10.1017/9781009401531

First published 2008
Reissued as OA 2023

• *A catalogue record for this publication is available from the British Library.*

ISBN 978-1-009-40149-4 Hardback
ISBN 978-1-009-40151-7 Paperback

Contents

Preface to the second edition

Scientific knowledge is a body of statements of varying degrees of certainty – some most unsure, some nearly sure, but none absolutely certain.

Richard Feynman

The book on Particle Detectors was originally published in German ('Teilchendetektoren') with the Bibliographisches Institut Mannheim in 1993. In 1996, it was translated and substantially updated by one of us (Claus Grupen) and published with Cambridge University Press. Since then many new detectors and substantial improvements of existing detectors have surfaced. In particular, the new proton collider under construction at CERN (the Large Hadron Collider LHC), the planning for new detectors at a future electron–positron linear collider, and experiments in astroparticle physics research require a further sophistication of existing and construction of novel particle detectors. With an ever increasing pace of development, the properties of modern detectors allow for high-precision measurements in fields like timing, spatial resolution, energy and momentum resolution, and particle identification.

Already in the past, electron–positron storage rings, like LEP at CERN, have studied electroweak physics and quantum chromodynamics at energies around the electroweak scale ($\approx 100\,\text{GeV}$). The measurement of lifetimes in the region of picoseconds required high spatial resolutions on the order of a few microns. The Large Hadron Collider and the Tevatron at Fermilab will hopefully be able to solve the long-standing question of the generation of masses by finding evidence for particles in the Higgs sector. Also the question of supersymmetry will be addressed by these colliders. Detectors for these enterprises require precision calorimetry and high spatial resolution as well as unanticipated time resolution and extreme selectivity of events, to cope with high backgrounds. Particles in crowded jets have to be identified to allow for the invariant-mass reconstruction of short-lived particles. Radiation hardness is certainly also a hot topic for detectors at hadron colliders.

Particle detection in astroparticle physics also presents a challenge. The origin of the highest-energy cosmic rays, even in spite of recent indications of possible correlations with active galactic nuclei, is still an unsolved problem. Detectors like in the Auger experiment or possibly also the giant IceCube array under construction in Antarctica will very likely find the sources of energetic cosmic rays either in our galaxy or beyond. Also the interaction mechanisms at very high energies, which are inaccessible at present and future accelerators and storage rings, will be attacked by measuring the shape and the elemental composition of the primary cosmic-ray spectrum beyond the expected Greisen cutoff, where energetic protons or nuclei are assumed to lose significant energy, e.g. in proton–photon collisions with the omnipresent blackbody radiation.

These modern developments in the field of particle detection are included in the second edition which is substantially updated compared to the first English edition. Also new results on modern micropattern detectors only briefly mentioned in the first edition and chapters on accelerators and neutrino detectors are included. The chapters on 'Electronics' and 'Data analysis' are completely rewritten.

We would like to mention that excellent books on particle detectors already exist. Without trying to be exhaustive we would like to mention the books of Kleinknecht [1], Fernow [2], Gilmore [3], Sauli [4], Tait [5], Knoll [6], Leo [7], Green [8], Wigmans [9], and Leroy and Rancoita [10]. There are also many excellent review articles in this field published in the literature.

We gratefully acknowledge the help of many colleagues. In particular, we would like to thank Helmuth Spieler for contributing the chapter on 'Electronics'. Archana Sharma has contributed some ideas for micropattern detectors and muon momentum measurement. Steve Armstrong assisted in rewriting the chapter on 'Data analysis'. Iskander Ibragimov very carefully transformed those figures which were recycled from the first edition, where they were just pasted in manually, into an electronic format. He also took care of the labelling of all figures to make them look uniform. T. Tsubo-yama, Richard Wigmans and V. Zhilich provided a number of figures, and A. Buzulutskov and Lev Shekhtman explained us several details concerning microstrip detectors. They also suggested a couple of relevant references. Some useful discussions with A. Bondar, A. Kuzmin, T. Ohshima, A. Vorobiov and M. Yamauchi were very helpful. Simon Eidelman and Tilo Stroh have carefully read the whole book and checked all problems. Tilo Stroh has also taken over the Herculean task to set the text in LaTeX, to improve the figures, to arrange the layout, and prepare a comprehensive index. This was of enormous help to us.

References

[1] K. Kleinknecht, *Detectors for Particle Radiation*, 2nd edition, Cambridge University Press (1998); *Detektoren für Teilchenstrahlung*, Teubner, Wiesbaden (2005)

[2] R. Fernow, *Introduction to Experimental Particle Physics*, Cambridge University Press (1989)

[3] R.S. Gilmore, *Single Particle Detection and Measurement*, Taylor and Francis, London (1992)

[4] F. Sauli (ed.), *Instrumentation in High Energy Physics*, World Scientific, Singapore (1992)

[5] W.H. Tait, *Radiation Detectors*, Butterworths, London (1980)

[6] G.F. Knoll, *Radiation Detection and Measurement*, 3rd edition, John Wiley & Sons Inc., New York (Wiley Interscience), New York (1999/2000)

[7] W.R. Leo *Techniques for Nuclear and Particle Physics Experiments*, Springer, Berlin (1987)

[8] D. Green, *The Physics of Particle Detectors*, Cambridge University Press (2000)

[9] R. Wigmans, *Calorimetry: Energy Measurement in Particle Physics*, Clarendon Press, Oxford (2000)

[10] C. Leroy & P.-G. Rancoita, *Principles of Radiation Interaction in Matter and Detection*, World Scientific, Singapore (November 2004)

Preface to the first edition

The basic motive which drives the scientist to new discoveries and understanding of nature is curiosity. Progress is achieved by carefully directed questions to nature, by experiments. To be able to analyse these experiments, results must be recorded. The most simple instruments are the human senses, but for modern questions, these natural detection devices are not sufficiently sensitive or they have a range which is too limited. This becomes obvious if one considers the human eye. To have a visual impression of light, the eye requires approximately 20 photons. A photomultiplier, however, is able to 'see' single photons. The dynamical range of the human eye comprises half a frequency decade (wavelengths from 400 nm to 800 nm), while the spectrum of electromagnetic waves from domestic current over radio waves, microwaves, infrared radiation, visible light, ultraviolet light, X-rays and gamma rays covers 23 frequency decades!

Therefore, for many questions to nature, precise measurement devices or detectors had to be developed to deliver objective results over a large dynamical range. In this way, the human being has sharpened his 'senses' and has developed new ones. For many experiments, new and special detectors are required and these involve in most cases not only just one sort of measurement. However, a multifunctional detector which allows one to determine all parameters at the same time does not exist yet.

To peer into the world of the microcosm, one needs microscopes. Structures can only be resolved to the size of the wavelength used to observe them; for visible light this is about 0.5 μm. The microscopes of elementary particle physicists are the present day accelerators with their detectors. Because of the inverse proportionality between wavelengths and momentum (de Broglie relation), particles with high momentum allow small structures to be investigated. At the moment, resolutions of the order of 10^{-17} cm can be reached, which is an improvement compared to the optical microscope of a factor of 10^{13}.

To investigate the macrocosm, the structure of the universe, energies in the ranges between some one hundred micro-electron-volts (μeV, cosmic microwave background radiation) up to 10^{20} eV (high energy cosmic rays) must be recorded. To master all these problems, particle detectors are required which can measure parameters like time, energy, momentum, velocity and the spatial coordinates of particles and radiation. Furthermore, the nature of particles must be identified. This can be achieved by a combination of a number of different techniques.

In this book, particle detectors are described which are in use in elementary particle physics, in cosmic ray studies, in high energy astrophysics, nuclear physics, and in the fields of radiation protection, biology and medicine. Apart from the description of the working principles and characteristic properties of particle detectors, fields of application of these devices are also given.

This book originated from lectures which I have given over the past 20 years. In most cases these lectures were titled 'Particle Detectors'. However, also in other lectures like 'Introduction to Radiation Protection', 'Elementary Particle Processes in Cosmic Rays', 'Gamma Ray Astronomy' and 'Neutrino Astronomy', special aspects of particle detectors were described. This book is an attempt to present the different aspects of radiation and particle detection in a comprehensive manner. The application of particle detectors for experiments in elementary particle physics and cosmic rays is, however, one of the main aspects.

I would like to mention that excellent books on particle detectors do already exist. In particular, I want to emphasise the four editions of the book of Kleinknecht [1] and the slightly out-of-date book of Allkofer [2]. But also other presentations of the subject deserve attention [3–25].

Without the active support of many colleagues and students, the completion of this book would have been impossible. I thank Dr U. Schäfer and Dipl. Phys. S. Schmidt for many suggestions and proposals for improvement. Mr R. Pfitzner and Mr J. Dick have carefully done the proof reading of the manuscript. Dr G. Cowan and Dr H. Seywerd have significantly improved my translation of the book into English. I thank Mrs U. Bender, Mrs C. Tamarozzi and Mrs R. Sentker for the production of a ready-for-press manuscript and Mr M. Euteneuer, Mrs C. Tamarozzi as well as Mrs T. Stöcker for the production of the many drawings. I also acknowledge the help of Mr J. Dick, Dipl. Phys.-Ing. K. Reinsch, Dipl. Phys. T. Stroh, Mr R. Pfitzner, Dipl. Phys. G. Gillessen and Mr Cornelius Grupen for their help with the computer layout of the text and the figures.

References

[1]   K. Kleinknecht, *Detektoren für Teilchenstrahlung*, Teubner, Stuttgart (1984, 1987, 1992); *Detectors for Particle Radiation*, Cambridge University Press, Cambridge (1986)

[2]   O.C. Allkofer, *Teilchendetektoren*, Thiemig, München (1971)

[3]   R. Fernow, *Introduction to Experimental Particle Physics*, Cambridge University Press, Cambridge (1989)

[4]   R.S. Gilmore, *Single Particle Detection and Measurement*, Taylor and Francis, London (1992)

[5]   F. Sauli (ed.), *Instrumentation in High Energy Physics*, World Scientific, Singapore (1992)

[6]   W.H. Tait, *Radiation Detectors*, Butterworths, London (1980)

[7]   W.R. Leo, *Techniques for Nuclear and Particle Physics Experiments*, Springer, Berlin (1987)

[8]   P. Rice-Evans, *Spark, Streamer, Proportional and Drift Chambers*, Richelieu Press, London (1974)

[9]   B. Sitar, G.I. Merson, V.A. Chechin & Yu.A. Budagov, *Ionization Measurements in High Energy Physics* (in Russian), Energoatomizdat, Moskau (1988)

[10] B. Sitar, G.I. Merson, V.A. Chechin & Yu.A. Budagov, *Ionization Measurements in High Energy Physics*, Springer Tracts in Modern Physics, Vol. 124, Springer, Berlin/Heidelberg (1993)

[11] T. Ferbel (ed.), *Experimental Techniques in High Energy Nuclear and Particle Physics*, World Scientific, Singapore (1991)

[12] C.F.G. Delaney & E.C. Finch, *Radiation Detectors*, Oxford Science Publications, Clarendon Press, Oxford (1992)

[13] R.C. Fernow, *Fundamental Principles of Particle Detectors*, Summer School on Hadron Spectroscopy, University of Maryland, 1988; BNL-Preprint, BNL-42114 (1988)

[14] G.F. Knoll, *Radiation Detection and Measurement*, John Wiley & Sons Inc., New York (1979)

[15] D.M. Ritson, *Techniques of High Energy Physics*, Interscience Publishers Inc., New York (1961)

[16] K. Siegbahn (ed.), *Alpha, Beta and Gamma-Ray Spectroscopy*, Vols. 1 and 2, Elsevier–North Holland, Amsterdam (1968)

[17] J.C. Anjos, D. Hartill, F. Sauli & M. Sheaff (eds.), *Instrumentation in Elementary Particle Physics*, World Scientific, Singapore (1992)

[18] G. Charpak & F. Sauli, High-Resolution Electronic Particle Detectors, *Ann. Rev. Nucl. Phys. Sci.* **34** (1984) 285–350

[19] W.J. Price, *Nuclear Radiation Detectors*, 2nd edition, McGraw-Hill, New York (1964)

[20] S.A. Korff, *Electron and Nuclear Counters*, 2nd edition, Van Nostrand, Princeton, New Jersey (1955)

[21] H. Neuert, *Kernphysikalische Meßverfahren zum Nachweis für Teilchen und Quanten*, G. Braun, Karlsruhe (1966)

[22] W. Stolz, *Messung ionisierender Strahlung: Grundlagen und Methoden*, Akademie-Verlag, Berlin (1985)

[23] E. Fenyves & O. Haimann, *The Physical Principles of Nuclear Radiation Measurements*, Akadémiai Kiadó, Budapest (1969)

[24] P.J. Ouseph, *Introduction to Nuclear Radiation Detectors*, Plenum Press, New York/London (1975)

[25] C.W. Fabjan, *Detectors for Elementary Particle Physics*, CERN-PPE-94-61 (1994)

Introduction

Every act of seeing leads to consideration, consideration to reflection,
reflection to combination, and thus it may be said that in every
attentive look on nature we already theorise.

Johann Wolfgang von Goethe

The development of particle detectors practically starts with the discovery
of radioactivity by Henri Becquerel in the year 1896. He noticed that
the radiation emanating from uranium salts could blacken photosensitive
paper. Almost at the same time X rays, which originated from materials
after the bombardment by energetic electrons, were discovered by Wilhelm
Conrad Röntgen.

The first nuclear particle detectors (X-ray films) were thus extremely
simple. Also the zinc-sulfide scintillators in use at the beginning of the
last century were very primitive. Studies of scattering processes – e.g. of
α particles – required tedious and tiresome optical registration of scintillation light with the human eye. In this context, it is interesting to note
that Sir William Crookes experimenting in 1903 in total darkness with
a very expensive radioactive material, radium bromide, first saw flashes
of light emitted from the radium salt. He had accidentally spilled a small
quantity of this expensive material on a thin layer of activated zinc sulfide
(ZnS). To make sure he had recovered every single speck of it, he used
a magnifying glass when he noticed emissions of light occurring around
each tiny grain of the radioactive material. This phenomenon was caused
by individual α particles emitted from the radium compound, striking the
activated zinc sulfide. The flashes of light were due to individual photons
caused by the interaction of α particles in the zinc-sulfide screen. A particle detector based on this effect, the *spinthariscope*, is still in use today
for demonstration experiments [1].

xx

Scintillations in the form of 'northern lights' (aurora borealis) had already been observed since long. As early as in 1733 this phenomenon was correctly interpreted as being due to radiation from the Sun (Jean-Jacques D'Ortous De Mairan). Without knowing anything about elementary particles, the atmosphere was realised to be a detector for solar electrons, protons and α particles. Also, already about 50 years before the discovery of *Cherenkov radiation*, Heaviside (1892) showed that charged particles moving faster than light emit an electromagnetic radiation at a certain angle with respect to the particle direction [2]. Lord Kelvin, too, maintained as early as 1901 that the emission of particles was possible at a speed greater than that of light [3, 4]. At the beginning of the twentieth century, in 1919, Madame Curie noticed a faint light emitted from concentrated solutions of radium in water thereby operating unknowingly the first Cherenkov detector. Similarly, Cherenkov radiation in water-cooled reactors or high-intensity radiation sources is fascinating, and sometimes extremely dangerous (e.g. in the Tokaimura nuclear reactor accident) to observe. The human eye can also act as Cherenkov detector, as the light flashes experienced by astronauts during their space mission with eyes closed have shown. These light emissions are caused by energetic primary cosmic rays passing through the vitreous body of the eye.

In the course of time the measurement methods have been greatly refined. Today, it is generally insufficient only to detect particles and radiation. One wants to identify their nature, i.e., one would like to know whether one is dealing, for example, with electrons, muons, pions or energetic γ rays. On top of that, an accurate energy and momentum measurement is often required. For the majority of applications an exact knowledge of the spatial coordinates of particle trajectories is of interest. From this information particle tracks can be reconstructed by means of optical (e.g. in spark chambers, streamer chambers, bubble and cloud chambers) or electronic (in multiwire proportional or drift chambers, micropattern or silicon pixel detectors) detection.

The trend of particle detection has shifted in the course of time from optical measurement to purely electronic means. In this development ever higher resolutions, e.g. of time (picoseconds), spatial reconstruction (micrometres), and energy resolutions (eV for γ rays) have been achieved. Early optical detectors, like cloud chambers, only allowed rates of one event per minute, while modern devices, like fast organic *scintillators*, can process data rates in the GHz regime. With GHz rates also new problems arise and questions of *radiation hardness* and *ageing* of detectors become an issue.

With such high data rates the electronic processing of signals from particle detectors plays an increasingly important rôle. Also the storage of data on magnetic disks or tapes and computer-aided preselection of data is already an integral part of complex detection systems.

Originally, particle detectors were used in cosmic rays and nuclear and particle physics. Meanwhile, these devices have found applications in medicine, biology, environmental science, oil exploration, civil engineering, archaeology, homeland security and arts, to name a few. While the most sophisticated detectors are still developed for particle physics and astroparticles, practical applications often require robust devices which also function in harsh environments.

Particle detectors have contributed significantly to the advancement of science. New detection techniques like cloud chambers, bubble chambers, multiwire proportional and drift chambers, and micropattern detectors allowed essential discoveries. The development of new techniques in this field was also recognised by a number of Nobel Prizes (C.T.R. Wilson, cloud chamber, 1927; P. Cherenkov, I. Frank, I. Tamm, Cherenkov effect, 1958; D. Glaser, bubble chamber, 1960; L. Alvarez, bubble-chamber analysis, 1968; G. Charpak, multiwire proportional chamber, 1992; R. Davis, M. Koshiba, neutrino detection, 2002).

In this book the chapters are ordered according to the object or type of measurement. However, most detectors are highlighted several times. First, their general properties are given, while in other places specific features, relevant to the dedicated subject described in special chapters, are discussed. The ordering principle is not necessarily unique because *solid-state detectors*, for example, in nuclear physics are used to make very precise energy measurements, but as solid-state strip or pixel detectors in elementary particle physics they are used for accurate track reconstruction.

The application of particle detectors in nuclear physics, elementary particle physics, in the physics of cosmic rays, astronomy, astrophysics and astroparticle physics as well as in biology and medicine or other applied fields are weighted in this book in a different manner. The main object of this presentation is the application of particle detectors in elementary particle physics with particular emphasis on modern fast high-resolution detector systems. This also includes astroparticle physics applications and techniques from the field of cosmic rays because these activities are very close to particle physics.

References

[1] http://www.unitednuclear.com/spinthariscope.htm
[2] O. Heaviside, *Electrical papers*, Vol. 2, Macmillan, London (1892) 490–9, 504–18

[3] Lord Kelvin, 'Nineteenth-Century Clouds over the Dynamical Theory of Heat and Light', Lecture to the Royal Institution of Great Britain, London, 27th April 1900; William Thomson, Lord Kelvin, 'Nineteenth-Century Clouds over the Dynamical Theory of Heat and Light', *The London, Edinburgh and Dublin Philosophical Magazine and Journal of Science* **2(6)** (1901) 1–40

[4] Pavel A. Cherenkov, 'Radiation of Particles Moving at a Velocity Exceeding that of Light, and Some of the Possibilities for Their Use in Experimental Physics', Nobel lecture, 11 December, 1958

1

Interactions of particles and radiation with matter

When the intervals, passages, connections, weights, impulses, collisions, movement, order, and position of the atoms interchange, so also must the things formed by them change.

Lucretius

Particles and radiation can be detected only through their interactions with matter. There are specific interactions for charged particles which are different from those of neutral particles, e.g. of photons. One can say that every interaction process can be used as a basis for a detector concept. The variety of these processes is quite rich and, as a consequence, a large number of detection devices for particles and radiation exist. In addition, for one and the same particle, different interaction processes at different energies may be relevant.

In this chapter, the main interaction mechanisms will be presented in a comprehensive fashion. Special effects will be dealt with when the individual detectors are being presented. The interaction processes and their cross sections will not be derived from basic principles but are presented only in their results, as they are used for particle detectors.

The main interactions of charged particles with matter are *ionisation* and *excitation*. For relativistic particles, *bremsstrahlung* energy losses must also be considered. Neutral particles must produce charged particles in an interaction that are then detected via their characteristic interaction processes. In the case of photons, these processes are the photoelectric effect, Compton scattering and pair production of electrons. The electrons produced in these *photon interactions* can be observed through their ionisation in the sensitive volume of the detector.

1

1.1   Interactions of charged particles

Charged particles passing through matter lose kinetic energy by *excitation* of bound electrons and by *ionisation*. Excitation processes like

$$e^- + \text{atom} \rightarrow \text{atom}^* + e^- \tag{1.1}$$
$$\hookrightarrow \text{atom} + \gamma$$

lead to low-energy photons and are therefore useful for particle detectors which can record this luminescence. Of greater importance are pure scattering processes in which incident particles transfer a certain amount of their energy to atomic electrons so that they are liberated from the atom.

The *maximum transferable kinetic energy* to an electron depends on the mass m_0 and the momentum of the incident particle. Given the momentum of the incident particle

$$p = \gamma m_0 \beta c \ , \tag{1.2}$$

where γ is the Lorentz factor ($= E/m_0 c^2$), $\beta c = v$ the velocity, and m_0 the rest mass, the maximum energy that may be transferred to an electron (mass m_e) is given by [1] (see also Problem 1.6)

$$E_{\text{kin}}^{\max} = \frac{2 m_e c^2 \beta^2 \gamma^2}{1 + 2\gamma m_e/m_0 + (m_e/m_0)^2} = \frac{2 m_e p^2}{m_0^2 + m_e^2 + 2 m_e E/c^2} \ . \tag{1.3}$$

In this case, it makes sense to give the kinetic energy, rather than total energy, since the electron is already there and does not have to be produced. The kinetic energy E_{kin} is related to the total energy E according to

$$E_{\text{kin}} = E - m_0 c^2 = c\sqrt{p^2 + m_0^2 c^2} - m_0 c^2 \ . \tag{1.4}$$

For low energies

$$2\gamma m_e/m_0 \ll 1 \tag{1.5}$$

and under the assumption that the incident particles are heavier than electrons ($m_0 > m_e$) Eq. (1.3) can be approximated by

$$E_{\text{kin}}^{\max} \approx 2 m_e c^2 \beta^2 \gamma^2 \ . \tag{1.6}$$

A particle (e.g. a muon, $m_\mu c^2 = 106\,\text{MeV}$) with a Lorentz factor of $\gamma = E/m_0 c^2 = 10$ corresponding to $E = 1.06\,\text{GeV}$ can transfer approximately $100\,\text{MeV}$ to an electron (mass $m_e c^2 = 0.511\,\text{MeV}$).

If one neglects the quadratic term in the denominator of Eq. (1.3), $(m_e/m_0)^2 \ll 1$, which is a good assumption for all incident particles except for electrons, it follows that

$$E_{\text{kin}}^{\max} = \frac{p^2}{\gamma m_0 + m_0^2/2m_e} \cdot \qquad (1.7)$$

For relativistic particles $E_{\text{kin}} \approx E$ and $pc \approx E$ holds. Consequently, the maximum transferable energy is

$$E^{\max} \approx \frac{E^2}{E + m_0^2 c^2/2m_e} \qquad (1.8)$$

which for muons gives

$$E^{\max} = \frac{E^2}{E + 11\,\text{GeV}} \cdot \qquad (1.9)$$

In the extreme relativistic case $\left(E \gg m_0^2 c^2/2m_e\right)$, the total energy can be transferred to the electron.

If the incident particle is an electron, these approximations are no longer valid. In this case, one gets, compare Eq. (1.3),

$$E_{\text{kin}}^{\max} = \frac{p^2}{m_e + E/c^2} = \frac{E^2 - m_e^2 c^4}{E + m_e c^2} = E - m_e c^2 \,, \qquad (1.10)$$

which is also expected in classical non-relativistic kinematics for particles of equal mass for a central collision.

1.1.1 Energy loss by ionisation and excitation

The treatment of the maximum transferable energy has already shown that incident electrons, in contrast to heavy particles $(m_0 \gg m_e)$, play a special rôle. Therefore, to begin with, we give the energy loss for 'heavy' particles. Following Bethe and Bloch [2–8]*, the average energy loss dE per length dx is given by

$$-\frac{\text{d}E}{\text{d}x} = 4\pi N_A r_e^2 m_e c^2 z^2 \frac{Z}{A} \frac{1}{\beta^2} \left(\ln \frac{2m_e c^2 \gamma^2 \beta^2}{I} - \beta^2 - \frac{\delta}{2} \right), \qquad (1.11)$$

---

* For the following considerations and formulae, not only the original literature but also secondary literature was used, mainly [1, 4–12] and references therein.

where

z – charge of the incident particle in units of the elementary charge

Z, A – atomic number and atomic weight of the absorber

m_e – electron mass

r_e – classical electron radius ($r_e = \frac{1}{4\pi\varepsilon_0} \cdot \frac{e^2}{m_e c^2}$ with ε_0 – permittivity of free space)

N_A – Avogadro number ($=$ number of atoms per gram atom) $= 6.022 \cdot 10^{23}\,\mathrm{mol}^{-1}$

I – mean excitation energy, characteristic of the absorber material, which can be approximated by

$$I = 16\,Z^{0.9}\,\mathrm{eV} \quad \text{for } Z > 1 \ .$$

To a certain extent, I also depends on the molecular state of the absorber atoms, e.g. $I = 15\,\mathrm{eV}$ for atomic and $19.2\,\mathrm{eV}$ for molecular hydrogen. For liquid hydrogen, I is $21.8\,\mathrm{eV}$.

δ – is a parameter which describes how much the extended transverse electric field of incident relativistic particles is screened by the charge density of the atomic electrons. In this way, the energy loss is reduced (*density effect*, 'Fermi plateau' of the energy loss). As already indicated by the name, this density effect is important in dense absorber materials. For gases under normal pressure and for not too high energies, it can be neglected.

For energetic particles, δ can be approximated by

$$\delta = 2\ln\gamma + \zeta \ ,$$

where ζ is a material-dependent constant.

Various approximations for δ and material dependences for parameters, which describe the density effect, are discussed extensively in the literature [9]. At very high energies

$$\delta/2 = \ln(\hbar\omega_\mathrm{p}/I) + \ln\beta\gamma - 1/2 \ ,$$

where $\hbar\omega_\mathrm{p} = \sqrt{4\pi N_e r_e^3}\,m_e c^2/\alpha = 28.8\sqrt{\varrho\,\langle Z/A\rangle}\,\mathrm{eV}$ is the plasma energy (ϱ in $\mathrm{g/cm}^3$), N_e the electron density, and α the fine-structure constant.

A useful constant appearing in Eq. (1.11) is

$$4\pi N_A r_e^2 m_e c^2 = 0.3071 \, \frac{\text{MeV}}{\text{g/cm}^2} \, . \tag{1.12}$$

In the logarithmic term of Eq. (1.11), the quantity $2m_e c^2 \gamma^2 \beta^2$ occurs in the numerator, which, according to Eq. (1.6), is identical to the maximum transferable energy. The average energy of electrons produced in the ionisation process in gases equals approximately the ionisation energy [2, 3].

If one uses the approximation for the maximum transferable energy, Eq. (1.6), and the shorthand

$$\kappa = 2\pi N_A r_e^2 m_e c^2 z^2 \cdot \frac{Z}{A} \cdot \frac{1}{\beta^2} \, , \tag{1.13}$$

the *Bethe–Bloch formula* can be written as

$$-\frac{\mathrm{d}E}{\mathrm{d}x} = 2\kappa \left(\ln \frac{E_{\text{kin}}^{\max}}{I} - \beta^2 - \frac{\delta}{2} \right) \, . \tag{1.14}$$

The energy loss $-\mathrm{d}E/\mathrm{d}x$ is usually given in units of $\text{MeV}/(\text{g/cm}^2)$. The length unit $\mathrm{d}x$ (in g/cm^2) is commonly used, because the energy loss per area density

$$\mathrm{d}x = \varrho \cdot \mathrm{d}s \tag{1.15}$$

with ϱ density (in g/cm^3) and $\mathrm{d}s$ length (in cm) is largely independent of the properties of the material. This length unit $\mathrm{d}x$ consequently gives the area density of the material.

Equation (1.11) represents only an approximation for the energy loss of charged particles by ionisation and excitation in matter which is, however, precise at the level of a few per cent up to energies of several hundred GeV. However, Eq. (1.11) cannot be used for slow particles, i.e., for particles which move with velocities which are comparable to those of atomic electrons or slower. For these velocities ($\alpha z \gg \beta \geq 10^{-3}$, $\alpha = \frac{e^2}{4\pi\varepsilon_0\hbar c}$: fine-structure constant) the energy loss is proportional to β. The energy loss of slow protons, e.g. in silicon, can be described by [10–12]

$$-\frac{\mathrm{d}E}{\mathrm{d}x} = 61.2 \, \beta \, \frac{\text{GeV}}{\text{g/cm}^2} \, , \quad \beta < 5 \cdot 10^{-3} \, . \tag{1.16}$$

Equation (1.11) is valid for all velocities

$$\beta \gg \alpha z \, . \tag{1.17}$$

Table 1.1. *Average energy loss of minimum-ionising particles in various materials [10–12]; gases for standard pressure and temperature*

Absorber	$\frac{dE}{dx}\big|_{min} \left[\frac{MeV}{g/cm^2}\right]$	$\frac{dE}{dx}\big|_{min} \left[\frac{MeV}{cm}\right]$
Hydrogen (H_2)	4.10	$0.37 \cdot 10^{-3}$
Helium	1.94	$0.35 \cdot 10^{-3}$
Lithium	1.64	0.87
Beryllium	1.59	2.94
Carbon (Graphite)	1.75	3.96
Nitrogen	1.82	$2.28 \cdot 10^{-3}$
Oxygen	1.80	$2.57 \cdot 10^{-3}$
Air	1.82	$2.35 \cdot 10^{-3}$
Carbon dioxide	1.82	$3.60 \cdot 10^{-3}$
Neon	1.73	$1.56 \cdot 10^{-3}$
Aluminium	1.62	4.37
Silicon	1.66	3.87
Argon	1.52	$2.71 \cdot 10^{-3}$
Titanium	1.48	6.72
Iron	1.45	11.41
Copper	1.40	12.54
Germanium	1.37	7.29
Tin	1.26	9.21
Xenon	1.25	$7.32 \cdot 10^{-3}$
Tungsten	1.15	22.20
Platinum	1.13	24.24
Lead	1.13	12.83
Uranium	1.09	20.66
Water	1.99	1.99
Lucite	1.95	2.30
Shielding concrete	1.70	4.25
Quartz (SiO_2)	1.70	3.74

Given this condition, the energy loss decreases like $1/\beta^2$ in the low-energy domain and reaches a broad minimum of ionisation near $\beta\gamma \approx 4$. Relativistic particles ($\beta \approx 1$), which have an energy loss corresponding to this minimum, are called *minimum-ionising particles* (MIPs). In light absorber materials, where the ratio $Z/A \approx 0.5$, the energy loss of minimum-ionising particles can be roughly represented by

$$-\left.\frac{dE}{dx}\right|_{min} \approx 2\ \frac{MeV}{g/cm^2}\ .\qquad (1.18)$$

In Table 1.1, the energy losses of minimum-ionising particles in different materials are given; for further values, see [10–12].

The energy loss increases again for $\gamma > 4$ (*logarithmic rise* or *relativistic rise*) because of the logarithmic term in the bracket of Eq. (1.11). The increase follows approximately a dependence like $2 \ln \gamma$.

The decrease of the energy loss at the ionisation minimum with increasing atomic number of the absorber originates mainly from the Z/A term in Eq. (1.11). A large fraction of the logarithmic rise relates to large energy transfers to few electrons in the medium (δ *rays* or knock-on electrons). Because of the density effect, the logarithmic rise of the energy loss saturates at high energies.

For heavy projectiles (e.g. like copper nuclei), the energy loss of slow particles is modified because, while being slowed down, electrons get attached to the incident nuclei, thereby decreasing their effective charge.

The energy loss by ionisation and excitation for muons in iron is shown in Fig. 1.1 [10, 11, 13].

The energy loss according to Eq. (1.11) describes only energy losses due to ionisation and excitation. At high energies, radiation losses become more and more important (see Sect. 1.1.5).

Figure 1.2 shows the ionisation energy loss for electrons, muons, pions, protons, deuterons and α particles in air [14].

Equation (1.11) gives only the average energy loss of charged particles by ionisation and excitation. For thin absorbers (in the sense of Eq. (1.15), average energy loss $\langle \Delta E \rangle \ll E_{\max}$), in particular, strong fluctuations around the average energy loss exist. The energy-loss distribution for thin absorbers is strongly asymmetric [2, 3].

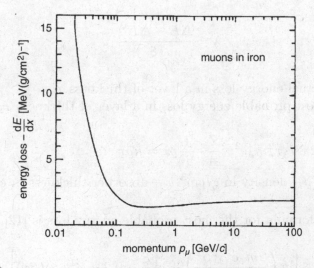

Fig. 1.1. Energy loss by ionisation and excitation for muons in iron and its dependence on the muon momentum.

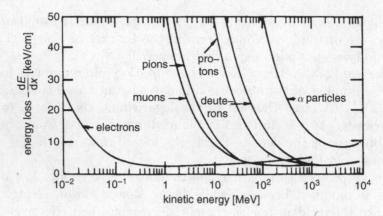

Fig. 1.2. Energy loss for electrons, muons, pions, protons, deuterons and α particles in air [14].

This behaviour can be parametrised by a *Landau distribution*. The Landau distribution is described by the inverse Laplace transform of the function s^s [15–18]. A reasonable approximation of the Landau distribution is given by [19–21]

$$L(\lambda) = \frac{1}{\sqrt{2\pi}} \cdot \exp\left[-\frac{1}{2}(\lambda + e^{-\lambda})\right] \ , \qquad (1.19)$$

where λ characterises the deviation from the *most probable energy loss*,

$$\lambda = \frac{\Delta E - \Delta E^{\mathrm{W}}}{\xi} \ , \qquad (1.20)$$

ΔE – actual energy loss in a layer of thickness x,
ΔE^{W} – most probable energy loss in a layer of thickness x,

$$\xi = 2\pi N_{\mathrm{A}} r_e^2 m_e c^2 z^2 \frac{Z}{A} \cdot \frac{1}{\beta^2} \varrho x = \kappa \varrho x \qquad (1.21)$$

(ϱ – density in g/cm^3, x – absorber thickness in cm).

The general formula for the most probable energy loss is [12]

$$\Delta E^{\mathrm{W}} = \xi \left[\ln\left(\frac{2 m_e c^2 \gamma^2 \beta^2}{I}\right) + \ln\frac{\xi}{I} + 0.2 - \beta^2 - \delta(\beta\gamma)\right] \ . \qquad (1.22)$$

For example, for argon and electrons of energies up to 3.54 MeV from a [106]Rh source the most probable energy loss is [19]

$$\Delta E^{\mathrm{W}} = \xi \left[\ln \left(\frac{2m_e c^2 \gamma^2 \beta^2}{I^2} \xi \right) - \beta^2 + 0.423 \right] \; . \qquad (1.23)$$

The most probable energy loss for minimum-ionising particles ($\beta\gamma = 4$) in 1 cm argon is $\Delta E^{\mathrm{W}} = 1.2\,\mathrm{keV}$, which is significantly smaller than the average energy loss of 2.71 keV [2, 3, 19, 22]. Figure 1.3 shows the energy-loss distribution of 3 GeV electrons in a thin-gap drift chamber filled with Ar/CH$_4$ (80:20) [23].

Experimentally, one finds that the actual energy-loss distribution is frequently broader than represented by the Landau distribution.

For thick absorber layers, the tail of the Landau distribution originating from high energy transfers, however, is reduced [24]. For very thick absorbers ($\frac{\mathrm{d}E}{\mathrm{d}x} \cdot x \gg 2m_e c^2 \beta^2 \gamma^2$), the energy-loss distribution can be approximated by a Gaussian distribution.

The energy loss $\mathrm{d}E/\mathrm{d}x$ in a compound of various elements i is given by

$$\frac{\mathrm{d}E}{\mathrm{d}x} \approx \sum_i f_i \left. \frac{\mathrm{d}E}{\mathrm{d}x} \right|_i \; , \qquad (1.24)$$

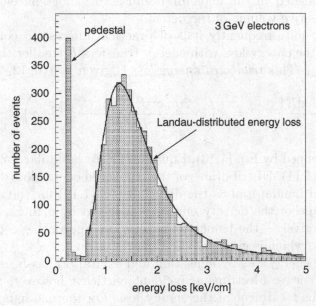

Fig. 1.3. Energy-loss distribution of 3 GeV electrons in a thin-gap drift chamber filled with Ar/CH$_4$ (80:20) [23].

where f_i is the mass fraction of the ith element and $\left.\frac{\mathrm{d}E}{\mathrm{d}x}\right|_i$, the average energy loss in this element. Corrections to this relation because of the dependence of the ionisation constant on the molecular structure can be safely neglected.

The energy transfers to ionisation electrons can be so large that these electrons can cause further ionisation. These electrons are called δ rays or knock-on electrons. The energy spectrum of knock-on electrons is given by [1, 10–12, 25]

$$\frac{\mathrm{d}N}{\mathrm{d}E_{\mathrm{kin}}} = \xi \cdot \frac{F}{E_{\mathrm{kin}}^2} \tag{1.25}$$

for $I \ll E_{\mathrm{kin}} \leq E_{\mathrm{kin}}^{\mathrm{max}}$.

F is a spin-dependent factor of order unity, if $E_{\mathrm{kin}} \ll E_{\mathrm{kin}}^{\mathrm{max}}$ [12]. Of course, the energy spectrum of knock-on electrons falls to zero if the maximum transferable energy is reached. This kinematic limit also constrains the factor F [1, 25]. The spin dependence of the spectrum of the knock-on electrons only manifests itself close to the maximum transferable energy [1, 25].

The strong fluctuations of the energy loss in thin absorber layers are quite frequently not observed by a detector. Detectors only measure the energy which is actually deposited in their sensitive volume, and this energy may not be the same as the energy lost by the particle. For example, the energy which is transferred to knock-on electrons may only be partially deposited in the detector because the knock-on electrons can leave the sensitive volume of the detector.

Therefore, quite frequently it is of practical interest to consider only that part of the energy loss with energy transfers E smaller than a given cut value E_{cut}. This *truncated energy loss* is given by [10–12, 26]

$$-\left.\frac{\mathrm{d}E}{\mathrm{d}x}\right|_{\leq E_{\mathrm{cut}}} = \kappa \left(\ln \frac{2m_e c^2 \beta^2 \gamma^2 E_{\mathrm{cut}}}{I^2} - \beta^2 - \delta \right) , \tag{1.26}$$

where κ is defined by Eq. (1.13). Equation (1.26) is similar, but not identical, to Eq. (1.11). Distributions of the truncated energy loss do not show a pronounced Landau tail as the distributions (1.19) for the mean value (1.11). Because of the density effect – expressed by δ in Eqs. (1.11) or (1.26), respectively – the truncated energy loss approaches a constant at high energies, which is given by the Fermi plateau.

So far, the energy loss by ionisation and excitation has been described for heavy particles. Electrons as incident particles, however, play a special rôle in the treatment of the energy loss. On the one hand, the total energy loss of electrons even at low energies (MeV range) is influenced by bremsstrahlung processes. On the other hand, the ionisation loss requires special treatment because the mass of the incident particle and the target electron is the same.

In this case, one can no longer distinguish between the primary and secondary electron after the collision. Therefore, the energy-transfer probability must be interpreted in a different manner. One electron after the collision receives the energy E_{kin} and the other electron the energy $E - m_e c^2 - E_{\text{kin}}$ (E is the total energy of the incident particle). All possible cases are considered if one allows the energy transfer to vary between 0 and $\frac{1}{2}(E - m_e c^2)$ and not up to $E - m_e c^2$.

This effect can be most clearly seen if in Eq. (1.11) the maximum energy transfer $E_{\text{kin}}^{\text{max}}$ of Eq. (1.6) is replaced by the corresponding expression for electrons. For relativistic particles, the term $\frac{1}{2}(E - m_e c^2)$ can be approximated by $E/2 = \frac{1}{2}\gamma m_e c^2$. Using $z = 1$, the ionisation loss of electrons then can be approximated by

$$-\frac{dE}{dx} = 4\pi N_A r_e^2 m_e c^2 \frac{Z}{A} \cdot \frac{1}{\beta^2} \left(\ln \frac{\gamma m_e c^2}{2I} - \beta^2 - \frac{\delta^*}{2} \right) , \qquad (1.27)$$

where δ^* takes a somewhat different value for electrons compared to the parameter δ appearing in Eq. (1.11). A more precise calculation considering the specific differences between incident heavy particles and electrons yields a more exact formula for the energy loss of electrons due to ionisation and excitation [27],

$$-\frac{dE}{dx} = 4\pi N_A r_e^2 m_e c^2 \frac{Z}{A} \cdot \frac{1}{\beta^2} \left[\ln \frac{\gamma m_e c^2 \beta \sqrt{\gamma - 1}}{\sqrt{2} I} \right.$$
$$\left. + \frac{1}{2}(1 - \beta^2) - \frac{2\gamma - 1}{2\gamma^2} \ln 2 + \frac{1}{16} \left(\frac{\gamma - 1}{\gamma} \right)^2 \right] . \qquad (1.28)$$

This equation agrees with the general Bethe–Bloch relation (1.11) within 10%–20%. It takes into account the kinematics of electron–electron collisions and also screening effects.

The treatment of the ionisation loss of positrons is similar to that of electrons if one considers that these particles are of equal mass, but not identical charge.

For completeness, we also give the ionisation loss of positrons [28]:

$$-\frac{dE}{dx} = 4\pi N_A r_e^2 m_e c^2 \frac{Z}{A} \frac{1}{\beta^2} \left\{ \ln \frac{\gamma m_e c^2 \beta \sqrt{\gamma - 1}}{\sqrt{2} I} \right.$$
$$\left. - \frac{\beta^2}{24} \left[23 + \frac{14}{\gamma + 1} + \frac{10}{(\gamma + 1)^2} + \frac{4}{(\gamma + 1)^3} \right] \right\} . \qquad (1.29)$$

Since positrons are antiparticles of electrons, there is, however, an additional consideration: if positrons come to rest, they will annihilate with an electron normally into two photons which are emitted anticollinearly. Both photons have energies of 511 keV in the centre-of-mass system, corresponding to the rest mass of the electrons. The *cross section for annihilation* in flight is given by [28]

$$\sigma(Z, E) = \frac{Z\pi r_e^2}{\gamma + 1} \left[\frac{\gamma^2 + 4\gamma + 1}{\gamma^2 - 1} \ln(\gamma + \sqrt{\gamma^2 - 1}) - \frac{\gamma + 3}{\sqrt{\gamma^2 - 1}} \right] . \qquad (1.30)$$

More details about the ionisation process of elementary particles, in particular, its spin dependence, can be taken from the books of Rossi and Sitar *et al.* [1–3].

1.1.2 Channelling

The energy loss of charged particles as described by the Bethe–Bloch formula needs to be modified for crystals where the collision partners are arranged on a regular lattice. By looking into a crystal it becomes immediately clear that the energy loss along certain crystal directions will be quite different from that along a non-aligned direction or in an amorphous substance. The motion along such channelling directions is governed mainly by coherent scattering on strings and planes of atoms rather than by the individual scattering off single atoms. This leads to anomalous energy losses of charged particles in crystalline materials [29].

It is obvious from the crystal structure that charged particles can only be channelled along a crystal direction if they are moving more or less parallel to crystal axes. The critical angle necessary for *channelling* is small (approx. 0.3° for $\beta \approx 0.1$) and decreases with energy. For the axial direction ($\langle 111 \rangle$, body diagonal) it can be estimated by

$$\psi \, [\text{degrees}] = 0.307 \cdot [z \cdot Z/(E \cdot d)]^{0.5} , \qquad (1.31)$$

where z and Z are the charges of the incident particle and the crystal atom, E is the particle's energy in MeV, and d is the interatomic spacing in Å. ψ is measured in degrees [30].

For protons ($z = 1$) passing through a silicon crystal ($Z = 14; d = 2.35$ Å), the critical angle for channelling along the direction-of-body diagonals becomes

$$\psi = 13 \, \mu\text{rad}/\sqrt{E \, [\text{TeV}]} . \qquad (1.32)$$

For planar channelling along the face diagonals ($\langle 110 \rangle$ axis) in silicon one gets [29]

$$\psi = 5 \, \mu\text{rad}/\sqrt{E \, [\text{TeV}]} . \qquad (1.33)$$

Of course, the channelling process also depends on the charge of the incident particle.

For a field inside a crystal of silicon atoms along the $\langle 110 \rangle$ crystal direction, one obtains a value of $1.3 \cdot 10^{10}$ V/cm. This field extends over macroscopic distances and can be used for the deflection of high-energy charged particles using bent crystals [30].

Channelled positive particles are kept away from a string of atoms and consequently suffer a relatively small energy loss. Figure 1.4 shows the energy-loss spectra for 15 GeV/c protons passing through a $740\,\mu$m thick germanium crystal [30]. The energy loss of channelled protons is lower by about a factor of 2 compared to random directions through the crystal.

1.1.3 Ionisation yield

The average energy loss by ionisation and excitation can be transformed into a number of electron–ion pairs produced along the track of a charged particle. One must distinguish between *primary ionisation*, that is the number of primarily produced electron–ion pairs, and the *total ionisation*. A sufficiently large amount of energy can be transferred to some primarily produced electrons so that they also can ionise (knock-on electrons). This secondary ionisation together with the primary ionisation forms the total ionisation.

The average energy required to form an electron–ion pair (W value) exceeds the ionisation potential of the gas because, among others, inner shells of the gas atoms can also be involved in the ionisation process,

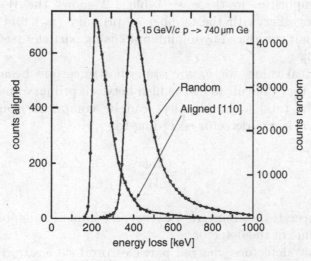

Fig. 1.4. The energy-loss spectra for 15 GeV/c protons passing through a $740\,\mu$m thick germanium crystal [30].

Table 1.2. *Compilation of some properties of gases. Given is the average effective ionisation potential per electron I_0, the average energy loss W per produced ion pair, the number of primary (n_p), and total (n_T) produced electron–ion pairs per cm at standard pressure and temperature for minimum-ionising particles [10, 11, 31–33]*

Gas	Density ϱ [g/cm^3]	I_0 [eV]	W [eV]	n_p [cm^{-1}]	n_T [cm^{-1}]
H_2	$8.99 \cdot 10^{-5}$	15.4	37	5.2	9.2
He	$1.78 \cdot 10^{-4}$	24.6	41	5.9	7.8
N_2	$1.25 \cdot 10^{-3}$	15.5	35	10	56
O_2	$1.43 \cdot 10^{-3}$	12.2	31	22	73
Ne	$9.00 \cdot 10^{-4}$	21.6	36	12	39
Ar	$1.78 \cdot 10^{-3}$	15.8	26	29	94
Kr	$3.74 \cdot 10^{-3}$	14.0	24	22	192
Xe	$5.89 \cdot 10^{-3}$	12.1	22	44	307
CO_2	$1.98 \cdot 10^{-3}$	13.7	33	34	91
CH_4	$7.17 \cdot 10^{-4}$	13.1	28	16	53
C_4H_{10}	$2.67 \cdot 10^{-3}$	10.8	23	46	195

and a fraction of the energy of the incident particle can be dissipated by excitation processes which do not lead to free electrons. The W value of a material is constant for relativistic particles and increases only slightly for low velocities of incident particles.

For gases, the W values are around 30 eV. They can, however, strongly depend on impurities in the gas. Table 1.2 shows the W values for some gases together with the number of primary (n_p) and total (n_T) electron–ion pairs produced by minimum-ionising particles (see Table 1.1) [10, 11, 31–33].

The numerical values for n_p are somewhat uncertain because experimentally it is very difficult to distinguish between primary and secondary ionisation. The total ionisation (n_T) can be computed from the total energy loss ΔE in the detector according to

$$n_T = \frac{\Delta E}{W} \, . \tag{1.34}$$

This is only true if the transferred energy is completely deposited in the sensitive volume of the detector.

In solid-state detectors, charged particles produce *electron–hole pairs*. For the production of an electron–hole pair on the average 3.6 eV in silicon and 2.85 eV in germanium are required. This means that the number of charge carriers produced in solid-state detectors is much larger compared to the production rate of electron–ion pairs in gases. Therefore, the statistical fluctuations in the number of produced charge carriers for a given energy loss is much smaller in solid-state detectors than in gaseous detectors.

The production of pairs of charge carriers for a given energy loss is a statistical process. If, on average, N charge-carrier pairs are produced one would naïvely expect this number to fluctuate according to Poisson statistics with an error of $\sqrt{N}$. Actually, the fluctuation around the average value is smaller by a factor $\sqrt{F}$ depending on the material; this was demonstrated for the first time by Fano [34]. If one considers the situation in detail, the origin of the *Fano factor* is clear. For a given energy deposit, the number of produced charge carriers is limited by energy conservation.

In the following, a formal justification for the Fano factor will be given [34, 35]. Let $E = E_{\text{total}}$ be the fixed energy deposited in a detector, e.g. by an X-ray photon or a stopping α particle. This energy is transferred in p steps to the detector medium, in general, in unequal portions E_p in each individual ionisation process. For each interaction step, m_p electron–ion pairs are produced. After N steps, the total energy is completely absorbed (Fig. 1.5).

Let

$m_p^{(e)} = \frac{E_p}{W}$ be the expected number of ionisations in the step p, and

$\overline{n}^{(e)} = \frac{E}{W}$ be the average expected number of the totally produced electron–ion pairs.

The quantity, which will finally describe the energy resolution, is

$$\sigma^2 = \left\langle (n - \overline{n})^2 \right\rangle \,, \tag{1.35}$$

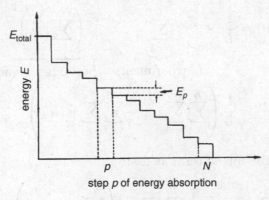

Fig. 1.5. Energy loss in N discrete steps with energy transfer E_p in the pth step [35].

where $\overline{n}$ is the average value over many experiments for fixed energy absorption:

$$\sigma^2 = \frac{1}{L} \sum_{k=1}^{L} (n_k - \overline{n})^2 \ . \tag{1.36}$$

That is, we perform L gedanken experiments, where in experiment k a total number n_k electron–ion pairs is produced. In experiment k the energy is transferred to the detector medium in N_k steps, where in the pth interval the number of produced electron–ion pairs is m_{pk};

$$n_k - \overline{n} = \sum_{p=1}^{N_k} m_{pk} - \frac{E}{W} = \sum_{p=1}^{N_k} m_{pk} - \frac{1}{W} \sum_{p=1}^{N_k} E_{pk} \ . \tag{1.37}$$

The second term in the sum constrains the statistical character of the charge-carrier production rate through energy conservation. Therefore, one would expect that the fluctuations are smaller compared to an unconstrained accidental energy-loss process.

The energy E is subdivided consequently into N_k discrete steps each with energy portion E_{pk}. If we introduce

$$\nu_{pk} = m_{pk} - \frac{E_{pk}}{W} \ , \tag{1.38}$$

it follows that

$$n_k - \overline{n} = \sum_{p=1}^{N_k} \nu_{pk} \ . \tag{1.39}$$

The variance for L experiments is given by

$$\sigma^2(n) = \frac{1}{L} \cdot \underbrace{\sum_{k=1}^{L}}_{L \text{ experiments}} \underbrace{\left(\sum_{p=1}^{N_k} \nu_{pk} \right)^2}_{\text{per experiment}} \ , \tag{1.40}$$

$$\sigma^2(n) = \frac{1}{L} \left(\sum_{k=1}^{L} \sum_{p=1}^{N_k} \nu_{pk}^2 + \sum_{k=1}^{L} \sum_{i \neq j}^{N_k} \nu_{ik} \nu_{jk} \right) \ . \tag{1.41}$$

Let us consider the mixed term at first:

$$\frac{1}{L} \sum_{k=1}^{L} \sum_{i \neq j}^{N_k} \nu_{ik} \nu_{jk} = \frac{1}{L} \sum_{k=1}^{L} \sum_{i=1}^{N_k} \nu_{ik} \left(\sum_{j=1}^{N_k} \nu_{jk} - \nu_{ik} \right) \ . \tag{1.42}$$

The last term in the bracket of Eq. (1.42) originates from the suppression of the product $\nu_{ik}\nu_{jk}$ for $i = j$, which is already contained in the quadratic terms.

For a given event k the average value

$$\overline{\nu}_k = \frac{1}{N_k} \sum_{j=1}^{N_k} \nu_{jk} \qquad (1.43)$$

can be introduced. Using this quantity, one gets

$$\frac{1}{L} \sum_{k=1}^{L} \sum_{i\neq j}^{N_k} \nu_{ik}\nu_{jk} = \frac{1}{L} \sum_{k=1}^{L} N_k \overline{\nu}_k (N_k \overline{\nu}_k - \overline{\nu}_k) \;. \qquad (1.44)$$

In this equation the last term ν_{ik} has been approximated by the average value $\overline{\nu}_k$. Under these conditions one obtains

$$\frac{1}{L} \sum_{k=1}^{L} \sum_{i\neq j}^{N_k} \nu_{ik}\nu_{jk} = \frac{1}{L} \sum_{k=1}^{L} N_k(N_k - 1)\overline{\nu}_k^2 = (\overline{N^2} - \overline{N})\overline{\nu}^2 \;, \qquad (1.45)$$

if one assumes that N_k and $\overline{\nu}_k$ are uncorrelated, and $\overline{\nu}_k = \overline{\nu}$, if N_k is sufficiently large.

The average value of ν, however, vanishes according to Eq. (1.38), consequently the second term in Eq. (1.41) does not contribute. The remaining first term gives

$$\sigma^2(n) = \frac{1}{L} \sum_{k=1}^{L} \sum_{p=1}^{N_k} \nu_{pk}^2 = \frac{1}{L} \sum_{k=1}^{L} N_k \overline{\nu_k^2} = \overline{N\nu^2} = \overline{N} \cdot \overline{(m_p - E_p/W)^2} \;.$$

$$\qquad (1.46)$$

In this case m_p is the actually measured number of electron–ion pairs in the energy-absorption step p with energy deposit E_p.

Remembering that $\overline{N} = \frac{\overline{n}}{\overline{m}_p}$, leads to

$$\sigma^2(n) = \frac{\overline{(m_p - E_p/W)^2}}{\overline{m}_p} \, \overline{n} \;. \qquad (1.47)$$

The variance of n consequently is

$$\sigma^2(n) = F \cdot \overline{n} \qquad (1.48)$$

with the Fano factor

$$F = \frac{\overline{(m_p - E_p/W)^2}}{\overline{m}_p} \;. \qquad (1.49)$$

Table 1.3. *Fano factors for typical detector materials at 300 K [35, 36]*

Absorber	F
$Ar + 10\% \ CH_4$	≈ 0.2
Si	0.12
Ge	0.13
GaAs	0.10
Diamond	0.08

As a consequence, the energy resolution is improved by the factor $\sqrt{F}$ compared to Poisson fluctuations. However, it must be remembered that one has to distinguish between the occasional very large fluctuations of the energy loss (Landau fluctuations) in thin absorber layers and the fluctuation of the number of produced electron–ion pairs for a given fixed well-defined energy loss. This last case is true for all particles which deposit their total energy in the sensitive volume of the detector.

Table 1.3 lists some Fano factors for various substances at 300 K [35, 36]. The improvement on the energy resolution can be quite substantial.

1.1.4   Multiple scattering

A charged particle traversing matter will be scattered by the Coulomb potentials of nuclei and electrons. In contrast to the ionisation energy loss which is caused by collisions with atomic electrons, multiple-scattering processes are dominated by deflections in the Coulomb field of nuclei. This leads to a large number of scattering processes with very low deviations from the original path. The distribution of scattering angles due to multiple *Coulomb scattering* is described by *Molière's theory* [10–12, 37]. For small scattering angles it is normally distributed around the average scattering angle $\Theta = 0$. Larger scattering angles caused by collisions of charged particles with nuclei are, however, more frequent than expected from a Gaussian distribution [38].

The root mean square of the projected *scattering-angle distribution* is given by [10–12]

$$\Theta_{\text{rms}}^{\text{proj.}} = \sqrt{\langle \Theta^2 \rangle} = \frac{13.6 \, \text{MeV}}{\beta cp} z \sqrt{\frac{x}{X_0}} \left[1 + 0.038 \ln(x/X_0) \right] , \qquad (1.50)$$

where p (in MeV/c) is the momentum, βc the velocity, and z the charge of the scattered particle. x/X_0 is the thickness of the scattering medium, measured in units of the radiation length (see Sect. 1.1.5) [1, 39, 40]

$$X_0 = \frac{A}{4\alpha N_A Z^2 r_e^2 \ln(183\, Z^{-1/3})} \, , \tag{1.51}$$

where Z and A are the atomic number and the atomic weight of the absorber, respectively.

Equation (1.50) is already an approximation. For most practical applications Eq. (1.50) can be further approximated for particles with $z = 1$ by

$$\Theta_{\text{rms}}^{\text{proj.}} = \sqrt{\langle \Theta^2 \rangle} \approx \frac{13.6\,\text{MeV}}{\beta c p} \sqrt{\frac{x}{X_0}} \, . \tag{1.52}$$

Equation (1.50) or (1.52) gives the root mean square of the projected distribution of the scattering angles. Such a projected distribution is, for example, of interest for detectors, which provide only a two-dimensional view of an event. The corresponding root mean square deviation for non-projected scattering angles is increased by factor $\sqrt{2}$ so that we have

$$\Theta_{\text{rms}}^{\text{space}} \approx \frac{19.2\,\text{MeV}}{\beta c p} \sqrt{\frac{x}{X_0}} \, . \tag{1.53}$$

1.1.5 Bremsstrahlung

Fast charged particles lose, in addition to their ionisation loss, energy by interactions with the Coulomb field of the nuclei of the traversed medium. If the charged particles are decelerated in the Coulomb field of the nucleus, a fraction of their kinetic energy will be emitted in form of photons (*bremsstrahlung*).

The energy loss by bremsstrahlung for high energies can be described by [1]

$$-\frac{\text{d}E}{\text{d}x} \approx 4\alpha \cdot N_A \cdot \frac{Z^2}{A} \cdot z^2 \left(\frac{1}{4\pi\varepsilon_0} \cdot \frac{e^2}{mc^2} \right)^2 \cdot E \ln\frac{183}{Z^{1/3}} \, . \tag{1.54}$$

In this equation

Z, A – are the atomic number and atomic weight of the medium,

z, m, E – are the charge number, mass and energy of the incident particle.

The bremsstrahlung energy loss of electrons is given correspondingly by

$$-\frac{\mathrm{d}E}{\mathrm{d}x} \approx 4\alpha N_\mathrm{A} \cdot \frac{Z^2}{A} r_e^2 \cdot E \, \ln \frac{183}{Z^{1/3}} \qquad (1.55)$$

if $E \gg m_e c^2 / \alpha Z^{1/3}$.

It should be pointed out that, in contrast to the ionisation energy loss, Eq. (1.11), the energy loss by bremsstrahlung is proportional to the energy of the particle and inversely proportional to the mass squared of the incident particles.

Because of the smallness of the electron mass, bremsstrahlung energy losses play an especially important rôle for electrons. For electrons ($z = 1$, $m = m_e$) Eq. (1.54) or Eq. (1.55), respectively, can be written in the following fashion:

$$-\frac{\mathrm{d}E}{\mathrm{d}x} = \frac{E}{X_0} \, . \qquad (1.56)$$

This equation defines the *radiation length* X_0. An approximation for X_0 has already been given by Eq. (1.51).

The proportionality

$$X_0^{-1} \propto Z^2 \qquad (1.57)$$

in Eq. (1.51) originates from the interaction of the incident particle with the Coulomb field of the target nucleus.

Bremsstrahlung, however, is also emitted in interactions of incident particles with the electrons of the target material. The cross section for this process follows closely the calculation of the bremsstrahlung energy loss on the target nucleus, the only difference being that for atomic target electrons the charge is always equal to unity, and therefore one obtains an additional contribution to the cross section, which is proportional to the number of target electrons, that is $\propto Z$. The cross section for bremsstrahlung must be extended by this term [9]. Therefore, the factor Z^2 in Eq. (1.51) must be replaced by $Z^2 + Z = Z(Z+1)$, which leads to a better description of the radiation length, accordingly,[†]

$$X_0 = \frac{A}{4\alpha N_\mathrm{A} Z(Z+1) r_e^2 \ln(183 \, Z^{-1/3})} \, \{\mathrm{g/cm^2}\} \, . \qquad (1.58)$$

In addition, one has to consider that the atomic electrons will screen the Coulomb field of the nucleus to a certain extent. If *screening effects*

[†] Units presented in curly brackets just indicate that the numerical result of the formula is given in the units shown in the brackets, i.e., in this case the radiation length comes out in g/cm^2.

are taken into account, the radiation length can be approximated by [10–12]

$$X_0 = \frac{716.4 \cdot A[\text{g/mol}]}{Z(Z+1)\ln(287/\sqrt{Z})} \ \text{g/cm}^2 \ . \tag{1.59}$$

The numerical results for the radiation length based on Eq. (1.59) deviate from those of Eq. (1.51) by a few per cent.

The radiation length X_0 is a property of the material. However, one can also define a radiation length for incident particles other than electrons. Because of the proportionality

$$X_0 \propto r_e^{-2} \tag{1.60}$$

and the relation

$$r_e = \frac{1}{4\pi\varepsilon_0} \cdot \frac{e^2}{m_e c^2} \ , \tag{1.61}$$

the 'radiation length', however, also has a dependence on the mass of the incident particle,

$$\tilde{X}_0 \propto m^2 \ . \tag{1.62}$$

The radiation lengths given in the literature, however, are always meant for electrons.

Integrating Eq. (1.54) or (1.56), respectively, leads to

$$E = E_0 \, e^{-x/X_0} \ . \tag{1.63}$$

This function describes the exponential attenuation of the *energy* of charged particles by radiation losses. Note the distinction from the exponential attenuation of the *intensity* of a photon beam passing through matter (see Sect. 1.2, Eq. (1.92)).

The radiation length of a mixture of elements or a compound can be approximated by

$$X_0 = \frac{1}{\sum_{i=1}^{N} f_i/X_0^i} \ , \tag{1.64}$$

where f_i are the mass fractions of the components with the radiation length X_0^i.

Energy losses due to bremsstrahlung are proportional to the energy while ionisation energy losses beyond the minimum of ionisation are proportional to the logarithm of the energy. The energy, where these two interaction processes for electrons lead to equal energy losses, is called the *critical energy E_c*,

$$-\frac{\mathrm{d}E}{\mathrm{d}x}(E_\mathrm{c})\bigg|_{\mathrm{ionisation}} = -\frac{\mathrm{d}E}{\mathrm{d}x}(E_\mathrm{c})\bigg|_{\mathrm{bremsstrahlung}} \qquad . \qquad (1.65)$$

The energy distribution of bremsstrahlung photons follows a $1/E_\gamma$ law (E_γ -- energy of the emitted photon). The photons are emitted preferentially in the forward direction ($\Theta_\gamma \approx m_e c^2/E$). In principle, the critical energy can be calculated from the Eqs. (1.11) and (1.54) using Eq. (1.65). Numerical values for the critical energy of electrons are given in the literature [9–11]. For solids the equation

$$E_\mathrm{c} = \frac{610\,\mathrm{MeV}}{Z + 1.24} \qquad (1.66)$$

describes the critical energies quite satisfactorily [41]. Similar parametrisations for gases, liquids and solids are given in [12]. The critical energy is related to the radiation length by

$$\left(\frac{\mathrm{d}E}{\mathrm{d}x}\right) \cdot X_0 \approx E_\mathrm{c} \quad . \qquad (1.67)$$

Table 1.4 lists the radiation lengths and critical energies for some materials [9–12]. The critical energy – as well as the radiation length – scales as the square of the mass of the incident particles. For muons ($m_\mu = 106\,\mathrm{MeV}/c^2$) in iron one obtains:

$$E_\mathrm{c}^\mu \approx E_\mathrm{c}^e \cdot \left(\frac{m_\mu}{m_e}\right)^2 = 890\,\mathrm{GeV} \quad . \qquad (1.68)$$

1.1.6 Direct electron-pair production

Apart from bremsstrahlung losses, additional energy-loss mechanisms come into play, particularly at high energies. Electron–positron pairs can be produced by virtual photons in the Coulomb field of the nuclei. For high-energy muons this energy-loss mechanism is even more important than bremsstrahlung losses. The energy loss by *trident production* (e.g. like μ+nucleus $\to \mu + e^+ + e^-$ +nucleus) is also proportional to the energy and can be parametrised by

$$-\frac{\mathrm{d}E}{\mathrm{d}x}\bigg|_{\mathrm{pair\ pr.}} = b_\mathrm{pair}(Z, A, E) \cdot E \quad ; \qquad (1.69)$$

Table 1.4. *Radiation lengths and critical energies for some absorber materials [9–12]. The values for the radiation lengths agree with Eq. (1.59) within a few per cent. Only the experimental value for helium shows a somewhat larger deviation. The numerical results for the critical energies of electrons scatter quite significantly in the literature. The effective values for Z and A of mixtures and compounds can be calculated for A by $A_{eff} = \sum_{i=1}^{N} f_i A_i$, where f_i are the mass fractions of the components with atomic weight A_i. Correspondingly, one obtains the effective atomic numbers using Eqs. (1.59) and (1.64). Neglecting the logarithmic Z dependence in Eq. (1.59), Z_{eff} can be calculated from $Z_{eff} \cdot (Z_{eff} + 1) = \sum_{i=1}^{N} f_i Z_i(Z_i + 1)$, where f_i are the mass fractions of the components with charge numbers Z_i. For the practical calculation of an effective radiation length of a compound one determines first the radiation length of the contributing components and then determines the effective radiation length according to Eq. (1.64)*

Material	Z	A	X_0 [g/cm^2]	X_0 [cm]	E_c [MeV]
Hydrogen	1	1.01	61.3	731 000	350
Helium	2	4.00	94	530 000	250
Lithium	3	6.94	83	156	180
Carbon	6	12.01	43	18.8	90
Nitrogen	7	14.01	38	30 500	85
Oxygen	8	16.00	34	24 000	75
Aluminium	13	26.98	24	8.9	40
Silicon	14	28.09	22	9.4	39
Iron	26	55.85	13.9	1.76	20.7
Copper	29	63.55	12.9	1.43	18.8
Silver	47	109.9	9.3	0.89	11.9
Tungsten	74	183.9	6.8	0.35	8.0
Lead	82	207.2	6.4	0.56	7.40
Air	7.3	14.4	37	30 000	84
SiO$_2$	11.2	21.7	27	12	57
Water	7.5	14.2	36	36	83

the $b(Z, A, E)$ parameter varies only slowly with energy for high energies. For 100 GeV muons in iron the energy loss due to *direct electron-pair production* can be described by [25, 42, 43]

$$-\left.\frac{dE}{dx}\right|_{\text{pair pr.}} = 3 \cdot 10^{-6} \cdot \frac{E}{\text{MeV}} \frac{\text{MeV}}{\text{g/cm}^2} , \qquad (1.70)$$

$$\text{i.e.} -\left.\frac{dE}{dx}\right|_{\text{pair pr.}} = 0.3 \frac{\text{MeV}}{\text{g/cm}^2} . \qquad (1.71)$$

The spectrum of total energy of directly produced electron–positron pairs at high energy transfers is steeper than the spectrum of bremsstrahlung photons. High fractional energy transfers are therefore dominated by bremsstrahlung processes [25].

1.1.7 Energy loss by photonuclear interactions

Charged particles can interact inelastically via virtual gauge particles (in this case, photons) with nuclei of the absorber material, thereby losing energy (nuclear interactions).

In the same way as for energy losses through bremsstrahlung or direct electron-pair production, the energy loss by *photonuclear interactions* is proportional to the particle's energy,

$$-\frac{dE}{dx}\bigg|_{\text{photonucl.}} = b_{\text{nucl.}}(Z, A, E) \cdot E \ . \tag{1.72}$$

For $100\,\text{GeV}$ muons in iron the energy-loss parameter b is given by $b_{\text{nucl.}} = 0.4 \cdot 10^{-6}\,\text{g}^{-1}\,\text{cm}^2$ [25], i.e.,

$$-\frac{dE}{dx}\bigg|_{\text{photonucl.}} = 0.04\,\frac{\text{MeV}}{\text{g/cm}^2}\ . \tag{1.73}$$

This energy loss is important for leptons and negligible for hadrons in comparison to direct nuclear interactions.

1.1.8 Total energy loss

In contrast to energy losses due to ionisation those by bremsstrahlung, direct electron-pair production and photonuclear interactions are characterised by large energy transfers with correspondingly large fluctuations. Therefore, it is somewhat problematic to speak of an average energy loss for these processes because extremely large fluctuations around this average value can occur [44, 45].

Nevertheless, the total energy loss of charged particles by the above mentioned processes can be parametrised by

$$-\frac{dE}{dx}\bigg|_{\text{total}} = -\frac{dE}{dx}\bigg|_{\text{ionisation}} -\frac{dE}{dx}\bigg|_{\text{brems.}} -\frac{dE}{dx}\bigg|_{\text{pair pr.}} -\frac{dE}{dx}\bigg|_{\text{photonucl.}}$$
$$= a(Z, A, E) + b(Z, A, E) \cdot E \ , \tag{1.74}$$

where $a(Z, A, E)$ describes the energy loss according to Eq. (1.11) and $b(Z, A, E)$ is the sum over the energy losses due to bremsstrahlung, direct electron-pair production and photonuclear interactions. The parameters

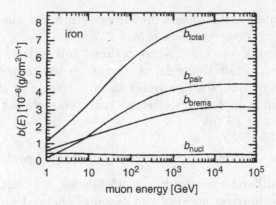

Fig. 1.6. Variation of the *b* parameters with energy for muons in iron. Plotted are the fractional energy losses by direct electron-pair production (b_{pair}), bremsstrahlung (b_{brems}), and photonuclear interactions (b_{nucl}), as well as their sum (b_{total}) [42].

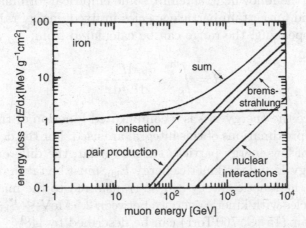

Fig. 1.7. Contributions to the energy loss of muons in iron [42].

a and *b* and their energy dependence for various particles and materials are given in the literature [46].

Figure 1.6 shows the *b parameters* and in Fig. 1.7 the various energy-loss mechanisms for muons in iron in their dependence on the muon energy are presented [42].

Up to energies of several hundred GeV the energy loss in iron due to ionisation and excitation is dominant. For energies in excess of several TeV direct electron-pair production and bremsstrahlung represent the main energy-loss processes. Photonuclear interactions contribute only at the 10% level. Since the energy loss due to these processes is proportional

to the muon's energy, this opens up the possibility of muon calorimetry
by means of energy-loss sampling [47].

The dominance of the energy-proportional interaction processes over
ionisation and excitation depends, of course, on the target material. For
uranium´this transition starts around several 100 GeV, while in hydrogen bremsstrahlung and direct electron-pair production prevail only at
energies in excess of 10 TeV.

1.1.9   Energy–range relations for charged particles

Because of the different energy-loss mechanisms, it is nearly impossible
to give a simple representation of the range of charged particles in matter. The definition of a range is in any case complicated because of the
fluctuations of the energy loss by catastrophic energy-loss processes, i.e.
by interactions with high energy transfers, and because of the multiple
Coulomb scattering in the material, all of which lead to substantial range
straggling. In the following, therefore, some empirical formulae are given,
which are valid for certain particle species in fixed energy ranges.

Generally speaking, the range can be calculated from:

$$R = \int_E^{m_0 c^2} \frac{\mathrm{d}E}{\mathrm{d}E/\mathrm{d}x} \ . \tag{1.75}$$

However, since the energy loss is a complicated function of the energy, in
most cases approximations of this integral are used. For the determination
of the range of low-energy particles, in particular, the difference between
the total energy E and the kinetic energy E_{kin} must be taken into account,
because only the kinetic energy can be transferred to the material.

For α particles with kinetic energies between $2.5\,\mathrm{MeV} \leq E_{\mathrm{kin}} \leq 20\,\mathrm{MeV}$
the range in air (15 °C, 760 Torr) can be described by [48]

$$R_\alpha = 0.31 (E_{\mathrm{kin}}/\mathrm{MeV})^{3/2}\,\mathrm{cm} \ . \tag{1.76}$$

For rough estimates of the range of α particles in other materials one can
use

$$R_\alpha = 3.2 \cdot 10^{-4} \frac{\sqrt{A/(\mathrm{g/mol})}}{\varrho/(\mathrm{g\,cm^{-3}})} \cdot R_{\mathrm{air}} \ \{\mathrm{cm}\} \tag{1.77}$$

(A atomic weight) [48]. The range of α particles in air is shown in Fig. 1.8.

For protons with kinetic energies between $0.6\,\mathrm{MeV} \leq E_{\mathrm{kin}} \leq 20\,\mathrm{MeV}$
the range in air [48] can be approximated by

$$R_p = 100 \cdot \left(\frac{E_{\mathrm{kin}}}{9.3\,\mathrm{MeV}} \right)^{1.8}\,\mathrm{cm} \ . \tag{1.78}$$

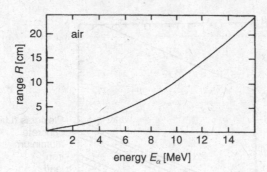

Fig. 1.8.   Range of α particles in air [48].

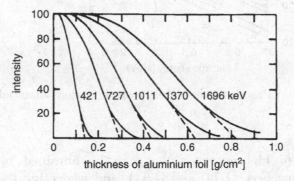

Fig. 1.9.   Absorption of electrons in aluminium [49, 50].

The range of low-energy electrons ($0.5\,\mathrm{MeV} \leq E_{\mathrm{kin}} \leq 5\,\mathrm{MeV}$) in aluminium is described [48] by

$$R_e = 0.526\,(E_{\mathrm{kin}}/\mathrm{MeV} - 0.094)\,\mathrm{g/cm^2}\ . \qquad (1.79)$$

Figure 1.9 shows the absorption of electrons in aluminium [49, 50]. Plotted is the fraction of electrons (with the energy E_{kin}), which penetrate through a certain absorber thickness.

This figure shows the difficulty in the definition of a range of a particle due to the pronounced range straggling, in this case mainly due to the fact that electrons will experience multiple scattering and will bremsstrahl in the absorber. For particles heavier than the electron the range is much better defined due to the reduced effect of multiple scattering ($\langle \Theta^2 \rangle \propto 1/p$). The extrapolation of the linear part of the curves shown in Fig. 1.9 to the intersection with the abscissa defines the *practical range* [50]. The range of electrons defined in this way is shown in Fig. 1.10 for various absorbers [50].

For higher energies the range of muons, pions and protons can be taken from Fig. 1.11 [12].

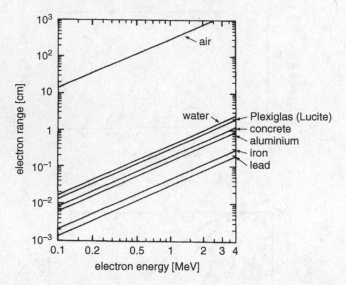

Fig. 1.10.   Practical range of electrons in various materials [50].

The range of high-energy muons can be obtained by integrating Eq. (1.75), using Eqs. (1.74) and (1.11), and neglecting the logarithmic term in Eq. (1.11). This leads to

$$R_\mu(E_\mu) = \frac{1}{b} \ln\left(1 + \frac{b}{a}E_\mu\right) \ . \tag{1.80}$$

For 1 TeV muons in iron Eq. (1.80) yields

$$R_\mu(1\,\text{TeV}) = 265\,\text{m} \ . \tag{1.81}$$

A numerical integration for the range of muons in rock (standard rock with $Z = 11$, $A = 22$) yields for $E_\mu > 10\,\text{GeV}$ [51]

$$R_\mu(E_\mu) = \left[\frac{1}{b}\ln(1 + \frac{b}{a}E_\mu)\right]\left(0.96\frac{\ln E_{\mu,\text{n}} - 7.894}{\ln E_{\mu,\text{n}} - 8.074}\right) \tag{1.82}$$

with $a = 2.2\,\frac{\text{MeV}}{\text{g/cm}^2}$, $b = 4.4 \cdot 10^{-6}\,\text{g}^{-1}\text{cm}^2$ and $E_{\mu,\text{n}} = E_\mu/\text{MeV}$. This energy–range dependence of muons in rock is shown in Fig. 1.12.

1.1.10   Synchrotron-radiation losses

There are further energy-loss processes of charged particles like *Cherenkov radiation, transition radiation* and *synchrotron radiation*. Cherenkov radiation and transition radiation will be discussed in those chapters where

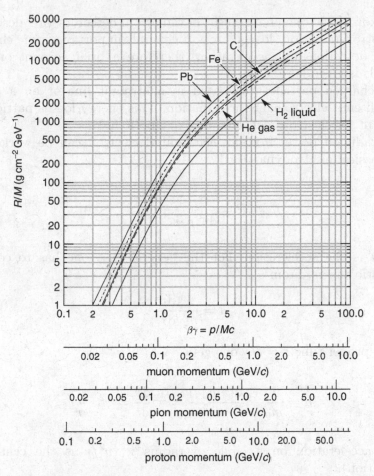

Fig. 1.11. Range of muons, pions and protons in liquid hydrogen, helium gas, carbon and lead [12].

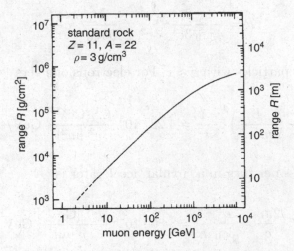

Fig. 1.12. Range of muons in rock [51].

Cherenkov detectors and transition-radiation detectors are described. Synchrotron-radiation losses are of general importance for charged-particle detection and acceleration, therefore a brief account on their essentials is given here.

Any charged particle accelerated in a straight line or on a curved path will emit electromagnetic radiation. This energy loss is particularly important for electrons deflected in a magnetic field.

The radiated power from an accelerated electron can be worked out from classical electrodynamics,

$$P = \frac{1}{4\pi\varepsilon_0}\frac{2e^2}{3c^3}a^2 \ , \tag{1.83}$$

where a is the acceleration. For the general case one has to consider relativistic effects. From

$$a = \frac{1}{m_0}\frac{\mathrm{d}p}{\mathrm{d}\tau} \tag{1.84}$$

and the proper time $\tau = t/\gamma$ one gets

$$a = \frac{1}{m_0}\cdot\gamma\frac{\mathrm{d}(\gamma m_0 v)}{\mathrm{d}t} = \gamma^2\frac{\mathrm{d}v}{\mathrm{d}t} = \gamma^2\cdot\frac{v^2}{r} \tag{1.85}$$

for an acceleration on a circle of radius r (v^2/r is the centrifugal acceleration).

This gives [40, 52]

$$P = \frac{1}{4\pi\varepsilon_0}\frac{2e^2}{3c^3}\gamma^4\frac{v^4}{r^2} = \frac{1}{6\pi\varepsilon_0}e^2 c\frac{\gamma^4}{r^2} \tag{1.86}$$

for relativistic particles with $v \approx c$. For electrons one gets

$$P = \frac{e^2 c}{6\pi\varepsilon_0}\left(\frac{E}{m_e c^2}\right)^4\cdot\frac{1}{r^2} = 4.22\cdot 10^3\frac{E^4\ [\mathrm{GeV}^4]}{r^2\ [\mathrm{m}^2]}\ \mathrm{GeV/s} \ . \tag{1.87}$$

The energy loss per turn in a circular accelerator is

$$\Delta E = P\cdot\frac{2\pi r}{c} = \frac{e^2}{3\varepsilon_0}\frac{\gamma^4}{r} = 8.85\cdot 10^{-5}\frac{E^4\ [\mathrm{GeV}^4]}{r\ [\mathrm{m}]}\ \mathrm{GeV} \ . \tag{1.88}$$

For the Large Electron–Positron collider LEP at CERN with a bending radius in the dipoles of 3100 m one obtains for a beam energy of 100 GeV

$$\Delta E = 2.85 \, \text{GeV per turn} ,$$ (1.89)

while for the Large Hadron Collider LHC for proton beam energies of 7 TeV in the LEP tunnel one has

$$\Delta E = 8.85 \cdot 10^{-5} \cdot \left(\frac{m_e}{m_p}\right)^4 \frac{E^4 \, [\text{GeV}^4]}{r \, [m]} \, \text{GeV} = 6 \cdot 10^{-6} \, \text{GeV} = 6 \, \text{keV} .$$ (1.90)

The emitted synchrotron photons have a broad energy spectrum with a characteristic (critical) energy of

$$E_c = \frac{3c}{2r} \hbar \gamma^3 .$$ (1.91)

They are emitted into a forward cone with opening angle $\propto \frac{1}{\gamma}$. In particular, for electron accelerators the synchrotron-radiation loss is a severe problem for high-energy electrons. Therefore, electron accelerators for $E \gg 100 \, \text{GeV}$ have to be linear instead of circular.

On the other hand, the synchrotron radiation from circular electron machines is used for other fields of physics like solid state or atomic physics, biophysics or medical physics. Here the high *brilliance* of these machines, often augmented by extra bending magnets (*undulators* and *wigglers*) provides excellent opportunities for structure analysis of a large variety of samples. Also the dynamical behaviour of fast biological processes can be investigated.

1.2 Interactions of photons

Photons are detected indirectly via interactions in the medium of the detector. In these processes charged particles are produced which are recorded through their subsequent ionisation in the sensitive volume of the detector. Interactions of photons are fundamentally different from ionisation processes of charged particles because in every photon interaction, the photon is either completely absorbed (*photoelectric effect, pair production*) or scattered through a relatively large angle (*Compton effect*). Since the absorption or scattering is a statistical process, it is impossible to define a range for γ rays. A photon beam is attenuated exponentially in matter according to

$$I = I_0 \, e^{-\mu x} .$$ (1.92)

The *mass attenuation coefficient* μ is related to the cross sections for the various interaction processes of photons according to

$$\mu = \frac{N_A}{A} \sum_i \sigma_i \ , \qquad (1.93)$$

where σ_i is the atomic cross section for the process i, A the atomic weight and N_A the Avogadro number.

The mass attenuation coefficient (according to Eq. (1.93) given per g/cm^2) depends strongly on the photon energy. For low energies $(100\,\text{keV} \geq E_\gamma \geq \text{ionisation energy})$ the photoelectric effect dominates,

$$\gamma + \text{atom} \ \rightarrow \ \text{atom}^+ + e^- \ . \qquad (1.94)$$

In the range of medium energies $(E_\gamma \approx 1\,\text{MeV})$ the Compton effect, which is the scattering of photons off quasi-free atomic electrons,

$$\gamma + e^- \ \rightarrow \ \gamma + e^- \ , \qquad (1.95)$$

has the largest cross section, and at higher energies $(E_\gamma \gg 1\,\text{MeV})$ the cross section for pair production dominates,

$$\gamma + \text{nucleus} \rightarrow \ e^+ + e^- + \text{nucleus} \ . \qquad (1.96)$$

The length x in Eq. (1.92) is an area density with the unit g/cm^2. If the length is measured in cm, the mass attenuation coefficient μ must be divided by the density ϱ of the material.

1.2.1 Photoelectric effect

Atomic electrons can absorb the energy of a photon completely, while – because of momentum conservation – this is not possible for free electrons. The absorption of a photon by an atomic electron requires a third collision partner which in this case is the atomic nucleus. The cross section for absorption of a photon of energy E_γ in the K shell is particularly large ($\approx 80\%$ of the total cross section), because of the proximity of the third collision partner, the atomic nucleus, which takes the recoil momentum. The total photoelectric cross section in the non-relativistic range away from the absorption edges is given in the non-relativistic *Born approximation* by [53]

$$\sigma_{\text{photo}}^{\text{K}} = \left(\frac{32}{\varepsilon^7}\right)^{1/2} \alpha^4 \cdot Z^5 \cdot \sigma_{\text{Th}}^e \ \{\text{cm}^2/\text{atom}\} \ , \qquad (1.97)$$

where $\varepsilon = E_\gamma/m_e c^2$ is the reduced photon energy and $\sigma_{\text{Th}}^e = \frac{8}{3}\pi\, r_e^2 = 6.65 \cdot 10^{-25}\,\text{cm}^2$ is the *Thomson cross section* for elastic scattering of photons on electrons. Close to the absorption edges, the energy dependence of the cross section is modified by a function $f(E_\gamma, E_\gamma^{\text{edge}})$. For higher energies ($\varepsilon \gg 1$) the energy dependence of the cross section for the photoelectric effect is much less pronounced,

$$\sigma_{\text{photo}}^{\text{K}} = 4\pi r_e^2 Z^5 \alpha^4 \cdot \frac{1}{\varepsilon} \;. \tag{1.98}$$

In Eqs. (1.97) and (1.98) the Z dependence of the cross section is approximated by Z^5. This indicates that the photon does not interact with an isolated atomic electron. Z-dependent corrections, however, cause σ_{photo} to be a more complicated function of Z. In the energy range between $0.1\,\text{MeV} \leq E_\gamma \leq 5\,\text{MeV}$ the exponent of Z varies between 4 and 5.

As a consequence of the photoelectric effect in an inner shell (e.g. of the K shell) the following secondary effects may occur. If the free place, e.g. in the K shell, is filled by an electron from a higher shell, the energy difference between those two shells can be liberated in the form of X rays of characteristic energy. The energy of characteristic X rays is given by *Moseley's law*,

$$E = Ry\,(Z-1)^2 \left(\frac{1}{n^2} - \frac{1}{m^2}\right) \;, \tag{1.99}$$

where Ry ($= 13.6\,\text{eV}$) is *Rydberg's constant* and n and m are the principal quantum numbers characterising the atomic shells. For a level transition from the L shell ($m = 2$) to the K shell ($n = 1$) one gets

$$E(\text{K}_\alpha) = \frac{3}{4} Ry\,(Z-1)^2 \;. \tag{1.100}$$

However, this energy difference can also be transferred to an electron of the *same* atom. If this energy is larger than the binding energy of the shell in question, a further electron can leave the atom (Auger effect, *Auger electron*). The energy of these Auger electrons is usually quite small compared to the energy of the primary photoelectrons.

If the photoionisation occurs in the K shell (binding energy B_K), and if the hole in the K shell is filled up by an electron from the L shell (binding energy B_L), the excitation energy of the atom ($B_\text{K} - B_\text{L}$) can be transferred to an L electron. If $B_\text{K} - B_\text{L} > B_\text{L}$, the L electron can leave the atomic shell with an energy $B_\text{K} - 2B_\text{L}$ as an Auger electron.

1.2.2 Compton effect

The Compton effect is the scattering of photons off quasi-free atomic electrons. In the treatment of this interaction process, the binding energy of

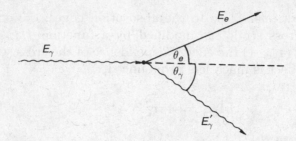

Fig. 1.13.   Definition of kinematic variables in Compton scattering.

the atomic electrons is neglected. The differential probability of Compton scattering $\phi_c(E_\gamma, E'_\gamma)\,\mathrm{d}E'_\gamma$ for $m_ec^2/2 < E'_\gamma < E_\gamma$ is given by the Klein–Nishina formula

$$\phi_c(E_\gamma, E'_\gamma)\,\mathrm{d}E'_\gamma = \pi r_e^2 \frac{N_A Z}{A} \frac{m_e c^2}{E_\gamma} \frac{\mathrm{d}E'_\gamma}{E'_\gamma} \left[1 + \left(\frac{E'_\gamma}{E_\gamma} \right)^2 - \frac{E'_\gamma}{E_\gamma} \sin^2 \theta_\gamma \right],$$

$$(1.101)$$

where θ_γ is the scattering angle of the photon in the laboratory system (see Fig. 1.13) and E_γ, E'_γ are the energies of the incident and scattered photon [54, 55]. The total cross section for Compton scattering per electron is given by [55]

$$\sigma_c^e = 2\pi r_e^2 \left[\left(\frac{1+\varepsilon}{\varepsilon^2} \right) \left\{ \frac{2(1+\varepsilon)}{1+2\varepsilon} - \frac{1}{\varepsilon} \ln(1+2\varepsilon) \right\} + \frac{1}{2\varepsilon} \ln(1+2\varepsilon) \right.$$
$$\left. - \frac{1+3\varepsilon}{(1+2\varepsilon)^2} \right] \{\mathrm{cm}^2/\mathrm{electron}\},$$

$$(1.102)$$

where

$$\varepsilon = \frac{E_\gamma}{m_e c^2}.$$

$$(1.103)$$

The angular and energy distributions of Compton electrons are discussed in great detail in R.D. Evans [56] and G. Hertz [48]. For the energy spectrum of Compton electrons one gets

$$\frac{\mathrm{d}\sigma_c^e}{\mathrm{d}E_{\mathrm{kin}}} = \frac{\mathrm{d}\sigma_c^e}{\mathrm{d}\Omega} \frac{2\pi}{\varepsilon^2 m_e c^2} \left[\frac{(1+\varepsilon)^2 - \varepsilon^2 \cos^2 \theta_e}{(1+\varepsilon)^2 - \varepsilon(2+\varepsilon)\cos^2 \theta_e} \right]^2,$$

$$(1.104)$$

where

$$\frac{\mathrm{d}\sigma_c^e}{\mathrm{d}\Omega} = \frac{r_e^2}{2} \left(\frac{E'_\gamma}{E_\gamma} \right)^2 \left[\frac{E_\gamma}{E'_\gamma} - \frac{E'_\gamma}{E_\gamma} - \sin^2 \theta_\gamma \right].$$

$$(1.105)$$

For Compton scattering off atoms the cross section is increased by the factor Z, because there are exactly Z electrons as possible scattering partners in an atom; consequently $\sigma_c^{\text{atomic}} = Z \cdot \sigma_c^e$.

At high energies the energy dependence of the Compton-scattering cross section can be approximated by [57]

$$\sigma_c^e \propto \frac{\ln \varepsilon}{\varepsilon} \; . \tag{1.106}$$

The ratio of scattered to incident photon energy is given by

$$\frac{E_\gamma'}{E_\gamma} = \frac{1}{1 + \varepsilon(1 - \cos \theta_\gamma)} \; . \tag{1.107}$$

For *backscattering* ($\theta_\gamma = \pi$) the energy transfer to the electron reaches a maximum value, leading to a ratio of scattered to incident photon energy of

$$\frac{E_\gamma'}{E_\gamma} = \frac{1}{1 + 2\varepsilon} \; . \tag{1.108}$$

The scattering angle of the electron with respect to the direction of the incident photon can be obtained from (see Problem 1.5)

$$\cot \theta_e = (1 + \varepsilon) \tan \frac{\theta_\gamma}{2} \; . \tag{1.109}$$

Because of momentum conservation the scattering angle of the electron, θ_e, can never exceed $\pi/2$.

In Compton-scattering processes only a fraction of the photon energy is transferred to the electron. Therefore, one defines an energy scattering cross section

$$\sigma_{\text{cs}} = \frac{E_\gamma'}{E_\gamma} \cdot \sigma_c^e \tag{1.110}$$

and subsequently an energy-absorption cross section

$$\sigma_{\text{ca}} = \sigma_c^e - \sigma_{\text{cs}} \; . \tag{1.111}$$

The latter is relevant for absorption processes and is related to the probability that an energy $E_{\text{kin}} = E_\gamma - E_\gamma'$ is transferred to the target electron.

In passing, it should be mentioned that in addition to the normal Compton scattering of photons on target electrons at rest, *inverse Compton scattering* also exists. In this case, an energetic electron collides with a low-energy photon and transfers a fraction of its kinetic energy to the photon which is blueshifted to higher frequencies. This inverse Compton-scattering process plays an important rôle, e.g. in astrophysics. Starlight photons (eV range) can be shifted in this way by collisions with energetic electrons into the X-ray (keV) or gamma (MeV) range. Laser photons backscattered from high-energy electron beams also provide energetic γ beams which are used in accelerator experiments [58].

Naturally, Compton scattering does not only occur with electrons, but also for other charged particles. For the measurement of photons in particle detectors, however, Compton scattering off atomic electrons is of special importance.

1.2.3   Pair production

The production of electron–positron pairs in the Coulomb field of a nucleus is only possible if the photon energy exceeds a certain threshold. This threshold energy is given by the rest masses of two electrons plus the recoil energy which is transferred to the nucleus. From energy and momentum conservation, this threshold energy can be calculated to be

$$E_\gamma \geq 2m_e c^2 + 2\frac{m_e^2}{m_{\text{nucleus}}} c^2 \; . \tag{1.112}$$

Since $m_{\text{nucleus}} \gg m_e$, the effective threshold can be approximated by

$$E_\gamma \geq 2m_e c^2 \; . \tag{1.113}$$

If, however, the electron–positron pair production proceeds in the Coulomb field of an electron, the threshold energy is

$$E_\gamma \geq 4m_e c^2 \; . \tag{1.114}$$

Electron–positron pair production in the Coulomb field of an electron is, however, strongly suppressed compared to pair production in the Coulomb field of the nucleus.

In the case that the nuclear charge is not screened by atomic electrons, (for low energies the photon must come relatively close to the nucleus to make pair production probable, which means that the photon sees only the 'naked' nucleus),

$$1 \ll \varepsilon < \frac{1}{\alpha Z^{1/3}} \; , \tag{1.115}$$

the pair-production cross section is given by [1]

$$\sigma_{\text{pair}} = 4\alpha r_e^2 Z^2 \left(\frac{7}{9} \ln 2\varepsilon - \frac{109}{54} \right) \; \{\text{cm}^2/\text{atom}\} \; ; \tag{1.116}$$

for complete screening of the nuclear charge, however, $\left(\varepsilon \gg \frac{1}{\alpha Z^{1/3}}\right)$ [1]

$$\sigma_{\text{pair}} = 4\alpha r_e^2 Z^2 \left(\frac{7}{9} \ln \frac{183}{Z^{1/3}} - \frac{1}{54}\right) \{\text{cm}^2/\text{atom}\} \ . \qquad (1.117)$$

(At high energies pair production can also proceed at relatively large impact parameters of the photon with a respect to the nucleus. But in this case the screening of the nuclear charge by the atomic electrons must be taken into account.)

For large photon energies, the pair-production cross section approaches an energy-independent value which is given by Eq. (1.117). Neglecting the small term $\frac{1}{54}$ in the bracket of this equation, this asymptotic value is given by

$$\sigma_{\text{pair}} \approx \frac{7}{9} \, 4\alpha \, r_e^2 Z^2 \ln \frac{183}{Z^{1/3}} \approx \frac{7}{9} \cdot \frac{A}{N_A} \cdot \frac{1}{X_0} \ , \qquad (1.118)$$

see Eq. (1.51).

The partition of the energy between the produced electrons and positrons is uniform at low and medium energies and becomes slightly asymmetric at large energies. The differential cross section for the creation of a positron of total energy between E_+ and $E_+ + dE_+$ with an electron of total energy E_- is given by [53]

$$\frac{d\sigma_{\text{pair}}}{dE_+} = \frac{\alpha r_e^2}{E_\gamma - 2m_e c^2} \cdot Z^2 \cdot f(\varepsilon, Z) \ \{\text{cm}^2/(\text{MeV} \cdot \text{atom})\} \ . \qquad (1.119)$$

$f(\varepsilon, Z)$ is a dimensionless, non-trivial function of ε and Z. The trivial Z^2 dependence of the cross section is, of course, already considered in a factor separated from $f(\varepsilon, Z)$. Therefore, $f(\varepsilon, Z)$ depends only weakly (logarithmically) on the atomic number of the absorber, see Eq. (1.117). $f(\varepsilon, Z)$ varies with Z only by few per cent [14]. The dependence of this function on the *energy-partition parameter*

$$x = \frac{E_+ - m_e c^2}{E_\gamma - 2m_e c^2} = \frac{E_+^{\text{kin}}}{E_{\text{pair}}^{\text{kin}}} \qquad (1.120)$$

for average Z values is shown in Fig. 1.14 for various parameters ε [14, 59, 60]. The curves shown in Fig. 1.14 do not just include the pair production on the nucleus, but also the pair-production probability on atomic electrons ($\propto Z$), so that the Z^2 dependence of the pair-production cross section, Eq. (1.119), is modified to $Z(Z+1)$ in a similar way as was argued when the electron-bremsstrahlung process was presented, see Eq. (1.58). The angular distribution of the produced electrons is quite narrow with a characteristic opening angle of $\Theta \approx m_e c^2/E_\gamma$.

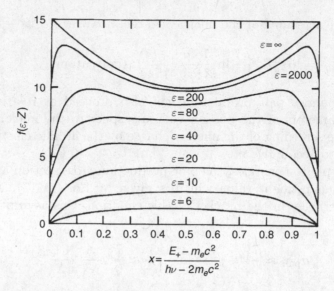

Fig. 1.14.   Form of the energy-partition function $f(\varepsilon, Z, x)$ with $\varepsilon = E_\gamma/m_e c^2$ as parameter. The total pair-production cross section is given by the area under the corresponding curve in units of $Z(Z+1)\alpha r_e^2$ [14, 59, 60].

1.2.4   Total photon absorption cross section

The total mass attenuation coefficient, which is related to the cross sections according to Eq. (1.93), is shown in Figs. 1.15–1.18 for the absorbers water, air, aluminium and lead [48, 56, 61, 62].

Since Compton scattering plays a special rôle for photon interactions, because only part of the photon energy is transferred to the target electron, one has to distinguish between the *mass attenuation coefficient* and the *mass absorption coefficient*. The mass attenuation coefficient μ_{cs} is related to the Compton-energy scattering cross section σ_{cs}, see Eq. (1.110), according to Eq. (1.93). Correspondingly, the mass absorption coefficient μ_{ca} is calculated from the energy absorption cross section σ_{ca}, Eq. (1.111) and Eq. (1.93). For various absorbers the Compton-scattering cross sections, or absorption coefficients shown in Figs. 1.15–1.18, have been multiplied by the atomic number of the absorber, since the Compton-scattering cross section, Eq. (1.102), given by the Klein–Nishina formula is valid per electron, but in this case, the atomic cross sections are required.

Ranges in which the individual photon interaction processes dominate, are plotted in Fig. 1.19 as a function of the photon energy and the atomic number of the absorber [14, 50, 53].

Further interactions of photons (photonuclear interactions, *photon–photon scattering*, etc.) are governed by extremely low cross sections.

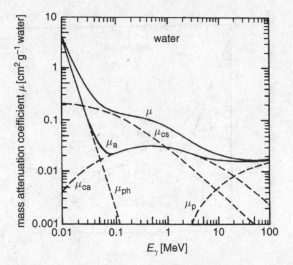

Fig. 1.15. Energy dependence of the mass attenuation coefficient μ and mass absorption coefficient μ_a for photons in water [48, 56, 61, 62]. μ_{ph} describes the photoelectric effect, μ_{cs} the Compton scattering, μ_{ca} the Compton absorption and μ_p the pair production. μ_a is the total mass absorption coefficient ($\mu_a = \mu_{ph} + \mu_p + \mu_{ca}$) and μ is the total mass attenuation coefficient ($\mu = \mu_{ph} + \mu_p + \mu_c$, where $\mu_c = \mu_{cs} + \mu_{ca}$).

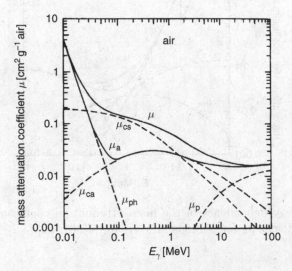

Fig. 1.16. Energy dependence of the mass attenuation coefficient μ and mass absorption coefficient μ_a for photons in air [48, 56, 61, 62].

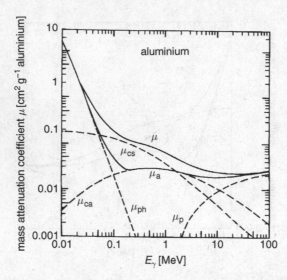

Fig. 1.17. Energy dependence of the mass attenuation coefficient μ and mass absorption coefficient μ_a for photons in aluminium [48, 56, 61, 62].

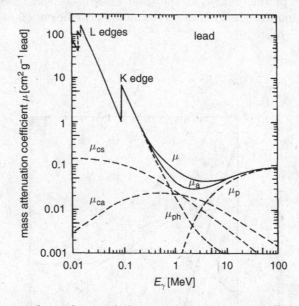

Fig. 1.18. Energy dependence of the mass attenuation coefficient μ and mass absorption coefficient μ_a for photons in lead [48, 56, 61, 62].

Therefore, these processes are of little importance for the detection of photons. However, these processes are of large interest in elementary particle physics and particle astrophysics.

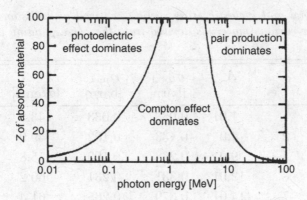

Fig. 1.19. Ranges in which the photoelectric effect, Compton effect and pair production dominate as a function of the photon energy and the target charge number Z [14, 50, 53].

1.3 Strong interactions of hadrons

Apart from the electromagnetic interactions of charged particles strong interactions may also play a rôle for particle detection. In the following we will sketch the strong interactions of hadrons.

In this case, we are dealing mostly with inelastic processes, where secondary strongly interacting particles are produced in the collision. The total cross section for proton–proton scattering can be approximated by a constant value of $50\,\mathrm{mb}$ ($1\,\mathrm{mb} = 10^{-27}\,\mathrm{cm}^2$) for energies ranging from $2\,\mathrm{GeV}$ to $100\,\mathrm{TeV}$. Both the elastic and inelastic part of the cross section show a rather strong energy dependence at low energies [12, 63],

$$\sigma_\text{total} = \sigma_\text{elastic} + \sigma_\text{inel} \ . \tag{1.121}$$

The specific quantity that characterises the inelastic processes is the average *interaction length* λ_I, which describes the absorption of hadrons in matter according to

$$N = N_0\, \mathrm{e}^{-x/\lambda_\mathrm{I}} \ . \tag{1.122}$$

The value of λ_I can be calculated from the inelastic part of the hadronic cross section as follows:

$$\lambda_\mathrm{I} = \frac{A}{N_\mathrm{A} \cdot \varrho \cdot \sigma_\text{inel}} \ . \tag{1.123}$$

If A is given in g/mol, N_A in mol^{-1}, ϱ in $\mathrm{g/cm}^3$ and the cross section in cm^2, then λ_I has the unit cm. The area density corresponding to $\lambda_\mathrm{I}\ \{\mathrm{cm}\}$

Table 1.5.   *Total and inelastic cross sections as well as collision and interaction lengths for various materials derived from the corresponding cross sections [10–12]*

Material	Z	A	σ_{total} [barn]	σ_{inel} [barn]	$\lambda_T \cdot \varrho$ [g/cm^2]	$\lambda_I \cdot \varrho$ [g/cm^2]
Hydrogen	1	1.01	0.0387	0.033	43.3	50.8
Helium	2	4.0	0.133	0.102	49.9	65.1
Beryllium	4	9.01	0.268	0.199	55.8	75.2
Carbon	6	12.01	0.331	0.231	60.2	86.3
Nitrogen	7	14.01	0.379	0.265	61.4	87.8
Oxygen	8	16.0	0.420	0.292	63.2	91.0
Aluminium	13	26.98	0.634	0.421	70.6	106.4
Silicon	14	28.09	0.660	0.440	70.6	106.0
Iron	26	55.85	1.120	0.703	82.8	131.9
Copper	29	63.55	1.232	0.782	85.6	134.9
Tungsten	74	183.85	2.767	1.65	110.3	185
Lead	82	207.19	2.960	1.77	116.2	194
Uranium	92	238.03	3.378	1.98	117.0	199

would be $\lambda_I \cdot \varrho$ {g/cm^2}. The *collision length* λ_T is related to the total cross section σ_{total} according to

$$\lambda_T = \frac{A}{N_A \cdot \varrho \cdot \sigma_{total}} \; . \tag{1.124}$$

Since $\sigma_{total} > \sigma_{inel}$, it follows that $\lambda_T < \lambda_I$.

The interaction and collision lengths for various materials are given in Table 1.5 [10–12].

Strictly speaking, the hadronic cross sections depend on the energy and vary somewhat for different strongly interacting particles. For the calculation of the interaction and collision lengths, however, the cross sections σ_{inel} and σ_{total} have been assumed to be energy independent and independent of the particle species (protons, pions, kaons, etc.).

For target materials with $Z \geq 6$ the interaction and collision lengths, respectively, are much larger than the radiation lengths X_0 (compare Table 1.4).

The definitions for λ_I and λ_T are not uniform in the literature.

The cross sections can be used to calculate the probabilities for interactions in a simple manner. If σ_N is the nuclear-interaction cross section (i.e. per nucleon), the corresponding probability for an interaction per g/cm^2 is calculated to be

$$\phi\{g^{-1}\,cm^2\} = \sigma_N \cdot N_A\,[mol^{-1}]/g \ , \tag{1.125}$$

where N_A is Avogadro's number. In the case that the atomic cross section σ_A is given, it follows that

$$\phi\{g^{-1}\,cm^2\} = \sigma_A \cdot \frac{N_A}{A} \ , \tag{1.126}$$

where A is the atomic weight.

1.4 Drift and diffusion in gases[‡]

Electrons and ions, produced in an ionisation process, quickly lose their energy by multiple collisions with atoms and molecules of a gas. They approach a thermal energy distribution, corresponding to the temperature of the gas.

Their average energy at room temperature is

$$\varepsilon = \frac{3}{2}kT = 40\,meV \ , \tag{1.127}$$

where k is the Boltzmann constant and T the temperature in Kelvin. They follow a Maxwell–Boltzmann distribution of energies like

$$F(\varepsilon) = const \cdot \sqrt{\varepsilon} \cdot e^{-\varepsilon/kT} \ . \tag{1.128}$$

The locally produced ionisation diffuses by multiple collisions corresponding to a Gaussian distribution

$$\frac{dN}{N} = \frac{1}{\sqrt{4\pi Dt}}\exp\left(-\frac{x^2}{4Dt}\right)dx \ , \tag{1.129}$$

where $\frac{dN}{N}$ is the fraction of the charge which is found in the length element dx at a distance x after a time t. D is the diffusion coefficient. For linear or volume diffusion, respectively, one obtains

$$\sigma_x = \sqrt{2Dt} \ , \tag{1.130}$$

$$\sigma_{vol} = \sqrt{3} \cdot \sigma_x = \sqrt{6Dt} \ . \tag{1.131}$$

[‡] Extensive literature to these processes is given in [2, 3, 12, 31, 32, 64–70].

Table 1.6. *Average mean free path λ_{ion}, diffusion constant D_{ion} and mobilities μ_{ion} of ions in some gases for standard pressure and temperature [32, 71]*

Gas	λ_{ion} [cm]	D_{ion} [cm^2/s]	$\mu_{\mathrm{ion}} \left[\frac{\mathrm{cm/s}}{\mathrm{V/cm}} \right]$
H$_2$	$1.8 \cdot 10^{-5}$	0.34	13.0
He	$2.8 \cdot 10^{-5}$	0.26	10.2
Ar	$1.0 \cdot 10^{-5}$	0.04	1.7
O$_2$	$1.0 \cdot 10^{-5}$	0.06	2.2

The *average mean free path* in the diffusion process is

$$\lambda = \frac{1}{N\sigma(\varepsilon)} \ , \tag{1.132}$$

where $\sigma(\varepsilon)$ is the energy-dependent collision cross section, and $N = \frac{N_A}{A}\varrho$ the number of molecules per unit volume. For noble gases one has $N = 2.69 \cdot 10^{19}$ molecules/cm^3 at standard pressure and temperature.

If the charge carriers are exposed to an electric field, an ordered drift along the field will be superimposed over the statistically disordered diffusion. A drift velocity can be defined according to

$$\vec{v}_{\mathrm{drift}} = \mu(E) \cdot \vec{E} \cdot \frac{p_0}{p} \ , \tag{1.133}$$

where

$\mu(E)$ – energy-dependent *charge-carrier mobility*,

$\vec{E}$ – electric field strength, and

p/p_0 – pressure normalised to standard pressure.

The statistically disordered transverse diffusion, however, is not influenced by the electric field.

The drift of free charge carriers in an electric field requires, however, that electrons and ions do not recombine and that they are also not attached to atoms or molecules of the medium in which the drift proceeds.

Table 1.6 contains numerical values for the average mean free path, the diffusion constant and the mobilities of ions [32, 71]. The mobility of ions does not depend on the field strength. It varies inversely proportional to the pressure, i.e. $\mu \cdot p \approx$ const [72, 73].

The corresponding quantity for electrons strongly depends on the energy of the electrons and thereby on the field strength. The mobilities

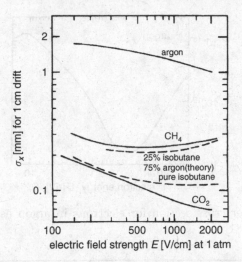

Fig. 1.20. Dependence of the root-mean-square deviation of an originally localised electron cloud after a drift of 1 cm in various gases [32, 74].

of electrons in gases exceed those of ions by approximately three orders of magnitude.

Figure 1.20 shows the root-mean-square deviation of an originally localised electron cloud for a drift of 1 cm [32, 74]. The width of the electron cloud $\sigma_x = \sqrt{2Dt}$ per 1 cm drift varies significantly with the field strength and shows characteristic dependences on the gas. For a gas mixture of argon (75%) and isobutane (25%) values around $\sigma_x \approx 200\,\mu$m are measured, which limit the spatial resolution of drift chambers. In principle, one has to distinguish between the *longitudinal diffusion* in the direction of the field and a *transverse diffusion* perpendicular to the electric field. The spatial resolution of drift chambers, however, is limited primarily by the longitudinal diffusion.

In a simple theory [75] the electron *drift velocity* can be expressed by

$$\vec{v}_{\text{drift}} = \frac{e}{m}\vec{E}\,\tau(\vec{E},\varepsilon) \;, \tag{1.134}$$

where $\vec{E}$ is the field strength and τ the time between two collisions, which in itself depends on $\vec{E}$. The collision cross section, and as a consequence also τ, depends strongly on the electron energy ε and passes through pronounced maxima and minima (*Ramsauer effect*). These phenomena are caused by interference effects, if the electron wavelength $\lambda = h/p$ (h – Planck's constant, p – electron momentum) approaches molecular dimensions. Of course, the electron energy and electric field strength are correlated. Figure 1.21 shows the Ramsauer cross section for electrons in argon as a function of the electron energy [76–81].

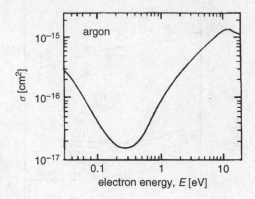

Fig. 1.21. Ramsauer cross section for electrons in argon as a function of the electron energy [76–81].

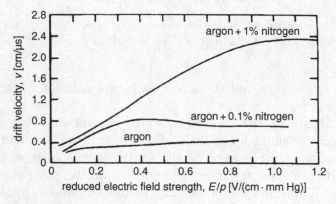

Fig. 1.22. Drift velocities of electrons in pure argon and in argon with minor additions of nitrogen [32, 76, 82, 83].

Even small contaminations of a gas can drastically modify the drift velocity (Fig. 1.22 [32, 76, 82, 83]).

Figure 1.23 shows the drift velocities for electrons in argon–methane mixtures [32, 84–86] and Fig. 1.24 those in argon–isobutane mixtures [32, 85, 87–89].

As an approximate value for high field strengths in argon–isobutane mixtures a typical value for the drift velocity of

$$v_{\mathrm{drift}} = 5\,\mathrm{cm/\mu s} \tag{1.135}$$

is observed. The dependence of the drift velocity on the field strength, however, may vary considerably for different gases [69, 85, 90]. Under comparable conditions the ions in a gas are slower by three orders of magnitude compared to electrons.

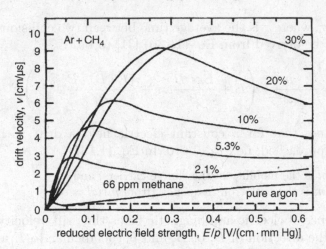

Fig. 1.23. Drift velocities for electrons in argon–methane mixtures [32, 84–86]. The percentage of methane is indicated on the curves.

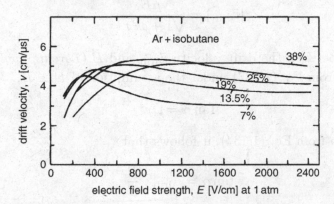

Fig. 1.24. Drift velocities for electrons in argon–isobutane mixtures [32, 85, 87–89]. The percentage of isobutane is indicated on the curves.

The drift velocity and, in general, the drift properties of electrons in gases are strongly modified in the presence of a magnetic field. In addition to the electric force, now the *Lorentz force* also acts on the charge carriers and forces the charge carriers into circular or spiral orbits.

The equation of motion for the free charge carriers reads

$$m\ddot{\vec{x}} = q\vec{E} + q \cdot \vec{v} \times \vec{B} + m\vec{A}(t) \; , \qquad (1.136)$$

where $m\vec{A}(t)$ is a time-dependent stochastic force, which has its origin in collisions with gas molecules. If one assumes that the time average of the product $m \cdot \vec{A}(t)$ can be represented by a velocity-proportional friction force $-m\vec{v}/\tau$, where τ is the average time between two collisions, the drift velocity can be derived from Eq. (1.136) [31] to be

$$\vec{v}_{\text{drift}} = \frac{\mu}{1 + \omega^2\tau^2}\left(\vec{E} + \frac{\vec{E} \times \vec{B}}{B}\omega\tau + \frac{(\vec{E} \cdot \vec{B}) \cdot \vec{B}}{B^2}\omega^2\tau^2\right) , \qquad (1.137)$$

if one assumes that for a constant electric field a drift with constant velocity is approached, i.e., $\dot{\vec{v}}_{\text{drift}} = 0$. In Eq. (1.137)

$\mu = e \cdot \tau/m$ is the mobility of the charge carriers, and
$\omega = e \cdot B/m$ is the cyclotron frequency (from $mr\omega^2 = evB$).

In the presence of electric and magnetic fields the drift velocity has components in the direction of $\vec{E}$, of $\vec{B}$, and perpendicular to $\vec{E}$ and $\vec{B}$ [91], see also Eq. (1.137). If $\vec{E} \perp \vec{B}$, the drift velocity $\vec{v}_{\text{drift}}$ along a line forming an angle α with the electric field can be derived from Eq. (1.137) to be

$$|\vec{v}_{\text{drift}}| = \frac{\mu E}{\sqrt{1 + \omega^2\tau^2}} . \qquad (1.138)$$

The angle between the drift velocity $\vec{v}_{\text{drift}}$ and $\vec{E}$ (*Lorentz angle*) can be calculated from Eq. (1.137) under the assumption of $\vec{E} \perp \vec{B}$,

$$\tan\alpha = \omega\tau ; \qquad (1.139)$$

if τ is taken from Eq. (1.134), it follows that

$$\tan\alpha = v_{\text{drift}} \cdot \frac{B}{E} . \qquad (1.140)$$

This result may also be derived if the ratio of the acting Lorentz force $e\vec{v} \times \vec{B}$ (with $\vec{v} \perp \vec{B}$) to the electric force $e\vec{E}$ is considered.

For $E = 500\,\text{V/cm}$ and a drift velocity in the electric field of $v_{\text{drift}} = 3.5\,\text{cm/\mu s}$, a drift velocity in a combined electric and magnetic field ($\vec{E} \perp \vec{B}$) is obtained from Eq. (1.138) for $B = 1.5\,\text{T}$ on the basis of these simple considerations to be

$$v(E = 500\,\text{V/cm}, B = 1.5\,\text{T}) = 2.4\,\text{cm/\mu s} ; \qquad (1.141)$$

correspondingly the Lorentz angle is calculated from Eq. (1.140) to be

$$\alpha = 46° , \qquad (1.142)$$

which is approximately consistent with the experimental findings and the results of a more exact calculation (Fig. 1.25) [32, 87].

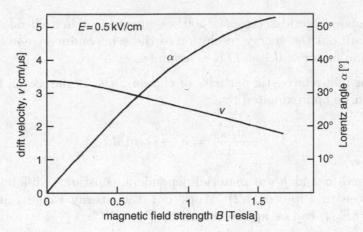

Fig. 1.25. Dependence of the electron drift velocity $\vec{v}_{\text{drift}}$ and the Lorentz angle α on the magnetic field for low electric field strengths (500 V/cm) in a gas mixture of argon (67.2%), isobutane (30.3%) and methylal (2.5%) [32, 87].

Small admixtures of electronegative gases (e.g. oxygen) considerably modify the drift behaviour due to electron attachment. For a 1% fraction of oxygen in argon at a drift field of 1 kV/cm the average mean free path of electrons for attachment is of the order 5 cm. Small admixtures of electronegative gases will reduce the charge signal and in case of strong electronegative gases (such as chlorine) operation of a drift chamber may be even impossible.

Because of the high density the effect of impurities is even more pronounced for liquefied gases. For liquid-noble-gas chambers the oxygen concentration must stay below the ppm ($\equiv 10^{-6}$) level. 'Warm' liquids, like tetramethylsilane (TMS) even require to reduce the concentration of electronegative impurities to below ppb ($\equiv 10^{-9}$).

1.5 Problems

1.1 The range of a 100 keV electron in water is about 200 μm. Estimate its stopping time.

1.2 The energy loss of TeV muons in rock can be parametrised by

$$-\frac{\mathrm{d}E}{\mathrm{d}x} = a + bE \ ,$$

where a stands for a parametrisation of the ionisation loss and the b term includes bremsstrahlung, direct electron-pair production and nuclear interactions ($a \approx 2\,\mathrm{MeV}/(\mathrm{g/cm^2})$, $b = 4.4 \cdot 10^{-6}\,(\mathrm{g/cm^2})^{-1}$) Estimate the range of a 1 TeV muon in rock.

1.3 Monoenergetic electrons of 500 keV are stopped in a silicon counter. Work out the energy resolution of the semiconductor detector if a Fano factor of 0.1 at 77 K is assumed.

1.4 For non-relativistic particles of charge z the Bethe–Bloch formula can be approximated by

$$-\frac{dE_{kin}}{dx} = a\,\frac{z^2}{E_{kin}}\ln(bE_{kin}) \;,$$

where a and b are material-dependent constants (different from those in Problem 1.2). Work out the energy–range relation if $\ln(bE_{kin})$ can be approximated by $(bE_{kin})^{1/4}$.

1.5 In *Compton telescopes* for astronomy or medical imaging one frequently needs the relation between the scattering angle of the electron and that of the photon. Work out this relation from momentum conservation in the scattering process.

1.6 The ionisation trail of charged particles in a gaseous detector is mostly produced by low-energy electrons. Occasionally, a larger amount of energy can be transferred to electrons (δ rays, knock-on electrons). Derive the maximum energy that a 100 GeV muon can transfer to a free electron at rest in a μe collision.

1.7 The production of δ rays can be described by the Bethe–Bloch formula. To good approximation the probability for δ-ray production is given by

$$\phi(E)\,dE = K\,\frac{1}{\beta^2}\,\frac{Z}{A}\cdot\frac{x}{E^2}\,dE \;,$$

where

$K = 0.154\,\mathrm{MeV}/(\mathrm{g/cm^2})$,

Z, A = atomic number and mass of the target,

x = absorber thickness in $\mathrm{g/cm^2}$.

Work out the probability that a 10 GeV muon produces a δ ray of more than $E_0 = 10\,\mathrm{MeV}$ in an 1 cm argon layer (gas at standard room temperature and pressure).

1.8 Relativistic particles suffer an approximately constant ionisation energy loss of about $2\,\mathrm{MeV}/(\mathrm{g/cm^2})$. Work out the *depth–intensity relation* of cosmic-ray muons in rock and estimate the intensity variation if a cavity of height $\Delta h = 1\,\mathrm{m}$ at a depth of 100 m were in the muon beam.

References

[1] B. Rossi, *High Energy Particles*, Prentice-Hall, Englewood Cliffs (1952)

[2] B. Sitar, G.I. Merson, V.A. Chechin & Yu.A. Budagov, *Ionization Measurements in High Energy Physics* (in Russian), Energoatomizdat, Moskau (1988)

[3] B. Sitar, G.I. Merson, V.A. Chechin & Yu.A. Budagov, *Ionization Measurements in High Energy Physics*, Springer Tracts in Modern Physics, Vol. 124, Springer, Berlin/Heidelberg (1993)

[4] H.A. Bethe, Theorie des Durchgangs schneller Korpuskularstrahlen durch Materie, *Ann. d. Phys.* **5** (1930) 325–400

[5] H.A. Bethe, Bremsformel für Elektronen mit relativistischen Geschwindigkeiten, *Z. Phys.* **76** (1932) 293–9

[6] F. Bloch, Bremsvermögen von Atomen mit mehreren Elektronen, *Z. Phys.* **81** (1933) 363–76

[7] R.M. Sternheimer & R.F. Peierls, General Expression for the Density Effect for the Ionization Loss of Charged Particles, *Phys. Rev.* **B3** (1971) 3681–92

[8] E.A. Uehling, Penetration of Heavy Charged Particles in Matter, *Ann. Rev. Nucl. Part. Sci.* **4** (1954) 315–50

[9] S. Hayakawa, *Cosmic Ray Physics*, John Wiley & Sons Inc. (Wiley Interscience) (1969)

[10] Particle Data Group, Review of Particle Properties, *Phys. Lett.* **239** (1990) 1–516

[11] Particle Data Group, Review of Particle Properties, *Phys. Rev.* **D45** (1992) 1–574; Particle Data Group, *Phys. Rev.* **D46** (1992) 5210-0 (Errata)

[12] Particle Data Group, Review of Particle Physics, S. Eidelman *et al.*, *Phys. Lett.* **B592 Vol. 1–4** (2004) 1–1109; W.-M. Yao *et al.*, *J. Phys.* **G33** (2006) 1–1232; http://pdg.lbl.gov

[13] C. Serre, *Evaluation de la Perte D'Energie et du Parcours de Particules Chargées Traversant un Absorbant Quelconque*, CERN **67-5** (1967)

[14] P. Marmier, *Kernphysik I*, Verlag der Fachvereine, Zürich (1977)

[15] L. Landau, On the Energy Loss of Fast Particles by Ionization, *J. Phys. USSR* **8** (1944) 201–5

[16] R.S. Kölbig, *Landau Distribution*, CERN Program Library G 110, CERN Program Library Section (1985)

[17] P.V. Vavilov, Ionization Losses of High Energy Heavy Particles, *Sov. Phys. JETP* **5** (1957) 749–51

[18] R. Werthenbach, *Elektromagnetische Wechselwirkungen von 200 GeV Myonen in einem Streamerrohr-Kalorimeter*, Diploma Thesis, University of Siegen (1987)

[19] S. Behrends & A.C. Melissinos, *Properties of Argon–Ethane/Methane Mixtures for Use in Proportional Counters*, University of Rochester, Preprint UR-776 (1981)

[20] J.E. Moyal, Theory of Ionization Fluctuations, *Phil. Mag.* **46** (1955), 263–80

[21] R.K. Bock *et al.* (eds.), *Formulae and Methods in Experimental Data Evaluation*, General Glossary, Vol. 1, European Physical Society, CERN/Geneva (1984) 1–231

[22] Y. Iga *et al.*, Energy Loss Measurements for Charged Particles and a New Approach Based on Experimental Results, *Nucl. Instr. Meth.* **213** (1983) 531–7

[23] K. Affholderbach *et al.*, Performance of the New Small Angle Monitor for BAckground (SAMBA) in the ALEPH Experiment at CERN, *Nucl. Instr. Meth.* **A410** (1998) 166–75

[24] S.I. Striganov, Ionization Straggling of High Energy Muons in Thick Absorbers, *Nucl. Instr. Meth.* **A322** (1992) 225–30

[25] C. Grupen, Electromagnetic Interactions of High Energy Cosmic Ray Muons, *Fortschr. der Physik* **23** (1976) 127–209

[26] U. Fano, Penetration of Photons, Alpha Particles and Mesons, *Ann. Rev. Nucl. Sci.* **13** (1963) 1–66

[27] G. Musiol, J. Ranft, R. Reif & D. Seeliger, *Kern- und Elementarteilchenphysik*, VCH Verlagsgesellschaft, Weinheim (1988)

[28] W. Heitler, *The Quantum Theory of Radiation*, Clarendon Press, Oxford (1954)

[29] S.P. Møller, *Crystal Channeling or How to Build a '1000 Tesla Magnet'*, CERN-94-05 (1994)

[30] D.S. Gemmell, Channeling and Related Effects in the Motion of Charged Particles through Crystals, *Rev. Mod. Phys.* **46** (1974) 129–227

[31] K. Kleinknecht, *Detektoren für Teilchenstrahlung*, Teubner, Stuttgart (1984, 1987, 1992); *Detectors for Particle Radiation*, Cambridge University Press, Cambridge (1986)

[32] F. Sauli, *Principles of Operation of Multiwire Proportional and Drift Chambers*, CERN-77-09 (1977) and references therein

[33] N.I. Koschkin & M.G. Schirkewitsch, *Elementare Physik*, Hanser, München/Wien (1987)

[34] U. Fano, Ionization Yield of Radiations. II. The Fluctuation of the Number of Ions, *Phys. Rev.* **72** (1947) 26–9

[35] A.H. Walenta, *Review of the Physics and Technology of Charged Particle Detectors*, Preprint University of Siegen SI-83-23 (1983)

[36] G. Lutz, *Semiconductor Radiation Detectors*, Springer, Berlin (1999)

[37] H.A. Bethe, Molière's Theory of Multiple Scattering, *Phys. Rev.* **89** (1953) 1256–66

[38] C. Grupen, *Physics for Particle Detection*, Siegen University publication ed. B. Wenclawiak, S. Wilnewski (2004); Proceedings of the 10 ICFA School on Instrumentation in Elementary Particle Physics Itacuruça, Rio de Janeiro 2003 (to be published 2007); www.cbpf.br/icfa2003/

[39] H.A. Bethe & W. Heitler, Stopping of Fast Particles and Creation of Electron Pairs, *Proc. R. Soc. Lond.* **A146** (1934) 83–112

[40] E. Lohrmann, *Hochenergiephysik*, Teubner, Stuttgart (1978, 1981, 1986, 1992)

[41] U. Amaldi, Fluctuations in Calorimetric Measurements, *Phys. Scripta* **23** (1981) 409–24

[42] W. Lohmann, R. Kopp & R. Voss, *Energy Loss of Muons in the Energy Range 1 – 10.000 GeV*, CERN-85-03 (1985)

[43] M.J. Tannenbaum, *Simple Formulas for the Energy Loss of Ultrarelativistic Muons by Direct Pair Production*, Brookhaven National Laboratory, BNL-44554 (1990)

[44] W.K. Sakumoto *et al.*, Measurement of TeV Muon Energy Loss in Iron, University of Rochester UR-1209 (1991); *Phys. Rev.* **D45** (1992) 3042–50

[45] K. Mitsui, Muon Energy Loss Distribution and its Applications to the Muon Energy Determination, *Phys. Rev.* **D45** (1992) 3051–60

[46] Lev I. Dorman, *Cosmic Rays in the Earth's Atmosphere and Underground*, Kluwer Academic Publishers, Dordercht (2004)

[47] R. Baumgart *et al.*, Interaction of 200 GeV Muons in an Electromagnetic Streamer Tube Calorimeter, *Nucl. Instr. Meth.* **A258** (1987) 51–7

[48] G. Hertz, *Lehrbuch der Kernphysik*, Bd. 1, Teubner, Leipzig (1966)

[49] J.S. Marshall & A.G. Ward, Absorption Curves and Ranges for Homogeneous β-Rays, *Canad. J. Res.* **A15** (1937) 39–41

[50] E. Sauter, *Grundlagen des Strahlenschutzes*, Siemens AG, Berlin/München (1971); *Grundlagen des Strahlenschutzes*, Thiemig, München (1982)

[51] A.G. Wright, A Study of Muons Underground and Their Energy Spectrum at Sea Level, Polytechnic of North London Preprint (1974); *J. Phys.* **A7** (1974) 2085–92

[52] R.K. Bock & A. Vasilescu, *The Particle Detector Briefbook*, Springer, Heidelberg (1998)

[53] P. Marmier & E. Sheldon, *Physics of Nuclei and Particles*, Vol. 1, Academic Press, New York (1969)

[54] E. Fenyves & O. Haimann, *The Physical Principles of Nuclear Radiation Measurements*, Akadémiai Kiadó, Budapest (1969)

[55] O. Klein & Y. Nishina, Über die Streuung von Strahlung durch freie Elektronen nach der neuen relativistischen Quantenmechanik von Dirac, *Z. Phys.* **52** (1929) 853–68

[56] R.D. Evans, *The Atomic Nucleus*, McGraw-Hill, New York (1955)

[57] W.S.C. Williams, *Nuclear and Particle Physics*, Clarendon Press, Oxford (1991)

[58] V. Telnov, *Photon Collider at TESLA*, hep-ex/0010033v4 (2000); *Nucl. Instr. Meth.* **A472** (2001) 43

[59] C. Grupen & E. Hell, Lecture Notes, *Kernphysik*, University of Siegen (1983)

[60] H.A. Bethe & J. Ashkin, Passage of Radiation through Matter, in E. Segrè (ed.), *Experimental Nucl. Phys.*, Vol. 1, John Wiley & Sons Inc. (Wiley Interscience), New York (1953) 166–201

[61] G.W. Grodstein, *X-Ray Attenuation Coefficients from 10 keV to 100 MeV*, Circ. Natl. Bur. Stand. No. 583 (1957)

[62] G.R. White, *X-ray Attenuation Coefficients from 10 keV to 100 MeV*, Natl. Bur. Standards (U.S.) Rept. 1003 (1952)

[63] Particle Data Group, Review of Particle Properties, *Phys. Lett.* **B111** (1982) 1–294

[64] P. Rice-Evans, *Spark, Streamer, Proportional and Drift Chambers*, Richelieu Press, London (1974)

[65] A. Andronic *et al.*, Drift Velocity and Gain in Argon and Xenon Based Mixtures, *Nucl. Instr. Meth.* **A523** (2004) 302–8

[66] P. Colas *et al.*, Electron Drift Velocity Measurements at High Electric Fields, DAPNIA-01-09 (2001) 10; *Nucl. Instr. Meth.* **A478** (2002) 215–19

[67] V. Palladino & B. Sadoulet, *Application of the Classical Theory of Electrons in Gases to Multiwire Proportional and Drift Chambers*, LBL-3013, UC-37, TID-4500-R62 (1974)

[68] W. Blum & L. Rolandi, *Particle Detection with Drift Chambers*, Springer Monograph XV, Berlin/New York (1993)

[69] A. Peisert & F. Sauli, *Drift and Diffusion in Gases: A Compilation*, CERN-84-08 (1984)

[70] L.G. Huxley & R.W. Crompton, *The Diffusion and Drift of Electrons in Gases*, John Wiley & Sons Inc. (Wiley Interscience), New York (1974)

[71] E.W. McDaniel & E.A. Mason, *The Mobility and Diffusion of Ions in Gases*, John Wiley & Sons Inc. (Wiley Interscience), New York (1973)

[72] M.P. Langevin, Sur la mobilité des ions dans les gaz, *C. R. Acad. Sci. Paris*, **134** (1903) 646–9; Recherches sur les gaz ionisés, *Ann. Chim. et Phys.*, **28** (1903) 233–89

[73] L.B. Loeb, *Basis Processes of Gaseous Electronics*, University of California Press, Berkeley (1961)

[74] V. Palladino & B. Sadoulet, Application of the Classical Theory of Electrons in Gases to Multiwire Proportional and Drift Chambers, *Nucl. Instr. Meth.* **128** (1975) 323–66

[75] J. Townsend, *Electrons in Gases*, Hutchinson, London (1947)

[76] S.C. Brown, *Basic Data of Plasma Physics*, MIT Press, Cambridge, MA. (1959), and John Wiley & Sons, Inc., New York (1959)

[77] C. Ramsauer & R. Kollath, Die Winkelverteilung bei der Streuung langsamer Elektronen an Gasmolekülen, *Ann. Phys.* **401** (1931) 756–68

[78] C. Ramsauer & R. Kollath, Über den Wirkungsquerschnitt der Edelgasmoleküle gegenüber Elektronen unterhalb 1 Volt, *Ann. Phys.* **395** (1929) 536–64

[79] C. Ramsauer, Über den Wirkungsquerschnitt der Gasmoleküle gegenüber langsamen Elektronen, *Ann. Phys.* **64** (1921) 513–40

[80] E. Brüche *et al.*, Über den Wirkungsquerschnitt der Edelgase Ar, Ne, He gegenüber langsamen Elektronen, *Ann. Phys.* **389** (1927) 279–91

[81] C.E. Normand, The Absorption Coefficient for Slow Electrons in Gases, *Phys. Rev.* **35** (1930) 1217–25

[82] L. Colli & U. Facchini, Drift Velocity of Electrons in Argon, *Rev. Sci. Instr.* **23** (1952) 39–42

[83] J.M. Kirshner & D.S. Toffolo, Drift Velocity of Electrons in Argon and Argon Mixtures, *J. Appl. Phys.* **23** (1952) 594–8

[84] H.W. Fulbright, Ionization Chambers in Nuclear Physics, in S. Flügge (ed.), *Handbuch der Physik*, Band **XLV**, Springer, Berlin (1958)

[85] J. Fehlmann & G. Viertel, *Compilation of Data for Drift Chamber Operation*, ETH-Zürich-Report (1983)

[86] W.N. English & G.C. Hanna, Grid Ionization Chamber Measurement of Electron Drift Velocities in Gas Mixtures, *Canad. J. Phys.* **31** (1953) 768–97

[87] A. Breskin *et al.*, Recent Observations and Measurements with High-Accuracy Drift Chambers, *Nucl. Instr. Meth.* **124** (1975) 189–214

[88] A. Breskin *et al.*, Further Results on the Operation of High-Accuracy Drift Chambers, *Nucl. Instr. Meth.* **119** (1974) 9–28

[89] G. Charpak & F. Sauli, High-Accuracy Drift Chambers and Their Use in Strong Magnetic Fields, *Nucl. Instr. Meth.* **108** (1973) 413–26

[90] J. Va'vra *et al.*, Measurement of Electron Drift Parameters for Helium and CF_4-Based Gases, *Nucl. Instr. Meth.* **324** (1993) 113–26

[91] T. Kunst *et al.*, Precision Measurements of Magnetic Deflection Angles and Drift Velocities in Crossed Electric and Magnetic Fields, *Nucl. Instr. Meth.* **A423** (1993) 127–40

2

Characteristic properties of detectors

Technical skill is the mastery of complexity while creativity is the
mastery of simplicity.

E. Christopher Zeeman

2.1   Resolutions and basic statistics

The criterion by which to judge the quality of a detector is its resolution
for the quantity to be measured (energy, time, spatial coordinates, etc.). If
a quantity with true value z_0 is given (e.g. the monoenergetic γ radiation
of energy E_0), the measured results z_{meas} of a detector form a *distribution
function $D(z)$* with $z = z_{\text{meas}} - z_0$; the *expectation value* for this quantity is

$$\langle z \rangle = \int z \cdot D(z)\, \mathrm{d}z \left/ \int D(z)\, \mathrm{d}z \right. , \qquad (2.1)$$

where the integral in the denominator normalises the distribution function. This normalised function is usually referred to as the *probability
density function (PDF)*.

The *variance* of the measured quantity is

$$\sigma_z^2 = \int (z - \langle z \rangle)^2 D(z)\, \mathrm{d}z \left/ \int D(z)\, \mathrm{d}z \right. . \qquad (2.2)$$

The integrals extend over the full range of possible values of the
distribution function.

As an example, the expectation value and the variance for a rectangular
distribution will be calculated. In a multiwire proportional chamber with
wire spacing δz, the coordinates of charged particles passing through the
chamber are to be determined. There is no drift-time measurement on

56

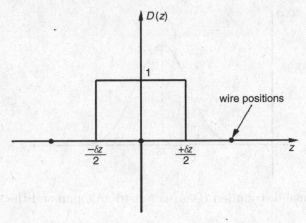

Fig. 2.1. Schematic drawing for the determination of the variance of a rectangular distribution.

the wires. Only a hit on a particular wire with number n_W is recorded (assuming only one hit per event) and its discrete coordinate, $z_{meas} = z_{in} + n_W \delta z$ is measured. The distribution function $D(z)$ is constant $= 1$ from $-\delta z/2$ up to $+\delta z/2$ around the wire which has fired, and outside this interval the distribution function is zero (see Fig. 2.1).

The expectation value for z is evidently zero ($\hat{=}$ position of the fired wire):

$$\langle z \rangle = \int_{-\delta z/2}^{+\delta z/2} z \cdot 1 \, dz \bigg/ \int_{-\delta z/2}^{+\delta z/2} dz = \frac{z^2}{2} \bigg|_{-\delta z/2}^{+\delta z/2} \bigg/ z \bigg|_{-\delta z/2}^{+\delta z/2} = 0 \; ; \quad (2.3)$$

correspondingly, the variance is calculated to be

$$\sigma_z^2 = \int_{-\delta z/2}^{+\delta z/2} (z - 0)^2 \cdot 1 \, dz \bigg/ \delta z = \frac{1}{\delta z} \int_{-\delta z/2}^{+\delta z/2} z^2 \, dz \quad (2.4)$$

$$= \frac{1}{\delta z} \frac{z^3}{3} \bigg|_{-\delta z/2}^{+\delta z/2} = \frac{1}{3 \delta z} \left(\frac{(\delta z)^3}{8} + \frac{(\delta z)^3}{8} \right) = \frac{(\delta z)^2}{12} \; , \quad (2.5)$$

which means

$$\sigma_z = \frac{\delta z}{\sqrt{12}} \; . \quad (2.6)$$

The quantities δz and σ_z have dimensions. The relative values $\delta z/z$ or σ_z/z, respectively, are dimensionless.

In many cases experimental results are normally distributed, corresponding to a distribution function (Fig. 2.2)

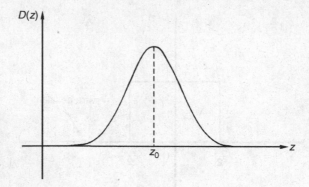

Fig. 2.2. Normal distribution (Gaussian distribution around the average value z_0).

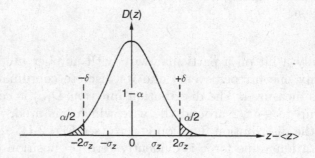

Fig. 2.3. Illustration of confidence levels.

$$D(z) = \frac{1}{\sigma_z\sqrt{2\pi}}\, e^{-(z-z_0)^2/2\sigma_z^2} . \qquad (2.7)$$

The variance determined according to Eq. (2.2) for this *Gaussian distribution* implies that 68.27% of all experimental results lie between $z_0 - \sigma_z$ and $z_0 + \sigma_z$. Within $2\sigma_z$ there are 95.45% and within $3\sigma_z$ there are 99.73% of all experimental results. In this way an interval ($[z_0 - \sigma_z, z_0 + \sigma_z]$) is defined which is called *confidence interval*. It corresponds to a *confidence level* of 68.27%. The value σ_z is usually referred to as a standard error or the standard deviation.

For the general definition we plot the normalised distribution function in its dependence on $z - \langle z \rangle$ (Fig. 2.3). For a normalised probability distribution with an expectation value $\langle z \rangle$ and root mean square deviation σ_z

$$1 - \alpha = \int_{\langle z \rangle - \delta}^{\langle z \rangle + \delta} D(z)\,\mathrm{d}z \qquad (2.8)$$

is the probability that the true value z_0 lies in the interval $\pm\delta$ around the measured quantity z or, equivalently: $100 \cdot (1-\alpha)\%$ of all measured values lie in an interval $\pm\delta$, centred on the average value $\langle z \rangle$.

As stated above, the choice of $\delta = \sigma_z$ for a Gaussian distribution leads to a *confidence interval*, which is called the *standard error*, and whose probability is $1-\alpha = 0.6827$ (corresponding to 68.27%). On the other hand, if a confidence level is given, the related width of the measurement interval can be calculated. For a confidence level of $1 - \alpha \cong 95\%$, one gets an interval width of $\delta = \pm 1.96\,\sigma_z$; $1 - \alpha \cong 99.9\%$ yields a width of $\delta = \pm 3.29\,\sigma_z$ [1]. In data analysis physicists deal very often with non-Gaussian distributions which provide a confidence interval that is asymmetric around the measured value. Consequently, this is characterised by asymmetric errors. However, even in this case the quoted interval of $\pm 1\sigma_z$ corresponds to the same confidence level, 68.27%. It should be noted that sometimes the confidence level is limited by only one border, while the other one extends to $+\infty$ or $-\infty$. In this case one talks about a lower or upper limit of the measured value set by the experiment.*

A frequently used quantity for a resolution is the half width of a distribution which can easily be read from the data or from a fit to it. The half width of a distribution is the *full width at half maximum* (FWHM). For normal distributions one gets

$$\Delta z(\text{FWHM}) = 2\sqrt{2\ln 2}\,\sigma_z = 2.3548\,\sigma_z \ . \tag{2.9}$$

The Gaussian distribution is a continuous distribution function. If one observes particles in detectors the events frequently follow a *Poisson distribution*. This distribution is asymmetric (negative values do not occur) and discrete.

For a mean value μ the individual results n are distributed according to

$$f(n,\mu) = \frac{\mu^n\,\mathrm{e}^{-\mu}}{n!} \ , \quad n = 0, 1, 2, \ldots \tag{2.10}$$

The expectation value for this distribution is equal to the mean value μ with a variance of $\sigma^2 = \mu$.

Let us assume that after many event-counting experiments the average value is three events. The probability to find, in an individual experiment, e.g. no event, is $f(0,3) = \mathrm{e}^{-3} = 0.05$ or, equivalently; if one finds no event in a single experiment, then the true value is smaller than or equal to 3 with a confidence level of 95%. For large values of n the Poisson distribution approaches the Gaussian.

---

* E.g., direct measurements on the electron-antineutrino mass from tritium decay yield a limit of less than 2 eV. From the mathematical point of view this corresponds to an interval from $-\infty$ to 2 eV. Then one says that this leads to an *upper limit* on the neutrino mass of 2 eV.

The determination of the efficiency of a detector represents a random experiment with only two possible outcomes: either the detector was efficient with probability p or not with probability $1 - p = q$. The probability that the detector was efficient exactly r times in n experiments is given by the binomial distribution (*Bernoulli distribution*)

$$f(n, r, p) = \binom{n}{r} p^r q^{n-r} = \frac{n!}{r!(n-r)!} p^r q^{n-r} \ . \tag{2.11}$$

The expectation value of this distribution is $\langle r \rangle = n \cdot p$ and the variance is $\sigma^2 = n \cdot p \cdot q$.

Let the efficiency of a detector be $p = 95\%$ for 100 triggers (95 particles were observed, 5 not). In this example the standard deviation (σ of the expectation value $\langle r \rangle$) is given by

$$\sigma = \sqrt{n \cdot p \cdot q} = \sqrt{100 \cdot 0.95 \cdot 0.05} = 2.18 \tag{2.12}$$

resulting in

$$p = (95 \pm 2.18)\% \ . \tag{2.13}$$

Note that with this error calculation the efficiency cannot exceed 100%, as is correct. Using a Poissonian error ($\pm\sqrt{95}$) would lead to a wrong result.

In addition to the distributions mentioned above some experimental results may not be well described by Gaussian, Poissonian or Bernoulli distributions. This is the case, e.g. for the energy-loss distribution of charged particles in thin layers of matter. It is obvious that a distribution function describing the energy loss must be asymmetric, because the minimum dE/dx can be very small, in principle even zero, but the maximum energy loss can be quite substantial up to the kinematic limit. Such a distribution has a Landau form. The Landau distribution has been described in detail in the context of the energy loss of charged particles (see Chap. 1).

The methods for the statistical treatment of experimental results presented so far include only the most important distributions. For low event rates Poisson-like errors lead to inaccurate limits. If, e.g., one genuine event of a certain type has been found in a given time interval, the experimental value which is obtained from the Poisson distribution, $n \pm \sqrt{n}$, in this case 1 ± 1, cannot be correct. Because, if one has found a genuine event, the experimental value can never be compatible with zero, also not within the error.

The statistics of small numbers therefore has to be modified, leading to the *Regener statistics* [2]. In Table 2.1 the $\pm 1\sigma$ limits for the quoted event numbers are given. For comparison the normal error which is the square root of the event rate is also shown.

Table 2.1. *Statistics of low numbers. Quoted are the ±1σ errors on the basis of the Regener statistics [2] and the ±1σ square root errors of the Poisson statistics*

lower limit		number of events	upper limit	
square root error	statistics of low numbers		statistics of low numbers	square root error
0	0	0	1.84	0
0	0.17	1	3.3	2
0.59	0.71	2	4.64	3.41
1.27	1.37	3	5.92	4.73
6.84	6.89	10	14.26	13.16
42.93	42.95	50	58.11	57.07

The determination of errors or confidence levels is even more complicated if one considers counting statistics with low event numbers in the presence of background processes which are detected along with searched-for events. The corresponding formulae for such processes are given in the literature [3–7].

A general word of caution, however, is in order in the statistical treatment of experimental results. The definition of statistical characteristics in the literature is not always consistent.

In the case of determination of resolutions or experimental errors, one is frequently only interested in relative quantities, that is, $\delta z/\langle z \rangle$ or $\sigma_z/\langle z \rangle$; one has to bear in mind that the average result of a number of experiments $\langle z \rangle$ must not necessarily be equal to the true value z_0. To obtain the relation between the experimental answer $\langle z \rangle$ and the true value z_0, the detectors must be calibrated. Not all detectors are linear, like

$$\langle z \rangle = c \cdot z_0 + d , \qquad (2.14)$$

where c, d are constants. Non-linearities such as

$$\langle z \rangle = c(z_0)z_0 + d \qquad (2.15)$$

may, however, be particularly awkward and require an exact knowledge of the calibration function (sometimes also called 'response function'). In many cases the *calibration parameters* are also time-dependent.

In the following some characteristic quantities of detectors will be discussed.

Energy resolutions, spatial resolutions and time resolutions are calculated as discussed above. Apart from the time resolution there are in addition a number of further *characteristic times* [8].

2.2   Characteristic times

The *dead time* τ_D is the time which has to pass between the registration of one set of incident particles and being sensitive to another set. The dead time, in which no further particles can be detected, is followed by a phase where particles can again be measured; however, the detector may not respond to the particle with full sensitivity. After a further time, the *recovery time* τ_R, the detector can again supply a signal of normal amplitude.

Let us illustrate this behaviour using the example of a Geiger–Müller counter (see Sect. 5.1.3) (Fig. 2.4). After the passage of the first particle the counter is completely insensitive for further particles for a certain time τ_D. Slowly, the field in the Geiger–Müller counter recovers so that for times $t > \tau_D$ a signal can again be recorded, although not at full amplitude. After a further time τ_R, the counter has recovered so that again the initial conditions are established.

The *sensitive time* τ_S is of importance for pulsed detectors. It is the time interval in which particles can be recorded, independent of whether these are correlated with the triggered event or not. If, for example, in an accelerator experiment the detector is triggered by a beam interaction (i.e. is made sensitive), usually a time window of defined length (τ_S) is opened, in which the event is recorded. If by chance in this time interval τ_S a cosmic-ray muon passes through the detector, it will also be recorded because the detector having been made sensitive once cannot distinguish at the trigger level between particles of interest and particles which just happen to pass through the detector in this time window.

The *readout time* is the time that is required to read the event, possibly into an electronic memory. For other than electronic registering (e.g. film), the readout time can be considerably long. Closely related to the readout time is the *repetition time*, which describes the minimum time which must pass between two subsequent events, so that they can be distinguished. The length of the repetition time is determined by the slowest element in the chain detector, readout and registering.

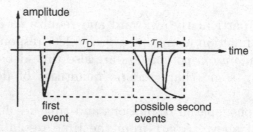

Fig. 2.4.   Illustration of dead and recovery times in a Geiger–Müller counter.

The *memory time* of a detector is the maximum allowed time delay between particle passage and trigger signal, which still yields a 50% efficiency.

The previously mentioned time resolution characterises the minimum time difference where two events can still be separated. This time resolution is very similar to the repetition time, the only difference being that the time resolution refers, in general, to an individual component of the whole detection system (e.g. only the front-end detector), while the repetition time includes all components. For example, the time resolution of a detector can be extremely short, but the whole speed can be lost by a slow readout.

The term *time resolution* is frequently used for the precision with which the arrival time of a particle in a detector can be recorded. The time resolution for individual events defined in this way is determined by the fluctuation of the rise time of the detector signal (see Chap. 14).

2.3 Dead-time corrections

Every particle detector has a *dead time* τ_D where no particles after an event can be recorded. The dead time can be as short as 1 ns in Cherenkov counters, but in Geiger–Müller tubes it can account for 1 ms.

If the count rate is N, the counter is dead for the fraction $N\tau_D$ of the time, i.e., it is only sensitive for the fraction $1 - N\tau_D$ of the measurement time. The *true count rate* – in the absence of dead-time effects – would then be

$$N_{\text{true}} = \frac{N}{1 - N\tau_D} \; . \tag{2.16}$$

Rate measurements have to be corrected, especially if

$$N\tau_D \ll 1 \tag{2.17}$$

is not guaranteed.

2.4 Random coincidences

Coincidence measurements, in particular for high count rates, can be significantly influenced by *chance coincidences*. Let us assume that N_1 and N_2 are the individual pulse rates of two counters in a twofold coincidence arrangement. For the derivation of the chance coincidence rate we assume that the two counters are independent and their count rates are given by Poisson statistics. The probability that counter 2 gives no signal in the time interval τ after a pulse in counter 1 can be derived from the Poisson distribution, see Eq. (2.10), to be

$$f(0, N_2) = \mathrm{e}^{-N_2\tau} . \tag{2.18}$$

Correspondingly, the chance of getting an uncorrelated count in this period is

$$P = 1 - \mathrm{e}^{-N_2\tau} . \tag{2.19}$$

Since normally $N_2\tau \ll 1$, one has

$$P \approx N_2\tau . \tag{2.20}$$

Because counter 2 can also have a signal before counter 1 within the resolving time of the coincidence circuit, the total random coincidence rate is [9, 10]

$$R_2 = 2N_1 N_2\tau . \tag{2.21}$$

If the signal widths of the two counters are different, one gets

$$R_2 = N_1 N_2(\tau_1 + \tau_2) . \tag{2.22}$$

In the general case of q counters with identical pulse widths τ the q-fold random coincidence rate is obtained to be [9, 10]

$$R_q = qN_1 N_2 \cdots N_q \tau^{q-1} . \tag{2.23}$$

To get coincidence rates almost free of random coincidences it is essential to aim for a high time resolution.

In practical situations a q-fold random coincidence can also occur, if $q - k$ counters are set by a true event and k counters have uncorrelated signals. The largest contribution mostly comes from $k = 1$:

$$R_{q,q-1} = 2(K_{q-1}^{(1)} \cdot N_1 + K_{q-1}^{(2)} \cdot N_2 + \cdots + K_{q-1}^{(q)} \cdot N_q) \cdot \tau , \tag{2.24}$$

where $K_{q-1}^{(i)}$ represents the rate of genuine $(q-1)$-fold coincidences when the counter i does not respond.

In the case of *majority coincidences* the following random coincidence rates can be determined: If the system consists of q counters and each counter has a counting rate of N, the number of random coincidences for p out of q stations is

$$R_p(q) = \binom{q}{p} pN^p \tau^{p-1} . \tag{2.25}$$

For $q = p = 2$ this reduces to

$$R_2(2) = 2N^2\tau \ , \tag{2.26}$$

as for the twofold chance coincidence rate. If the counter efficiency is high it is advisable to use a coincidence level with p not much smaller than q to reduce the chance coincidence rate.

2.5 Efficiencies

A very important characteristic of each detector is its efficiency, that is, the probability that a particle which passes through the detector is also seen by it. This *efficiency ε* can vary considerably depending on the type of detector and radiation. For example, γ rays are measured in gas counters with probabilities on the order of a per cent, whereas charged particles in scintillation counters or gas detectors are seen with a probability of 100%. Neutrinos can only be recorded with extremely low probabilities ($\approx 10^{-18}$ for MeV neutrinos in a massive detector).

In general, efficiency and resolution of a detector are strongly correlated. Therefore one has to find an optimum for these two quantities also under consideration of possible backgrounds. If, for example, in an experiment with an energy-loss, Cherenkov, or transition-radiation detector a pion–kaon separation is aimed at, this can in principle be achieved with a low *misidentification probability*. However, for a small misidentification probability one has to cut into the distribution to get rid of the unwanted particle species. This inevitably results in a low efficiency: one cannot have both high efficiency and high two-particle resolution at the same time (see Chaps. 9 and 13).

The efficiency of a detector can be measured in a simple experiment (Fig. 2.5). The detector whose unknown efficiency ε has to be determined is placed between two trigger counters with efficiencies ε_1 and ε_2; one must make sure that particles which fulfil the trigger requirement, which in this case is a twofold coincidence, also pass through the sensitive volume of the detector under investigation.

The twofold coincidence rate is $R_2 = \varepsilon_1 \cdot \varepsilon_2 \cdot N$, where N is the number of particles passing through the detector array. Together with the threefold coincidence rate $R_3 = \varepsilon_1 \cdot \varepsilon_2 \cdot \varepsilon \cdot N$, the efficiency of the detector in question is obtained as

$$\varepsilon = \frac{R_3}{R_2} \ . \tag{2.27}$$

If one wants to determine the error on the efficiency ε one has to consider that R_2 and R_3 are correlated and that we are dealing in this case

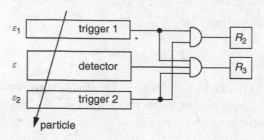

Fig. 2.5. A simple experiment for the determination of the efficiency of a detector.

with Bernoulli statistics. Therefore, the absolute error on the threefold coincidence rate is given by, see Eq. (2.12),

$$\sigma_{R_3} = \sqrt{R_2 \cdot \varepsilon(1 - \varepsilon)} \ , \tag{2.28}$$

and the relative error of the threefold coincidence rate, normalised to the number of triggers R_2, is

$$\frac{\sigma_{R_3}}{R_2} = \sqrt{\frac{\varepsilon(1 - \varepsilon)}{R_2}} \ . \tag{2.29}$$

If the efficiency is small ($R_3 \ll R_2$, $\varepsilon \ll 1$), Eq. (2.28) reduces to

$$\sigma_{R_3} \approx \sqrt{R_3} \ . \tag{2.30}$$

In case of a high efficiency ($R_3 \approx R_2$, $1 - \varepsilon \ll 1$, i.e. $\varepsilon \approx 1$) the error can be approximated by

$$\sigma_{R_3} \approx \sqrt{R_2 - R_3} \ . \tag{2.31}$$

In these extreme cases Poisson-like errors can be used as an approximation.

If an experimental setup consists of n detector stations, frequently only a *majority coincidence* is asked for, i.e., one would like to know the efficiency that k or more out of the n installed detectors have seen a signal. If the single detector efficiency is given by ε, the efficiency for the majority coincidence, ε_{M}, is worked out to be

$$\varepsilon_{\mathrm{M}} = \varepsilon^k(1 - \varepsilon)^{n-k}\binom{n}{k} + \varepsilon^{k+1}(1 - \varepsilon)^{n-(k+1)}\binom{n}{k+1} + \cdots$$

$$+ \varepsilon^{n-1}(1 - \varepsilon)\binom{n}{n-1} + \varepsilon^n \ . \tag{2.32}$$

The first term is motivated as follows: to have exactly k detectors efficient one gets the efficiency ε^k, but in addition the other $(n-k)$ detectors are inefficient leading to $(1-\varepsilon)^{n-k}$. However, there are $\binom{n}{k}$ possibilities to pick k counters out of n stations. Hence the product of multiplicities is multiplied by this number. The other terms can be understood along similar arguments.

The efficiency of a detector normally also depends on the point where the particle has passed through the detector (homogeneity, uniformity), on the angle of incidence (isotropy), and on the time delay with respect to the trigger.

In many applications of detectors it is necessary to record many particles at the same time. For this reason, the *multiparticle efficiency* is also of importance. The multiparticle efficiency can be defined as the probability that exactly N particles are registered if N particles have simultaneously passed through the detector. For normal spark chambers the multitrack efficiency defined this way decreases rapidly with increasing N, while for scintillation counters it will probably vary very little with N. The multiparticle efficiency for drift chambers can also be affected by the way the readout is done ('single hit' where only one track is recorded or 'multiple hit' where many tracks (up to a preselected maximum) can be analysed).

In modern tracking systems (e.g. time-projection chambers) the multitrack efficiency is very high. This is also necessary if many particles in jets must be resolved and properly reconstructed, so that the invariant mass of the particle that has initiated the jet can be correctly worked out. In time-projection chambers in heavy-ion experiments as many as 1000 tracks must be reconstructed to allow for an adequate event interpretation. Figure 2.6 shows the final state of a head-on collision of two gold nuclei at a centre-of-mass energy of 130 GeV in the time-projection chamber of the STAR experiment [11]. Within these dense particle bundles also decays of short-lived particles must be identified. This is in particular also true for tracking detectors at the Large Hadron Collider (LHC), where a good *multitrack reconstruction efficiency* is essential so that rare and interesting events (like the Higgs production and decay) are not missed. The event shown in Fig. 2.6, however, is a little misleading in the sense that it represents a two-dimensional projection of a three-dimensional event. Overlapping tracks in this projection might be well separated in space thereby allowing track reconstruction.

The multitrack efficiency in such an environment can, however, be influenced by problems of *occupancy*. If the density of particle tracks is getting too high – this will for sure occur in tracking devices close to the interaction point – different tracks may occupy the same readout element. If

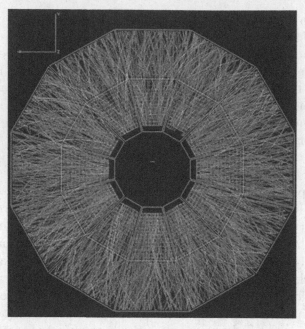

Fig. 2.6. A reconstructed Au + Au collision in the STAR time-projection chamber at a centre-of-mass energy of 130 GeV [11].

the two-track resolution of a detector is denoted by Δx, and two particles or more have mutual distances less than Δx, track coordinates will be lost, which will eventually lead to a problem in track reconstruction efficiency if too many coordinates are affected by this limitation. This can only be alleviated if the *pixel size* for a readout segment is decreased. This implies an increased number of readout channels associated with higher costs. For inner trackers at high-luminosity colliders the question of occupancy is definitely an issue.

Event-reconstruction capabilities might also suffer from the deterioration of detector properties in *harsh radiation environments* (*ageing*). A limited *radiation hardness* can lead to *gain losses* in wire chambers, increase in dark currents in semiconductor counters, or reduction of transparency for scintillation or Cherenkov counters. Other factors limiting the performance are, for example, related to events overlapping in time. Also a possible gain drift due to temperature or pressure variation must be kept under control. This requires an *on-line monitoring* of the relevant detector parameters which includes a measurement of ambient conditions and the possibility of on-line calibration by the injection of standard pulses into the readout system or using known and well-understood processes to monitor the stability of the whole detector system (*slow control*).

2.6 Problems

2.1 The thickness of an aluminium plate, x, is to be determined by the absorption of ^{137}Cs γ rays. The count rate N in the presence of the aluminium plate is 400 per 10 seconds, and without absorber it is 576 in 10 seconds. The mass attenuation coefficient for Al is $\mu/\varrho = (0.07 \pm 0.01)\,(\text{g/cm}^2)^{-1}$. Calculate the thickness of the foil and the total error.

2.2 Assume that in an experiment at the LHC one expects to measure 10 neutral Higgs particles of mass $115\,\text{GeV}/c^2$ in hundred days of running. Use the Poisson statistics to determine the probability of detecting

> 5 Higgs particles in 100 days,
> 2 particles in 10 days,
> no Higgs particle in 100 days.

2.3 A pointlike radioactive γ-ray source leads to a count rate of $R_1 = 90\,000$ per second in a GM counter at a distance of $d_1 = 10\,\text{cm}$. At $d_2 = 30\,\text{cm}$ one gets $R_2 = 50\,000$ per second. What is the dead time of the GM counter, if absorption effects in the air can be neglected?

References

[1] Particle Data Group, Review of Particle Physics, S. Eidelman *et al.*, *Phys. Lett.* **B592 Vol. 1–4** (2004) 1–1109; W.-M. Yao *et al.*, *J. Phys.* **G33** (2006) 1–1232; http://pdg.lbl.gov

[2] V.H. Regener, Statistical Significance of Small Samples of Cosmic Ray Counts, *Phys. Rev.* **84** (1951) 161–2

[3] Particle Data Group, Review of Particle Properties, *Phys. Lett.* **239** (1990) 1–516

[4] G. Zech, Upper Limits in Experiments with Background or Measurement Errors, *Nucl. Instr. Meth.* **277** (1989) 608–10; *Nucl. Instr. Meth.* **A398** (1997) 431–3

[5] O. Helene, Upper Limit of Peak Area, *Nucl. Instr. Meth.* **212** (1983) 319–22

[6] S. Brandt, *Datenanalyse*, 4. Auflage; Spektrum Akademischer Verlag, Heidelberg/Berlin (1999); *Data Analysis: Statistical and Computational Methods for Scientists and Engineers*, 3rd edition, Springer, New York (1998)

[7] G. Cowan, *Statistical Data Analysis*, Oxford Science Publications, Clarendon Press, Oxford (1998)

[8] O.C. Allkofer, W.D. Dau & C. Grupen, *Spark Chambers*, Thiemig, München (1969)

[9]   E. Fenyves & O. Haimann, *The Physical Principles of Nuclear Radiation Measurements*, Akadémiai Kiadó, Budapest (1969)
[10] L. Jánossy, *Cosmic Rays*, Clarendon Press, Oxford (1948)
[11] www.np.ph.bham.ac.uk/research/heavyions1.htm

3

Units of radiation measurements and radiation sources

Ketchup left overnight on dinner plates has a longer half-life than radioactive waste.

Wes Smith

3.1 Units of radiation measurement

Many measurements and tests with detectors are made with radioactive sources. Radiation aspects are also an issue at any accelerator and, in particular, at hadron colliders. Even at neutrino factories the radiation levels can be quite high. Basic knowledge of the units of radiation measurement and the biological effects of radiation are therefore useful [1–5].

Let us assume that there are initially N_0 nuclei of a certain radioactive element. The number will decrease in the course of time t due to decay according to

$$N = N_0 \, e^{-t/\tau} , \tag{3.1}$$

where τ is the *lifetime* of the radioisotope. One has to distinguish between the lifetime and the *half-life* $T_{1/2}$. The half-life can be calculated from Eq. (3.1) as

$$N(t = T_{1/2}) = \frac{N_0}{2} = N_0 \, e^{-T_{1/2}/\tau} , \tag{3.2}$$

$$T_{1/2} = \tau \cdot \ln 2 . \tag{3.3}$$

The decay constant of the radioactive element is

$$\lambda = \frac{1}{\tau} = \frac{\ln 2}{T_{1/2}} . \tag{3.4}$$

The activity of a source gives the number of decays per unit time,

$$A = -\frac{dN}{dt} = \frac{1}{\tau}N = \lambda N \ . \tag{3.5}$$

The unit of the activity is *Becquerel* (Bq). 1 Bq means 1 decay per second. (In passing it should be mentioned that the physical quantity with the dimension s^{-1} already has a name: Hertz! However, this unit Hz is mostly used for periodic phenomena, while Bq is used for statistically distributed events.) The unit Bq supersedes the old unit Curie (Ci). Historically 1 Ci was the activity of 1 g of radium,

$$1\,\text{Ci} = 3.7 \cdot 10^{10}\,\text{Bq} \tag{3.6}$$

or

$$1\,\text{Bq} = 27 \cdot 10^{-12}\,\text{Ci} = 27\,\text{pCi} \ . \tag{3.7}$$

1 Bq is a very small unit of the activity. The radioactivity of the human body amounts to about 7500 Bq, mainly due to ^{14}C, ^{40}K and ^{232}Th.

The activity in Bq does not say very much about possible biological effects. These are related to the energy which is deposited per unit mass by a radioactive source.

The *absorbed dose D* (absorbed energy per mass unit)

$$D = \frac{1}{\varrho}\frac{dW}{dV} \tag{3.8}$$

(dW – absorbed energy; ϱ – density; dV – unit of volume) is measured in *Grays* (1 Gray = 1 J/kg). The old cgs unit rad (**r**öntgen **a**bsorbed **d**ose, 1 rad = 100 erg/g) is related to Gray according to

$$1\,\text{Gy} = 100\,\text{rad} \ . \tag{3.9}$$

Gray and rad describe only the physical energy absorption, and do not take into account any biological effect. Since, however, α-, β-, γ- and neutron-emitting sources have different biological effects for the same energy absorption, a *relative biological effectiveness* (RBE) is defined. The absorbed dose D_γ obtained from the exposure to γ or X rays serves as reference. The absorbed dose of an arbitrary radiation which yields the same biological effect as D_γ leads to the definition of the relative biological effectiveness as

$$D_\gamma = RBE \cdot D \ . \tag{3.10}$$

The RBE factor has a complicated dependence on the radiation field, the radiation energy and the dose rate. For practical reasons, therefore,

Table 3.1. *Radiation weighting factors w_R*

Radiation and energy range	Radiation weighting factor w_R
Photons, all energies	1
Electrons and muons, all energies	1
Neutrons $E_n < 10\,\text{keV}$	5
$10\,\text{keV} \le E_n \le 100\,\text{keV}$	10
$100\,\text{keV} < E_n \le 2\,\text{MeV}$	20
$2\,\text{MeV} < E_n \le 20\,\text{MeV}$	10
$E_n > 20\,\text{MeV}$	5
Protons, except recoil protons, $E > 2\,\text{MeV}$	5
α particles, nuclear fragments, heavy nuclei	20

a *radiation weighting factor w_R* (formerly called *quality factor*) is introduced. The absorbed dose D multiplied by this weighting factor is called *equivalent dose H*. The unit of the equivalent dose is 1 *Sievert* (Sv),

$$H\{\text{Sv}\} = w_R \cdot D\,\{\text{Gy}\} \ . \tag{3.11}$$

The weighting factor has the unit Sv/Gy. The old cgs unit rem ($H\{\text{rem}\} = w_R \cdot D\{\text{rad}\}$, rem = **r**öntgen **e**quivalent **m**an) is related to Sievert according to

$$1\,\text{Sv} = 100\,\text{rem} \ . \tag{3.12}$$

The radiation weighting factors w_R are listed in Table 3.1.

According to Table 3.1, neutrinos do not present a radiation hazard. This is certainly true for natural neutrino sources, however, the high flux of energetic neutrinos from future neutrino factories might present a radiation problem.

It should be mentioned that the biological effect of radiation is also influenced by, for example, the time sequence of absorption (e.g. fractionated irradiation), the energy spectrum of radiation, or the question whether the irradiated person has been sensitised or desensitised by a pharmaceutical drug.

The biological effect also depends on which particular part of the human body is irradiated. To take account of this effect a further *tissue weighting factor w_T* is introduced leading to a general expression for the effective equivalent dose

$$H_{\text{eff}} = \sum_T w_T\, H_T \ , \tag{3.13}$$

Table 3.2.   *Tissue weighting factors w_R*

Organ or tissue	Tissue weighting factor w_T
Gonads	0.20
Red bone marrow	0.12
Colon	0.12
Lung	0.12
Stomach	0.12
Bladder	0.05
Chest	0.05
Liver	0.05
Oesophagus	0.05
Thyroid gland	0.05
Skin	0.01
Bone surface	0.01
Other organs or tissue	0.05

where the sum extends over those irradiated parts of the human body which have received the doses H_T. The tissue weighting factors are listed in Table 3.2.

The most general form of the effective equivalent dose is therefore

$$H_{\text{eff}} = \sum_T w_T \sum_R w_R D_{T,R} , \qquad (3.14)$$

where the sums run over the partial body doses received in different radiation fields properly weighted by the radiation and tissue weighting factors.

The *equivalent whole-body dose rate* from pointlike radiation sources can be calculated from the relation

$$\dot{H} = \Gamma \frac{A}{r^2} , \qquad (3.15)$$

where A is the activity (in Bq) and r the distance from the source in metres. Γ is the dose constant which depends on the radiation field and the radioisotope. Specific γ-ray dose constants ($\Gamma_\gamma = 8.46 \cdot 10^{-14} \frac{\text{Sv} \cdot \text{m}^2}{\text{Bq} \cdot \text{h}}$ for ^{137}Cs) and β-ray dose constants ($\Gamma_\beta = 2.00 \cdot 10^{-11} \frac{\text{Sv} \cdot \text{m}^2}{\text{Bq} \cdot \text{h}}$ for ^{90}Sr) are listed in the literature [4].

Apart from these units, there is still another one describing the quantity of produced charge, which is the *Röntgen* (R). One Röntgen is the radiation dose for X-ray or γ radiation which produces, under normal conditions, one electrostatic charge unit (esu) of electrons and ions in $1\,\mathrm{cm}^3$ of dry air.

The charge of an electron is $1.6 \cdot 10^{-19}\,\mathrm{C}$ or $4.8 \cdot 10^{-10}\,\mathrm{esu}$. (The esu is a cgs unit with $1\,\mathrm{esu} = \frac{1}{3 \cdot 10^9}\,\mathrm{C}$.) If one electrostatic charge unit is produced, the number of generated electrons per cm^3 is given by

$$N = \frac{1}{4.8 \cdot 10^{-10}} = 2.08 \cdot 10^9 \ . \qquad (3.16)$$

If the unit Röntgen is transformed into an ion charge per kg, it gives

$$1\,\mathrm{R} = \frac{N \cdot q_e \,\{\mathrm{C}\}}{m_{\mathrm{air}}(1\,\mathrm{cm}^3)\,\{\mathrm{kg}\}} = \frac{1\,\mathrm{esu}}{m_{\mathrm{air}}(1\,\mathrm{cm}^3)\,\{\mathrm{kg}\}} \ , \qquad (3.17)$$

where q_e is the electron charge in Coulomb, $m_{\mathrm{air}}(1\,\mathrm{cm}^3)$ is the mass of $1\,\mathrm{cm}^3$ air; consequently

$$1\,\mathrm{R} = 2.58 \cdot 10^{-4}\,\mathrm{C/\,kg} \quad \text{for air} \ . \qquad (3.18)$$

If Röntgen has to be converted to an absorbed dose, one has to consider that the production of an electron–ion pair in air requires an energy of about $W = 34\,\mathrm{eV}$,

$$1\,\mathrm{R} = N \cdot \frac{W}{m_{\mathrm{air}}} = 0.88\,\mathrm{rad} = 8.8\,\mathrm{mGy} \ . \qquad (3.19)$$

To obtain a feeling for these abstract units, it is quite useful to establish a natural scale by considering the radiation load from the environment.

The radioactivity of the human body amounts to about $7500\,\mathrm{Bq}$, mainly caused by the radioisotope $^{14}\mathrm{C}$ and the potassium isotope $^{40}\mathrm{K}$. The average radioactive load (at sea level) by cosmic radiation ($\approx 0.3\,\mathrm{mSv/a}$)*, by terrestrial radiation ($\approx 0.5\,\mathrm{mSv/a}$) and by incorporation of radioisotopes (inhalation $\approx 1.1\,\mathrm{mSv/a}$, ingestion $\approx 0.3\,\mathrm{mSv/a}$) are all of approximately the same order of magnitude, just as the radiation load caused by civilisation ($\approx 1.0\,\mathrm{mSv/a}$), which is mainly caused by X-ray diagnostics and treatment and by exposures in nuclear medicine. The total annual per capita dose consequently is about $3\,\mathrm{mSv}$.

The natural radiation load, of course, depends on the place where one lives; it has a typical fluctuation corresponding to a factor of two. The radiation load caused by civilisation naturally has a much larger fluctuation. The average value in this case results from relatively high doses obtained by few persons.

---

* a (Latin) = annum = year.

The *lethal whole-body dose* (50% mortality in 30 days without medical treatment) is 4 Sv (= 400 rem).

The International Commission for Radiological Protection (ICRP) has recommended a limit for the whole-body dose for persons working in controlled areas of 20 mSv/a (= 2 rem/a) which has been adopted in most national radiation protection regulations. The ICRP has also proposed *exemption limits* for the handling of radioactive sources (e.g. 10^4 Bq for ^{137}Cs) and *clearance levels* for discharging radioactive material from radiation areas (e.g. 0.5 Bq/g for solid or liquid material containing ^{137}Cs). A radiation officer has to be installed whose responsibility is to watch that the various radiation protection regulations are respected.

3.2   Radiation sources

There is a large variety of radiation sources which can be used for detector tests. Historically, radioactive sources were the first ones to be employed. In β decay electrons or positrons with continuous energy spectra are produced. In β^- decay a neutron inside the nucleus decays according to

$$n \rightarrow p + e^- + \bar{\nu}_e \qquad (3.20)$$

while in positron decay a proton of the radioactive element undergoes the transformation

$$p \rightarrow n + e^+ + \nu_e . \qquad (3.21)$$

The electron-capture reaction

$$p + e^- \rightarrow n + \nu_e \qquad (3.22)$$

mostly leads to excited states in the daughter nucleus. The excited nucleus is a source of monochromatic γ rays or monoenergetic electrons which originate from the K or the L shell if the nuclear excitation energy is directly transferred to atomic electrons. The energies of these *conversion electrons* are $E_\mathrm{ex} - E_\mathrm{binding}$ where E_ex is the nuclear excitation energy and E_binding the binding energy in the respective atomic shell. As a consequence of internal conversion or other processes which liberate electrons from atomic shells, *Auger electrons* may be emitted. This happens when the excitation energy of the atomic shell is transferred to an electron of the outer shells, which then can leave the atom. If, e.g., the excitation energy of a nucleus liberates a K-shell electron, the free electron state can be filled up by an L-shell electron. The excitation-energy difference of the atomic shells, $E_{\mathrm{K}} - E_{\mathrm{L}}$, can be emitted as either characteristic K_α X ray or can directly be transferred to an L electron which then gets the energy $E_{\mathrm{K}} - 2 \cdot E_{\mathrm{L}}$. Such an electron is called an Auger electron. Conversion electrons are typically in the MeV range while Auger electrons are in the keV region.

In most cases β decays do not reach the ground state of the daughter nucleus. The excited daughter nucleus de-excites by γ-ray emission. There is a large selection of γ-ray emitters covering the energy region from keV to several MeV. γ rays can also come from *annihilation*

$$e^+ + e^- \rightarrow \gamma + \gamma \tag{3.23}$$

which provides monoenergetic γ rays of 511 keV or from other annihilation reactions.

γ rays from a wide energy spectrum or X rays can be produced in bremsstrahlung reactions where a charged particle (mostly electrons) is decelerated in the Coulomb potential of a nucleus, as it is typical in an X-ray tube. If charged particles are deflected in a magnetic field, synchrotron photons (*magnetic bremsstrahlung*) are emitted.

Sometimes, also heavily ionising particles are required for detector tests. For this purpose α rays from radioactive sources can be used. Because of the short range of α particles ($\approx 4\,\mathrm{cm}$ in air), the sources must be very close to the detector or even integrated into the sensitive volume of the detector.

For tests of radiation hardness of detectors one frequently also has to use neutron beams. *Radium–beryllium sources* provide neutrons in the MeV region. In these sources α particles from $^{226}\mathrm{Ra}$ decay interact with beryllium according to

$$\alpha + {}^9\mathrm{Be} \rightarrow {}^{12}\mathrm{C} + n \ . \tag{3.24}$$

Neutrons can also be produced in photonuclear reactions.

In Table 3.3 some α-, β- and γ-ray emitters, which are found to be quite useful for detector tests, are listed [6–8]. (For β-ray emitters the maximum energies of the continuous energy spectra are given; *EC* means electron capture, mostly from the K shell.)

If gaseous detectors are to be tested, an $^{55}\mathrm{Fe}$ source is very convenient. The $^{55}\mathrm{Fe}$ nucleus captures an electron from the K shell leading to the emission of characteristic X rays of manganese of 5.89 keV. X rays or γ rays do not provide a trigger. If one wants to test gaseous detectors with triggered signals, one should look for electron emitters with an electron energy as high as possible. Energetic electrons have a high range making

Table 3.3. *A compilation of useful radioactive sources along with their characteristic properties [5–12]*

Radio-isotope	Decay mode/ branching fraction	$T_{1/2}$	Energy of radiation	
			β, α	γ
$^{22}_{11}$Na	β^+ (89%)	2.6 a	β_1^+ 1.83 MeV (0.05%)	1.28 MeV
	EC (11%)		β_2^+ 0.54 MeV (90%)	0.511 MeV (annihilation)
$^{55}_{26}$Fe	EC	2.7 a		Mn X rays 5.89 keV (24%) 6.49 keV (2.9%)
$^{57}_{27}$Co	EC	267 d		14 keV (10%) 122 keV (86%) 136 keV (11%)
$^{60}_{27}$Co	β^-	5.27 a	β^- 0.316 MeV (100%)	1.173 MeV (100%) 1.333 MeV (100%)
$^{90}_{38}$Sr	β^-	28.5 a	β^- 0.546 MeV (100%)	
$\rightarrow$ $^{90}_{39}$Y	β^-	64.8 h	β^- 2.283 MeV (100%)	
$^{106}_{44}$Ru	β^-	1.0 a	β^- 0.039 MeV (100%)	
$\rightarrow$ $^{106}_{45}$Rh	β^-	30 s	β_1^- 3.54 MeV (79%) β_2^- 2.41 MeV (10%) β_3^- 3.05 MeV (8%)	0.512 MeV (21%) 0.62 MeV (11%)
$^{109}_{48}$Cd	EC	1.27 a	monoenergetic conversion electrons 63 keV (41%) 84 keV (45%)	88 keV (3.6%) Ag X rays
$^{137}_{55}$Cs	β^-	30 a	β_1^- 0.514 MeV (94%) β_2^- 1.176 MeV (6%)	0.662 MeV (85%)
$^{207}_{83}$Bi	EC	32.2 a	monoenergetic conversion electrons 0.482 MeV (2%) 0.554 MeV (1%) 0.976 MeV (7%) 1.048 MeV (2%)	0.570 MeV (98%) 1.063 MeV (75%) 1.770 MeV (7%)
$^{241}_{95}$Am	α	433 a	α 5.443 MeV (13%) α 5.486 MeV (85%)	60 keV (36%) Np X rays

it possible to penetrate the detector and also a trigger counter. ^{90}Y produced in the course of ^{90}Sr decay has a maximum energy of 2.28 MeV (corresponding to $\approx$ 4 mm aluminium). An Sr/Y radioactive source has the convenient property that almost no γ rays, which are hard to shield, are emitted. If one wants to achieve even higher electron energies, one can use a ^{106}Rh source, it being a daughter element of ^{106}Ru. The electrons of this source with a maximum energy of 3.54 MeV have a range of $\approx$ 6.5 mm in aluminium. The *electron capture* (*EC*) emitter ^{207}Bi emits monoenergetic conversion electrons, and is therefore particularly well suited for an energy calibration and a study of the energy resolution of detectors. A compilation of commonly used radioactive sources along with their characteristic properties is given in Table 3.3. The *decay-level schemes* of these sources are presented in Appendix 5.

If higher energies are required, or more penetrating radiation, one can take advantage of test beams at accelerators or use muons from cosmic radiation.

In these test beams almost any particle with well-defined momentum and charge (electrons, muons, pions, kaons, protons, ...) can be provided. These beams are mostly produced in interactions of energetic protons in a target. A suitable test-beam equipment consisting of momentum-selection magnets, beam-defining scintillators and Cherenkov counters for tagging special particle species can tailor the secondary beam to the needs of the experimenter. If no particle accelerator is at hand, the omnipresent cosmic rays provide an attractive alternative – albeit at relatively low rates – for detector tests.

The flux of cosmic-ray muons through a horizontal area amounts to approximately $1/(\text{cm}^2 \cdot \text{min})$ at sea level. The muon flux per solid angle from near vertical directions through a horizontal area is $8 \cdot 10^{-3}\,\text{cm}^{-2}\,\text{s}^{-1}\,\text{sr}^{-1}$ [6, 13].

The angular distribution of muons roughly follows a $\cos^2\theta$ law, where θ is the zenith angle measured with respect to the vertical direction. Muons account for 80% of all charged cosmic-ray particles at sea level.

3.3 Problems

3.1 Assume that some piece of radioactive material has a nearly constant gamma activity of 1 GBq. Per decay a total energy of 1.5 MeV is liberated. Work out the daily absorbed dose, if the ionising radiation is absorbed in a mass of $m = 10\,\text{kg}$?

3.2 In an accident in a nuclear physics laboratory a researcher has inhaled dust containing the radioactive isotope ^{90}Sr, which led to a dose rate of $1\,\mu Sv/h$ in his body. The physical half-life of ^{90}Sr is 28.5 years, the biological half-life is only 80 days. How long does it take this dose rate to decay to a level of $0.1\,\mu Sv/h$?

3.3 Consider a pocket dosimeter with a chamber volume of $2.5\,cm^3$ and a capacitance of $7\,pF$. Originally it had been charged to a voltage of $200\,V$. After a visit in a nuclear power plant it only showed a voltage of $170\,V$. What was the received dose?

The density of air is $\dot{\varrho}_L = 1.29 \cdot 10^{-3}\,g/cm^3$.

3.4 In a reactor building (volume $V_1 = 4000\,m^3$) a tritium concentration of $100\,Bq/m^3$ has been measured. The tritium originated from the containment area of volume $500\,m^3$. Work out the original tritium concentration and the total activity.

3.5 Assume that in a certain working area a ^{60}Co concentration in the air of $1\,Bq/m^3$ exists. Based on a respiratory annual volume of $8000\,m^3$ this would lead to an intake of $8000\,Bq$ in this environment. What sort of amount of ^{60}Co would this cobalt activity correspond to $(T_{1/2}(^{60}Co) = 5.24\,a$, mass of a ^{60}Co nucleus $m_{Co} = 1 \cdot 10^{-22}\,g)$?

3.6 A large shielded shipping container (mass $m = 120$ tons) with an inventory activity of $10^{17}\,Bq$ will warm up as a consequence of the emitted ionising radiation. Assume that $10\,MeV$ are liberated per decay which is transferred to the container for a period of 24 hours without any losses. What would be the corresponding temperature increase if the shipping container is made from iron, and if it had originally a temperature of $20\,°C$ (specific heat of iron: $c = 0.452\,kJ/(kg\,K)$)?

3.7 The absorption coefficient for $50\,keV$ X rays in aluminium is $\mu = 0.3\,(g/cm^2)^{-1}$. Work out the thickness of an aluminium shielding which reduces the radiation level by a factor of $10\,000$.

3.8 How does the radiation dose received in a four-week holiday in the high mountains $(3000\,m)$ compare to the radiation load caused by an X-ray of the human chest in an X-ray mass screening?

3.9 ^{137}Cs is stored in a human with a biological half-life of about 111 days $(T_{1/2}^{phys} = 30\,a)$. Assume that a certain quantity of ^{137}Cs corresponding to an activity of $4 \cdot 10^6\,Bq$ is incorporated due to a radiation accident. Work out the ^{137}Cs content of the radiation worker after a period of three years.

3.10 Assume that during the mounting of a radiation facility for medical tumour irradiation a 10 Ci ^{60}Co source falls down and is almost immediately recovered by a technician with his naked, unprotected hand. Work out the partial body dose and also estimate a value for the whole-body dose (exposure time $\approx$ 60 seconds for the hands and 5 minutes for the whole body).

3.11 A nuclear physics laboratory had been contaminated with a radioactive isotope. The *decontamination procedure* had an efficiency of $\varepsilon = 80\%$. After three decontamination procedures a remaining *surface contamination* of $512\,\text{Bq/cm}^2$ was still measured. Work out the initial contamination! By how much did the third decontamination procedure reduce the surface contamination?

   If the level of contamination had to be suppressed to $1\,\text{Bq/cm}^2$, how many procedures would have been required?

References

[1]   G.F. Knoll, *Radiation Detection and Measurement*, 3rd edition, John Wiley & Sons Inc., New York (Wiley Interscience), New York (1999/2000)

[2]   James E. Martin, *Physics for Radiation Protection*, John Wiley & Sons Inc., New York (2000)

[3]   E. Pochin, *Nuclear radiation: Risks and Benefits*, Clarendon Press, Oxford (1983)

[4]   C. Grupen, *Grundkurs Strahlenschutz*, Springer, Berlin (2003)

[5]   Alan Martin, Samuel A. Harbison, *An Introduction to Radiation Protection*, 3rd edition, Chapman and Hall, London (1987)

[6]   Particle Data Group, Review of Particle Physics, S. Eidelman *et al.*, *Phys. Lett.* **B592 Vol. 1–4** (2004) 1–1109; W.-M. Yao *et al.*, *J. Phys.* **G33** (2006) 1–1232; http://pdg.lbl.gov

[7]   Edgardo Browne & Richard B. Firestone, *Table of Radioactive Isotopes*, John Wiley & Sons Inc., New York (1986)

[8]   Richard B. Firestone & Virginia S. Shirley, *Table of Isotopes*, 8th edition, Wiley Interscience, New York (1998)

[9]   C. M. Lederer, *Table of Isotopes*, John Wiley & Sons Inc., New York (1978)

[10] H. Landolt & R. Börnstein, *Atomkerne und Elementarteilchen*, Vol. 5, Springer, Berlin (1952)

[11] Landolt–Börnstein, *Group I Elementary Particles, Nuclei and Atoms*, Springer, Berlin (2004), also as on-line version, see www.springeronline.com

[12] R.C. Weast & M.J. Astle (eds.), *Handbook of Chemistry and Physics*, CRC Press, Boca Raton, Florida (1979, 1987)

[13] C. Grupen, *Astroparticle Physics*, Springer, New York (2005)

4

Accelerators

The 'microscopes' of the particle physicist are enormous particle
accelerators.

American Institute of Physics

Accelerators are in use in many different fields, such as particle accelerators in nuclear and elementary particle physics, in nuclear medicine for tumour treatment, in material science, e.g. in the study of elemental composition of alloys, and in food preservation. Here we will be mainly concerned with accelerators for particle physics experiments [1–5]. Other applications of particle accelerators are discussed in Chapter 16.

Historically Röntgen's X-ray cathode-ray tube was an accelerator for electrons which were accelerated in a static electric field up to several keV. With electrostatic fields one can accelerate charged particles up to the several-MeV range.

In present-day accelerators for particle physics experiments much higher energies are required. The particles which are accelerated must be charged, such as electrons, protons or heavier ions. In some cases – in particular for colliders – also antiparticles are required. Such particles like positrons or antiprotons can be produced in interactions of electrons or protons. After identification and momentum selection they are then transferred into the accelerator system [6].

Accelerators can be linear or circular. *Linear accelerators* (Fig. 4.1) are mostly used as injectors for *synchrotrons*, where the magnetic guiding field is increased in a synchronous fashion with the increasing momentum so that the particle can stay on the same orbit. The guiding field is provided by magnetic dipoles where the Lorentz force keeps the particles on track. Magnetic quadrupoles provide a focussing of the beam. Since quadrupoles focus the beam in only one direction and defocus it in the perpendicular direction, one has to use pairs of quadrupoles to

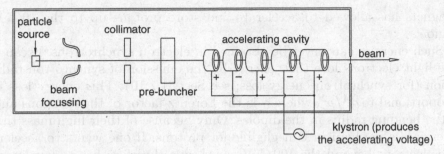

Fig. 4.1. Sketch of a linear accelerator. Particles emitted from the source are focussed and collimated. The continuous particle flow from the source is transformed into a discontinuous bunched beam which is steered into an accelerating cavity. The cavity is powered by a klystron.

achieve an overall focussing effect. The particles gain their energy in cavities which are fed by a radiofrequency generation (e.g. klystrons), which means that they are accelerated in an alternating electromagnetic field. *Field gradients* of more than $10\,\mathrm{MeV/m}$ can be achieved. Since the particles propagate on a circular orbit, they see the accelerating gradient on every revolution and thereby can achieve high energies. In addition to dipoles and quadrupoles there are usually also sextupoles and correction coils for beam steering. Position and beam-loss monitors are required for beam diagnostics, adjustments and control (Fig. 4.2). It almost goes without saying that the particles have to travel in an evacuated beam pipe, so that they do not lose energy by ionising collisions with gas molecules.

The maximum energy which can be achieved for protons is presently limited by the magnetic guiding field strength in synchrotrons and available resources. The use of large bending radii and superconducting

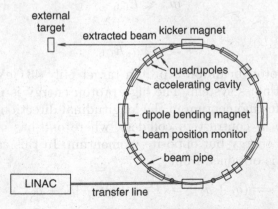

Fig. 4.2. Schematic layout of a synchrotron; LINAC – linear accelerator. The kicker magnet extracts the particle beam from the synchrotron.

magnets has allowed to accelerate and store protons up to the 10 TeV region.

Such energies can never be obtained in electron synchrotrons, because the light electrons lose their energy by the emission of synchrotron radiation (for synchrotron energy loss, see Sect. 1.1.10). This energy loss is proportional to γ^4/ρ^2, where γ is the Lorentz factor of the electrons and ρ the bending radius in the dipoles. Only because of their high mass this energy-loss mechanism is negligible for protons. If one wants to accelerate electrons beyond the 100 GeV range, one therefore has to use linear accelerators. With present-day technology a linear accelerator for electrons with a maximum beam energy of several hundred giga-electron-volts must have a length of ≈ 15 km, so that a linear e^+e^- collider would have a total length of ≈ 30 km [7].

In the past particle physics experiments were mostly performed in the *fixed-target mode*. In this case the accelerated particle is ejected from the synchrotron and steered into a fixed target, where the target particles except for the Fermi motion are at rest. The advantage of this technique is that almost any material can be used as target. With the target density also the interaction probability can be controlled. The disadvantage is that most of the kinetic energy of the projectile cannot be used for particle production since the centre-of-mass energy for the collision is relatively low. If $q_p = (E_{\text{lab}}, \vec{p}_{\text{lab}})$ and $q_{\text{target}} = (m_p, 0)$ are the four-momenta of the accelerated proton and the proton of the target, respectively, the centre-of-mass energy $\sqrt{s}$ in a collision with a target proton at rest is worked out to be

$$s = (q_p + q_{\text{target}})^2 = E_{\text{lab}}^2 + 2m_p E_{\text{lab}} + m_p^2 - p_{\text{lab}}^2 = 2m_p E_{\text{lab}} + 2m_p^2 \; . \quad (4.1)$$

Since for high energies

$$m_p \ll E_{\text{lab}} \; , \quad (4.2)$$

one has

$$\sqrt{s} = \sqrt{2m_p E_{\text{lab}}} \quad . \quad (4.3)$$

For a 1 TeV proton beam on a proton target only 43 GeV are available in the centre-of-mass system. The high proton energy is used to a large extent to transfer momentum in the longitudinal direction.

This is quite in contrast to colliders, where one has counterrotating beams of equal energy but opposite momentum. In this case the centre-of-mass energy is obtained from

$$s = (q_1 + q_2)^2 = (E_1 + E_2)^2 - |\vec{p}_1 + \vec{p}_2|^2 \; , \quad (4.4)$$

where q_1, q_2 are the four-momenta of the colliding particles. If the beams are of the same energy – when they travel in the same beam pipe such as in *particle–antiparticle colliders* this is always true – and if $\vec{p}_2 = -\vec{p}_1$, one gets

$$s = 4\,E^2 \qquad (4.5)$$

or

$$\sqrt{s} = 2\,E \ . \qquad (4.6)$$

In this case the full energy of the beams is made available for particle production. These conditions are used in *proton–antiproton* or *electron–positron colliders*. It is also possible to achieve this approximately for pp or e^-e^- collisions, however, at the expense of having to use two vacuum beam pipes, because in this case the colliding beams of equal charge must travel in opposite directions while in $p\bar{p}$ and e^+e^- machines both particle types can propagate in opposite directions in the same beam pipe. There is, however, one difference between pp or e^-e^- colliders on the one hand and $p\bar{p}$ or e^+e^- machines on the other hand: because of baryon- and lepton-number conservation the beam particles from pp or e^-e^- colliders – or equivalent baryonic or leptonic states – will also be present in the final state, so that not the full centre-of-mass energy is made available for particle production. For e^+e^- colliders one has also the advantage that the final-state particle production starts from a *well-defined quantum state*.

If particles other than protons or electrons are required as beam particles, they must first be made in collisions. Pions and kaons and other strongly interacting particles are usually produced in proton–nucleon collisions, where the secondary particles are momentum selected and identified. Secondary pion beams can also provide muons in their decay ($\pi^+ \rightarrow \mu^+ + \nu_\mu$). At high enough energies these muons can even be transferred into a collider ring thereby making $\mu^+\mu^-$ collisions feasible. *Muon colliders* have the advantage over electron–positron colliders that – due to their higher mass – they suffer much less synchrotron-radiation energy loss.

In high flux proton accelerators substantial neutrino fluxes can be provided which allow a study of neutrino interactions. Muon colliders also lead to intense neutrino fluxes in their decay which can be used in *neutrino factories*.

Almost all types of long-lived particles (π, K, Λ, Σ, ...) can be prepared for secondary fixed-target beams. In electron machines photons can be produced by bremsstrahlung allowing the possibility of $\gamma\gamma$ colliders as byproduct of linear e^+e^- colliders.

An important parameter in accelerator experiments is the number of events that one can expect for a particular reaction. For fixed-target experiments the *interaction rate* ϕ depends on the rate of beam particles n hitting the target, the cross section for the reaction under study, σ, and the target thickness d according to

$$\phi = \sigma \cdot N_A \; [\text{mol}^{-1}]/\text{g} \cdot \varrho \cdot n \cdot d \; \{\text{s}^{-1}\} \; , \tag{4.7}$$

where σ is the cross section per nucleon, N_A Avogadro's number, d the target thickness (in cm), and ϱ the density of the target material (in g/cm^3). Equation (4.7) can be rewritten as

$$\phi = \sigma L \; , \tag{4.8}$$

where L is called *luminosity*.

In collider experiments the situation is more complicated. Here one beam represents the target for the other. The interaction rate in this case is related to the luminosity of the collider which is a measure of the number of particles per cm^2 and s. If N_1 and N_2 are the numbers of particles in the colliding beams and σ_x and σ_y are the transverse beam dimensions, the luminosity L is related to these parameters by

$$L \propto \frac{N_1 N_2}{\sigma_x \sigma_y} \; . \tag{4.9}$$

It is relatively easy to determine N_1 and N_2. The measurement of the transverse beam size is more difficult. For a high interaction rate the two particle beams must of course completely overlap at the interaction point. The precise measurement of all parameters which enter into the luminosity determination cannot be performed with the required accuracy. Since, however, the luminosity is related to the interaction rate ϕ by Eq. (4.8), a process of well-known cross section σ can be used to fix the luminosity.

In e^+e^- colliders the well-understood QED process

$$e^+e^- \rightarrow e^+e^- \tag{4.10}$$

(*Bhabha scattering*, see Fig. 4.3), with a large cross section, can be precisely measured. Since this cross section is known theoretically with high precision, the e^+e^- luminosity can be accurately determined ($\delta L/L \ll 1\%$).

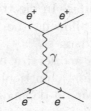

Fig. 4.3. Feynman diagram for Bhabha scattering in t-channel exchange.

The luminosity determination in pp or $p\bar{p}$ colliders is more difficult. One could use the elastic scattering for calibration purposes or the W and/or Z production. In Z production one can rely on the decay $Z \to \mu^+\mu^-$. Since the cross section for Z production and the branching ratio into muon pairs are well known, the luminosity can be derived from the number of muon pairs recorded.

In $\gamma\gamma$ colliders the QED process

$$\gamma\gamma \to e^+e^- \tag{4.11}$$

could be the basis for the luminosity measurement. Unfortunately, this process is only sensitive to one spin configuration of the two photons so that further processes (like the radiative process $\gamma\gamma \to e^+e^-\gamma$) must be used to determine the total luminosity.

If energies beyond the reach of earthbound accelerators are required, one has to resort to *cosmic accelerators* [8–10]. Experiments with cosmic-ray particles are always fixed-target experiments. To obtain centre-of-mass energies beyond $10\,\mathrm{TeV}$ in pp collisions with cosmic-ray protons one has to use cosmic-ray energies of

$$E_{\mathrm{lab}} \geq \frac{s}{2m_p} \approx 50\,\mathrm{PeV}(= 5 \cdot 10^{16}\,\mathrm{eV}) \; . \tag{4.12}$$

Since one has no command over the cosmic-ray beam, one must live with the low intensity of cosmic-ray particles at the high energies.

4.1 Problems

4.1 At the Large Hadron Collider the centre-of-mass energy of the two head-on colliding protons is $14\,\mathrm{TeV}$. How does this compare to a cosmic-ray experiment where an energetic proton collides with a proton at rest?

4.2 A betatron essentially works like a transformer. The current in an evacuated beam pipe acts as a secondary winding. The primary coil induces a voltage

$$U = \int \vec{E} \cdot \mathrm{d}\vec{s} = |\vec{E}| \cdot 2\pi R = -\frac{\mathrm{d}\phi}{\mathrm{d}t} = -\pi R^2 \frac{\mathrm{d}B}{\mathrm{d}t} \; .$$

While the induction increases by $\mathrm{d}B$, the accelerated electron gains an energy

$$\mathrm{d}E = e\,\mathrm{d}U = e|\vec{E}|\,\mathrm{d}s = e \cdot \frac{1}{2}R\frac{\mathrm{d}B}{\mathrm{d}t}\,\mathrm{d}s = e\frac{R}{2}v\,\mathrm{d}B \; , \quad v = \frac{\mathrm{d}s}{\mathrm{d}t} \; . \tag{4.13}$$

If the electron could be forced to stay on a closed orbit, it would gain the energy

$$E = e\frac{R}{2}\int_0^B v\,\mathrm{d}B\;.$$

To achieve this a guiding field which compensates the centrifugal force is required. Work out the relative strength of this steering field in relation to the accelerating time-dependent field B.

4.3 A possible uncontrolled beam loss in a proton storage ring might cause severe damage. Assume that a beam of 7 TeV protons ($N_p = 2\cdot10^{13}$) is dumped into a stainless-steel pipe of 3 mm thickness over a length of 3 m. The lateral width of the beam is assumed to be 1 mm. The 3 mm thick beam pipe absorbs about 0.3% of the proton energy. What happens to the beam pipe hit by the proton beam?

4.4 The LEP dipoles allow a maximum field of $B = 0.135$ T. They cover about two thirds of the 27 km long storage ring. What is the maximum electron energy that can be stored in LEP?

For LHC 10 T magnets are foreseen. What would be the maximum storable proton momentum?

4.5 Quadrupoles are used in accelerators for beam focussing. Let z be the direction of the beam. If ℓ is the length of the bending magnet, the bending angle α is

$$\alpha = \frac{\ell}{\rho} = \frac{e\,B_y}{p}\cdot\ell\;.$$

To achieve a focussing effect, this bending angle must be proportional to the beam excursion in x:

$$\alpha \propto x \;\Rightarrow\; B_y\cdot\ell \propto x\;;$$

for symmetry reasons: $B_x\cdot\ell \propto y$.

Which magnetic potential fulfils these conditions, and what is the shape of the surface of the quadrupole magnet?

References

[1] D.A. Edwards & M.J. Syphers, *An Introduction to the Physics of High Energy Accelerators*, John Wiley & Sons Inc., New York (1993)

[2] A.W. Chao & M. Tigner (eds.), *Handbook of Accelerator Physics and Engineering*, World Scientific, Singapore (1999)

[3] K. Wille, *The Physics of Particle Accelerators – An Introduction*, Oxford University Press, Oxford (2001)

[4] S.Y. Lee, *Accelerator Physics*, World Scientific Publishing Co., Singapore (1999)

[5] E.J.N. Wilson, *An Introduction to Particle Accelerators*, Oxford University Press, Oxford (2001)

[6] M. Reiser, *Theory and Design of Charged Particle Beams*, John Wiley & Sons Inc., New York (1994)

[7] M. Sands, *The Physics of Electron Storage Rings – An Introduction*, SLAC-R-121, UC-28 (ACC), 1970

[8] C. Grupen, *Astroparticle Physics*, Springer, New York (2005)

[9] T.K. Gaisser, *Cosmic Rays and Particle Physics*, Cambridge University Press, Cambridge (1991)

[10] T.K. Gaisser & G.P. Zank (eds.), *Particle Acceleration in Cosmic Plasmas*, AIP Conference Proceedings, No. 264 (1992) and Vol. 264, Springer, New York (1997)

5

Main physical phenomena used for particle detection and basic counter types

What we observe is not nature itself, but nature exposed to our method of questioning.

Werner Heisenberg

A particular type of detector does not necessarily make only one sort of measurement. For example, a segmented calorimeter can be used to determine particle tracks; however, the primary aim of such a detector is to measure the energy. The main aim of drift chambers is a measurement of particle trajectories but these devices are often used for particle identification by ionisation measurements. There is a number of such examples.

This chapter considers the main physical principles used for particle detection as well as the main types of counters (detector elements). The detectors intended for the measurement of certain particle characteristics are described in the next chapters. A brief introduction to different types of detectors can be found in [1].

5.1 Ionisation counters

5.1.1 *Ionisation counters without amplification*

An *ionisation counter* is a gaseous detector which measures the amount of ionisation produced by a charged particle passing through the gas volume. Neutral particles can also be detected by this device via secondary charged particles resulting from the interaction of the primary ones with electrons or nuclei. Charged particles are measured by separating the charge-carrier pairs produced by their ionisation in an electric field and guiding the

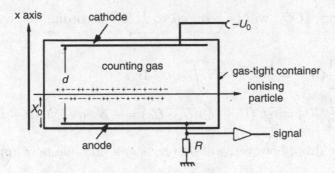

Fig. 5.1. Principle of operation of a planar ionisation chamber.

ionisation products to the anode or cathode, respectively, where corresponding signals can be recorded. If a particle is totally absorbed in an *ionisation chamber*, such a detector type measures its energy [2, 3].

In the simplest case an ionisation chamber consists of a pair of parallel electrodes mounted in a gas-tight container that is filled with a gas mixture which allows electron and ion drift. A voltage applied across the electrodes produces a homogeneous electric field.

In principle the counting gas can also be a liquid or even a solid (solid-state ionisation chamber). The essential properties of ionisation chambers are not changed by the phase of the counting medium.

Let us assume that a charged particle is incident parallel to the electrodes at a distance x_0 from the anode (Fig. 5.1). Depending on the particle type and energy, it produces along its track an ionisation, where the average energy required for the production of an electron–ion pair, W, is characteristic of the gas (see Table 1.2).

The voltage U_0 applied to the electrodes provides a uniform electric field

$$|\vec{E}| = E_x = U_0/d \ . \tag{5.1}$$

In the following we will assume that the produced charge is completely collected in the electric field and that there are no *secondary ionisation* processes or *electron capture* by possible electronegative gas admixtures.

The parallel electrodes of the ionisation chamber, acting as a capacitor with capacitance C, are initially charged to the voltage U_0. To simplify the consideration let us assume that the load resistor R is very large so that the capacitor can be considered to be independent. Suppose N charge-carrier pairs are produced along the particle track at a distance x_0 from the anode. The drifting charge carriers induce an electric charge on the electrodes which leads to certain change of the voltage, ΔU. Thereby the stored energy $\frac{1}{2}CU_0^2$ will be reduced to $\frac{1}{2}CU^2$ according to the following equations:

$$\frac{1}{2}CU^2 = \frac{1}{2}CU_0^2 - N\int_{x_0}^{x} qE_x\,dx\;, \tag{5.2}$$

$$\frac{1}{2}CU^2 - \frac{1}{2}CU_0^2 = \frac{1}{2}C(U+U_0)(U-U_0) = -N\cdot q\cdot E_x\cdot(x-x_0)\;. \tag{5.3}$$

The voltage drop, however, will be very small and one may approximate

$$U+U_0\approx 2U_0\;,\quad U-U_0=\Delta U\;. \tag{5.4}$$

Using $E_x = U_0/d$ one can work out ΔU with the help of Eq. (5.3),

$$\Delta U = -\frac{N\cdot q}{C\cdot d}(x-x_0)\;. \tag{5.5}$$

The signal amplitude ΔU has contributions from fast-moving electrons and the slowly drifting ions. If v^+ and v^- are the constant *drift velocities* of ions and electrons while $+e$ and $-e$ are their charges, one obtains

$$\Delta U^+ = -\frac{Ne}{Cd}v^+\Delta t\;,$$

$$\Delta U^- = -\frac{N(-e)}{Cd}(-v^-)\Delta t\;, \tag{5.6}$$

where Δt is the drift time. For ions $0 < \Delta t < T^+ = (d-x_0)/v^+$ while for electrons $0 < \Delta t < T^- = x_0/v^-$. It should be noted that electrons and ions cause contributions of the same sign since these carriers have opposite charges and opposite drift directions.

Because of $v^- \gg v^+$, the signal amplitude will initially rise linearly up to

$$\Delta U_1 = \frac{Ne}{Cd}\cdot(-x_0) \tag{5.7}$$

(the electrons will arrive at the anode, which is at $x=0$, at the time T^-) and then will increase more slowly by the amount which originates from the movement of ions,

$$\Delta U_2 = -\frac{Ne}{Cd}(d-x_0)\;. \tag{5.8}$$

Therefore the total signal amplitude, that is reached at $t=T^+$, is

$$\Delta U = \Delta U_1 + \Delta U_2 = -\frac{Ne}{Cd}x_0 - \frac{Ne}{Cd}(d-x_0) = -\frac{N\cdot e}{C}\;. \tag{5.9}$$

This result can also be derived from the equation describing the charge on a capacitor, $\Delta Q = -N \cdot e = C \cdot \Delta U$, which means that, independent of the construction of the ionisation chamber, the charge Q on the capacitor is reduced by the collected ionisation ΔQ, and this leads to a voltage amplitude of $\Delta U = \Delta Q/C$.

These considerations are only true if the charging resistor is infinitely large or, more precisely,

$$RC \gg T^-, T^+ .\tag{5.10}$$

When $RC \neq \infty$, expressions (5.6) should be modified,

$$\Delta U^+ = -\frac{Ne}{d} v^+ R(1 - e^{-\Delta t/RC}) ,$$

$$\Delta U^- = -\frac{N(-e)}{d}(-v^-)R(1 - e^{-\Delta t/RC}) .\tag{5.11}$$

In practical cases RC is usually large compared to T^-, but smaller than T^+. In this case one obtains [4]

$$\Delta U = -\frac{Ne}{Cd}x_0 - \frac{Ne}{d}v^+ R(1 - e^{-\Delta t/RC}) ,\tag{5.12}$$

which reduces to Eq. (5.9) if $RC \gg T^+ = (d - x_0)/v^+$.

For electric field strengths of $500\,\mathrm{V/cm}$ and typical drift velocities of $v^- = 5\,\mathrm{cm/\mu s}$, collection times for electrons of $2\,\mu s$ and for ions of about $2\,\mathrm{ms}$ are obtained for a drift path of $10\,\mathrm{cm}$. If the time constant $RC \gg 2\,\mathrm{ms}$, the signal amplitude is independent of x_0.

For many applications this is much too long. If one restricts oneself to the measurement of the electron component, which can be done by differentiating the signal, the total amplitude will not only be smaller, but also depend on the point in which the ionisation is produced, see Eq. (5.7).

This disadvantage can be overcome by mounting a grid between the anode and cathode (*Frisch grid* [5]). If the charged particle enters the larger volume between the grid and cathode, the produced charge carriers will first drift through this region which is shielded from the anode. Only when electrons penetrate through the grid, the signal on the working resistor R will rise. Ions will not produce any signal on R because their effect is screened by the grid. Consequently, this type of ionisation chamber with a Frisch grid measures only the electron signal which, in this configuration, is independent of the ionisation production region, as long as it is between the grid and the cathode.

Ionisation counters of this type are well suited for the detection of low-energy heavy particles. For example, $5\,\mathrm{MeV}$ α particles will deposit all their energy in a counter of $4\,\mathrm{cm}$ thickness filled with argon. Since

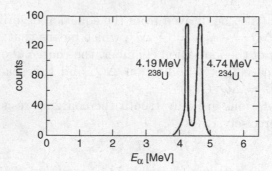

Fig. 5.2.   Pulse-height spectrum of α particles emitted from a ^{234}U/^{238}U isotope mixture recorded with a Frisch grid ionisation chamber [4].

$W \approx 26\,\mathrm{eV}$ for argon (see Table 1.2), the total number of electron–ion pairs will be

$$N = 5 \cdot 10^6\,\mathrm{eV}/26\,\mathrm{eV} = 1.9 \cdot 10^5 \; . \tag{5.13}$$

Assuming a capacity of $C = 10\,\mathrm{pF}$ we obtain the amplitude of the signal due to electrons as $\Delta U \approx 3\,\mathrm{mV}$ which can be easily measured with rather simple electronics.

Figure 5.2 shows the pulse-height spectrum of α particles emitted from a mixture of radioisotopes ^{234}U and ^{238}U recorded by a Frisch grid ionisation chamber [4]. ^{234}U emits α particles with energies of 4.77 MeV (72%) and 4.72 MeV (28%), while ^{238}U emits mainly α particles of energy 4.19 MeV. Although the adjacent α energies of the ^{234}U isotope cannot be resolved, one can, however, clearly distinguish between the two different uranium isotopes.

Ionisation chambers can also be used in the spectroscopy of particles of higher charge because in this case the deposited energies are in general larger compared to those of singly charged minimum-ionising particles. And indeed, minimum-ionising particles passing the same 4 cm of argon deposit only about 11 keV which provides about 400 pairs. To detect such a small signal is a very difficult task!

Apart from planar ionisation counters, *cylindrical ionisation counters* are also in use. Because of the cylindrical arrangement of the electrodes, the electric field in this case is no longer constant but rather rises like $1/r$ to the anode wire (see, for example, the famous book [6]):

$$\vec{E} = \frac{\tau}{2\pi\varepsilon_0 r}\frac{\vec{r}}{r} \; , \tag{5.14}$$

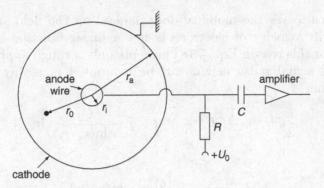

Fig. 5.3.   Principle of operation of a cylindrical ionisation counter.

where τ is the linear charge density on the wire. The potential distribution is obtained by integration:

$$U = U(r_i) - \int_{r_i}^{r} E(r)\, \mathrm{d}r \ . \tag{5.15}$$

Here r_a – radius of cylindrical cathode, r_i – anode-wire radius (Fig. 5.3). By taking into account the boundary condition $U(r_i) = U_0$, $U(r_a) = 0$, Formulae (5.15), (5.14) provide $U(r)$ and $E(r)$ using the intermediate $C_\tau = 2\pi\varepsilon_0/\ln(r_a/r_i)$ for the capacitance per unit length of the counter and $U_0 = \tau/C_\tau$:

$$U(r) = \frac{U_0 \ln(r/r_a)}{\ln(r_i/r_a)} \ , \quad |\vec{E}(r)| = \frac{U_0}{r\ln(r_a/r_i)} \ . \tag{5.16}$$

The field-dependent drift velocity can no longer assumed to be constant. The drift time of electrons is obtained by

$$T^- = \int_{r_0}^{r_i} \frac{\mathrm{d}r}{v^-(r)} \ , \tag{5.17}$$

if the ionisation has been produced locally at a distance r_0 from the counter axis (e.g. by the absorption of an X-ray photon). The drift velocity can be expressed by the mobility $\mu(\vec{v}^- = \mu^- \cdot \vec{E})$, and in the approximation that the mobility does not depend on the field strength one obtains $(\vec{v} \| (-\vec{E}))$,

$$T^- = -\int_{r_0}^{r_i} \frac{\mathrm{d}r}{\mu^- \cdot E} = -\int_{r_0}^{r_i} \frac{\mathrm{d}r}{\mu^- \cdot U_0} r\ln(r_a/r_i)$$
$$= \frac{\ln(r_a/r_i)}{2\mu^- \cdot U_0}(r_0^2 - r_i^2) \ . \tag{5.18}$$

In practical cases the mobility does depend on the field strength, so that the drift velocity of electrons is not a linear function of the field strength. For this reason Eq. (5.18) presents only a rough approximation. The related signal pulse height can be computed in a way similar to Eq. (5.2) from

$$\frac{1}{2}CU^2 = \frac{1}{2}CU_0^2 - N\int_{r_0}^{r_i} q \cdot \frac{U_0}{r\ln(r_a/r_i)}\,\mathrm{d}r \qquad (5.19)$$

to

$$\Delta U^- = -\frac{Ne}{C\ln(r_a/r_i)}\ln(r_0/r_i) \qquad (5.20)$$

with $q = -e$ for drifting electrons and C – the detector capacitance. It may clearly be seen that the signal pulse height in this case depends only logarithmically on the production point of ionisation.

The signal contribution due to the drift of the positive ions is obtained similarly,

$$\Delta U^+ = -\frac{Ne}{C}\frac{\ln(r_a/r_0)}{\ln(r_a/r_i)}\ . \qquad (5.21)$$

The ratio of pulse heights originating from ions and electrons, respectively, is obtained as

$$\frac{\Delta U^+}{\Delta U^-} = \frac{\ln(r_a/r_0)}{\ln(r_0/r_i)}\ . \qquad (5.22)$$

Assuming that the ionisation is produced at a distance $r_a/2$ from the anode wire, one gets

$$\frac{\Delta U^+}{\Delta U^-} = \frac{\ln 2}{\ln(r_a/2r_i)}\ . \qquad (5.23)$$

Since $r_a \gg r_i$, we obtain

$$\Delta U^+ \ll \Delta U^- , \qquad (5.24)$$

i.e., for all practical cases (homogeneous illumination of the chamber assumed) the largest fraction of the signal in the cylindrical ionisation chamber originates from the movement of electrons. For typical values of $r_a = 1\,\mathrm{cm}$ and $r_i = 15\,\mu\mathrm{m}$ the signal ratio is

$$\Delta U^+/\Delta U^- = 0.12 . \qquad (5.25)$$

The pulse duration from ionisation chambers varies in a wide range depending on the gas mixtures (e.g., 80% Ar and 20% CF_4 provides very fast pulses, $\approx 35\,\mathrm{ns}$) [7]. The length of a tube ionisation chamber is almost unlimited, for example, a detector in the form of a gas dielectric cable with a length of 3500 m was used as a beam-loss monitor at SLAC [8].

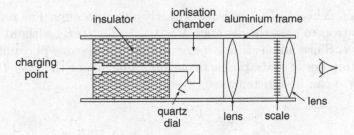

Fig. 5.4. Construction of an ionisation pocket dosimeter.

For radiation-protection purposes ionisation chambers are frequently used in a *current mode*, rather than a *pulse mode*, for monitoring the personal radiation dose. These *ionisation dosimeters* usually consist of a cylindrical air capacitor. The capacitor is charged to a voltage U_0. The charge carriers which are produced in the capacitor under the influence of radiation will drift to the electrodes and partially discharge the capacitor. The voltage reduction is a measure for the absorbed dose. The directly readable *pocket dosimeters* (Fig. 5.4) are equipped with an electrometer. The state of discharge can be read at any time using a built-in optics [9, 10].

5.1.2 Proportional counters

In ionisation chambers the primary ionisation produced by the incident particle is merely collected via the applied electric field. If, however, the field strength in some region of the counter volume is high, an electron can gain enough energy between two collisions to ionise another atom. Then the number of charge carriers increases. In cylindrical chambers the maximum field strength is around the thin-diameter anode wires due to the $1/r$ dependence of the electric field, see Eq. (5.16). The physics of electrical discharges in gases was developed by J.S. Townsend [11] and a good introduction is presented in [12, 13]. The signal amplitude is increased by the gas amplification factor A; therefore one gets, see Eq. (5.9),

$$\Delta U = -\frac{eN}{C} \cdot A \ . \tag{5.26}$$

The energy gain between two collisions is

$$\Delta E_{\text{kin}} = eE \cdot \lambda_0 \ , \tag{5.27}$$

assuming that the field strength $\vec{E}$ does not change over the mean free path length λ_0. To consider the multiplication process let us take a simple model. When the electron energy ΔE_{kin} at the collision is lower than a certain threshold, I_{ion}, the electron loses its energy without ionisation, while, when $\Delta E_{kin} > I_{ion}$, ionisation nearly always occurs. The probability for an electron to pass the distance $\lambda > \lambda_{ion} = I_{ion}/(eE)$ without collision is $e^{-\lambda_{ion}/\lambda_0}$. Since an electron experiences $1/\lambda_0$ collisions per unit length, the total number of ionisation acts per unit length – or the *first Townsend coefficient* – can be written as

$$\alpha = \frac{1}{\lambda_0} e^{-\lambda_{ion}/\lambda_0} . \qquad (5.28)$$

Taking into account the inverse proportionality of λ_0 to the gas pressure p, this can be rewritten as

$$\frac{\alpha}{p} = a \cdot e^{\frac{b}{E/p}} , \qquad (5.29)$$

where a and b are constants. In spite of its simplicity, this model reasonably describes the observed dependence when a and b are determined from experiment.

The first Townsend coefficient for different gases is shown in Fig. 5.5 for noble gases, and in Fig. 5.6 for argon with various additions of organic vapours. The first Townsend coefficient for argon-based gas mixtures at high electric fields can be taken from literature [14, 15].

If N_0 primary electrons are produced, the number of particles, $N(x)$, at the point x is calculated from

$$dN(x) = \alpha N(x) \, dx \qquad (5.30)$$

to be

$$N(x) = N_0 \, e^{\alpha x} . \qquad (5.31)$$

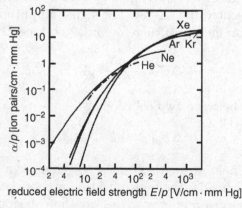

Fig. 5.5.   First Townsend coefficient for some noble gases [15–18].

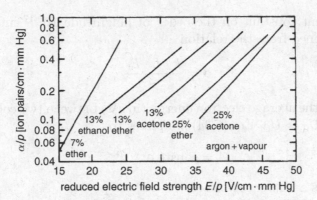

Fig. 5.6. First Townsend coefficient for argon with some organic vapour admixtures [16, 19, 20].

The first Townsend coefficient α depends on the field strength $\vec{E}$ and thereby on the position x in the gas counter. Therefore, more generally, it holds that

$$N(x) = N_0 \cdot e^{\int \alpha(x)\,dx} , \qquad (5.32)$$

where the gas amplification factor is given by

$$A = \exp\left\{ \int_{r_k}^{r_i} \alpha(x)\,dx \right\} . \qquad (5.33)$$

The lower integration limit is fixed by the distance r_k from the centre of the gas counter, where the electric field strength exceeds the critical value E_k from which point on charge-carrier multiplication starts. The upper integration limit is the anode-wire radius r_i.

The proportional range of a counter is characterised by the fact that the *gas amplification factor* A takes a constant value. As a consequence, the measured signal is proportional to the produced ionisation. Gas amplification factors of up to 10^6 are possible in the proportional mode. Typical gas amplifications are rather in the range between 10^4 up to 10^5.

If U_{th} is the threshold voltage for the onset of the proportional range, the gas amplification factor expressed by the detector parameters can be calculated to be [16]

$$A = \exp\left\{ 2\sqrt{\frac{kLCU_0 r_i}{2\pi\varepsilon_0}} \left[\sqrt{\frac{U_0}{U_{th}}} - 1 \right] \right\} ; \qquad (5.34)$$

where U_0 – applied anode voltage; $C = \frac{2\pi\varepsilon_0}{\ln r_a/r_i}$ – capacitance per unit length of the counter; L – number of atoms/molecules per unit volume ($\frac{N_A}{V_{mol}} = 2.69 \cdot 10^{19}/cm^3$) at normal pressure and temperature; k is a gas-dependent constant on the order of magnitude $10^{-17}\,\mathrm{cm^2/V}$, which can be obtained from the relation

$$\alpha = \frac{k \cdot L \cdot E_e}{e} \ , \tag{5.35}$$

where E_e is the average electron energy (in eV) between two collisions [16]. In the case $U_0 \gg U_{\mathrm{th}}$ Eq. (5.34) simplifies to

$$A = \mathrm{const} \cdot \mathrm{e}^{U_0/U_{\mathrm{ref}}} \ , \tag{5.36}$$

where U_{ref} is a reference voltage.

Equation (5.36) shows that the gas amplification rises exponentially with the applied anode-wire voltage. The detailed calculation of the gas amplification is difficult [11, 21–30], However, it can be measured quite easily. Let N_0 be the number of primary charge carriers produced in the proportional counter which, for example, have been created by the absorption of an X-ray photon of energy E_γ ($N_0 = E_\gamma/W$, where W is the average energy that is required for the production of one electron–ion pair). The integration of the current at the output of the proportional counter leads to the gas-amplified charge

$$Q = \int i(t)\,\mathrm{d}t \ , \tag{5.37}$$

which is again given by the relation $Q = e \cdot N_0 \cdot A$. From the current integral and the known primary ionisation N_0 the gas amplification A can be easily obtained.

At high field collisions of electrons with atoms or molecules can cause not only ionisation but also excitation. De-excitation is often followed by photon emission. The previous considerations are only true as long as photons produced in the course of the avalanche development are of no importance. These photons, however, will produce further electrons by the photoelectric effect in the gas or at the counter wall, which affect the avalanche development. Apart from gas-amplified primary electrons, secondary avalanches initiated by the photoelectric processes must also be taken into account. For the treatment of the gas amplification factor with inclusion of photons we will first derive the number of produced charge carriers in different generations.

In the first generation, N_0 primary electrons are produced by the ionising particle. These N_0 electrons are gas amplified by a factor A. If γ is the probability that one photoelectron per electron is produced in the avalanche, an additional number of $\gamma(N_0 A)$ photoelectrons is produced via photoprocesses. These, however, are again gas amplified so that in the second generation $(\gamma N_0 A) \cdot A = \gamma N_0 A^2$ gas-amplified photoelectrons the anode wire, which again create $(\gamma N_0 A^2)\gamma$ further photoelectrons in the gas amplification process, which again are gas amplified themselves. The gas amplification A_γ under inclusion of photons, therefore, is obtained from

$$N_0 A_\gamma = N_0 A + N_0 A^2 \gamma + N_0 A^3 \gamma^2 + \cdots$$

$$= N_0 A \cdot \sum_{k=0}^{\infty} (A\gamma)^k = \frac{N_0 A}{1 - \gamma A} \qquad (5.38)$$

to be

$$A_\gamma = \frac{A}{1 - \gamma A} . \qquad (5.39)$$

The factor γ, which determines the gas amplification under inclusion of photons, is also called the *second Townsend coefficient*.

As the number of produced charges increases, they begin to have an effect on the external applied field and saturation effects occur. For $\gamma A \to 1$ the signal amplitude will be independent of the primary ionisation. The proportional or, rather, the saturated proportional region is limited by gas amplification factors around $A_\gamma = 10^8$.

The process of *avalanche formation* takes place in the immediate vicinity of the anode wire (Fig. 5.7). One has to realise that half of the total produced charge appears at the last step of the avalanche! The mean free paths of electrons are on the order of μm so that the total avalanche formation process according to Eq. (5.31) requires only about 10–$20\,\mu$m. As a consequence, the effective production point of the charge (start of the avalanche process) is

$$r_0 = r_\mathrm{i} + k \cdot \lambda_0 , \qquad (5.40)$$

where k is the number of mean free paths which are required for the avalanche formation.

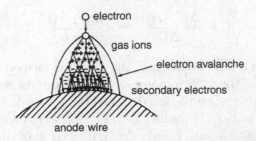

Fig. 5.7. Illustration of the avalanche formation on an anode wire in a proportional counter. By lateral diffusion a drop-shaped avalanche develops.

The ratio of signal amplitudes which are caused by the drift of positive ions or electrons, respectively, is determined to be, see Eq. (5.22),

$$\frac{\Delta U^+}{\Delta U^-} = \frac{-\frac{Ne}{C}\frac{\ln(r_a/r_0)}{\ln(r_a/r_i)}}{-\frac{Ne}{C}\frac{\ln(r_0/r_i)}{\ln(r_a/r_i)}} = \frac{\ln(r_a/r_0)}{\ln(r_0/r_i)} = R \ . \tag{5.41}$$

The gas amplification factor cancels in this ratio because equal numbers of electrons and ions are produced.

With typical values of $r_a = 1\,\text{cm}$, $r_i = 20\,\mu\text{m}$ and $k\lambda = 10\,\mu\text{m}$ this ratio is $R \approx 14$, which implies that in the proportional counter the signal on the anode wire is caused mainly by ions drifting slowly away from the wire and not by electrons which quickly drift in the direction of the wire.

The *rise time* of the electron signal can be calculated from Eq. (5.18). For electron mobilities in the range between $\mu^- = 100$ and $1000\,\text{cm}^2/\text{V s}$, an anode voltage of several hundred volts and typical detector dimensions as given above, the rise time is on the order of nanoseconds. The total ion drift time T^+ can be found analogously to Formula (5.18),

$$T^+ = \frac{\ln(r_a/r_i)}{2\mu^+ \cdot U_0}(r_a^2 - r_0^2) \ . \tag{5.42}$$

For the counter dimensions given above, $U_0 = 1000\,\text{V}$ and an ion mobility at normal pressure equal to $\mu^+ = 1.5\,\text{cm}^2/\text{V s}$, the ion drift time T^+ is about $2\,\text{ms}$.

On the other hand, the time dependence of the signal induced by the motion of ions, $\Delta U^+(t)$, is quite non-linear. The voltage drop caused by the drift of the ions created near the anode wire ($r \approx r_i$) to the point r_1 is, see Formula (5.21),

$$\Delta U^+(r_i, r_1) = -\frac{Ne}{C}\frac{\ln(r_1/r_i)}{\ln(r_a/r_i)} \ ; \tag{5.43}$$

and the ratio of this value to the total ion amplitude is

$$R = \frac{\Delta U^+(r_i, r_1)}{\Delta U^+} = \frac{\ln(r_1/r_i)}{\ln(r_a/r_i)} \ . \tag{5.44}$$

One can note that a large fraction of the signal is formed when ions move only a small part of the way from anode to cathode. As an example, let us calculate the R value when the ion drifts from the anode ($r = r_i = 20\,\mu\text{m}$) to the distance $r_1 = 10\,r_i$. Formula (5.44) gives $R \approx 0.4$ while the time that the ions require for this path is only $\Delta t^+(r_i, 10\,r_i) \approx 0.8\,\mu\text{s}$. It means that by differentiating the signal with an RC combination (as it is illustrated by Fig. 5.8) one can obtain a reasonably high and rather fast signal.

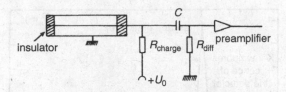

Fig. 5.8.   Readout of a proportional counter.

Fig. 5.9.   Photographic reproduction of an electron avalanche [13, 16, 31, 32]. The photo shows the form of the avalanche. It was made visible in a cloud chamber (see Chap. 6) by droplets which had condensed on the positive ions.

Of course, if $R_{\mathrm{diff}} \cdot C \approx 1\,\mathrm{ns}$ is chosen, one can even resolve the time structure of the ionisation in the proportional counter.

Raether was the first to photograph *electron avalanches* (Fig. 5.9, [13, 16, 31, 32]). In this case, the avalanches were made visible in a cloud chamber by droplets which had condensed on the positive ions. The size of the luminous region of an avalanche in a proportional chamber is rather small compared to different gas-discharge operation modes, such as in Geiger–Müller or streamer tubes.

Proportional counters are particularly suited for the spectroscopy of X rays. Figure 5.10 shows the energy spectrum of 59.53 keV X-ray photons

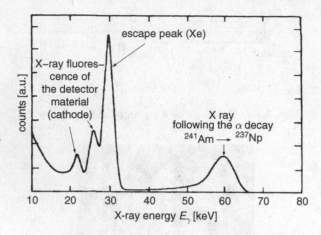

Fig. 5.10. Energy spectrum of 59.53 keV X-ray photons which are emitted in the α decay of ^{241}Am, measured in a xenon proportional counter [33].

which are emitted in the α decay $^{241}_{95}\text{Am} \rightarrow {}^{237}_{93}\text{Np}^* + \alpha$ from the excited neptunium nucleus. The spectrum was measured in a xenon proportional counter. The characteristic X-ray lines of the detector material and the Xe *escape peak* are also seen [33]. The escape peak is the result of the following process. The incident X rays ionise the Xe gas in most cases in the K shell. The resulting photoelectron only gets the X-ray energy minus the binding energy in the K shell. If the gap in the K shell is filled up by electrons from outer shells, X rays characteristic of the gas may be emitted. If these characteristic X rays are also absorbed by the photoelectric effect in the gas, a total-absorption peak is observed; if the characteristic X rays leave the counter undetected, the escape peak is formed (see also Sect. 1.2.1).

Proportional counters can also be used for X-ray imaging. Special electrode geometries allow one- or two-dimensional readout with high resolution for X-ray synchrotron-radiation experiments which also work at high rates [34, 35]. The electronic imaging of ionising radiation with limited avalanches in gases has a wide field of application ranging from cosmic-ray and elementary particle physics to biology and medicine [36].

The energy resolution of proportional counters is limited by the fluctuations of the charge-carrier production and their multiplication. Avalanche formation is localised to the point of ionisation in the vicinity of the anode wire. It *does not* propagate along the anode wire.

5.1.3 Geiger counters

The increase of the field strength in a proportional counter leads to a copious production of photons during the avalanche formation. As a

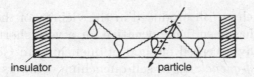

Fig. 5.11. Schematic representation of the transverse avalanche propagation along the anode wire in a Geiger counter.

consequence, the probability to produce further new electrons by the photoelectric effect increases. This photoelectric effect can also occur at points distant from the production of the primary avalanche. These electrons liberated by the photoelectric effect will initiate new avalanches whereby the discharge will propagate along the anode wire [37, 38] (Fig. 5.11).

The probability of photoelectron production per electron, γ, in the original avalanche becomes so large that the total number of charge carriers produced by various secondary and tertiary avalanches increases rapidly. As a consequence, the proportionality between the signal and the primary ionisation gets lost. The domain in which the liberated amount of charge does not depend on the primary ionisation is called the *Geiger mode*. The signal only depends on the applied voltage. In this mode of operation the signal amplitude corresponds to a charge signal of 10^8 up to 10^{10} electrons per primary produced electron.

After the passage of a particle through a Geiger counter (also called *Geiger–Müller counter* [39]) a large number of charge carriers are formed all along the anode wire. The electrons are quickly drained by the anode, however, the ions form a kind of flux tube which is practically stationary. The positive ions migrate with low velocities to the cathode. Upon impact with the electrode they will liberate, with a certain probability, new electrons, thereby starting the discharge anew.

Therefore, the discharge must be interrupted. This can be achieved if the charging resistor R is chosen to be so large that the momentary anode voltage $U_0 - IR$ is smaller than the threshold value for the Geiger mode (quenching by resistor).

Together with the total capacitance C the time constant RC has to be chosen in such a way that the voltage reduction persists until all positive ions have arrived at the cathode. This results in times on the order of magnitude of milliseconds, which strongly impairs the rate capability of the counter.

It is also possible to lower the applied external voltage to a level below the threshold for the Geiger mode for the ion drift time. This will, however, also cause long dead times. These can be reduced if the polarity of the electrodes is interchanged for a short time interval, thereby draining the positive ions, which are all produced in the vicinity of the anode wire, to the anode which has been made negative for a very short period.

A more generally accepted method of quenching in Geiger counters is the method of *self-quenching*. In self-quenching counters a quench gas is admixed to the counting gas which is in most cases a noble gas. Hydrocarbons like methane (CH_4), ethane (C_2H_6), isobutane (iC_4H_{10}), alcohols like ethyl alcohol (C_2H_5OH) or methylal ($CH_2(OCH_3)_2$), or halides like ethyl bromide are suitable as quenchers. These additions will absorb photons in the ultraviolet range (wavelength 100–200 nm) thereby reducing their range to a few wire radii ($\approx 100\,\mu m$). The transverse propagation of the discharge proceeds only along and in the vicinity of the anode wire because of the short range of the photons. The photons have no chance to liberate electrons from the cathode by the photoelectric effect because they will be absorbed before they can reach the cathode.

After a flux tube of positive ions has been formed along the anode wire, the external field is reduced by this space charge by such an amount that the avalanche development comes to an end. The positive ions drifting in the direction of the cathode will collide on their way with quench-gas molecules, thereby becoming neutralised,

$$Ar^+ + CH_4 \to Ar + CH_4^+ \ . \tag{5.45}$$

The molecule ions, however, have insufficient energy to liberate electrons from the cathode upon impact. Consequently, the discharge stops by itself. The charging resistor, therefore, can be chosen to be smaller, with the result that time constants on the order of $1\,\mu s$ are possible.

Contrary to the proportional mode, the discharge propagates along the whole anode wire in the Geiger mode. Therefore, it is impossible to record two charged particles in one Geiger tube at the same time. This is only achievable if the lateral propagation of the discharge along the anode wire can be interrupted. This can be accomplished by stretching insulating fibres perpendicular to the anode wire or by placing small droplets of insulating material on the anode wire. In these places the electric field is so strongly modified that the avalanche propagation is stopped. This locally *limited Geiger mode* allows the simultaneous registration of several particles on one anode wire. However, it has the disadvantage that the regions close to the fibres are inefficient for particle detection. The inefficient zone is typically 5 mm wide. The readout of simultaneous particle passages in the limited Geiger range is done via segmented cathodes.

5.1.4 Streamer tubes

In Geiger counters the fraction of counting gas to quenching gas is typically 90:10. The anode wires have diameters of $30\,\mu m$ and the anode voltage is around 1 kV. If the fraction of the quenching gas is considerably increased, the lateral propagation of the discharge along the anode wire can be completely suppressed. One again obtains, as in the proportional counter, a localised discharge with the advantage of large signals (gas amplification $\geq 10^{10}$ for sufficiently high anode voltages), which can be processed without any additional preamplifiers. These *streamer tubes* (Iarocci tubes, also developed by D.M. Khazins) [40–43] are operated with 'thick' anode wires between $50\,\mu$m and $100\,\mu$m diameter. Gas mixtures with $\leq 60\%$ argon and $\geq 40\%$ isobutane can be used. Streamer tubes operated with pure isobutane also proved to function well [44]. In this mode of operation the transition from the proportional range to streamer mode proceeds avoiding the Geiger discharges.

Figure 5.12 shows the amplitude spectra from a cylindrical counter with anode-wire diameter $100\,\mu$m, filled with argon/isobutane in proportion 60:40 under irradiation of electrons from a ^{90}Sr source [45]. At relatively low voltages of 3.2 kV small proportional signals caused by electrons are seen. At higher voltages (3.4 kV) for the first time streamer signals

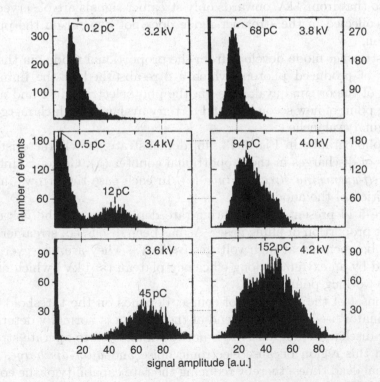

Fig. 5.12. Amplitude spectra of charge signals in a streamer tube. With increasing anode voltage the transition from the proportional to the streamer mode is clearly visible [45].

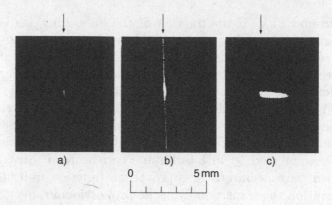

Fig. 5.13. Gas discharges in (a) a proportional counter, (b) a Geiger counter and (c) a self-quenching streamer tube; the arrows indicate the position of the anode wire [46].

with distinctly higher amplitudes also occur along with the proportional signals. For even higher voltages the proportional mode completely disappears, so that from 4 kV onwards only streamer signals are observed. The charge collected in the streamer mode does not depend on the primary ionisation.

The streamer mode develops from the proportional mode via the large number of produced photons which are re-absorbed in the immediate vicinity of the original avalanche via the photoelectric effect and are the starting point of new secondary and tertiary avalanches which merge with the original avalanche.

The photographs in Fig. 5.13 [46] demonstrate the characteristic differences of discharges in the proportional counter (a), Geiger counter (b) and a *self-quenching streamer tube* (c). In each case the arrows indicate the position of the anode wire.

Figure 5.14 presents the counting-rate dependence on the voltage for different proportion of filling gases. As has been discussed, streamer tubes have to be operated at high voltages ($\approx 5\,\mathrm{kV}$). They are, however, characterised by an extremely long efficiency plateau ($\approx 1\,\mathrm{kV}$) which enables a stable working point.

The onset of the efficiency, of course, depends on the threshold of the discriminator used. The upper end of the plateau is normally determined by after-discharges and noise. It is not recommended to operate streamer tubes in this region because electronic noise and *after-discharges* cause additional dead times, thereby reducing the rate capability of the counter.

If 'thick' anode wires are used, the avalanche is caused mostly by only one primary electron and the discharge is localised to the side of the anode wire which the electron approaches. The signals can be directly measured

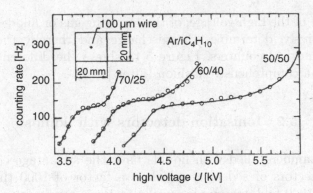

Fig. 5.14. Dependence of the counting rate on the high voltage in a streamer tube [45].

on the anode wire. Additionally or alternatively, one can also record the signals induced on the cathodes. A segmentation of the cathodes allows the determination of the track position along the anode wire.

Because of the simple mode of operation and the possibility of multi-particle registration on one anode wire, streamer tubes are an excellent candidate for sampling elements in calorimeters. A fixed charge signal Q_0 is recorded per particle passage. If a total charge Q is measured in a streamer tube, the number of equivalent particles passing is calculated to be $N = Q/Q_0$.

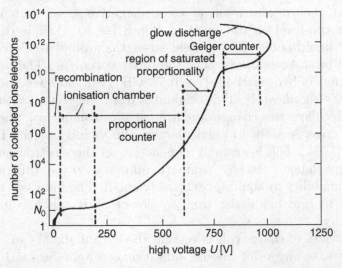

Fig. 5.15. Characterisation of the modes of operation of cylindrical gas detectors (after [16]). When the high voltage is increased beyond the Geiger regime (for counters with small-diameter anode wires), a *glow discharge* will develop and the voltage breaks down. This will normally destroy the counter.

The choice of the high voltage, of the counting gas or anode-wire diameter, respectively, determines the discharge and thereby the operation mode of cylindrical counters. Figure 5.15 shows the different regions of operation in a comprehensive fashion (after [16]).

5.2   Ionisation detectors with liquids

Ionisation chambers filled with liquids have the advantage compared to gas-filled detectors of a density which is a factor of 1000 times higher. It implies a 1000-fold energy absorption in these media for a relativistic particle and the photon-detection efficiency increases by the same factor. Therefore, ionisation chambers filled with liquids are excellent candidates for sampling and homogeneous-type calorimeters [47–51].

The average energy for the production of an electron–ion pair in liquid argon (LAr) is 24 eV, and in liquid xenon (LXe) it is 16 eV. A technical disadvantage, however, is related to the fact that noble gases only become liquid at low temperatures. Typical temperatures of operation are 85 K for LAr, 117 K for LKr and 163 K for LXe. Liquid gases are homogeneous and therefore have excellent counting properties. Problems may, however, arise with electronegative impurities which must be kept at an extremely low level because of the slow drift velocities in the high-density liquid counting medium. To make operation possible, the absorption length λ_{ab} of electrons must be comparable to the electrode distance. This necessitates that the concentration of electronegative gases such as O_2 be reduced to the level on the order of 1 ppm ($\equiv 10^{-6}$). The drift velocity in pure liquid noble gases at field strengths around 10 kV/cm, which are typical for *LAr counters*, is of the order 0.4 cm/µs. The addition of small amounts of hydrocarbons (e.g. 0.5% CH_4) can, however, increase the drift velocity significantly. This originates from the fact that the admixture of molecular gases changes the average electron energy. The electron scattering cross section, in particular, in the vicinity of the Ramsauer minimum [17, 52–56], is strongly dependent on the electron energy. So, small energy changes can have dramatic influence on the drift properties.

The ion mobility in liquids is extremely small. The induced charge due to the ion motion has a rise time so slow that it can hardly be used electronically.

The processes of charge collection and the output signal can be considered in the same way as for gaseous ionisation counters (Sect. 5.1.1). Often the integration time in the readout electronics is chosen much shorter than the electron drift time. This decreases the pulse height but makes the signal faster and reduces the dependence on the point of ionisation production.

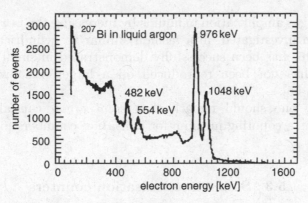

Fig. 5.16. Energy spectrum of conversion electrons from the isotope ^{207}Bi in a
liquid-argon chamber [57]. The spectrum also shows the Compton edges of the
570 keV and 1064 keV photons.

Figure 5.16 shows the energy spectrum of conversion electrons from
^{207}Bi, recorded with a liquid-argon ionisation chamber. The ^{207}Bi nuclei
decay by electron capture into excited states of lead nuclei. De-excitation
occurs by the emission of 570 keV and 1064 keV photons or by transfer
of this excitation energy to the electrons at K and L shells of lead (see
Table 3.3, Appendix 5). Thus, two K and L line pairs corresponding to the
nuclear level transitions of 570 keV and 1064 keV are seen in the spectrum.
The liquid-argon chamber separates the K and L electrons relatively well
and achieves a resolution of $\sigma_E = 11$ keV [57].

The operation of *liquid-noble-gas ionisation chambers* requires cryogenic equipment. This technical disadvantage can be overcome by the
use of *'warm' liquids*. The requirements for such 'warm' liquids, which
are already in the liquid state at room temperature, are considerable:
they must possess excellent drift properties and they must be extremely
free of electronegative impurities (< 1 ppb). The molecules of the 'warm'
liquid must have a high symmetry (i.e. a near spherical symmetry) to
allow favourable drift properties. Some organic substances like tetramethylsilane (TMS) or tetramethylpentane (TMP) are suitable as 'warm'
liquids [49, 58–61].

Attempts to obtain higher densities, in particular, for the application of
liquid ionisation counters in calorimeters, have also been successful. This
can be achieved, for example, if the silicon atom in the TMS molecule is
replaced by lead or tin (tetramethyltin (TMT) [62] or tetramethyllead).
The flammability and toxicity problems associated with such materials
can be handled in practice, if the liquids are sealed in vacuum-tight con-
tainers. These 'warm' liquids show excellent *radiation hardness*. Due to
the high fraction of hydrogen they also allow for compensation of signal
amplitudes for electrons and hadrons in calorimeters (see Chap. 8).

Obtaining gas amplification in liquids by increasing the working voltage has also been investigated, in a fashion similar to cylindrical ionisation chambers. This has been successfully demonstrated in small prototypes; however, it has not been reproduced on a larger scale with full-size detectors [63–65].

In closing, one should remark that *solid argon* can also be used successfully as a counting medium for ionisation chambers [66].

5.3   Solid-state ionisation counters

Solid-state detectors are essentially ionisation chambers with solids as a counting medium. Because of their high density compared to gaseous detectors, they can absorb particles of correspondingly higher energy. Charged particles or photons produce *electron–hole pairs* in a crystal. An electric field applied across the crystal allows the produced charge carriers to be collected.

The operating principle of solid-state detectors can be understood from the *band model of solids*. An introduction to the band theory of solids can be found, for example, in [67]. In the frame of this theory, the discrete electron energy levels of individual atoms or ions within a whole crystal are merged forming energy bands, as it is shown in Fig. 5.17. According to the Pauli exclusion principle, each band can contain only a finite number of electrons. So, some low energy bands are fully filled with electrons while the high energy bands are empty, at least at low temperature. The lowest partially filled or empty band is called *conduction band* while the highest fully filled band is referred to as *valence band*. The gap between the top of the valence band, V_V, and the bottom of the conduction band, V_C, is called *forbidden band* or *energy gap* with a width of $E_g = V_C - V_V$.

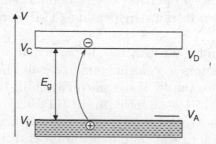

Fig. 5.17.   Band structure of solid-state material. V_V and V_C are the top of valence band and bottom of the conduction band; E_g – forbidden gap; V_A and V_D – acceptor and donor levels.

When the 'conduction band' is partially filled, electrons can move easily under the influence of an electric field, hence, this solid is a conductor. Such a material cannot be used as an ionisation counter. The solids which have basically empty conduction bands are divided conventionally into insulators (specific resistivity 10^{14}–$10^{22}\,\Omega\,cm$ at room temperature) and semiconductors (10^{9}–$10^{-2}\,\Omega\,cm$). The electric charge in these materials is carried by electrons which have been excited to the conduction band from the valence band. The corresponding vacancies in the valence band are called *holes* and are able to drift in the electric field as well. The main difference between insulators and semiconductors lies in the value of E_{g}. For insulators it is typically $E > 3\,eV$ while for semiconductors it is in the range of $1\,eV$.

Insulators are not widely used as ionisation counters. The main reasons are the low hole mobility in most of such crystals as well as the necessity of using very high-purity crystals. *Impurities* can create deep traps in wide-gap solids causing polarisation of the crystal under irradiation. The common solid-state ionisation counters are based on semiconductors.

The *specific resistivity* of the material is determined as

$$\varrho = \frac{1}{e(n\mu_e + p\mu_p)} , \qquad (5.46)$$

where n and p are electron and hole concentrations, respectively, while μ_e and μ_p are their mobilities and e is the elementary charge.

In a pure semiconductor the electron concentration, n, is equal to the hole concentration, p. These values can be approximated by the expression [68, 69]

$$n = p \approx 5 \cdot 10^{15}\,(T[\mathrm{K}])^{3/2}\,\mathrm{e}^{-E_\mathrm{g}/(2kT)} , \qquad (5.47)$$

where T is the temperature in K. For silicon with a band gap of $E_{\mathrm{g}} = 1.07\,eV$ Eq. (5.47) results in $n \approx 2 \cdot 10^{10}\,cm^{-3}$ at $T = 300\,K$. Taking $\mu_e = 1300\,cm^2\,s^{-1}\,V^{-1}$ and $\mu_p = 500\,cm^2\,s^{-1}\,V^{-1}$ one gets an estimation for the specific resistivity, $\varrho \approx 10^5\,\Omega\,cm$. For Ge ($E_{\mathrm{g}} = 0.7\,eV$, $\mu_e = 4000\,cm^2\,s^{-1}\,V^{-1}$, $\mu_p = 2000\,cm^2\,s^{-1}\,V^{-1}$) the specific resistivity is about one order of magnitude lower. The impurities, even at low level, which almost always exist in the material, can substantially decrease these values.

Thus, semiconductors are characterised by a relatively high dark current. To suppress this, usually multilayer detectors containing layers with different properties are built. Electron and hole concentrations in these layers are intentionally changed by special doping.

Germanium and silicon have four electrons in the outer shell. If an atom with five electrons in the outer shell, like phosphorus or arsenic, is incorporated to the crystal lattice, the fifth electron of the impurity atom is only weakly bound and forms a *donor level* V_D that is just under but very close to the conduction band (see Fig. 5.17). Typically, the difference $V_C - V_D$ is in the range of 0.05 eV and this electron can easily be lifted to the conduction band. Material with an impurity of this type has a high concentration of free electrons and is therefore referred to as *n-type semiconductor*.

If trivalent electron acceptor impurities like boron or indium are added to the lattice, one of the silicon bonds remains incomplete. This *acceptor level*, which is about 0.05 eV above the edge of the valence band (V_A in Fig. 5.17), tries to attract one electron from a neighbouring silicon atom creating a hole in the valence band. This type of material with high concentration of free holes is referred to as *p-type semiconductor*.

Let us consider the phenomena of a *pn* junction at the interface of two semiconductors of p and n type. The electrons of the n-type semiconductor diffuse into the p type, and the holes from the p type to the n-type region. This leads to the formation of a space-charge distribution shown in Fig. 5.18. The positive charge in the n-type region and the negative charge in p-type area provide the electric field which draws the free electrons and holes to the opposite direction out of the region of the electric field. Thus, this area has a low concentration of free carriers and is called the *depletion region* or *depletion layer*. When no external voltage is applied, the diffusion of carriers provides a contact potential, U_c, which is typically $\approx 0.5\,\mathrm{V}$.

The *pn junction* has the properties of a diode. At a 'direct' bias, when an external positive voltage is applied to the p region, the depletion area shortens causing a large direct current. At *reverse bias*, when an external positive voltage is applied to the n region, the depletion layer increases. A detailed consideration of the physics of *pn* junctions is given, for example, in [69] and its application to semiconductor detectors can be found in [68]. Electron–hole pairs released by photons interacting in the depletion area or by charged particles crossing the depletion layer are separated by the electric field and the carriers are collected by the electrodes inducing a current pulse. It is worth mentioning that electron–hole pairs created beyond the depletion layer do not produce an electric pulse since the electric field outside the *pn* junction is negligible due to the high charge-carrier concentration there.

Thus, a semiconductor device with a *pn* junction can be used as an ionisation detector. The total charge collected by this detector is proportional to the energy deposited in the depletion layer. Usually, one of the two semiconductor layers (p or n) has a much higher carrier concentration than the other. Then the depletion region extends practically all over the

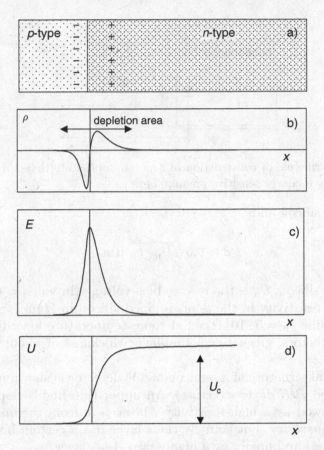

Fig. 5.18. (a) Working principle of a *pn* semiconductor counter; (b) space-charge distribution including all kinds of charge carriers: free electrons and holes, fixed positive non-compensating ions, electrons captured at acceptor levels; (c) electric field; (d) potential distribution. When no external voltage is applied, the maximum potential is equal to the contact voltage U_c.

area with low carrier concentration and, hence, high resistivity. The width of the depletion area, d, in this case can be expressed as [68]

$$d = \sqrt{2\varepsilon(U + U_c)\mu\varrho_d} \, , \tag{5.48}$$

where U is the external *reverse-bias voltage*, ε the dielectric constant of the material ($\varepsilon = 11.9\,\varepsilon_0 \approx 1\,\text{pF/cm}$), ϱ_d the specific resistivity of the low-doped semiconductor, and μ the mobility of the main carriers in the low-doped area. This expression leads to

$$d \approx 0.3\sqrt{U_n \cdot \varrho_p} \;\mu\text{m} \tag{5.49}$$

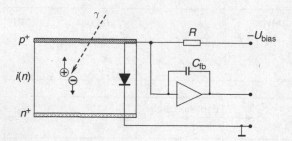

Fig. 5.19. Principle of construction of a p–i–n solid state detector along with its readout by a charge-sensitive preamplifier.

for p-doped silicon and

$$d \approx 0.5\sqrt{U_\mathrm{n} \cdot \varrho_\mathrm{n}}\ \mu\mathrm{m} \tag{5.50}$$

for n-doped silicon. U_n is the reverse-bias voltage (in volts, $= U/\mathrm{V}$), $\varrho_\mathrm{p,n}$ the specific resistivity in the p- or n-doped silicon in $\Omega\,\mathrm{cm}$ ($= \varrho_\mathrm{d}/\Omega\,\mathrm{cm}$). A typical value $\varrho_\mathrm{n} = 5 \cdot 10^3\,\Omega\,\mathrm{cm}$, at room temperature for n-type silicon used for detectors, gives a depletion-layer thickness of about $350\,\mu\mathrm{m}$ at $V = 100\,\mathrm{V}$.

The typical structure of a semiconductor detector is shown in Fig. 5.19 (the so-called *PIN diode structure*). An upper thin highly doped p layer (p^+) is followed by a high-resistivity i layer (i is from *intrinsically conducting* or *insulator*, but actually the i layer has a certain but very low n or p doping) and finally by a highly doped n^+ layer.*

In the example presented in Fig. 5.19 the pn junction appears at the p^+–$i(n)$ border and extends over the whole $i(n)$ area up to the n^+ layer that plays the rôle of an electrode. Usually, the upper p^+ layer is shielded by a very thin SiO_2 film.

Since *semiconductor diodes* do not have an intrinsic amplification, the output signal from this device is quite small. For example, a minimum-ionising particle crossing a depletion layer of $300\,\mu\mathrm{m}$ thickness produces about $3 \cdot 10^4$ electron–hole pairs corresponding to only $4.8 \cdot 10^{-15}\,\mathrm{C}$ of collected charge. Therefore, the processing of signals from solid-state detectors requires the use of low-noise *charge-sensitive amplifiers*, as shown in Fig. 5.19, followed by a shaper (see Chap. 14). To suppress electronics noise the integration time is usually relatively rather long – from hundreds of ns to tens of μs.

---

* Normally, only a very low concentration of dopant atoms is needed to modify the conduction properties of a semiconductor. If a comparatively small number of dopant atoms is added (concentration $\approx 10^{-8}$), the doping concentration is said to be low, or light, denoted by n^- or p^-. If a much higher number is required ($\approx 10^{-4}$) then the doping is referred to as heavy, or high, denoted by n^+ or p^+.

The charge collection time for such a detector can be estimated by taking an average field strength of $E = 10^3 \, \text{V/cm}$ and charge-carrier mobilities of $\mu = 10^3 \, \text{cm}^2/\text{V s}$:

$$t_s = \frac{d}{\mu E} \approx 3 \cdot 10^{-8} \, \text{s} \; . \tag{5.51}$$

The shape of the signal can be found in a similar way as for gaseous ionisation detectors (see Sect. 5.1.1), but for semiconductors the difference between mobilities of holes and electrons is only a factor of 2 to 3 in contrast to gases where ions are by 3 orders of magnitude less mobile than electrons. Therefore, the signal in semiconductor detectors is determined by both types of carriers and the collected charge does not depend on the point of where the ionisation was produced.

For α and electron spectroscopy the depletion layer in semiconductor counters should be very close to the surface to minimise the energy loss in an inactive material. The *surface-barrier detectors* meet this requirement. These detectors are made of an n-conducting silicon crystal by a special treatment of its surface producing a super-thin p-conducting film. A thin evaporated gold layer of several µm thickness serves as a high-voltage contact. This side is also used as an entrance window for charged particles.

Semiconductor counters with depletion areas up to 1 mm are widely used for α-, low-energy β-, and X-ray detection and spectroscopy. Detectors of this type can be operated at room temperature as well as under cooling for dark-current suppression. In particle physics at high energies silicon detectors are typically used as high-resolution tracking devices in the form of strip, pixel or voxel counters (see Chaps. 7 and 13).

However, for gamma and electron spectroscopy in the MeV range as well as for α and proton energy measurements in the 10–100 MeV range the thickness of the depletion area should be much larger. To achieve this one should use a material with high intrinsic resistivity, as is seen from Formulae (5.48), (5.49), (5.50). One way of increasing the resistivity is cooling the device (see Formula (5.47)).

In the early 1960s high-resistivity-compensated silicon and germanium became available. In these materials the net free charge-carrier concentration was reduced by drifting lithium into p-conducting, e.g. boron-doped, silicon. Lithium has only one electron in the outer shell and is therefore an electron donor, since this outer electron is only weakly bound. Lithium atoms are allowed to diffuse into the p-conducting crystal at a temperature of about 400 °C. Because of their small size, reasonable diffusion velocities are obtained with lithium atoms. A region is formed in which the number of boron ions is compensated by the lithium ions. This technology provides material with a specific resistivity of $3 \cdot 10^5 \, \Omega \, \text{cm}$ in the depletion layer, which is approximately equal to the intrinsic conductivity of silicon without any impurities. In this way, p–i–n structures can be produced with relatively thin p and n regions and with i zones up to 5 mm.

From the early 1980s on *high-purity germanium crystals* (HPGe) with impurity concentrations as low as 10^{10} cm^{-3} became available. Nowadays, HPGe detectors have almost replaced the Ge(Li) type. HPGe detectors have the additional advantage that they only have to be cooled during operation, while Ge(Li) detectors must be permanently cooled to prevent the lithium ions from diffusing out of the intrinsically conducting region. At present HPGe detectors with an area up to 50 cm^2 and a thickness of the sensitive layer up to 5 cm are commercially available, the largest coaxial-type HPGe detector has a diameter and a length of about 10 cm [70]. Usually, all Ge detectors operate at liquid-nitrogen temperatures, i.e. at ≈ 77 K.

The energy resolution of semiconductor detectors can be approximated by the combination of three terms:

$$\sigma_E = \sqrt{\sigma_{\text{eh}}^2 + \sigma_{\text{noise}}^2 + \sigma_{\text{col}}^2} \,, \tag{5.52}$$

where σ_{eh} is the statistical fluctuation of the number of electron–hole pairs, σ_{noise} the contribution of electronics noise, and σ_{col} the contribution of the non-uniformity of the charge collection efficiency and other technical effects.

For solid-state counters, just as with gaseous detectors, the statistical fluctuation of the number of produced charge carriers is smaller than Poissonian fluctuations, $\sigma_{\text{P}} = \sqrt{n}$. The shape of a monoenergetic peak is somewhat asymmetric and narrower than a Gaussian distribution. The Fano factor F (measurements for silicon and germanium give values from 0.08 to 0.16 [68], see also Chap. 1) modifies the Gaussian variance σ_{P}^2 to $\sigma^2 = F\sigma_{\text{P}}^2$, so that the electron–hole-pair statistics contribution to the energy resolution – because E is proportional to n – can be represented by

$$\frac{\sigma_{\text{eh}}(E)}{E} = \frac{\sqrt{F\sigma_{\text{P}}^2}}{n} = \frac{\sqrt{n}\sqrt{F}}{n} = \frac{\sqrt{F}}{\sqrt{n}} \,. \tag{5.53}$$

Since the number of electron–hole pairs is $n = E/W$, where W is the average energy required for the production of one charge-carrier pair, one obtains

$$\frac{\sigma(E)}{E} = \frac{\sqrt{F \cdot W}}{\sqrt{E}} \,. \tag{5.54}$$

The properties of commonly used semiconductors are presented in Table 5.1.

Figure 5.20 presents the energy spectrum of photons from a ^{60}Co radioactive source, as measured with a HPGe detector (Canberra GC

Table 5.1. *Properties of commonly used semiconductors [68, 71]*

Characteristic property	Si	Ge
Atomic number	14	32
Atomic weight	28.09	72.60
Density in g/cm^3	2.33	5.32
Dielectric constant	12	16
Energy gap at 300 K in eV	1.12	0.67
Energy gap at 0 K in eV	1.17	0.75
Charge carrier density at 300 K in cm^{-3}	$1.5 \cdot 10^{10}$	$2.4 \cdot 10^{13}$
Resistivity at 300 K in Ω cm	$2.3 \cdot 10^5$	47
Electron mobility at 300 K in cm^2/V s	1350	3900
Electron mobility at 77 K in cm^2/V s	$2.1 \cdot 10^4$	$3.6 \cdot 10^4$
Hole mobility at 300 K in cm^2/V s	480	1900
Hole mobility at 77 K in cm^2/V s	$1.1 \cdot 10^4$	$4.2 \cdot 10^4$
Energy per e–h pair at 300 K in eV	3.62	≈ 3 for HPGe[a]
Energy per e–h pair at 77 K in eV	3.76	2.96
Fano factor[b] at 77 K	≈ 0.15	≈ 0.12

[a] For room-temperature operation High Purity Germanium (HPGe) is required.
[b] The value of the Fano factor shows a large scatter in different publications, see [68].

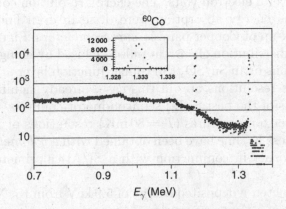

Fig. 5.20. Gamma spectrum from a ^{60}Co source measured by a HPGe detector (by courtesy of V. Zhilich). The peaks correspond to the 1.17 MeV and 1.33 MeV ^{60}Co γ lines, while the two shoulders in the central part of the spectrum are caused by Compton edges related to the full-absorption lines (see Table 3.3 and Appendix 5).

2518) [72]. The two ^{60}Co γ lines, 1.17 MeV and 1.33 MeV, are clearly seen, and the energy resolution at $E_\gamma = 1.33$ MeV is about 2 keV (FWHM) or FWHM/$E_\gamma \approx 1.5 \cdot 10^{-3}$. The theoretical limitation imposed by electron–hole-pair statistics on the energy resolution for this case can be estimated from Formula (5.54). Values of $W \approx 3$ eV and $F \approx 0.1$ lead to

$$\sigma(E)/E \approx 4.7 \cdot 10^{-4} \ , \ \ \text{FWHM}/E = 2.35 \cdot \sigma(E)/E \approx 1.1 \cdot 10^{-3} \quad (5.55)$$

for $E_\gamma = 1.33$ MeV. This result is not very far from the experimentally obtained detector parameters.

Although only silicon and germanium semiconductor detectors are discussed here, other materials like gallium arsenide (GaAs) [73, 74], cadmium telluride (CdTe) and cadmium–zinc telluride [75] can be used for particle detectors in the field of nuclear and elementary particle physics.

To compare the spectroscopic properties of solid-state detectors with other counters (see Sects. 5.1, 5.2, and 5.4), we have to note that the average energy required for the creation of an electron–hole pair is only $W \approx 3$ eV. This parameter, according to Formulae (5.53) and (5.54), provides in principle the limitation for the energy resolution. For gases and liquid noble gases W is approximately 10 times larger, $W \approx 20$–30 eV, while for scintillation counters the energy required to produce one photoelectron at the photosensor is in the range of 50–100 eV. In addition, one cannot gain anything here from the Fano factor ($F \approx 1$).

Semiconductor counters are characterised by quantum transitions in the range of several electron volts. The energy resolution could be further improved if the energy absorption were done in even finer steps, such as by the break-up of Cooper pairs in *superconductors*. Figure 5.21 shows the amplitude distribution of current pulses, caused by manganese K$_\alpha$ and K$_\beta$ X-ray photons in an Sn/SO$_x$/Sn tunnel-junction layer at $T = 400$ mK. The obtainable resolutions are, in this case, already significantly better than the results of the best Si(Li) semiconductor counters [76].

For even lower temperatures ($T = 80$ mK) resolutions of 17 eV FWHM for the manganese K$_\alpha$ line have been obtained with a *bolometer* made from a HgCdTe absorber in conjunction with a Si/Al calorimeter (Fig. 5.22) [77, 78].

With a bolometer, a deposited energy of 5.9 keV from K$_\alpha$ X rays is registered by means of a temperature rise. These microcalorimeters must have an extremely low heat capacity, and they have to be operated at cryogenic temperatures. In most cases they consist of an absorber with a relatively large surface (some millimetres in diameter), which is coupled to a semiconductor thermistor. The deposited energy is collected in the absorber part, which forms, together with the thermistor readout, a totally absorbing calorimeter. Such two-component bolometers allow one to obtain

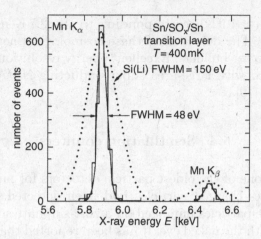

Fig. 5.21.   Amplitude distribution of Mn K_α and Mn K_β X-ray photons in an Sn/SO$_x$/Sn tunnel-junction layer. The dotted line shows the best obtainable resolution with a Si(Li) semiconductor detector for comparison [76].

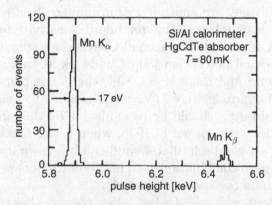

Fig. 5.22.   Amplitude distribution of 5.9 keV and 6.47 keV X rays from the Mn K_α and K_β lines in a bolometer consisting of a HgCdTe absorber and a Si/Al calorimeter. The K_α line corresponds to a transition from the L into the K shell, the K_β line to a transition from the M into the K shell [78].

excellent energy resolution, but they cannot, at the moment, process high rates of particles since the decay time of the thermal signals is on the order of 20 μs. Compared to standard calorimetric techniques, which are based on the production and collection of ionisation electrons, bolometers have the large advantage that they can in principle also detect weakly or non-ionising particles such as slow magnetic monopoles, weakly interacting massive particles (WIMPs), astrophysical neutrinos, or, for example, primordial neutrino radiation as remnant from the Big Bang with energies around $0.2\,\text{meV}$ ($\approx 1.9\,\text{K}$), corresponding to the $2.7\,\text{K}$ microwave background radiation. The detection of these cosmological neutrinos is a real challenge for detector builders. Excellent energy resolution for X rays also has been obtained with large-area superconducting Nb/Al–AlO$_x$/Al/Nb *tunnel junctions* [79].

5.4   Scintillation counters

A *scintillator* is one of the oldest particle detectors for nuclear radiation. In the early times charged particles had been detected by light flashes emitted when the particles impinged on a zinc-sulphate screen. This light was registered with the naked eye. It has been reported that the sensitivity of the human eye can be significantly increased by a cup of strong coffee possibly with a small dose of strychnine.

After a longer period of accommodation in complete darkness, the human eye is typically capable of recognising approximately 15 photons as a light flash, if they are emitted within one tenth of a second and if their wavelength is matched to the maximum sensitivity of the eye.

The time span of a tenth of a second corresponds roughly to the time constant of the visual perception [80]. Chadwick [81] refers occasionally to a paper by Henri and Bancels [82, 83], where it is mentioned that an energy deposit of approximately $3\,\text{eV}$, corresponding to a single photon in the green spectral range, should be recognisable by the human eye [84].

New possibilities were opened in 1948, when it was found that crystals of *sodium iodide* are good scintillators and can be grown up to a large size [85]. These crystals in combination with photomultipliers were successfully exploited for gamma-ray spectroscopy [86].

The measurement principle of scintillation counters has remained essentially unchanged. The function of a scintillator is twofold: first, it should convert the excitation of, e.g., the crystal lattice caused by the energy loss of a particle into visible light; and, second, it should transfer this light either directly or via a light guide to an optical receiver (photomultiplier, photodiode, etc.) [87–89]. Reference [87] gives a detailed review of physical principles and characteristics of scintillation detectors.

The disadvantage of such indirect detection is that a much larger energy is required for the generation of one photoelectron than it is necessary for the creation of one electron–hole pair in solid-state ionisation detectors. We have to compare an amount of about $50\,\text{eV}$ for the best scintillation counters with $3.65\,\text{eV}$ for silicon detectors. But this drawback is compensated by the possibility to build a detector of large size and mass, up to tens of metres and hundreds of tons at relatively low cost of the scintillation material.

The main scintillator characteristics are: scintillation efficiency, light output, emission spectrum and decay time of the scintillation light. The *scintillation efficiency* ε_{sc} is defined as the ratio of the energy of the emitted photons to the total energy absorbed in the scintillator. The *light output* L_{ph} is measured as the number of photons per 1 MeV of energy absorbed in the scintillator. The *emission spectrum* usually has a maximum (sometimes more than one) at a characteristic wavelength λ_{em}. For light collection the index of refraction $n(\lambda)$ and the light attenuation length λ_{sc} are important. The scintillation flash is characterised by a fast rise followed by a much longer exponential decay with a *decay time* τ_D characteristic of the scintillation material. Often, more than one exponential component is required to describe the light pulse shape. In that case several decay times $\tau_{D,i}$ are needed to describe the trailing edge of the pulse.

Scintillator materials can be inorganic crystals, organic compounds, liquids and gases. The scintillation mechanism in these scintillator materials is fundamentally different.

Inorganic scintillators are mostly crystals, pure ($Bi_4Ge_3O_{12}$, BaF_2, CsI, etc.) or doped with small amounts of other materials ($NaI(Tl)$, $CsI(Tl)$, $LiI(Eu)$, etc.) [90, 91].

The *scintillation mechanism* in inorganic substances can be understood by considering the energy bands in crystals. Since the scintillator must be transparent for the emitted light, the number of free electrons in the conduction band should be small and the gap between valence and conduction bands should be wide enough, at least several eV. Halide crystals, which are most commonly used, are insulators. The valence band is completely occupied, but the conduction band is normally empty (Fig. 5.23). The energy difference between both bands amounts to about 3 eV to 10 eV.

Electrons are transferred from the valence band to the conduction band by the energy deposited by an incident charged particle or γ ray. In the conduction band they can move freely through the crystal lattice. In this excitation process a hole remains in the valence band. The electron can

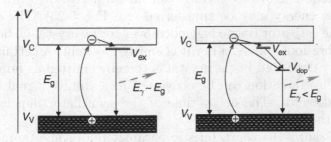

Fig. 5.23. Energy bands in a pure (left) and doped (right) crystal.

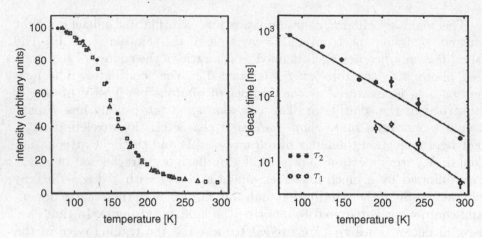

Fig. 5.24.   Light output (left) and decay time (right) for a pure CsI crystal [92]. Two curves in the right-hand figure correspond to two decay-time constants.

recombine with the hole or create a bound state with a hole called exciton. The *exciton level* V_{ex} is slightly below the lower edge of the conduction band, V_C. The exciton migrates in the crystal for some time and then can be de-excited in a collision with a phonon or it just recombines emitting a photon corresponding to its excitation energy E_{ex}. At room temperature the probability of photon emission is low while at cryogenic temperatures this mechanism mainly defines the exciton lifetime. So, the scintillation efficiency becomes quite high for pure alkali–halide crystals at low temperatures. Figure 5.24 shows the temperature dependence of the light output and the decay time of pure CsI [92]. It can be seen from the figure that the light output increases at low temperature while the decay time of the light flash becomes longer.

To improve the scintillation efficiency at room temperature, dopant impurities, which act as *activator centres*, are deliberately introduced into the crystal lattice. These impurities are energetically localised between the valence and the conduction band thereby creating additional energy levels V_{dop}. Excitons or free electrons can hit an activator centre whereby their binding energy may be transferred (see Fig. 5.23). The excitation energy of the activator centre is handed over to the crystal lattice in form of lattice vibrations (phonons) or it is emitted as light. A certain fraction of the energy deposited in the crystal is thereby emitted as luminescence radiation. This radiation can be converted to a voltage signal by a photosensitive detector. The decay time of the scintillator depends on the lifetimes of the excited levels.

Table 5.2 shows the characteristic parameters of some inorganic scintillators [93–98]. As it can be seen from the table, inorganic scintillators

Table 5.2. *Characteristic parameters of some inorganic scintillators [93–98]*

Scintillator	Density ϱ [g/cm^3]	X_0 [cm]	τ_D [ns]	L_{ph}, N_{ph} [per MeV]	λ_{em} [nm]	$n(\lambda_{em})$
NaI(Tl)	3.67	2.59	230	$3.8 \cdot 10^4$	415	1.85
LiI(Eu)	4.08	2.2	1400	$1 \cdot 10^4$	470	1.96
CsI	4.51	1.85	30	$2 \cdot 10^3$	315	1.95
CsI(Tl)	4.51	1.85	1000	$5.5 \cdot 10^4$	550	1.79
CsI(Na)	4.51	1.85	630	$4 \cdot 10^4$	420	1.84
$Bi_4Ge_3O_{12}$ (BGO)	7.13	1.12	300	$8 \cdot 10^3$	480	2.15
BaF_2	4.88	2.1	0.7	$2.5 \cdot 10^3$	220	1.54
			630	$6.5 \cdot 10^3$	310	1.50
$CdWO_4$	7.9	1.06	5000	$1.2 \cdot 10^4$	540	2.35
			20 000		490	
$PbWO_4$ (PWO)	8.28	0.85	10/30	70–200	430	2.20
$Lu_2SiO_5(Ce)$ (LSO)	7.41	1.2	12/40	$2.6 \cdot 10^4$	420	1.82

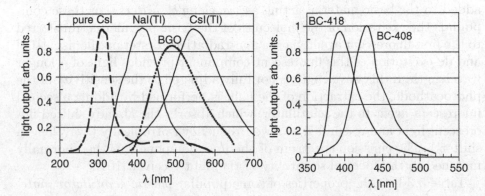

Fig. 5.25. The luminescence spectra of some scintillation crystals and plastic scintillators [94, 98]. The curves are arbitrarily scaled.

have decay times and light outputs in a wide range. Some of the scintillators are widely used in high energy physics experiments as well as in nuclear spectroscopy while others are still under study [91, 99].

The *luminescence spectra* of some inorganic scintillation crystals are shown in Fig. 5.25 in comparison to plastic-scintillator spectra, which are usually more narrow.

Table 5.3.   *Characteristic parameters of some organic scintillators [87, 93, 94, 102, 103]*

Scintillator	base	density ϱ [g/cm^3]	τ_D [ns]	L_{ph}, N_{ph} [per MeV]	λ_{em} [nm]	$n(\lambda_{em})$
Anthracene		1.25	30	16 000	440	1.62
BC-408 (BICRON)	PVT	1.032	2.1	10 000	425	1.58
BC-418 (BICRON)	PVT	1.032	1.5	11 000	391	1.58
UPS-89 (AMCRYS-H)	PS	1.06	2.4	10 000	418	1.60
UPS-91F (AMCRYS-H)	PS	1.06	0.6	6 500	390	1.60

Organic scintillators are polymerised plastics, liquids or sometimes also crystals, although the latter are rarely used at present. Plastic scintillation materials most widely used now are usually based on polymers having benzene rings in their molecular structure. Such materials luminesce after charged-particle energy deposition. However, the emitted light is in the ultraviolet range and the absorption length of this light is quite short: the fluorescent agent is opaque for its own light. To obtain light output in the maximum-sensitivity wavelength range of the photomultiplier (typically 400 nm) one or two (sometimes even three) fluorescent agents are added to the basic material acting as *wavelength shifters*. For these compounds the excitation of the molecules of the basic polymer is transferred to the first fluorescent agent via the non-radiative Förster mechanism [100] and de-excitation of this fluorescent component provides light of a longer wavelength. If this wavelength is not fully adjusted to the sensitivity of the photocathode, the extraction of the light is performed by adding a second fluorescent agent to the scintillator, which absorbs the already shifted fluorescent light and re-emits it at lower frequency isotropically ('wavelength shifter'). The emission spectrum of the final component is then normally matched to the spectral sensitivity of the light receiver [101].

Table 5.3 lists the properties of some popular *plastic scintillator materials* in comparison with the organic crystal anthracene. The best plastic scintillators are based on polyvinyltoluene (PVT, polymethylstyrene) and polystyrene (PS, polyvinylbenzene). Sometimes a non-scintillating base, like PMMA (polymethyl methacrylate, 'Plexiglas' or 'Perspex'), is used with an admixture ($\approx 10\%$) of naphthalene. This scintillator is cheaper than the PVT- or PS-based ones and has a good transparency for its own light, but the light output is typically a factor of two lower than that of the best materials. Organic scintillators are characterised by short decay times, which lie typically in the nanosecond range.

The active components in an organic scintillator are either dissolved in an organic liquid or are mixed with an organic material to form a polymerising structure. In this way liquid or plastic scintillators can be produced in almost any geometry. In most cases scintillator sheets of 1 mm up to 30 mm thickness are made.

All solid and liquid scintillators are characterised by a non-linear response to the energy deposition at very high ionisation density [87, 104, 105]. For example, the scintillation signal from a CsI(Tl) crystal normalised to the deposited energy is about a factor of two lower for a 5 MeV α particle than for MeV γ quanta or relativistic charged particles.

For organic scintillators Birks [87] suggested a semi-empirical model describing the light-output degradation at high ionisation density dE/dx,

$$L = L_0 \frac{1}{1 + k_\mathrm{B} \cdot dE/dx} \, , \qquad (5.56)$$

where dE/dx is the ionisation loss in $MeV/(g/cm^2)$ and k_B is *Birks' constant* which is characteristic for the scintillation material used. Typically, k_B is in the range $(1–5) \cdot 10^{-3} \, g/(cm^2 \, MeV)$.

Gas scintillation counters use light which is produced when charged particles excite atoms in interactions and these atoms subsequently decay into the ground state by light emission [87, 106]. The lifetime of the excited levels lies in the nanosecond range. Because of the low density, the light yield in gas scintillators is relatively low. However, liquid argon (LAr), krypton (LKr) and xenon (LXe) were found to be efficient scintillators [47, 107]. For example, the light output of liquid xenon is about the same as that of a NaI(Tl) crystal while the decay time is about 20 ns only. However, the maximum-emission wavelength is 174 nm (128 nm for LAr and 147 nm for LKr), which makes detection of this light very difficult, especially taking into account the necessity of the cryogenic environment.

The scintillation counter has to have high *light collection efficiency* and *uniform response* over its volume. To achieve this the light attenuation in a crystal should be small and the light attenuation length λ_sc becomes a very important characteristic of the scintillator.

Usually the scintillation light is collected from one or two faces of the counter by a photosensor surface S_out, which is much smaller than the total surface S_tot of the counter. For counters of not too large size ($\approx 5 \, \mathrm{cm}$ or less) the best collection efficiency is obtained when all the surface except an output window is diffusively reflecting. A fine powder of magnesium oxide or porous Teflon film can be used as an effective reflector. For counters of approximately equal dimensions (e.g., close to spherical or cubic) the light collection efficiency η_C can be estimated by a simple formula (see Problem 5.4),

$$\eta_\mathrm{C} = \frac{1}{1 + \mu/q} \, , \qquad (5.57)$$

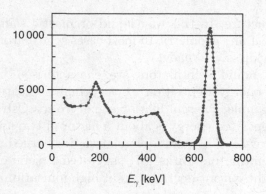

Fig. 5.26.   Energy spectrum measured with a CsI(Tl) counter exposed to 662 keV γ rays from a ^{137}Cs radioactive source. The crystal of dimension $2 \cdot 2 \cdot 2 \, \text{cm}^3$ is viewed by a $1 \, \text{cm}^2$ silicon photodiode. The energy resolution, FWHM/E_γ, is about 6%.

where $q = S_{\text{out}}/S_{\text{tot}}$ and μ is the absorption coefficient for diffusive reflection. For crystals of medium size it is possible to reach $\mu \approx 0.05$–0.02 [108, 109], which provides a light collection efficiency of $\eta_\text{C} \approx 60\%$–$70\%$.

Figure 5.26 shows a typical energy spectrum measured with a CsI(Tl) counter exposed to 662 keV γ rays from a ^{137}Cs radioactive source. The rightmost peak corresponds to full absorption of the photon (see Chap. 1), commonly called *photopeak*. The *Compton edge* at the end of the flat Compton continuum is to the left of the photopeak. Another peak observed at 184 keV is produced by photons backscattered from the surrounding material into the detector when they get absorbed by the photoelectric effect. The energy of this peak corresponds to the difference between full absorption and the Compton edge. The energy resolution is dominated by statistical fluctuations of the number of photoelectrons, N_{pe}, generated in the photosensor by the scintillation light and by electronics noise. The number of photoelectrons, N_{pe}, can be calculated from

$$N_{\text{pe}} = L_{\text{ph}} \cdot E_{\text{dep}} \cdot \eta_\text{C} \cdot Q_\text{s} \, , \qquad (5.58)$$

where E_{dep} is the deposited energy, L_{ph} the number of light photons per 1 MeV of energy absorbed in the scintillator, η_C the light collection efficiency, and Q_s the quantum efficiency of the photosensor. Then the energy resolution is given by the formula

$$\sigma_{E_{\text{dep}}}/E_{\text{dep}} = \sqrt{\frac{f}{N_{\text{pe}}} + \left(\frac{\sigma_\text{e}}{E_{\text{dep}}}\right)^2 + \Delta^2} \, . \qquad (5.59)$$

Here f is the so-called 'excess noise factor' describing the statistical fluctuations introduced by the photosensor, σ_e is the noise of the read-out electronics, and Δ stands for other contributions like non-linear scintillator response, non-uniformity of the light collection, etc.

For scintillation counters of large sizes, especially having the shape of long bars or sheets, the optimal way of light collection is to use the effect of internal (total) reflection. To achieve this all surfaces of the scintillator must be carefully polished. Let us consider a scintillator of parallelepiped-like shape with a photosensor attached to one of its faces with perfect optical contact. The light that does not fulfil the condition of internal reflection leaves the counter through one of the five faces while the remaining light is collected by the photosensor. Assuming an uniform angular distribution of scintillation photons, the amount of light leaving the scintillator through each face is given by the formula

$$\frac{\Delta I}{I_{\text{tot}}} = \frac{1 - \cos\beta_{\text{ir}}}{2} = \frac{1}{2}\left(1 - \sqrt{1 - \frac{1}{n^2}}\right) \,, \qquad (5.60)$$

where β_{ir} is the angle for internal reflection, n the index of refraction of the scintillator, I_{tot} the total surface area of the counter, and ΔI the area of one of the faces. Then the light collection efficiency is

$$\eta_{\text{C}} = 1 - 5 \cdot \frac{\Delta I}{I_{\text{tot}}} \,. \qquad (5.61)$$

For counters of large size the loss of scintillation light due to bulk absorption or surface scattering is not negligible. High-quality scintillators have light attenuation lengths including both effects of about 2 m.

Normally large-area scintillators are read out with several photomultipliers. The relative pulse heights of these photomultipliers can be used to determine the particle's point of passage and thereby enable a correction of the measured light yield for absorption effects.

Plastic scintillators used in detectors are usually in the form of scintillator plates. The scintillation light emerges from the edges of these plates and has to be guided to a photomultiplier and also matched to the usually circular geometry of the photosensing device. This matching is performed with *light guides*. In the most simple case (Fig. 5.27) the light

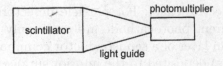

Fig. 5.27. Light readout with a 'fish-tail' light guide.

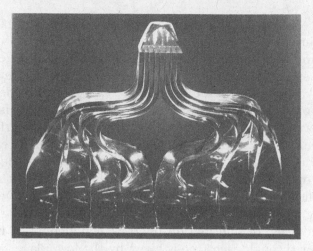

Fig. 5.28.   Photograph of an adiabatic light guide [110].

is transferred via a triangular light guide (*fish-tail*) to the photocathode of a photomultiplier. A complete light transfer, i.e. light transfer without any losses, using fish-tail light guides is impossible. Only by using complicated light guides can the end face of a scintillator plate be imaged onto the photocathode without appreciable loss of light (*adiabatic light guides*). Figure 5.28 shows the working principle of an adiabatic light guide ($dQ = 0$, i.e. no loss of light). Individual parts of the light-guide system can be only moderately bent because otherwise the light, which is normally contained in the light guide by internal reflection, will be lost at the bends.

The scintillator end face cannot be focussed without light loss onto a photocathode with a smaller area because of *Liouville's theorem*, which states: 'The volume of an arbitrary phase space may change its form in the course of its temporal and spatial development, its size, however, remains constant.'

5.5   Photomultipliers and photodiodes

The most commonly used instrument for the measurement of fast light signals is the *photomultiplier* (PM). Light in the visible or ultraviolet range – e.g. from a scintillation counter – liberates electrons from a photocathode via the photoelectric effect. For particle detectors photomultipliers with a semi-transparent photocathode are commonly used. This photocathode is a very thin layer of a semiconductor compound ($SbCs$, $SbKCs$, $SbRbKCs$ and others) deposited to the interior surface of the transparent input window.

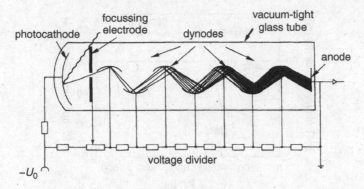

Fig. 5.29. Working principle of a photomultiplier. The electrode system is mounted in an evacuated glass tube. The photomultiplier is usually shielded by a mu-metal cylinder made from high-permeability material against stray magnetic fields (e.g. the magnetic field of the Earth).

For most counters a negative high voltage is applied to the photocathode, although for some types of measurements the opposite way (where a positive high voltage is applied to the anode) is recommended. Photoelectrons are focussed by an electric guiding field onto the first dynode, which is part of the multiplication system. The anode is normally at ground potential. The voltage between the photocathode and anode is subdivided by a chain of resistors. This voltage divider supplies the dynodes between the photocathode and anode so that the applied negative high voltage is subdivided linearly (Fig. 5.29). Detailed descriptions of photomultiplier operation and applications can be found in [111, 112].

An important parameter of a photomultiplier is its *quantum efficiency*, i.e. the mean number of photoelectrons produced per incident photon. For the most popular bialkali cathodes (Cs–K with Sb) the quantum efficiency reaches values around 25% for a wavelength of about 400 nm. It is worth to note that in the last years photomultiplier tubes with GaAs and GaInAsP photocathodes having quantum efficiencies up to 50% became commercially available. However, these devices are up to now not in frequent use and they do have some limitations.

Figure 5.30 shows the quantum efficiency for bialkali cathodes as a function of the wavelength [111]. The quantum efficiency decreases for short wavelengths because the transparency of the photomultiplier window decreases with increasing frequency, i.e. shorter wavelength. The range of efficiency can only be extended to shorter wavelengths by using UV-transparent quartz windows.

The dynodes must have a high *secondary-electron emission coefficient* (BeO or Mg–O–Cs). For electron energies from around 100 eV up to 200 eV, which correspond to typical acceleration voltages between

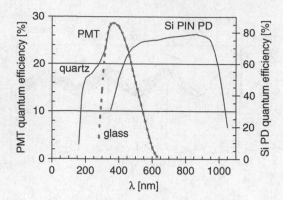

Fig. 5.30.   Quantum efficiency of a bialkali cathode as function of the wavelength [111] in comparison to a silicon PIN photodiode [113]. Note that the quantum efficiencies for the photomultiplier and the silicon photodiode are marked with different scales at opposite sides of the figure.

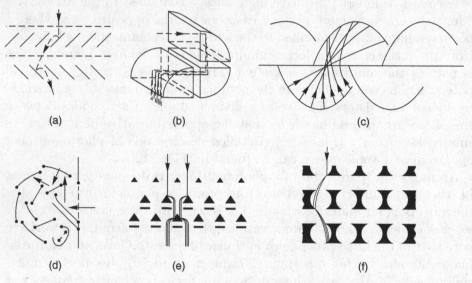

Fig. 5.31.   Some dynode system configurations: (a) venetian blind, (b) box, (c) linear focussing, (d) circular cage, (e) mesh and (f) foil [111].

two dynodes, approximately three to five secondary electrons are emitted [111]. Various types of geometries for dynode systems are shown in Fig. 5.31. For an n-dynode photomultiplier with a secondary emission coefficient g, the current amplification is given by

$$A = g^n .$$
(5.62)

For typical values of $g = 4$ and $n = 12$ one obtains $A = 4^{12} \approx 1.7 \cdot 10^7$.

The charge arriving at the anode for one photoelectron,

$$Q = eA \approx 2.7 \cdot 10^{-12}\,\mathrm{C}\ , \tag{5.63}$$

is collected within approximately 5 ns leading to an anode current of

$$i = \frac{\mathrm{d}Q}{\mathrm{d}t} \approx 0.5\,\mathrm{mA}\ . \tag{5.64}$$

If the photomultiplier is terminated with a $50\,\Omega$ resistor, a voltage signal of

$$\Delta U = R \cdot \frac{\mathrm{d}Q}{\mathrm{d}t} \approx 27\,\mathrm{mV} \tag{5.65}$$

is obtained.

Thus one photoelectron can be firmly detected. Figure 5.32 shows the pulse-height distribution for a single-photoelectron signal of a photomultiplier with a linear-focussing dynode system. The ratio of the maximum and minimum values of this distribution is called 'peak-to-valley ratio' and reaches about 3. The peak width is mostly determined by the Poisson statistics of the secondary electrons emitted from the first dynode. The left part of the spectrum is caused by the thermoemission from the first dynode and from electronics noise.

The contribution of the photomultiplier to the overall counter energy resolution is determined by photoelectron statistics, non-uniformity of the quantum efficiency and photoelectron collection efficiency over the photocathode, and the excess noise factor f, see Eq. (5.59). The term Δ

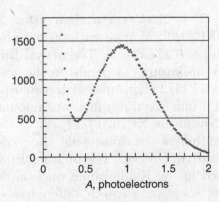

Fig. 5.32. Anode pulse distribution for a single-photoelectron signal for a photomultiplier with a linear-focussing dynode system.

is usually negligible for photomultiplier tubes. The excess noise factor for a photomultiplier is given by

$$f = 1 + \frac{1}{g_1} + \frac{1}{g_1 g_2} + \cdots + \frac{1}{g_1 g_2 \cdots g_n} \approx 1 + \frac{1}{g_1} \ , \qquad (5.66)$$

where g_i is the gain at the ith dynode.

The *rise time* of the photomultiplier signal is typically 1–3 ns. This time has to be distinguished from the time required for electrons to traverse the photomultiplier. This *transit time* depends on the phototube type and varies typically from 10 ns to 40 ns.

The *time jitter* in the arrival time of electrons at the anode poses a problem for reaching a high time resolution. Two main sources of the time jitter are the variation in the velocity of the photoelectrons and the difference of the path lengths from the production point of the photoelectrons to the first dynode which can be subject to large fluctuations.

The time jitter (or transit-time spread, TTS) caused by different velocities of photoelectrons can easily be estimated. If s is the distance between photocathode and the first dynode, then the time t_1 when an electron with initial kinetic energy T reaches the first dynode can be found from the expression

$$s = \frac{1}{2} \frac{eE}{m} t_1^2 + t_1 \cdot \sqrt{2T/m} \ , \qquad (5.67)$$

where E is the electric field strength and m the electron mass. Then one can estimate the difference between t_1 for photoelectrons at $T = 0$ and those with an average kinetic energy T,

$$\delta t = \frac{\sqrt{2mT}}{eE} \ . \qquad (5.68)$$

For $T = 1\,\text{eV}$ and $E = 200\,\text{V/cm}$ a time jitter of $\delta t = 0.17\,\text{ns}$ is obtained. For a fast XP2020 PM tube with 50 mm photocathode diameter this spread is $\sigma_{\text{TTS}} = 0.25\,\text{ns}$ [114]. The arrival-time difference based on path-length variations strongly depends on the size and shape of the photocathode. For an XP4512 phototube with planar photocathode and a cathode diameter of 110 mm this time difference amounts to $\sigma_{\text{TTS}} = 0.8\,\text{ns}$ in comparison to 0.25 ns for an XP2020 [114].

For large photomultipliers the achievable time resolution is limited essentially by path-length differences. The photomultipliers with a 20-inch cathode diameter used in the Kamiokande nucleon decay and neutrino experiment [115, 116] show path-length differences of up to 5 ns. For this phototube the distance between photocathode and first dynode is so large that the Earth's magnetic field has to be well shielded so that the

Fig. 5.33. Photograph of an 8-inch photomultiplier (type R 4558) [117].

photoelectrons can reach the first dynode. Figure 5.33 shows a photograph of an 8-inch photomultiplier [117].

To obtain position sensitivity, the anode of a photomultiplier tube can be subdivided into many independent pads or it can be built as a set of strips (or two layers of crossed strips) [112]. To preserve the position information the dynode system has to transfer the image from the photocathode with minimal distortions. To meet this condition, the dynode system of this device should be placed very close to the cathode. It can be made as a set of layers of fine mesh or foils. The anode pixel size can be $2 \times 2\,\text{mm}^2$ at a pitch of 2.5 mm. Such photomultiplier tubes are used in gamma cameras for medical applications [118] as well as in high energy physics experiments [119, 120].

The path-length fluctuations can be significantly reduced in photomultipliers (*channel plates*) with microchannel plates as multiplication system (MCP-PMT). The principle of operation of such channel plates is shown in Fig. 5.34 [112]. A voltage of about 1000 V is applied to a thin glass tube (diameter 6–50 µm, length 1–5 mm) which is coated on the inside with a resistive layer. Incident photons produce photoelectrons on a photocathode or on the inner wall of the microchannel. These are, like in the normal phototube, multiplied at the – in this case – continuous dynode. Channel plates contain a large number (10^4 to 10^7) of such channels which are implemented as holes in a lead-glass plate. A microphotographic record of such channels with a diameter of 12.5 µm [121] is shown in Fig. 5.34. A photomultiplier tube with a single MCP provides a gain up to 10^3–10^4. To obtain a higher gain, two or three MCP in series can be incorporated into the MCP-PMT.

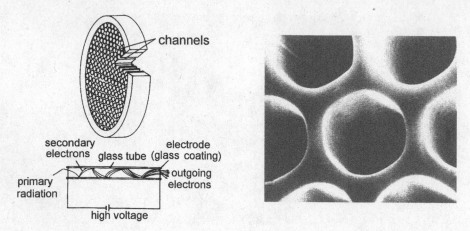

Fig. 5.34.   Working principle of a channel plate [112] (left) and microphotograph of microchannels (right) [121].

Because of the short mean path lengths of electrons in the longitudinal electric field, path-length fluctuations are drastically reduced compared to a normal photomultiplier. Transit-time differences of about 30 ps for multiplication factors between 10^5 and 10^6 are obtained [122].

While normal photomultipliers practically cannot be operated in magnetic fields (or if so, only heavily shielded), the effect of magnetic fields on channel plates is comparatively small. This is related to the fact that in channel plates the distance between cathode and anode is much shorter. There are, however, recent developments of conventional photomultipliers with transparent wire-mesh dynodes, which can withstand moderate magnetic fields.

A problem with channel plates is the flux of positive ions produced by electron collisions with the residual gas in the channel plate that migrate in the direction of the photocathode. The lifetime of channel plates would be extremely short if the positive ions were not prevented from reaching the photocathode. By use of extremely thin aluminium windows of ≈ 7 nm thickness (transparent for electrons) mounted between photocathode and channel plate, the positive ions are absorbed. In this way, the photocathode is shielded against the ion bombardment.

A very promising photosensor is the hybrid photomultiplier tube (HPMT) [123, 124]. This device has only a photocathode and a silicon PIN diode as an anode. A high voltage, up to 15–20 kV, is applied to the gap between the photocathode and PIN diode. The diode of 150–300 µm thickness is fully depleted under reverse-bias voltage. Photoelectrons accelerated by the electric field penetrate a very thin (about 500 Å) upper contact layer, get to the depleted area, and produce multiple electron–hole pairs. The gain of this device can reach 5000. Figure 5.35

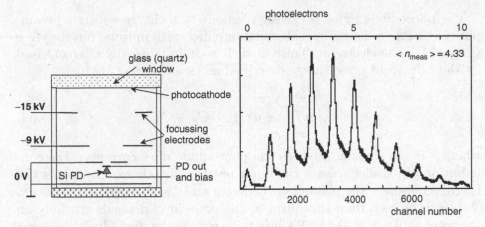

Fig. 5.35. The layout of the hybrid PMT and the amplitude spectrum measured with light flashes from scintillation fibres. Each peak corresponds to a certain number of photoelectrons emerging from the photocathode [125].

shows the layout of a HPMT and the amplitude spectrum of light flashes measured with a HPMT [125]. The peaks which correspond to signals with a certain number of photoelectrons are distinctively seen (compare with Fig. 5.32 for a usual PM tube).

Recent developments of modern electronics made it possible to use low-gain photosensors for particle detectors. These are one- or two-dynode PM tubes (*phototriodes* and *phototetrodes*) [126–128] as well as *silicon photodiodes*. The main reasons to use these devices are their low sensitivity or insensitivity to magnetic fields, their compactness, better stability and lower price.

Semiconductor photodetectors are known for a long time and the possibility to use silicon photodiodes (PD) for particle detection in combination with large-size scintillation crystals CsI(Tl), NaI(Tl), BGO was demonstrated in 1982–5 [129, 130].

The main operation principles as well as the structure of PIN photodiodes are very similar to that for a silicon particle detector described in Sect. 5.3 (see Fig. 5.19). The difference is that the layers in front of the depletion region should be transparent for the light to be detected. A photodiode contains a very thin layer of highly doped p^+ silicon followed by a layer of moderately doped n-Si of 200–500 μm thickness (called i layer) and ending with a highly doped n^+-Si layer. A SiO_2 film is mounted on top of the p^+ layer. The whole structure is attached to a ceramic substrate and covered with a transparent window.

The photon enters the depletion area, penetrates to the i layer where the bias voltage U_b is applied, and creates an electron–hole pair that is separated by the electric field which exists in this area.

The photodiode signal is usually read out by a charge-sensitive preamplifier (CSPA) followed by a shaping amplifier with optimal filtering (see Chap. 14). The electronics noise of such a chain is usually characterised by the equivalent noise charge described as (see also Sect. 14.9)

$$\sigma_Q = \sqrt{2eI_{\mathrm{D}}\tau + a\tau + \frac{b}{\tau}(C_{\mathrm{p}} + C_{\mathrm{fb}})^2} \, , \qquad (5.69)$$

where τ is the shaping time, I_{D} the photodiode dark current, a the contribution of parallel noise of the input preamplifier chain, b describes the thermal noise, C_{p} the photodiode capacity and C_{fb} the feedback capacity.

As can be seen from this formula, the noise level depends crucially on the total capacity at the CSPA input. To reduce the photodiode capacity, the depletion layer should be extended. On the other hand, the photodiode dark current is related to the electron–hole pairs produced in the depletion layer by thermal excitation. Hence the dark current should be proportional to the depleted-area volume (in reality part of this current is due to a surface component). At present, photodiodes used for particle detectors have an area of 0.5–4 cm^2 and i-layer thicknesses of 200–500 μm. The dark current of these devices is 0.5–3 nA/cm^2 while the capacity is about 50 pF/cm^2.

A specific feature of the semiconductor photodiode performance is the so-called *nuclear counter effect*, i.e. the possibility of electron–hole production not only by photons but also by charged particles crossing the *pn* junction. This effect should be taken into account when designing a detector system using photodiodes. On the other hand, X-ray absorption provides an opportunity for a direct counter calibration. For that we can irradiate a photodiode by X rays of known energy from a radioactice source, e.g. ^{241}Am. An example of such a spectrum is presented in Fig. 5.36. The number of electron–hole pairs corresponding to the 60 keV peak is easily calculated taking into account $W_{\mathrm{Si}} = 3.65$ eV.

Since the operation principles of a solid-state ionisation counter are similar to a gaseous one, it is a natural idea to use photodiodes in the proportional mode. The first successful devices of this type, suitable for usage in scintillation counters, were developed in 1991–3 [131–133]. At present, these avalanche photosensors are used rather widely [134].

The principle of operation of *avalanche photodiodes* (APDs) is illustrated in Fig. 5.37. This device has a complex doping profile which provides a certain area with high electric field. Photons penetrate several microns into a *p*-silicon layer before they create an electron–hole pair. A weak electric field existing in this area separates the pair and causes the electrons to drift to the *pn* junction which exhibits a very high field strength. Here the electron can gain enough energy to create new

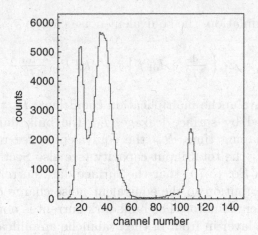

Fig. 5.36. Pulse-height distribution taken with a $1\,\mathrm{cm}^2$ Si photodiode at room temperature exposed to X rays from an ^{241}Am source. The rightmost peak corresponds to photons of $60\,\mathrm{keV}$ energy while the second, broad peak results from an overlap of further non-resolved γ-ray and X-ray lines in the range of 15–$30\,\mathrm{keV}$. For details see also Appendix 5, Fig. A5.10.

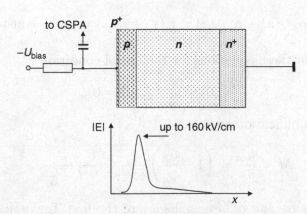

Fig. 5.37. The layout of an avalanche photodiode together with the electric field-strength distribution.

electron–hole pairs. Due to impact ionisation such a device can provide an amplification of up to 1000 for existing avalanche photodiodes.

Since the avalanche multiplication is a statistical process, the statistical fluctuation of the collected charge is higher than that determined from the Poisson spread of the number of initial photons, $\sigma = \sqrt{f/n}$, where f is the 'excess noise factor'.

As first approximation, the equivalent noise charge σ_q is expressed by

$$\sigma_q^2 = 2e \left(\frac{I_{\mathrm{ds}}}{M^2} + I_{\mathrm{db}} f \right) \tau + 4kT R_{\mathrm{en}} \frac{C_{\mathrm{tot}}^2}{M^2} \frac{1}{\tau} \ , \tag{5.70}$$

where M is the avalanche multiplication coefficient, I_{ds} the dark-current component caused by surface leakage, I_{db} the bulk dark-current component, τ the shaping time, R_{en} the equivalent noise resistance of the amplifier and C_{tot} the total input capacity (see also Sect. 14.9).

It is clear from Eq. (5.70) that the surface dark current does not give a significant contribution to the equivalent noise charge due to the large factor M in the denominator. The bulk dark current is normally quite low due to the thin p layer in front of the avalanche amplification region.

Let us consider the simplest APD model assuming that an avalanche occurs in a uniform electric field in a layer $0 < x < d$. Both electrons and holes can produce new pairs but the energy threshold for holes is much higher due to their larger effective mass. Denoting the probabilities of ionisation per unit drift length as α_e and α_p, respectively, we arrive at the following equations for electron (i_e) and hole (i_p) currents:

$$\frac{\mathrm{d} i_e(x)}{\mathrm{d} x} = \alpha_e i_e(x) + \alpha_p i_p(x) \ , \quad i_e(x) + i_p(x) = i_{\mathrm{tot}} = \mathrm{const} \ . \tag{5.71}$$

The solution of these equations for the initial conditions

$$i_e(0) = i_0 \ , \quad i_p(d) = 0 \tag{5.72}$$

leads to an amplification of

$$M = \frac{i_{\mathrm{tot}}}{i_0} = \left(1 - \frac{\alpha_p}{\alpha_e} \right) \frac{1}{e^{-(\alpha_e - \alpha_p)d} - \frac{\alpha_p}{\alpha_e}} \ . \tag{5.73}$$

The quantities α_e and α_p are analogues of the first Townsend coefficient (see Sect. 5.1.2), and they increase with increasing field strength. The gain rises according to Eq. (5.73). When α_p becomes sufficiently high to fulfil the condition

$$e^{-(\alpha_e - \alpha_p)d} = \frac{\alpha_p}{\alpha_e} \ , \tag{5.74}$$

the APD will break down.

The gain and dark current in their dependence on the bias voltage for a $5 \times 5\,\mathrm{mm}^2$ APD produced by 'Hamamatsu Photonics' are shown in Fig. 5.38a. The noise level versus gain is presented in Fig. 5.38b.

In the bias-voltage range where $\alpha_p \ll \alpha_e$, the APD can be considered as a multistage multiplier with a multiplication coefficient at each stage

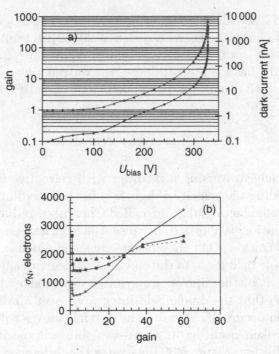

Fig. 5.38. (a) The gain (▲) and dark current (◆) in their dependence on the bias voltage for a $5 \times 5\,\text{mm}^2$ APD produced by 'Hamamatsu Photonics'; (b) The noise level versus gain for different shaping times: $2\,\mu\text{s}$ (●), $0.25\,\mu\text{s}$ (■) and $0.1\,\mu\text{s}$ (▲).

equal to 2. For this case one can derive a value for the excess noise factor of $f = 2$.

A more detailed theory of processes in APDs was developed in [135, 136]. This theory gives an expression for the excess noise factor of

$$f = K_{\text{eff}}M + (2 - 1/M)(1 - K_{\text{eff}}) , \qquad (5.75)$$

where K_{eff} is a constant on the order of 0.01.

Avalanche photodiodes have two important advantages in comparison to photodiodes without amplification: a much lower nuclear counter effect and a much higher radiation tolerance [133, 134]. Both of these features originate from the small thickness of the p layer in front of the avalanche area. The quantum efficiency of an APD is basically close to that for a normal PIN photodiode.

When the bias voltage approaches the breakdown threshold, the APD reaches a regime similar to the Geiger mode (see Sect. 5.1.3). As in the Geiger regime the output signals do not depend on the amount of light at the input, but rather are limited by the resistance and capacity of the device. However, this mode became the basis for another promising photosensor – the so-called *silicon photomultiplier*. Such a device consists of a set of pixel Geiger APDs with a size of 20–50 μm built on a common substrate with the total area of 0.5–1 mm². When the total number of photons in a light flash is not too large, the output pulse is proportional to this number with a multiplication coefficient of $\approx 10^6$ [137].

5.6 Cherenkov counters

A charged particle, traversing a medium with refractive index n with a velocity v exceeding the velocity of light c/n in that medium, emits a characteristic electromagnetic radiation, called Cherenkov radiation [138, 139]. Cherenkov radiation is emitted because the charged particle polarises atoms along its track so that they become electric dipoles. The time variation of the dipole field leads to the emission of electromagnetic radiation. As long as $v < c/n$, the dipoles are symmetrically arranged around the particle path, so that the dipole field integrated over all dipoles vanishes and no radiation occurs. If, however, the particle moves with $v > c/n$, the symmetry is broken resulting in a non-vanishing dipole moment, which leads to the radiation. Figure 5.39 illustrates the difference in polarisation for the cases $v < c/n$ and $v > c/n$ [140, 141].

The contribution of Cherenkov radiation to the energy loss is small compared to that from ionisation and excitation, Eq. (1.11), even for minimum-ionising particles. For gases with $Z \geq 7$ the energy loss by Cherenkov radiation amounts to less than 1% of the ionisation loss of minimum-ionising particles. For light gases (He, H) this fraction amounts to about 5% [21, 22].

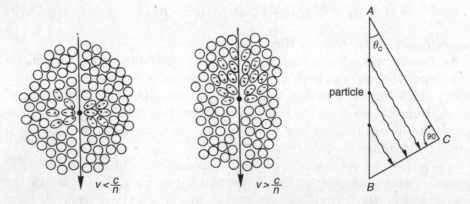

Fig. 5.39.   Illustration of the Cherenkov effect [140, 141] and geometric determination of the Cherenkov angle.

The angle between the emitted Cherenkov photons and the track of the charged particle can be obtained from a simple argument (Fig. 5.39). While the particle has travelled the distance $AB = t\beta c$, the photon has advanced by $AC = t \cdot c/n$. Therefore one obtains

$$\cos\theta_c = \frac{c}{n\beta c} = \frac{1}{n\beta} \ . \tag{5.76}$$

For the emission of Cherenkov radiation there is a *threshold effect*. Cherenkov radiation is emitted only if $\beta > \beta_c = \frac{1}{n}$. At threshold, Cherenkov radiation is emitted in the forward direction. The *Cherenkov angle* increases until it reaches a maximum for $\beta = 1$, namely

$$\theta_c^{max} = \arccos\frac{1}{n} \ . \tag{5.77}$$

Consequently, Cherenkov radiation of wavelength λ requires $n(\lambda) > 1$. The maximum emission angle, θ_c^{max}, is small for gases ($\theta_c^{max} \approx 1.4°$ for air) and becomes large for condensed media (about 45° for usual glass).

For fixed energy, the threshold Lorentz factor depends on the mass of the particle. Therefore, the measurement of Cherenkov radiation is well suited for particle-identification purposes.

The number of Cherenkov photons emitted per unit path length with wavelengths between λ_1 and λ_2 is given by

$$\frac{\mathrm{d}N}{\mathrm{d}x} = 2\pi\alpha z^2 \int_{\lambda_1}^{\lambda_2} \left(1 - \frac{1}{(n(\lambda))^2\beta^2}\right) \frac{\mathrm{d}\lambda}{\lambda^2} \ , \tag{5.78}$$

for $n(\lambda) > 1$, where z is the electric charge of the particle producing Cherenkov radiation and α is the fine-structure constant.

Neglecting the dispersion of the medium (i.e. n independent of λ) leads to

$$\frac{\mathrm{d}N}{\mathrm{d}x} = 2\pi\alpha z^2 \cdot \sin^2\theta_c \cdot \left(\frac{1}{\lambda_1} - \frac{1}{\lambda_2}\right) \ . \tag{5.79}$$

For the optical range ($\lambda_1 = 400\,\mathrm{nm}$ and $\lambda_2 = 700\,\mathrm{nm}$) one obtains for singly charged particles ($z = 1$)

$$\frac{\mathrm{d}N}{\mathrm{d}x} = 490 \sin^2\theta_c \,\mathrm{cm}^{-1} \ . \tag{5.80}$$

Figure 5.40 shows the number of Cherenkov photons emitted per unit path length for various materials as a function of the velocity of the particle [142].

The photon yield can be increased by up to a factor of two or three if the photons emitted in the ultraviolet range can also be detected. Although

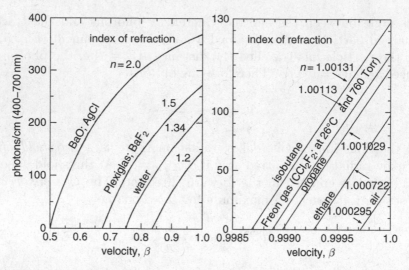

Fig. 5.40. Number of produced Cherenkov photons per unit path length for various materials as a function of the particle velocity [142].

the spectrum of emitted Cherenkov photons exhibits a $1/\lambda^2$ dependence, see Eq. (5.78), Cherenkov photons are not emitted in the X-ray range because in this region the index of refraction is $n = 1$, and therefore the condition for Cherenkov emission cannot be fulfilled.

To obtain the correct number of photons produced in a Cherenkov counter, Eq. (5.78) must be integrated over the region for which $\beta \cdot n(\lambda) > 1$. Also the response function of the light collection system must be taken into account to obtain the number of photons arriving at the photon detector.

All transparent materials are candidates for Cherenkov radiators. In particular, Cherenkov radiation is emitted in all scintillators and in the light guides which are used for the readout. The scintillation light, however, is approximately 100 times more intense than the Cherenkov light. A large range of indices of refraction can be covered by the use of solid, liquid or gaseous radiators (Table 5.4).

Ordinary liquids have indices of refraction greater than ≈ 1.33 (H_2O) and gases have n less than about 1.002 (pentane). Although gas Cherenkov counters can be operated at high pressure, thus increasing the index of refraction, the substantial gap between $n = 1.33$ and $n = 1.002$ cannot be bridged in this way.

By use of *silica aerogels*, however, it has become feasible to cover this missing range of the index of refraction. Aerogels are phase mixtures from m (SiO_2) and $2m$ (H_2O) where m is an integer. Silica aerogels form a porous structure with pockets of air. The diameter of the air bubbles

Table 5.4. *Compilation of Cherenkov radiators [1, 143]. The index of refraction for gases is for 0 °C and 1 atm (STP). Solid sodium is transparent for wavelengths below 2000 Å [144, 145]*

Material	$n-1$	β threshold	γ threshold
Solid sodium	3.22	0.24	1.029
Diamond	1.42	0.41	1.10
Flint glass (SFS1)	0.92	0.52	1.17
Lead fluoride	0.80	0.55	1.20
Aluminium oxide	0.76	0.57	1.22
Lead glass	0.67	0.60	1.25
Polystyrene	0.60	0.63	1.28
Plexiglas (Lucite)	0.48	0.66	1.33
Borosilicate glass (Pyrex)	0.47	0.68	1.36
Lithium fluoride	0.39	0.72	1.44
Water	0.33	0.75	1.52
Liquid nitrogen	0.205	0.83	1.79
Silica aerogel	0.007–0.13	0.993–0.884	8.46–2.13
Pentane (STP)	$1.7 \cdot 10^{-3}$	0.9983	17.2
CO_2 (STP)	$4.3 \cdot 10^{-4}$	0.9996	34.1
Air (STP)	$2.93 \cdot 10^{-4}$	0.9997	41.2
H_2 (STP)	$1.4 \cdot 10^{-4}$	0.99986	59.8
He (STP)	$3.3 \cdot 10^{-5}$	0.99997	123

in the aerogel is small compared to the wavelength of the light so that the light 'sees' an average index of refraction between the air and the solid forming the aerogel structure. Silica aerogels can be produced with densities between $0.1\,\mathrm{g/cm^3}$ and $0.6\,\mathrm{g/cm^3}$ [1, 101, 146] and indices of refraction between 1.01 and 1.13. There is a simple relation between the aerogel density (in $\mathrm{g/cm^3}$) and the index of refraction [147, 148]:

$$n = 1 + 0.21 \cdot \varrho \; [\mathrm{g/cm^3}] \; . \tag{5.81}$$

The Cherenkov effect is used for particle identification in threshold detectors as well as in detectors which exploit the angular dependence of the radiation. These *differential Cherenkov counters* provide in fact a direct measurement of the particle velocity. The working principle of a differential Cherenkov counter which accepts only particles in a certain velocity range is shown in Fig. 5.41 [4, 149–151].

All particles with velocities above $\beta_{\min} = 1/n$ are accepted. With increasing velocity the Cherenkov angle increases and finally reaches the critical angle for internal reflection, θ_{t}, in the radiator so that no light can

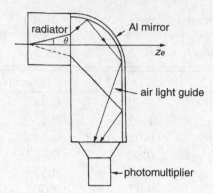

Fig. 5.41. Working principle of a differential (Fitch-type) Cherenkov counter [149–151].

escape into the air light guide. The critical angle for internal reflection can be computed from Snell's law of refraction to be

$$\sin\theta_t = \frac{1}{n} \, .$$

(5.82)

Because

$$\cos\theta = \sqrt{1 - \sin^2\theta} = \frac{1}{n\beta}$$

(5.83)

the maximum detectable velocity is

$$\beta_{max} = \frac{1}{\sqrt{n^2 - 1}} \, .$$

(5.84)

For polystyrene ($n = 1.6$) β_{min} is 0.625 and β_{max} is equal to 0.80. In this way, such a differential Cherenkov counter selects a velocity window of about $\Delta\beta = 0.17$. If the optical system of a differential Cherenkov counter is optimised, so that chromatic aberrations are corrected (DISC counter, DIScriminating Cherenkov counter [152]), a velocity resolution of $\Delta\beta/\beta = 10^{-7}$ can be achieved. The main types of Cherenkov detectors are discussed in Chap. 9.

5.7 Transition-radiation detectors (TRD)

Below Cherenkov threshold, charged particles can also emit electromagnetic radiation. This radiation is emitted in those cases where charged particles traverse the boundary between media with different dielectric properties [153]. This occurs, for example, when a charged particle enters a dielectric through a boundary from the vacuum or from air, respectively.

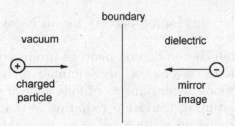

Fig. 5.42.   Illustration of the production of transition radiation at boundaries.

The energy loss by *transition radiation* represents only a negligibly small contribution to the total energy loss of charged particles.

A charged particle moving towards a boundary forms together with its mirror charge an electric dipole, whose field strength varies in time, i.e. with the movement of the particle (Fig. 5.42). The field strength vanishes when the particle enters the medium. The time-dependent dipole electric field causes the emission of electromagnetic radiation.

The emission at boundaries can be understood in such a way that although the electric displacement $\vec{D} = \varepsilon\varepsilon_0\vec{E}$ varies continuously in passing through the boundary, the electric field strength does not [154–156].

The energy radiated from a single boundary (transition from vacuum to a medium with dielectric constant ε) is proportional to the Lorentz factor of the incident charged particle [157–159]:

$$S = \frac{1}{3}\alpha z^2\hbar\omega_{\mathrm{p}}\gamma \ , \ \ \hbar\omega_{\mathrm{p}} = \sqrt{4\pi N_e r_e^3}m_e c^2/\alpha \ , \qquad (5.85)$$

where N_e is the electron density in the material, r_e is classical electron radius, and $\hbar\omega_{\mathrm{p}}$ is the *plasma energy*. For commonly used plastic radiators (styrene or similar materials) one has

$$\hbar\omega_{\mathrm{p}} \approx 20\,\mathrm{eV} \ . \qquad (5.86)$$

The radiation yield drops sharply for frequencies

$$\omega > \gamma\omega_{\mathrm{p}} \ . \qquad (5.87)$$

The number of emitted transition-radiation photons with energy $\hbar\omega$ higher than a certain threshold $\hbar\omega_0$ is

$$N_\gamma(\hbar\omega > \hbar\omega_0) \approx \frac{\alpha z^2}{\pi}\left[\left(\ln\frac{\gamma\hbar\omega_{\mathrm{p}}}{\hbar\omega_0} - 1\right)^2 + \frac{\pi^2}{12}\right] \ . \qquad (5.88)$$

At each interface the emission probability for an X-ray photon is on the order of $\alpha = 1/137$.

The number of transition-radiation photons produced can be increased if the charged particle traverses a large number of boundaries, e.g. in porous media or periodic arrangements of foils and air gaps.

The attractive feature of transition radiation is that the energy radiated by transition-radiation photons increases with the Lorentz factor γ (i.e. the energy) of the particle, and is not proportional only to its velocity [160, 161]. Since most processes used for particle identification (energy loss by ionisation, time of flight, Cherenkov radiation, etc.) depend on the velocity, thereby representing only very moderate identification possibilities for relativistic particles ($\beta \to 1$), the γ-dependent effect of transition radiation is extremely valuable for particle identification at high energies.

An additional advantage is the fact that transition-radiation photons are emitted in the X-ray range [162]. The increase of the radiated energy in transition radiation proportional to the Lorentz factor originates mainly from the increase of the average energy of X-ray photons and much less from the increase of the radiation intensity. In Fig. 5.43 the average energy of transition-radiation photons is shown in its dependence on the electron momentum for a typical radiator [152].

The angle of emission of transition-radiation photons is inversely proportional to the Lorentz factor,

$$\theta = \frac{1}{\gamma_{\text{particle}}} \, .$$

(5.89)

For periodical arrangements of foils and gaps, interference effects occur, which produce an effective threshold behaviour at a value of $\gamma \approx 1000$

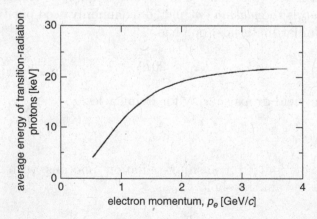

Fig. 5.43. Typical dependence of the average energy of transition-radiation photons on the electron momentum for standard-radiator arrangements [152].

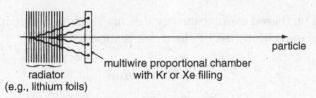

Fig. 5.44. Working principle of a transition-radiation detector.

[163, 164], i.e., for particles with $\gamma < 1000$ almost no transition-radiation photons are emitted.

A typical arrangement of a *transition-radiation detector* (TRD) is shown in Fig. 5.44. The TRD is formed by a set of foils consisting of a material with an atomic number Z as low as possible. Because of the strong dependence of the photoabsorption cross section on $Z (\sigma_{\text{photo}} \propto Z^5)$ the transition-radiation photons would otherwise not be able to escape from the radiator. The transition-radiation photons have to be recorded in a detector with a high efficiency for X-ray photons. This requirement is fulfilled by a multiwire proportional chamber filled with krypton or xenon, i.e. gases with high atomic number for an effective absorption of X rays.

In the set-up sketched in Fig. 5.44 the charged particle also traverses the photon detector, leading to an additional energy deposit by ionisation and excitation. This energy loss is superimposed onto the energy deposit by transition radiation. Figure 5.45 shows the energy-loss distribution in a transition-radiation detector for highly relativistic electrons for the case that (a) the radiator has gaps and (b) the radiator has no gaps (dummy). In both cases the amount of material in the radiators is the same. In the first case, because of the gaps, transition-radiation photons are emitted,

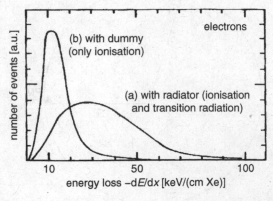

Fig. 5.45. Typical energy-loss distribution for high-energy electrons in a transition-radiation detector with radiator and with dummy radiator [152].

leading to an increased average energy loss of electrons, while in the second case only the ionisation loss of electrons is measured [152].

5.8   Problems

5.1 A Geiger–Müller counter (dead time $500\,\mu s$) measures in a strong radiation field a count rate of $1\,kHz$. What is the dead-time corrected true rate?

5.2 In practical situations the energy measurement in a semiconductor counter is affected by some dead layer of thickness d, which is usually unknown, in front of the sensitive volume. How can d be determined experimentally?

5.3 The Cherenkov angle is normally derived to be related to the particle velocity β and index of refraction n according to

$$\cos\Theta = \frac{1}{n\beta} \ .$$

This, however, neglects the recoil of the emitted Cherenkov photon on the incident particle. Determine the exact relation for the Cherenkov angle considering the recoil effect.

5.4 A detector may consist of a spherical vessel of radius R filled with liquid scintillator which is read out by a photomultiplier tube with photocathode area $S_p \ll S_{tot} = 4\pi R^2$ (Fig. 5.46). All inner surface of the vessel except the output window is covered with a diffusive

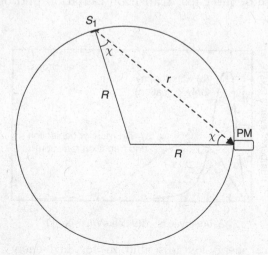

Fig. 5.46.   Illustration for Problem 5.4.

reflector of efficiency $(1 - \mu)$. Estimate the light collection efficiency if the diffusive reflection is governed by Lambert's law [165]: $dJ = (J_0/\pi) \cdot \cos \chi \cdot d\Omega$, where J_0 is the total amount of the reflected light and χ is the angle between the observation line and the normal to the surface. The detector is irradiated uniformly.

References

[1] Particle Data Group, Review of Particle Physics, S. Eidelman *et al.*, *Phys. Lett.* **B592 Vol. 1–4** (2004) 1–1109; W.-M. Yao *et al.*, *J. Phys.* **G33** (2006) 1–1232; http://pdg.lbl.gov

[2] B. Rossi & H. Staub, *Ionization Chambers and Counters*, McGraw-Hill, New York (1949)

[3] D.M. Wilkinson, *Ionization Chambers and Counters*, Cambridge University Press, Cambridge (1950)

[4] O.C. Allkofer, *Teilchendetektoren*, Thiemig, München (1971)

[5] O.R. Frisch, *British Atomic Energy Report* **BR-49** (1944)

[6] R. Feynman, *The Feynman Lectures on Physics*, Vol. 2, Addison-Wesley, Reading, MA (1964)

[7] D. McCormick, *Fast Ion Chambers for SLC*, SLAC-Pub-6296 (1993)

[8] M. Fishman & D. Reagan, The SLAC Long Ion Chamber For Machine Protection, *IEEE Trans. Nucl. Sci.* **NS-14** (1967) 1096–8

[9] E. Sauter, *Grundlagen des Strahlenschutzes*, Siemens AG, Berlin/ München (1971); *Grundlagen des Strahlenschutzes*, Thiemig, München (1982)

[10] S.E. Hunt, *Nuclear Physics for Engineers and Scientists*, John Wiley & Sons Inc. (Wiley Interscience), New York (1987)

[11] J.S. Townsend, *Electricity in Gases*, Clarendon Press, Oxford (1915)

[12] S.C. Brown, *Introduction to Electrical Discharges in Gases*, John Wiley & Sons Inc. (Wiley Interscience), New York (1966)

[13] H. Raether, *Electron Avalanches and Breakdown in Gases*, Butterworth, London (1964)

[14] A. Sharma & F. Sauli, A Measurement of the First Townsend Coefficient in Ar-based Mixtures at High Fields, *Nucl. Instr. Meth.* **A323** (1992) 280–3

[15] A. Sharma & F. Sauli, First Townsend Coefficient Measured in Argon Based Mixtures at High Fields, *Nucl. Instr. Meth.* **A334** (1993) 420–4

[16] F. Sauli, *Principles of Operation of Multiwire Proportional and Drift Chambers*, CERN-77-09 (1977) and references therein

[17] S.C. Brown, *Basic Data of Plasma Physics*, MIT Press, Cambridge, MA (1959), and John Wiley & Sons Inc., New York (1959)

[18] A. von Engel, Ionization in Gases by Electrons in Electric Fields, in S. Flügge (ed.), *Handbuch der Physik*, Elektronen-Emission; Gasentladungen I, Bd. **XXI**, Springer, Berlin (1956)

[19]   A. Arefev *et al.*, *A Measurement of the First Townsend Coefficient in CF_4, CO_2, and CF_4/CO_2-Mixtures at High, Uniform Electric Field*, RD5 Collaboration, CERN-PPE-93-082 (1993)

[20]   E. Bagge & O.C. Allkofer, Das Ansprechvermögen von Parallel-Platten Funkenzählern für schwach ionisierende Teilchen, *Atomenergie* **2** (1957) 1–17

[21]   B. Sitar, G.I. Merson, V.A. Chechin & Yu.A. Budagov, *Ionization Measurements in High Energy Physics* (in Russian), Energoatomizdat, Moskau (1988)

[22]   B. Sitar, G.I. Merson, V.A. Chechin & Yu.A. Budagov, *Ionization Measurements in High Energy Physics*, Springer Tracts in Modern Physics, Vol. 124, Springer, Berlin/Heidelberg (1993)

[23]   L.G. Huxley & R.W. Crompton, *The Diffusion and Drift of Electrons in Gases*, John Wiley & Sons Inc. (Wiley Interscience), New York (1974)

[24]   T.Z. Kowalski, Generalised Parametrization of the Gas Gain in Proportional Counters, *Nucl. Instr. Meth.* **A243** (1986) 501–4; On the Generalised Gas Gain Formula for Proportional Counters, *Nucl. Instr. Meth.* **A244** (1986) 533–6; Measurement and Parametrization of the Gas Gain in Proportional Counters, *Nucl. Instr. Meth.* **A234** (1985) 521–4

[25]   T. Aoyama, Generalised Gas Gain Formula for Proportional Counters, *Nucl. Instr. Meth.* **A234** (1985) 125–31

[26]   H.E. Rose & S.A. Korff, Investigation of Properties of Proportional Counters, *Phys. Rev.* **59** (1941) 850–9

[27]   A. Williams & R.I. Sara, Parameters Effecting the Resolution of a Proportional Counter, *Int. J. Appl. Radiation Isotopes* **13** (1962) 229–38

[28]   M.W. Charles, Gas Gain Measurements in Proportional Counters, *J. Phys.* **E5** (1972) 95–100

[29]   A. Zastawny, Gas Amplification in a Proportional Counter with Carbon Dioxide, *J. Sci. Instr.* **43** (1966) 179–81; *Nukleonika* **11** (1966) 685–90

[30]   L.G. Kristov, Measurement of the Gas Gain in Proportional Counters, *Doklady Bulg. Acad. Sci.* **10** (1947) 453–7

[31]   L.B. Loeb, *Basis Processes of Gaseous Electronics*, University of California Press, Berkeley (1961)

[32]   G.A. Schröder, Discharge in Plasma Physics, in S.C. Haydon (ed.), *Summer School Univ. of New England*, The University of New England, Armidale (1964)

[33]   M. Salehi, *Nuklididentifizierung durch Halbleiterspektrometer*, Diploma Thesis, University of Siegen (1990)

[34]   V. Aulchenko *et al.*, Fast, parallax-free, one-coordinate X-ray detector OD-3, *Nucl. Instr. Meth.* **A405** (1998) 269–73

[35]   G.C. Smith *et al.*, High Rate, High Resolution, Two-Dimensional Gas Proportional Detectors for X-Ray Synchrotron Radiation Experiments, *Nucl. Instr. Meth.* **A323** (1992) 78–85

[36]   G. Charpak, *Electronic Imaging of Ionizing Radiation with Limited Avalanches in Gases*, Nobel-Lecture 1992, CERN-PPE-93-25 (1993); *Rev. Mod. Phys.* **65** (1993) 591–8

[37]   H. Geiger, Method of Counting α and β-Rays, *Verh. d. Deutsch. Phys. Ges.* **15** (1913) 534–9

[38]   E. Rutherford & H. Geiger, An Electrical Method of Counting the Number of α-Particles from Radio-active Substances, *Proc. R. Soc. Lond.* **81** (1908) 141–61

[39]   H. Geiger & W. Müller, Das Elektronenzählrohr, *Z. Phys.* **29** (1928) 839–41; Technische Bemerkungen zum Elektronenzählrohr, *Z. Phys.* **30** (1929) 489–93

[40]   G.D. Alekseev, D.B. Khazins & V.V. Kruglov, Selfquenching Streamer Discharge in a Wire Chamber, *Lett. Nuovo Cim.* **25** (1979) 157–60; *Fiz. Elem. Chast. Atom. Yadra* **13** (1982) 703–48; Investigation of Selfquenching Streamer Discharge in a Wire Chamber, *Nucl. Instr. Meth.* **177** (1980) 385–97

[41]   E. Iarocci, Plastic Streamer Tubes and Their Applications in High Energy Physics, *Nucl. Instr. Meth.* **217** (1983) 30–42

[42]   G. Battistoni *et al.*, Operation of Limited Streamer Tubes, *Nucl. Instr. Meth.* **164** (1979) 57–66

[43]   G.D. Alekseev, Investigation of Self-Quenching Streamer Discharge in a Wire Chamber, *Nucl. Instr. Meth.* **177** (1980) 385–97

[44]   R. Baumgart *et al.*, The Response of a Streamer Tube Sampling Calorimeter to Electrons, *Nucl. Instr. Meth.* **A239** (1985) 513–17; Performance Characteristics of an Electromagnetic Streamer Tube Calorimeter, *Nucl. Instr. Meth.* **A256** (1987) 254–60; Interactions of 200 GeV Muons in an Electromagnetic Streamer Tube Calorimeter, *Nucl. Instr. Meth.* **A258** (1987) 51–7; Test of an Iron/Streamer Tube Calorimeter with Electrons and Pions of Energy between 1 and 100 GeV, *Nucl. Instr. Meth.* **A268** (1988) 105–11

[45]   R. Baumgart *et al.*, Properties of Streamers in Streamer Tubes, *Nucl. Instr. Meth.* **222** (1984) 448–57

[46]   *Dubna: Self Quenching Streamers Revisited*, CERN-Courier, **21(8)** (1981) 358

[47]   T. Doke, A Historical View on the R&D for Liquid Rare Gas Detectors, *Nucl. Instr. Meth.* **A327** (1993) 113–18

[48]   T. Doke (ed.), Liquid Radiation Detectors, *Nucl. Instr. Meth.* **A327** (1993) 3–226

[49]   T.S. Virdee, Calorimeters Using Room Temperature and Noble Liquids, *Nucl. Instr. Meth.* **A323** (1992) 22–33

[50]   A.A. Grebenuk, Liquid Noble Gas Calorimeters for KEDR and CMD-2M Detectors, *Nucl. Instr. Meth.* **A453** (2000) 199–204

[51]   M. Jeitler, The NA48 Liquid-Krypton Calorimeter, *Nucl. Instr. Meth.* **A494** (2002) 373–7

[52]   C. Ramsauer & R. Kollath, Die Winkelverteilung bei der Streuung langsamer Elektronen an Gasmolekülen, *Ann. Phys.* **401** (1931) 756–68

[53]   C. Ramsauer & R. Kollath, Über den Wirkungsquerschnitt der Edelgasmoleküle gegenüber Elektronen unterhalb 1 Volt, *Ann. Phys.* **395** (1929) 536–64

[54] C. Ramsauer, Über den Wirkungsquerschnitt der Gasmoleküle gegenüber langsamen Elektronen, *Ann. Phys.* **64** (1921) 513–40

[55] E. Brüche *et al.*, Über den Wirkungsquerschnitt der Edelgase Ar, Ne, He gegenüber langsamen Elektronen, *Ann. Phys.* **389** (1927) 279–91

[56] C.E. Normand, The Absorption Coefficient for Slow Electrons in Gases, *Phys. Rev.* **35** (1930) 1217–25

[57] D. Acosta *et al.*, Advances in Technology for High Energy Subnuclear Physics. Contribution of the LAA Project, *Riv. del Nuovo Cim.* **13(10–11)** (1990) 1–228; and G. Anzivino *et al.*, The LAA Project, *Riv. del Nuovo Cim.* **13(5)** (1990) 1–131

[58] J. Engler, H. Keim & B. Wild, Performance Test of a TMS Calorimeter, *Nucl. Instr. Meth.* **A252** (1986) 29–34

[59] M.G. Albrow *et al.*, Performance of a Uranium/Tetramethylpentane Electromagnetic Calorimeter, *Nucl. Instr. Meth.* **A265** (1988) 303–18

[60] K. Ankowiak *et al.*, Construction and Performance of a Position Detector for the UA1 Uranium-TMP Calorimeter, *Nucl. Instr. Meth.* **A279** (1989) 83–90

[61] M. Pripstein, *Developments in Warm Liquid Calorimetry*, LBL-30282, Lawrence-Berkeley Laboratory (1991); B. Aubert *et al.*, *Warm Liquid Calorimetry*, Proc. 25th Int. Conf. on High Energy Physics, Singapore, Vol. 2 (1991) 1368–71; B. Aubert *et al.*, A Search for Materials Compatible with Warm Liquids, *Nucl. Instr. Meth.* **A316** (1992) 165–73

[62] J. Engler, Liquid Ionization Chambers at Room Temperatures, *J. Phys. G: Nucl. Part. Phys.* **22** (1996) 1–23

[63] G. Bressi *et al.*, Electron Multiplication in Liquid Argon on a Tip Array, *Nucl. Instr. Meth.* **A310** (1991) 613–7

[64] R.A. Muller *et al.*, Liquid Filled Proportional Counter, *Phys. Rev. Lett.* **27** (1971) 532–6

[65] S.E. Derenzo *et al.*, *Liquid Xenon-Filled Wire Chambers for Medical Imaging Applications*, LBL-2092, Lawrence-Berkeley Laboratory (1973)

[66] E. Aprile, K.L. Giboni & C. Rubbia, *A Study of Ionization Electrons Drifting Large Distances in Liquid and Solid Argon*, Harvard University Preprint (May 1985)

[67] C. Kittel, *Introduction to Solid State Physics*, 8th edition, Wiley Interscience, New York (2005), and *Einführung in die Festkörperphysik*, Oldenbourg, München/Wien (1980)

[68] G.F. Knoll, *Radiation Detection and Measurement*, 3rd edition, John Wiley & Sons Inc., New York (Wiley Interscience), New York (1999/2000)

[69] S.M. Sze, *Physics of semiconductor devices*, 2nd edition, Wiley, New York (1981)

[70] P. Sangsingkeow *et al.*, Advances in Germanium Detector Technology, *Nucl. Instr. Meth.* **A505** (2003) 183–6

[71] G. Bertolini & A. Coche (eds.), *Semiconductor Detectors*, Elsevier–North Holland, Amsterdam (1968)

[72] V. Zhilich, private communication

[73] S.P. Beaumont *et al.*, Gallium Arsenide Microstrip Detectors for Charged Particles, *Nucl. Instr. Meth.* **A321** (1992) 172–9

[74] A.I. Ayzenshtat *et al.*, GaAs as a Material for Particle Detectors, *Nucl. Instr. Meth.* **A494** (2002) 120–7

[75] R.H. Redus *et al.*, Multielement CdTe Stack Detectors for Gamma-Ray Spectroscopy, *IEEE Trans. Nucl. Sci.* **51(5)** (2004) 2386–94

[76] A.H. Walenta, Strahlungsdetektoren – Neuere Entwicklungen und Anwendungen, *Phys. Bl.* **45** (1989) 352–6

[77] R. Kelley, D. McCammon *et al.*, *High Resolution X-Ray Spectroscopy Using Microcalorimeters*, NASA-Preprint LHEA 88-026 (1988); publ. in 'X-Ray Instrumentation in Astronomy', Proc. S.P.I.E. 982 (1988) 219–24; D. McCammon *et al.*, Cryogenic microcalorimeters for high resolution spectroscopy: Current status and future prospects, *Nucl. Phys.* **A527** (1991) 821C–4C

[78] F. Cardone & F. Celani, Rivelatori a Bassa Temperatura e Superconduttori per la Fisica delle Particelle di Bassa Energia, *Il Nuovo Saggiatore* **6(3)** (1990) 51–61

[79] A. Matsumura *et al.*, High Resolution Detection of X-Rays with a Large Area Nb/Al − Al O_x/Al/Nb Superconducting Tunnel Injection, *Nucl. Instr. Meth.* **A309** (1991) 350–2

[80] G. Hertz, *Lehrbuch der Kernphysik*, Bd. 1, Teubner, Leipzig (1966)

[81] J. Chadwick, Observations Concerning Artificial Disintegration of Elements, *Phil. Mag.* **7(2)** (1926) 1056–61

[82] V. Henri & J. Larguier des Bancels, Photochimie de la Rétine, *Journ. de Physiol. et de Pathol. Gén.* **XIII** (1911) 841–58; Anwendung der physikalisch-chemischen Untersuchungsmethoden auf das Studium verschiedener allgemein-biologischer Erscheinungen *Journ. de Physiol. et de Pathol. Gén.* **2** (Mars 1904)

[83] D.A. Baylor, Photoreceptor Signals and Vision: Proctor Lecture. *Investigative Ophthalmology and Visual Science*, **28** (1987) 34–49

[84] K.W.F. Kohlrausch, Radioaktivität, in W. Wien & F. Harms (eds.), *Handbuch der Experimentalphysik*, Band **15**, Akademische Verlagsanstalt, Leipzig (1928)

[85] R. Hofstadter, Alkali Halide Scintillation Counters, *Phys. Rev.* **74** (1948) 100–1; and Erratum: Alkali Halide Scintillation Counters, *Phys. Rev.* **74** (1948) 628

[86] R. Hofstadter & J.A. McIntyre, Gamma-Ray Spectroscopy with Crystals of NaI(Tl), *Nucleonics* **7** (1950) 32–7

[87] J.B. Birks, *The Theory and Practice of Scintillation Counting*, Pergamon Press, Oxford (1964, 1967); *Scintillation Counters*, Pergamon Press, Oxford (1953)

[88] K.D. Hildenbrand, *Scintillation Detectors*, Darmstadt GSI-Preprint GSI 93-18 (1993)

[89] E.B. Norman, *Scintillation Detectors*, LBL-31371, Lawrence-Berkeley Laboratory (1991)

[90] R. Hofstadter, *Twenty-Five Years of Scintillation Counting*, IEEE Scintillation and Semiconductor Counter Symposium, Washington DC, HEPL Report No. 749, Stanford University (1974)

[91] R. Novotny, Inorganic Scintillators: A Basic Material for Instrumentation in Physics, *Nucl. Instr. Meth.* **A537** (2005) 1–5

[92] C. Amsler *et al.*, Temperature Dependence of Pure CsI: Scintillation Light Yield and Decay Time, *Nucl. Instr. Meth.* **A480** (2002) 494–500

[93] Scintillation materials & Detectors, Catalogue of Amcrys-H, Kharkov, Ukraine (2000)

[94] Scintillation Detectors, Crismatec, Catalogue (1992)

[95] I. Holl, E. Lorenz & G. Mageras, A Measurement of the Light Yield of Some Common Inorganic Scintillators, *IEEE Trans. Nucl. Sci.* **35(1)** (1988) 105–9

[96] C.L. Melcher & J.S. Schweitzer, A Promising New Scintillator: Cerium-doped Luthetium Oxyorthosilicate, *Nucl. Instr. Meth.* **A314** (1992) 212–14

[97] M. Moszynski *et al.*, Large Size LSO:Ce and YSO:Ce Scintillators for 50 MeV Range γ-ray Detector, *IEEE Trans. Nucl. Sci.* **47**, No. 4 (2000) 1324–8

[98] M.E. Globus & B.V. Grinyov, Inorganic Scintillation Crystals: New and Traditional Materials, Kharkov (Ukraine), Akta (2000)

[99] C.L. Melcher, Perspectives on the Future Development of New Scintillators, *Nucl. Instr. Meth.* **A537** (2005) 6–14

[100] T. Förster, Zwischenmolekulare Energiewanderung und Fluoreszenz, *Ann. Phys.* **2** (1948) 55–67

[101] K. Kleinknecht, *Detektoren für Teilchenstrahlung*, Teubner, Stuttgart (1984, 1987, 1992); *Detectors for Particle Radiation*, Cambridge University Press, Cambridge (1986)

[102] Saint-Gobain, Crystals and Detectors, Organic Scintillators; General Characteristics and Technical Data; www.detectors.saint-gobain.com

[103] 'Physical Properties of Plastic Scintillators', www.amcrys-h.com/plastics.htm, www.amcrys-h.com/organics.htm

[104] R.B. Murray & A. Meyer, Scintillation Response of Activated Inorganic Crystals to Various Charged Particles, *Phys. Rev.* **122** (1961) 815–26

[105] W. Mengesha *et al.*, Light Yield Nonproportionality of CsI(Tl), CsI(Na) and YAP, *IEEE Trans. Nucl. Sci.* **45(3)** (1998) 456–60

[106] B.A. Dolgoshein & B.U. Rodionov, The Mechanism of Noble Gas Scintillation, *Elementary Particles and Cosmic Rays*, No. 2, Sect. 6.3, Atomizdat, Moscow (1969)

[107] T. Doke & K. Masuda, Present Status of Liquid Rare Gas Scintillation Detectors and Their New Application to Gamma-ray Calorimetry, *Nucl. Instr. Meth.* **A420** (1999) 62–80

[108] B. Mouellic, *A Comparative Study of the Reflectivity of Several Materials Used for the Wrapping of Scintillators in Particle Physics Experiments*, CERN-PPE-94-194, CERN (1994)

[109] B.J. Pichler *et al.*, Production of a Diffuse Very High Reflectivity Material for Light Collection in Nuclear Detectors, *Nucl. Instr. Meth.* **A442** (2000) 333–6

[110] Nuclear Enterprises, *Scintillation Materials*, Edinburgh (1977)

[111] Photomultiplier Tubes: Principles & Applications, Re-edited by S.-O. Flyct & C. Marmonier, Photonis, Brive, France (2002)

[112] Hamamatsu Photonics K.K., Editorial Committee, *Photomultipliers Tubes, Basics and Applications*, 2nd edition, Hamamatsu (1999)

[113] *Si Photodiode*, Catalogue, Hamamatsu Photonics K.K., Solid State Division, Hamamatsu (2002)

[114] Photomultiplier tubes, Catalogue Photonis, Brive, France (2000)

[115] K.S. Hirata *et al.*, Observation of a Neutrino Burst from the Supernova SN 1987 A, *Phys. Rev. Lett.* **58** (1987) 1490–3

[116] K.S. Hirata *et al.*, *Observation of ^8B Solar Neutrinos in the Kamiokande II Detector*, Inst. f. Cosmic Ray Research, ICR-Report 188-89-5 (1989)

[117] Hamamatsu Photonics K.K., *Measure Weak Light from Indeterminate Sources with New Hemispherical PM*, CERN-Courier, **21(4)** (1981) 173; private communication by Dr. H. Reiner, Hamamatsu Photonics, Germany

[118] K. Blazek *et al.*, YAP Multi-Crystal Gamma Camera Prototype, *IEEE Trans. Nucl. Sci.* **42(5)** (1995) 1474–82

[119] J. Bibby *et al.*, Performance of Multi-anode Photomultiplier Tubes for the LHCb RICH detectors, *Nucl. Instr. Meth.* **A546** (2005) 93–8

[120] E. Aguilo *et al.*, Test of Multi-anode Photomultiplier Tubes for the LHCb Scintillator Pad Detector, *Nucl. Instr. Meth.* **A538** (2005) 255–64

[121] Philips, *Imaging: From X-Ray to IR*, CERN-Courier, **23(1)** (1983) 35

[122] M. Akatsu *et al.*, MCP-PMT Timing Property for Single Photons, *Nucl. Instr. Meth.* **A528** (2004) 763–75

[123] G. Anzivino *et al.*, Review of the Hybrid Photo Diode Tube (HPD) an Advanced Light Detector for Physics, *Nucl. Instr. Meth.* **A365** (1995) 76–82

[124] C. D'Ambrosio & H. Leutz, Hybrid Photon Detectors, *Nucl. Instr. Meth.* **A501** (2003) 463–98

[125] C. D'Ambrosio *et al.*, Gamma Spectroscopy and Optoelectronic Imaging with Hybrid Photon Detector, *Nucl. Instr. Meth.* **A494** (2003) 186–97

[126] D.N. Grigoriev *et al.*, Study of a Calorimeter Element Consisting of a Csi(Na) Crystal and a Phototriode, *Nucl. Instr. Meth.* **A378** (1996) 353–5

[127] P.M. Beschastnov *et al.*, The Results of Vacuum Phototriodes Tests, *Nucl. Instr. Meth.* **A342** (1994) 477–82

[128] P. Checchia *et al.*, Study of a Lead Glass Calorimeter with Vacuum Phototriode Readout, *Nucl. Instr. Meth.* **A248** (1986) 317–25

[129] H. Grassman, E. Lorenz, H.G. Mozer & H. Vogel, Results from a CsI(Tl) Test Calorimeter with Photodiode Readout between 1 GeV and 20 GeV, *Nucl. Instr. Meth.* **A235** (1985) 319–25

[130] H. Dietl *et al.*, Reformance of BGO Calorimeter with Photodiode Readout and with Photomultiplier Readout at Energies up to 10 GeV, *Nucl. Instr. Meth.* **A235** (1985) 464–74

[131] S.J. Fagen (ed.), *The Avalanche Photodiode Catalog*, Advanced Photonix Inc., Camarillo, CA93012, USA (1992)

[132] E. Lorenz *et al.*, *Test of a Fast, Low Noise Readout of Pure CsI Crystals with Avalanche Photodiodes*, Proc. 4th Int. Conf. on Calorimetry in High-energy Physics, La Biodola, Italy (19–25 September 1993) 102–6

[133] E. Lorenz *et al.*, Fast Readout of Plastic and Crystal Scintillators by Avalanche Photodiodes, *Nucl. Instr. Meth.* **A344** (1994) 64–72

[134] I. Britvich *et al.*, Avalanche Photodiodes Now and Possible Developments, *Nucl. Instr. Meth.* **A535** (2004) 523–7

[135] R.J. McIntyre, Multiplication Noise in Uniform Avalanche Diodes, *IEEE Trans. on Electron Devices* **ED-13** (1966) 164–8

[136] P.P. Webb, R.J. McIntyre & J. Conradi, Properties of Avalanche Photodiodes, *RCA Review* **35** (1974) 234–78

[137] A.N. Otte *et al.*, A Test of Silicon Photomultipliers as Readout for PET, *Nucl. Instr. Meth.* **A545** (2005) 705–15

[138] P.A. Cherenkov, Visible Radiation Produced by Electrons Moving in a Medium with Velocities Exceeding that of Light, *Phys. Rev.* **52** (1937) 378–9

[139] P.A. Cherenkov, *Radiation of Particles Moving at a Velocity Exceeding that of Light, and some of the Possibilities for Their Use in Experimental Physics*, and I.M. Frank, *Optics of Light Sources Moving in Refractive Media*, and I.E. Tamm, *General Characteristics of Radiations Emitted by Systems Moving with Super Light Velocities with some Applications to Plasma Physics*, Nobel Lectures 11 December, 1958, publ. in *Nobel Lectures in Physics 1942–62*, Elsevier Publ. Comp., New York (1964) 426–40 (Cherenkov), 471–82 (Tamm) and 443–68 (Frank)

[140] P. Marmier & E. Sheldon, *Physics of Nuclei and Particles*, Vol. 1, Academic Press, New York (1969)

[141] J.V. Jelley, *Cherenkov Radiation and its Applications*, Pergamon Press, London/New York (1958)

[142] C. Grupen & E. Hell, Lecture Notes, *Kernphysik*, University of Siegen (1983)

[143] R.C. Weast & M.J. Astle (eds.), *Handbook of Chemistry and Physics*, CRC Press, Boca Raton, Florida (1979, 1987)

[144] M. Born & E. Wolf, *Principles of Optics*, Pergamon Press, New York (1964)

[145] N.W. Ashcroft & N.D. Mermin, *Solid State Physics*, Holt-Saunders, New York (1976)

[146] A.F. Danilyuk *et al.*, Recent Results on Aerogel Development for Use in Cherenkov Counters, *Nucl. Instr. Meth.* **A494** (2002) 491–4

[147] G. Poelz & R. Reithmuller, Preparation of Silica Aerogel for Cherenkov Counters, *Nucl. Instr. Meth.* **195** (1982) 491–503

[148] E. Nappi, Aerogel and its Applications to RICH Detectors, ICFA Instrumentation Bulletin and SLAC-JOURNAL-ICFA-17-3, *Nucl. Phys. B Proc. Suppl.* **61** (1998) 270–6

[149] H. Bradner & D.A. Glaser, *Methods of Particle Identification for High energy Physics Experiments*, 2nd United Nations International Conference on Peaceful Uses of Atomic Energy, A/CONF.15/P/730/Rev. 1 (1958)

[150] C. Biino *et al.*, A Glass Spherical Cherenkov Counter Based on Total Internal Reflection, *Nucl. Instr. Meth.* **A295** (1990) 102–8

[151] V. Fitch & R. Motley, Mean Life of K^+ Mesons, *Phys. Rev.* **101** (1956) 496–8; Lifetime of τ^+ Mesons, *Phys. Rev.* **105** (1957) 265–6

[152] C.W. Fabjan & H.G. Fischer, *Particle Detectors*, CERN-EP-80-27 (1980)

[153] V.L. Ginzburg & V.N. Tsytovich, *Transition Radiation and Transition Scattering*, Inst. of Physics Publishing, Bristol (1990)

[154] A. Bodek *et al.*, Observation of Light Below Cherenkov Threshold in a 1.5 Meter Long Integrating Cherenkov Counter, *Z. Phys.* **C18** (1983) 289–306

[155] W.W.M. Allison & J.H. Cobb, Relativistic Charged Particle Identification by Energy Loss, *Ann. Rev. Nucl. Sci.* **30** (1980) 253–98

[156] W.W.M. Allison & P.R.S. Wright, *The Physics of Charged Particle Identification: dE/dx, Cherenkov and Transition Radiation*, Oxford University Preprint OUNP 83-35 (1983)

[157] Particle Data Group, R.M. Barnett *et al.*, *Phys. Rev.* **D54** (1996) 1–708; *Eur. Phys. J.* **C3** (1998) 1–794; *Eur. Phys. J.* **C15** (2000) 1–878

[158] R.C. Fernow, *Brookhaven Nat. Lab. Preprint* **BNL-42114** (1988)

[159] S. Paul, *Particle Identification Using Transition Radiation Detectors*, CERN-PPE-91-199 (1991)

[160] B. Dolgoshein, Transition Radiation Detectors, *Nucl. Instr. Meth.* **A326** (1993) 434–69

[161] V.L. Ginzburg & I.M. Frank, Radiation of a Uniformly Moving Electron due to its Transitions from One Medium into Another, *JETP* **16** (1946) 15–29

[162] G.M. Garibian, *Macroscopic Theory of Transition Radiation*, Proc. 5th Int. Conf. on Instrumentation for High Energy Physics, Frascati (1973) 329–33

[163] X. Artru *et al.*, Practical Theory of the Multilayered Transition Radiation Detector, *Phys. Rev.* **D12** (1975) 1289–306

[164] J. Fischer *et al.*, Lithium Transition Radiator and Xenon Detector Systems for Particle Identification at High Energies, JINR-Report D13-9164, Dubna (1975); *Nucl. Instr. Meth.* **127** (1975) 525–45

[165] A.S. Glassner, Surface Physics for Ray Tracing, in A.S. Glassner (ed.), *Introduction to Ray Tracing*, Academic Press, New York (1989) 121–60

6

Historical track detectors

A man should look for what is, and not for what he thinks should be.

Albert Einstein

In this chapter some historical particle detectors will be briefly described. These are mainly optical devices that have been used in the early days of cosmic rays and particle physics. Even though some of these detectors have been 'recycled' for recent elementary particle physics experiments, like nuclear emulsions for the discovery of the tau neutrino (ν_τ) or bubble chambers with holographic readout for the measurement of short-lived hadrons, these optical devices are nowadays mainly integrated into demonstration experiments in exhibitions or employed as eye-catchers in lobbies of physics institutes (like spark chambers or diffusion cloud chambers).

6.1   Cloud chambers

The *cloud chamber* ('Wilson chamber') is one of the oldest detectors for track and ionisation measurement [1–4]. In 1932 Anderson discovered the positron in cosmic rays by operating a cloud chamber in a strong magnetic field (2.5 T). Five years later Anderson, together with Neddermeyer, discovered the muon again in a cosmic-ray experiment with cloud chambers.

A cloud chamber is a container filled with a gas–vapour mixture (e.g. air–water vapour, argon–alcohol) at the vapour saturation pressure. If a charged particle traverses the cloud chamber, it produces an ionisation trail. The lifetime of positive ions produced in the ionisation process in the chamber gas is relatively long ($\approx$ ms). Therefore, after the passage of the

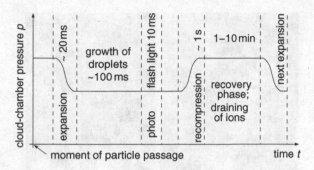

Fig. 6.1. Expansion cycle in a cloud chamber [5].

particle a trigger signal, for example, can be derived from a coincidence of scintillation counters, which initiates a fast expansion of the chamber. By means of adiabatic expansion the temperature of the gas mixture is lowered and the vapour gets supersaturated. It condenses on seeds, which are represented by the positive ions yielding droplets marking the particle trajectory. The track consisting of droplets is illuminated and photographed. A complete expansion cycle in a cloud chamber is shown in Fig. 6.1 [5].

The characteristic times, which determine the length of a cycle, are the lifetime of condensation nuclei produced by the ionisation ($\approx 10\,\mathrm{ms}$), the time required for the droplets to grow to a size where they can be photographed ($\approx 100\,\mathrm{ms}$), and the time which has to pass after the recording of an event until the chamber is recycled to be ready for the next event. The latter time can be very long since the sensitive volume of the chamber must be cleared of the slowly moving positive ions. In addition, the cloud chamber must be transformed into the initial state by recompression of the gas–vapour mixture.

In total, cycle times from 1 min up to 10 min can occur, limiting the application of this chamber type to rare events in the field of cosmic rays.

Figure 6.2 shows electron cascades initiated by cosmic-ray muons in a multiplate cloud chamber [6, 7].

A *multiplate cloud chamber* is essentially a sampling calorimeter with photographic readout (see Chap. 8 on 'Calorimetry'). The introduction of lead plates into a cloud chamber, which in this case was used in an extensive-air-shower experiment, serves the purpose of obtaining an electron/hadron/muon separation by means of the different interaction behaviour of these elementary particles.

In contrast to the *expansion cloud chamber*, a diffusion cloud chamber is permanently sensitive. Figure 6.3 shows schematically the construction of a *diffusion cloud chamber* [5, 8–11]. The chamber is, like the expansion cloud chamber, filled with a gas–vapour mixture. A constant temperature

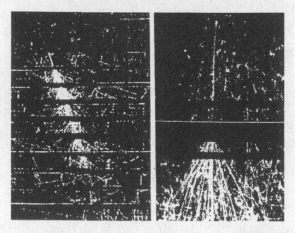

Fig. 6.2. Electromagnetic cascades in the core of an extensive air shower initiated by cosmic-ray muons (presumably via muon bremsstrahlung) in a multiplate cloud chamber [6, 7].

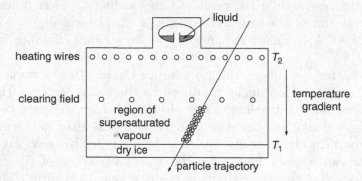

Fig. 6.3. Schematic representation of the construction of a diffusion cloud chamber [5].

gradient provides a region where the vapour is in a permanently supersaturated state. Charged particles entering this region produce a trail automatically without any additional trigger requirement. Zone widths (i.e. regions in which trails can form) with supersaturated vapour of 5 cm to 10 cm can be obtained. A clearing field removes the positive ions from the chamber.

The advantage of permanent sensitivity is obtained at the expense of small sensitive volumes. Since the chamber cannot be triggered, all events, even background events without interest, are recorded.

Because of the long repetition time for triggered cloud chambers and the disadvantage of photographic recording, this detector type is rarely used nowadays.

6.2 Bubble chambers

Bubble chambers [12–17], like cloud chambers, belong to the class of visual detectors and, therefore, require optical recording of events. This method of observation includes the tedious analysis of bubble-chamber pictures which certainly limits the possible statistics of experiments. However, the bubble chamber allows the recording and reconstruction of events of high complexity with high spatial resolution. Therefore, it is perfectly suited to study rare events (e.g. neutrino interactions); still, the bubble chamber has now been superseded by detectors with a purely electronic readout.

In a bubble chamber the liquid (H_2, D_2, Ne, C_3H_8, Freon, etc.) is held in a pressure container close to the boiling point. Before the expected event the chamber volume is expanded by retracting a piston. The expansion of the chamber leads to a reduction in pressure thereby exceeding the boiling temperature of the bubble-chamber liquid. If in this *superheated liquid state* a charged particle enters the chamber, bubble formation sets in along the particle track.

The positive ions produced by the incident particles act as nuclei for bubble formation. The lifetime of these nuclei is only 10^{-11} s to 10^{-10} s. This is too short to trigger the expansion of the chamber by the incoming particles. For this reason the superheated state has to be reached *before* the arrival time of the particles. Bubble chambers, however, can be used at accelerators where the arrival time of particles in the detector is known and, therefore, the chamber can be expanded in time (*synchronisation*).

In the superheated state the bubbles grow until the growth is stopped by a termination of the expansion. At this moment the bubbles are illuminated by light flashes and photographed. Figure 6.4 shows the principle of operation of a bubble chamber [5, 8]. The inner walls of the container have to be extremely smooth so that the liquid 'boils' only in those places where bubble formation should occur, namely, along the particle trajectory, and not on the chamber walls.

Depending on the size of the chamber repetition times down to 100 ms can be obtained with bubble chambers.

The bubble-chamber pressure before expansion is several atmospheres. To transform the gases into the liquid state they generally must be strongly cooled. Because of the large amount of stored gases, the experiments with hydrogen bubble chambers can be potentially dangerous because of the possible formation of explosive oxyhydrogen gas, if the chamber gas leaks from the bubble chamber. Also operation with organic liquids, which must be heated for the operation, represents a risk because of their flammability. Bubble chambers are usually operated in a high magnetic field (several Tesla). This allows to measure particle momenta

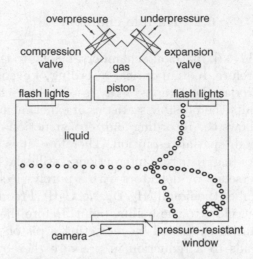

Fig. 6.4. Schematic construction of a bubble chamber [5, 8].

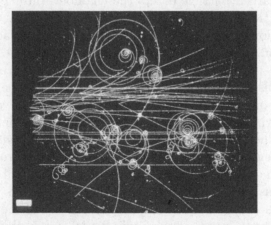

Fig. 6.5. Tracks of charged particles in a bubble chamber. Also seen are δ electrons produced by interactions of the incident particles in the bubble-chamber liquid, which are spiralling in the transverse magnetic field [18].

with high precision since the spatial resolution of bubble chambers is excellent. Furthermore, the bubble density along the track is proportional to the energy loss $\mathrm{d}E/\mathrm{d}x$ by ionisation. For $p/m_0c = \beta\gamma \ll 4$ the energy loss can be approximated by

$$\frac{\mathrm{d}E}{\mathrm{d}x} \propto \frac{1}{\beta^2} \; . \tag{6.1}$$

If the momentum of the particle is known and if the velocity is determined from an energy-loss measurement, the particle can be identified.

Figure 6.5 shows tracks of charged particles in a bubble chamber. One can see the decay of a neutral particle producing a 'V' (presumably

Table 6.1. *Characteristic properties of bubble-chamber liquids [5, 19]*

Bubble-chamber filling	Boiling point T [K]	Vapour pressure [bar]	Density [g/cm^3]	Radiation length X_0 [cm]	Nuclear interaction length λ_I [cm]
^{4}He	3.2	0.4	0.14	1027	437
^{1}H$_2$	26	4	0.06	1000	887
D$_2$	30	4.5	0.14	900	403
^{20}Ne	36	7.7	1.02	27	89
C$_3$H$_8$	333	21	0.43	110	176
CF$_3$Br (Freon)	303	18	1.5	11	73

$K^0 \rightarrow \pi^+ + \pi^-$) and δ electrons spiralling in the transverse magnetic field.

For the investigation of photoproduction on protons naturally, the best choice is a pure hydrogen filling. Results on photoproduction off neutrons can be obtained from D$_2$ fillings, because no pure neutron liquid exists (maybe with the exception of neutron stars). The photonuclear cross section on neutrons can be determined according to

$$\sigma(\gamma, n) = \sigma(\gamma, d) - \sigma(\gamma, p) \ . \tag{6.2}$$

If, e.g., the production of neutral pions is to be investigated, a bubble-chamber filling with a small radiation length X_0 is required, because the π^0 decays in two photons which have to be detected via the formation of electromagnetic showers. In this case, xenon or Freon can be chosen as chamber gas.

Table 6.1 lists some important gas fillings for bubble chambers along with their characteristic parameters [5, 19].

If one wants to study nuclear interactions with bubble chambers, the nuclear interaction length λ_I should be as small as possible. In this case heavy liquids like Freon are indicated.

Bubble chambers are an excellent device if the main purpose of the experiment is to analyse complex and rare events. For example, the Ω^- – after first hints from experiments in cosmic rays – could be unambiguously discovered in a bubble-chamber experiment.

In recent times the application of bubble chambers has, however, been superseded by other detectors like electronic devices. The reasons for this originate from some serious intrinsic drawbacks of the bubble chamber listed as follows:

- Bubble chambers cannot be triggered.

- They cannot be used in storage-ring experiments because it is difficult to achieve a 4π geometry with this type of detector. Also the 'thick' entrance windows required for the pressure container prevents good momentum resolution because of multiple scattering.

- For high energies the bubble chamber is not sufficiently massive to stop the produced particles. This precludes an electron and hadron calorimetry – not to mention the difficult and tedious analysis of these cascades – because shower particles will escape from the detector volume.

- The identification of muons with momenta above several GeV/c in the bubble chamber is impossible because they look almost exactly like pions as far as the specific energy loss is concerned. Only by use of additional detectors (external muon counters) a π/μ separation can be achieved.

- The lever arm of the magnetic field is generally insufficient for an accurate momentum determination of high-momentum particles.

- Experiments which require high statistics are not really practical because of the time-consuming analysis of bubble-chamber pictures.

However, bubble chambers are still used in experiments with external targets (fixed-target experiments) and in non-accelerator experiments. Because of their high intrinsic spatial resolution of several micrometre, bubble chambers can serve as vertex detectors in these experiments [20, 21].

To be able to measure short lifetimes in bubble chambers the size of the bubbles must be limited. This means that the event under investigation must be photographed relatively soon after the onset of bubble formation when the bubble size is relatively small, thereby guaranteeing a good spatial resolution and, as a consequence, also good time resolution. In any case the bubble size must be small compared to the decay length of the particle.

By use of the technique of holographic recording a three-dimensional event reconstruction can be achieved [22]. With these high-resolution bubble chambers, e.g., the lifetimes of short-lived particles can be determined precisely. For a spatial resolution of $\sigma_x = 6\,\mu$m decay-time measurement errors of

$$\sigma_\tau = \frac{\sigma_x}{c} = 2 \cdot 10^{-14}\,\text{s} \tag{6.3}$$

are reachable.

Bubble chambers have contributed significantly to the field of high-energy hadron collisions and neutrino interactions [23].

6.3 Streamer chambers

In contrast to streamer *tubes*, which represent a particular mode of operation for special cylindrical counters, streamer *chambers* are large-volume detectors in which events are normally recorded photographically [24–28]. In streamer chambers the volume between two planar electrodes is filled with a counting gas. After passage of a charged particle, a high-voltage pulse of high amplitude, short rise time and limited duration is applied to the electrodes. Figure 6.6 sketches the principle of operation of such a detector.

In the most frequent mode of operation, particles are incident approximately perpendicular to the electric field into the chamber. Each individual ionisation electron will start an avalanche in the homogeneous, very strong electric field in the direction of the anode. Since the electric field is time-dependent (amplitude of the high-voltage pulse $\approx 500\,\text{kV}$, rise and decay time $\approx 1\,\text{ns}$, pulse duration: several ns), the avalanche formation is interrupted after the decay time of the high-voltage pulse. The high amplitude of the voltage pulse leads to large gas amplifications ($\approx 10^8$) like in streamer tubes; however, the streamers can only extend over a very small region of space. Naturally, in the course of the avalanche development large numbers of gas atoms are excited and subsequently de-excite, leading to light emission. Luminous streamers are formed. Normally, these streamers are not photographed in the side view as sketched in Fig. 6.6, but through one electrode which can be made from a transparent wire mesh. In this projection the longish streamers appear as luminous dots characterising the track of the charged particle.

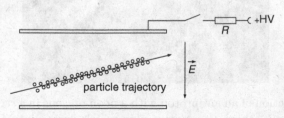

Fig. 6.6.   Principle of construction of a streamer chamber.

The art of the operation of streamer chambers lies in the production of a high-voltage signal with the required properties. The rise time must be extremely short (ns), otherwise the leading edge of the pulse would displace the ionisation electrons from the original track. A slow leading edge of the pulse would act as a clearing field resulting in a displacement of the particle track. Streamer development proceeds in very large electric fields ($\approx 30\,\text{kV/cm}$). It must, however, be interrupted after a short time, so that the streamers will not grow too large or even produce a spark. Streamers that are too large imply a poor spatial resolution. A suitable high-voltage pulse can be obtained using a Marx generator connected by a suitable circuit (*transmission line*, Blumlein circuit, spark gaps) to the streamer chamber providing short signals of high amplitude [24, 26].

For fast repetition rates the large number of electrons produced in the course of streamer formation poses a problem. It would take too long a time to remove these electrons from the chamber volume by means of a clearing field. Therefore, electronegative components are added to the counting gas to which the electrons are attached. *Electronegative quenchers* like SF_6 or SO_2 have proven to be good. These quenchers allow cycle times of several $100\,\text{ms}$. The positive ions produced during streamer formation do not present a problem because they can never start new streamer discharges due to their low mobility.

Streamer chambers provide pictures of excellent quality. Also targets can be mounted in the chamber to obtain the interaction vertex in the sensitive volume of the detector.

Figure 6.7 shows the interaction of an antiproton with a neon nucleus in a streamer chamber in which – among others – a positive pion is produced.

Fig. 6.7. Interaction of an antiproton with a neon nucleus in a streamer chamber producing – among others – a positive pion which decays into a muon yielding eventually a positron which escapes from the chamber [18].

This π^+ spirals anticlockwise and decays into a muon, which also spirals in the transverse magnetic field and eventually decays into a positron, which escapes from the chamber [18].

In a different mode of operation of the streamer chamber, particles are incident within $\pm 30°$ with respect to the electric field into the detector. Exactly as mentioned before, very short streamers will develop which now, however, merge into one another and form a plasma channel along the particle track. (This variant of the streamer chamber is also called *track spark chamber* [8, 28].) Since the high-voltage pulse is very short, no spark between the electrodes develops. Consequently, only a very low current is drawn from the electrodes [5, 8, 24].

Streamer chambers are well suited for the recording of complex events; they have, however, the disadvantage of a time-consuming analysis.

6.4 Neon-flash-tube chamber

The *neon-flash-tube chamber* is also a discharge chamber [17, 29–32]. Neon- or neon/helium-filled glass tubes (at atmospheric pressure), glass spheres (*Conversi tubes*) or polypropylene-extruded plastic tubes with rectangular cross section are placed between two metal electrodes (Fig. 6.8).

After a charged particle has passed through the neon-flash-tube stack, a high-voltage pulse is applied to the electrodes that initiates a gas discharge in those tubes, which have been passed by the particle. This gas discharge propagates along the total length of the tube and leads to a glow discharge in the whole tube. Typical tube lengths are around 2 m with diameters between 5 mm and 10 mm. The glow discharge can be intensified by afterpulsing with high voltage so that the flash tubes can be photographed end on. But purely electronic recording with the help of pickup electrodes at the faces of the neon tubes can also be applied ('Ayre–Thompson technique' [33, 34]). These pickup electrodes supply large signals which can be directly processed without additional preamplifiers.

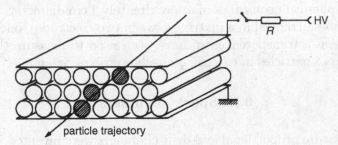

Fig. 6.8. Working principle of a neon-flash-tube chamber.

Fig. 6.9.   Shower of parallel muons in a flash-tube chamber [35, 36].

Depending on the tube diameter, spatial resolutions of several millimetres can be obtained. The memory time of this detector lies in the range around $20\,\mu s$; the dead time, however, is rather long at 30–1000 ms. For reasons of geometry, caused by the tube walls, the efficiency of one layer of neon flash tubes is limited to $\approx 80\%$. To obtain three-dimensional coordinates crossed layers of neon flash tubes are required.

Because of the relatively long dead time of this detector it was mainly used in cosmic-ray experiments, in the search for nucleon decay or in neutrino experiments. Figure 6.9 shows a shower of parallel cosmic-ray muons in a neon-flash-tube chamber [35, 36]. In Fig. 6.10 a single muon track is seen in an eight-layer stack of polypropylene-extruded plastic tubes [37].

A variant of the neon flash tube is the spherical Conversi tube [30, 31]. These are spherical neon tubes of approximately 1 cm diameter. Layers of Conversi tubes arranged in a matrix between two electrodes, one of which being made as a transparent grid, have been used to measure the lateral distribution of particles in extensive-air-shower experiments [8, 38].

6.5   Spark chambers

Before multiwire proportional and drift chambers were invented, the *spark chamber* was the most commonly used track detector which could be triggered ([8, 17, 39–43], and references in [17]).

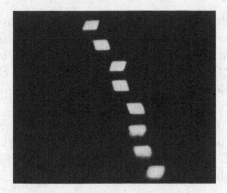

Fig. 6.10. Single muon track in a stack of polypropylene-extruded plastic tubes. Such extruded plastic tubes are very cheap since they are normally used as packing material. Because they have not been made for particle tracking, their structure is somewhat irregular, which can clearly be seen [37].

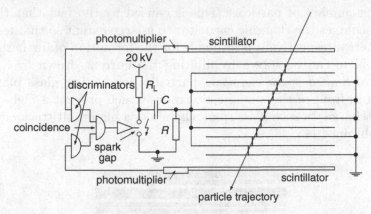

Fig. 6.11. Principle of operation of a multiplate spark chamber.

In a spark chamber a number of parallel plates are mounted in a gas-filled volume. Typically, a mixture of helium and neon is used as counting gas. Alternatingly, the plates are either grounded or connected to a high-voltage supply (Fig. 6.11). The high-voltage pulse is normally triggered to every second electrode by a coincidence between two scintillation counters placed above and below the spark chamber. The gas amplification is chosen in such a way that a spark discharge occurs at the point of the passage of the particle. This is obtained for gas amplifications between 10^8 and 10^9. For lower gas amplifications sparks will not develop, while for larger gas amplifications sparking at unwanted positions (e.g. at spacers which separate the plates) can occur. The discharge channel follows the electric field. Up to an angle of $30°$ the conducting plasma channel can, however, follow the particle trajectory [8] as in the track spark chamber.

Between two discharges the produced ions are removed from the detector volume by means of a *clearing field*. If the time delay between the passage of the particle and the high-voltage signal is less than the memory time of about $100\,\mu s$, the efficiency of the spark chamber is close to 100%. A clearing field, of course, removes also the primary ionisation from the detector volume. For this reason the time delay between the passage of the particle and the application of the high-voltage signal has to be chosen as short as possible to reach full efficiency. Also the rise time of the high-voltage pulse must be short because otherwise the leading edge acts as a clearing field before the critical field strength for spark formation is reached.

Figure 6.12 shows the track of a cosmic-ray muon in a multiplate spark chamber [5, 44].

If several particles penetrate the chamber simultaneously, the probability that all particles will form a spark trail decreases drastically with increasing number of particles. This is caused by the fact that the first spark discharges the charging capacitor to a large extent so that less voltage or energy, respectively, is available for the formation of further sparks. This problem can be solved by limiting the current drawn by a spark. In *current-limited spark chambers* partially conducting glass plates are mounted in front of the metallic electrodes which prevent a high-current spark discharge. In such glass spark chambers a high multitrack efficiency can be obtained [45, 46].

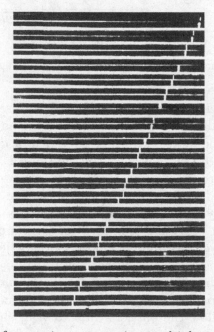

Fig. 6.12.   Track of a cosmic-ray muon in a multiplate spark chamber [44].

Apart from the photographic recording in spark chambers, which has to be made stereoscopically to allow three-dimensional event reconstruction, a purely electronic readout is also possible.

If the electrodes are made from layers of wires, the track coordinate can be obtained like in the multiwire proportional chamber by identifying the discharged wire. This method would require a large number of wires to obtain a high spatial resolution. On the other hand, the track reconstruction can be simplified with the help of a *magnetostrictive readout*.

The spark discharge represents a time-dependent current dI/dt. The current signal propagates along the chamber wire and reaches a *magnetostrictive delay line* stretched perpendicular to the chamber wires. This magnetostrictive delay line is positioned directly on the chamber wires without having ohmic contact to them. The current signal, along with its associated time-dependent magnetic field $d\vec{H}/dt$, produces in the magnetostrictive delay line a magnetostriction, i.e. a local variation of the length, which propagates in time and space with its characteristic velocity of sound. In a pickup coil at the end of the magnetostrictive delay line the mechanical signal of magnetostriction is converted back into a time-dependent magnetic-field signal $d\vec{H}/dt$ leading to a detectable voltage pulse. The measurement of the propagation time of the sound wave on the magnetostrictive delay line can be used to identify the number and hence the spatial coordinate of the discharged wire. Typical sound velocities of $\approx 5\,\mathrm{km/s}$ lead to spatial resolutions on the order of $200\,\mu\mathrm{m}$ [8, 47].

A somewhat older method of identifying discharged wires in wire spark chambers uses *ferrite cores* to localise the discharged wire. In this method each chamber wire runs through a small ferrite core [8]. The ferrite core is in a well-defined state. A discharging spark-chamber wire causes the ferrite core to flip. The state of ferrite cores is recorded by a readout wire. After reading out the event the flipped ferrite cores are reset into the initial state by a reset wire.

For all spark-chamber types a clearing field is necessary to remove the positive ions from the detector volume. This causes dead times of several milliseconds.

6.6 Nuclear emulsions

Tracks of charged particles in *nuclear emulsions* can be recorded by the photographic method [48–52]. Nuclear emulsions consist of fine-grained silver-halide crystals (AgBr and AgCl), which are embedded in a gelatine substrate. A charged particle produces a latent image in the emulsion. Due to the free charge carriers liberated in the ionisation process, some halide molecules are reduced to metallic silver in the emulsion.

In the subsequent development process the silver-halide crystals are chemically reduced. This affects preferentially those microcrystals (nuclei) which are already disturbed and partly reduced. These are transformed into elemental silver. The process of fixation dissolves the remaining silver halide and removes it. Thereby the charge image, which has been transformed into elemental silver particles, remains stable.

The evaluation of the emulsion is usually done under a microscope by eye, but it can also be performed by using a Charge-Coupled Device (CCD) camera and a semi-automatic pattern-recognition device. Fully automated emulsion-analysis systems have also been developed [53].

The sensitivity of the nuclear emulsion must be high enough so that the energy loss of minimum-ionising particles is sufficient to produce individual silver-halide microcrystals along the track of a particle. Usually commercially available photoemulsions do not have this property. Furthermore, the silver grains which form the track and also the silver-halide microcrystals must be sufficiently small to enable a high spatial resolution. The requirements of high sensitivity and low grain size are in conflict and, therefore, demand a compromise. In most nuclear emulsions the silver grains have a size of $0.1\,\mu$m to $0.2\,\mu$m and so are much smaller than in commercial films (1–$10\,\mu$m). The mass fraction of the silver halide (mostly AgBr) in the emulsion amounts to approximately 80%.

A typical thickness of emulsions is 20 to 1000 microns with sizes up to $50\,$cm $\times\ 50\,$cm. In the developing, fixing, washing and drying process extreme care must be taken not to lose the intrinsically high space resolution. In particular, a possible shrinking of the emulsion must be well understood.

Because of the high density of the emulsion ($\varrho = 3.8\,$g/cm^3) and the related short radiation length ($X_0 = 2.9\,$cm), stacks of nuclear emulsions are perfectly suited to detect electromagnetic cascades. On the other hand, hadron cascades hardly develop in such stacks because of the much larger nuclear interaction length ($\lambda_\mathrm{I} = 35\,$cm).

The efficiency of emulsions for single or multiple particle passages is close to 100%. Emulsions are permanently sensitive but they cannot be triggered. They have been, and still are, in use in many cosmic-ray experiments [51]. They are, however, also suited for accelerator experiments as vertex detectors with high spatial resolution ($\sigma_x \approx 2\,\mu$m) for the investigation of decays of short-lived particles.

The high resolution of complex events with large multiplicities is clearly shown in the interaction of a 228.5 GeV uranium nucleus in a nuclear emulsion (Fig. 6.13) [54].

Figure 6.14 shows the ionisation profile of an iron nucleus and a heavy nucleus ($Z \approx 90$) in a nuclear emulsion [55].

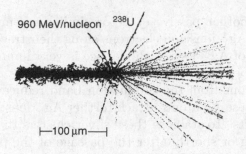

960 MeV/nucleon ^{238}U

├─100 μm─┤

Fig. 6.13. Interaction of an uranium nucleus of energy 228.5 GeV in a nuclear emulsion [54].

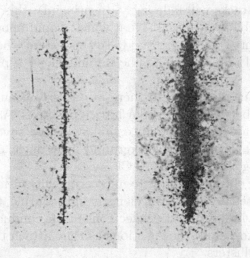

Fig. 6.14. Ionisation profile of an iron nucleus and a heavy nucleus ($Z \approx 90$) in a nuclear emulsion [55].

Occasionally, nuclear emulsions are used in accelerator experiments where extremely high spatial resolution is needed for rare events, like in the search for the tau neutrino (see Chap. 10 on 'Neutrino detectors').

Among others, the emulsion technique has contributed significantly in past decades to the fields of cosmic-ray physics, high-energy heavy-ion collisions, hypernuclear physics, neutrino oscillations, and to the study of charm and bottom particles [56].

6.7 Silver-halide crystals

The disadvantage of nuclear emulsions is that the sensitive volume of the detector is usually very small. The production of large-area $AgCl$ crystals has been made possible several years ago. This has allowed the construction of another passive detector similar to emulsions. Charged particles produce Ag^+ ions and electrons along their track in a AgCl crystal. The mobility of Ag^+ ions in the lattice is very limited. They usually occupy positions between regular lattice atoms thereby forming a lattice defect. Free electrons from the conduction band reduce the Ag^+ ions to metallic silver. These Ag atoms attach further Ag^+ ions: the formation of silver clusters starts. To stabilise these silver clusters the crystal must be illuminated during or shortly after the passage of the particle providing the free electrons required for the reduction of the Ag^+ ions (storage or *conservation of particle tracks*). This is frequently done by using light with a wavelength of around 600 nm [57]. If this illumination is not done during data taking, the tracks will fade away. In principle, this illumination can also be triggered by an external signal which would allow the separation of interesting events from background. To this extent, the event recording in an AgCl crystal – in contrast to nuclear emulsions or plastic detectors – can be triggered.

Even in the unirradiated state there are a certain number of Ag ions occupying places between the regular lattice positions. A small admixture of cadmium chloride serves to reduce this unwanted silver concentration. This minimises the formation of background silver nuclei on lattice defects, thereby decreasing the 'noise' in AgCl crystals.

To allow the Ag clusters to grow to microscopically visible size, the AgCl crystal is irradiated by short-wavelength light during the development process. This provides further free electrons in the conduction band which in turn will help to reduce the Ag^+ ions as they attach themselves to the already existing clusters.

This process of track amplification (*decoration*) produces a stable track which can then be evaluated under a microscope.

Silver-chloride detectors show – just like plastic detectors – a certain threshold effect. The energy loss of relativistic protons is too small to produce tracks which can be developed in the crystal. The AgCl detector, however, is well suited to measure tracks of heavy nuclei ($Z \geq 3$).

The tedious evaluation of nuclear tracks under a microscope can be replaced by automatic pattern-recognition methods similar to those which are in use for nuclear emulsions and plastic detectors [58–60]. The spatial resolution which can be achieved in AgCl crystals is comparable to that of nuclear emulsions.

6.8   X-ray films

Emulsion chambers, i.e. stacks of nuclear emulsions when used in cosmic-ray experiments, are frequently equipped with additional large-area *X-ray films* [61–63]. The main differences of the X-ray films from nuclear emulsions are the low grain size from 50 nm to 200 nm (0.05–0.2 microns) at a thickness of 7–20 microns [64]. These industrial X-ray films allow the detection of high-energy electromagnetic cascades (see Chap. 8 on 'Calorimetry') and the determination of the energy of electrons or photons initiating these cascades by photometric methods. This is done by constructing a stack of X-ray films alternating with thin lead sheets. The longitudinal and lateral development of electromagnetic cascades can be inferred from the structure of the darkness in the X-ray films.

X-ray films employed in cosmic-ray experiments are mainly used for the detection of photons and electrons in the TeV range. Hadronic cascades are harder to detect in stacks with X-ray films. They can, however, be recorded via the π^0 fraction in the hadron shower ($\pi^0 \to \gamma\gamma$). This is related to the fact that photons and electrons initiate narrowly collimated cascades producing dark spots on the X-ray film, whereas hadronic cascades, because of the relatively large transverse momenta of secondary particles, spread out over a larger area on the film thus not exceeding the threshold required for a blackening of the film.

Saturation effects in the region of the maximum of shower development (central blackening) cause the relation between the deposited energy E and the photometrically measured blackening D not to be linear [64]. For typical X-ray films which are used in the TeV range one gets

$$D \propto E^{0.85} . \tag{6.4}$$

The radial distribution of the blackening allows the determination of the point of particle passage with relatively high precision.

6.9 Thermoluminescence detectors

Thermoluminescence detectors are used in the field of radiation protection [65–67] and also in cosmic-ray experiments.

Particle detection in thermoluminescence detectors is based on the fact that in certain crystals ionising radiation causes electrons to be transferred from the valence band to the conduction band where they may occupy stable energy states [68]. In the field of radiation protection, media, which retain the dose information, such as manganese- or titanium-activated calcium-fluoride (CaF_2) or lithium-fluoride (LiF) crystals, are used. The stored energy caused by irradiating the crystal is proportional to the absorbed dose. Heating the thermoluminescence dosimeter to a temperature between 200 °C and 400 °C can liberate this energy by emission of photons. The number of produced photons is proportional to the absorbed energy dose.

In cosmic-ray experiments thermoluminescence films (similar to X-ray films) are used for the measurement of high-energy electromagnetic cascades. A thermoluminescence detector is made by coating a glass or metal surface with a layer of thermoluminescent powder. The smaller the grain size of microcrystals on the film the better the spatial resolution that can be reached. The ionising particles in the electron cascade produce stable thermoluminescence centres. The determination of where the energy is deposited on the film can be achieved by scanning the film with an infrared laser. During the process of scanning the intensity of emitted photons must be measured with a photomultiplier. If the spatial resolution is not limited by the radial extension of the laser spot, resolutions on the order of a few micrometres can be obtained [69].

Apart from doped calcium-fluoride or lithium-fluoride crystals and storage phosphors, which are commonly used in the field of radiation protection, cosmic-ray experiments utilise mainly $BaSO_4$, Mg_2SiO_4 and $CaSO_4$ as thermoluminescent agents. While thermoluminescence dosimeters measure the integrated absorbed energy dose, in cosmic-ray experiments the measurement of individual events is necessary.

In such experiments thermoluminescence films are stacked similar to X-ray films or emulsions alternatingly with lead-absorber sheets. The hadrons, photons or electrons to be detected initiate hadronic or electromagnetic cascades in the thermoluminescence calorimeter. Neutral pions produced in hadronic cascades decay relatively quickly (in $\approx 10^{-16}$ s) into two photons thereby initiating electromagnetic subcascades.

In contrast to hadronic cascades with a relatively large lateral width, the energy in electromagnetic cascades is deposited in a relatively small region thereby enabling a recording of these showers. That is why electromagnetic cascades are directly measured in such a detector type while hadronic cascades are detected only via their π^0 content. Thermoluminescence detectors exhibit an energy threshold for the detection of particles. This threshold is approximately 1 TeV per event in europium-doped $BaSO_4$ films [69].

6.10   Radiophotoluminescence detectors

Silver-activated phosphate glass, after having been exposed to ionising radiation, emits fluorescence radiation in a certain frequency range if irradiated by ultraviolet light. The intensity of the fluorescence radiation is a measure for the energy deposition by the ionising radiation. The Ag^+ ions produced by ionising particles in the glass represent stable photoluminescence centres. Reading out the energy deposition with ultraviolet light does not erase the information of the energy loss in the detector [68]. *Yokota glass* is mostly used in these phosphate-glass detectors. It consists of 45% $AlPO_3$, 45% $LiPO_3$, 7.3% $AgPO_3$ and 2.7% B_2O_3 and has a typical density of $2.6\,g/cm^3$ for a silver mass fraction of 3.7%. Such phosphate-glass detectors are mainly used in the field of radiation protection for dosimetric measurements.

By scanning a two-dimensional radiophotoluminescence sheet with a UV laser, it is possible to determine the spatial dependence of the energy deposit by measuring the position-dependent fluorescence light yield. If individual events are recorded, a threshold energy on the order of 1 TeV is required just as in thermoluminescence detectors. The spatial resolution that can be obtained is limited also in this case by the resolution of the scanning system.

6.11 Plastic detectors

Particles of high electric charge destroy the local structure in a solid along their tracks. This local destruction can be intensified by etching and thereby made visible. Solids such as inorganic crystals, glasses, plastics, minerals or even metals can be used for this purpose [70–74]. The damaged parts of the material react with the etching agent more intensively than the undamaged material and characteristic *etch cones* will be formed.

If the *etching process* is not interrupted, the etch cones starting from the surface of the plastic will merge and form a hole at the point of the particle track. The etching procedure will also remove some part of the surface material.

Figure 6.15 [75] shows a microphotograph of tracks made in a CR-39 *plastic nuclear-track detector* exposed on board the Mir Space Station during a NASA mission. The width of the track at the centre is approximately 15 μm [75].

For inclined incidence the etch cones exhibit an elliptical form.

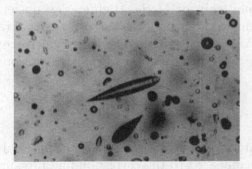

Fig. 6.15. Microphotograph of cosmic-ray tracks in a plastic nuclear-track detector. Typical track widths are on the order of 10 μm [75].

The determination of the energy of heavy ions is frequently done in stacks containing a large number of foils. The radiation damage of the material – just as the energy loss of charged particles – is proportional to the square of their charge, and depends also on the velocity of the particles.

Plastic detectors show a threshold effect: the minimum radiation damage caused by protons and α particles is frequently insufficient to produce etchable tracks. The detection and measurement of heavy ions, e.g. in primary cosmic rays $(Z \geq 3)$, will consequently not be disturbed by a high background of protons and α particles. The size of the etch cones (for a fixed etching time) is a measure of the energy loss of the particles. It allows, therefore, if the velocity of the particles is known, a determination of the charge of the nuclei. A stack of plastic detectors, flown in a balloon in a residual atmosphere of several grams per square centimetre, thus permits a determination of the elemental abundance in primary cosmic rays.

Plastic detectors are also utilised in the search for magnetic monopoles which, according to theory, should cause strong ionisation. Such experiments can also be performed on proton storage rings because the high background of singly charged particles does not impair the search for monopoles due to the threshold behaviour of the plastic material.

In a similar way to plastic detectors, minerals also conserve a local radiation damage over a long period of time. This leads to the possibility of dating uranium-containing minerals by counting the number of spontaneous fission events. If the minerals are time calibrated in this way, the number of tracks initiated by cosmic radiation in these minerals indicates that the intensity of cosmic rays has not varied significantly $(\leq 10\%)$ over the past 10^6 years [76, 77].

The evaluation of plastic detectors under the microscope is very tiresome. The information on particle tracks in a plastic sheet can, however, also be digitised by means of a CCD camera looking through a microscope onto the foil. The digitised event is subsequently processed with a programme for automatic pattern reconstruction [74].

A nuclear detector with super-high spatial resolution is provided, for example, by a small chip of MoS_2. High-energy nuclei penetrating the MoS_2 sample produce craters on its surface due to local radiation damage. Analysing these craters by scanning tunnelling microscopy allows spatial resolutions on the order of $10\,\text{Å}$ and two-track resolutions of $30\text{–}50\,\text{Å}$ [78].

6.12   Problems

6.1 The saturation vapour pressure p_r over a spherical surface of radius r is larger compared to the corresponding pressure p_∞ over a planar surface. To achieve good quality tracks in a cloud chamber one has to aim for $p_r/p_\infty = 1.001$. What kind of droplet size results from this condition in a cloud chamber with air and water vapour (air and alcohol vapour)?

The surface tension for water (alcohol) is $72.8\,\mathrm{dyn/cm}$ ($22.3\,\mathrm{dyn/cm}$).

6.2 In a discharge chamber the gas multiplication is characterised by the first Townsend coefficient α which describes the increase of the electron number dn/dx per primary electron over the distance dx.

$$dn = \alpha n \, dx \ .$$

On the other hand, some of the electrons can be attached to electronegative gases in the chamber (attachment coefficient β).

Work out the number of electrons and negative ions on a multiplication distance d and the total charge increase $(n_e + n_{\mathrm{ion}})/n_0$, where n_0 is the primary ionisation ($n_0 = 100/\mathrm{cm}, \alpha = 20/\mathrm{cm}, \beta = 2/\mathrm{cm}, d = 1\,\mathrm{cm}$).

6.3 In a nuclear emulsion ($X_0 = 5\,\mathrm{cm}$) of $500\,\mathrm{\mu m}$ thickness an average projected scattering angle for electrons is found to be $\sqrt{\langle\theta^2\rangle} = 5°$. Work out the electron momentum.

References

[1] C.T.R. Wilson, On a Method of Making Visible the Paths of Ionizing Particles, *Proc. R. Soc. Lond.* **A85** (1911) 285–8; Expansion Apparatus for Making Visible the Tracks of Ionizing Particles in Gases: Results Obtained, *Proc. R. Soc. Lond.* **A87** (1912) 277–92

[2] C.T.R. Wilson, Uranium Rays and Condensation of Water Vapor, *Cambridge Phil. Soc. Proc.* **9** (1897) 333–6; *Phil. Trans. R. Soc. Lond.* **189** (1897) 265–8; On the Condensation Nuclei Produced in Gases by the Action of Röntgen Rays, Uranium Rays, Ultra-Violet Light, and Other Agents, *Proc. R. Soc. Lond.* **64** (1898–9), 127–9

[3] C.M. York, Cloud Chambers, in S. Flügge (ed.), *Handbuch der Physik*, Band **XLV**, Springer, Berlin (1958) 260–313

[4] G.D. Rochester & J.G. Wilson, *Cloud Chamber Photographs of Cosmic Radiation*, Pergamon Press, London (1952)

[5] O.C. Allkofer, *Teilchendetektoren*, Thiemig, München (1971)

[6] W. Wolter, private communication (1969)

[7] U. Wiemken, *Untersuchungen zur Existenz von Quarks in der Nähe der Kerne Großer Luftschauer mit Hilfe einer Nebelkammer*, Ph.D. Thesis, University of Kiel; U. Wiemken, Diploma Thesis, University of Kiel (1972); K. Sauerland, private communication (1993)

[8] O.C. Allkofer, W.D. Dau & C. Grupen, *Spark Chambers*, Thiemig, München (1969)

[9]   A. Langsdorf, A Continuously Sensitive Cloud Chamber, *Phys. Rev.* **49** (1936) 422–34; *Rev. Sci. Instr.* **10** (1939) 91–103

[10]   V.K. Ljapidevski, Die Diffusionsnebelkammer, *Fortschr. der Physik* **7** (1959) 481–500

[11]   E.W. Cowan, Continuously Sensitive Diffusion Cloud Chamber, *Rev. Sci. Instr.* **21** (1950) 991–6

[12]   D.A. Glaser, Some Effects of Ionizing Radiation on the Formation of Bubbles in Liquids, *Phys. Rev.* **87** (1952) 665

[13]   D.A. Glaser, Bubble Chamber Tracks of Penetrating Cosmic Ray Particles, *Phys. Rev.* **91** (1953) 762–3

[14]   D.A. Glaser, Progress Report on the Development of Bubble Chambers, *Nuovo Cim. Suppl.* **2** (1954) 361–4

[15]   D.A. Glaser, The Bubble Chamber, in S. Flügge (ed.), *Handbuch der Physik*, Band **XLV**, Springer, Berlin (1958) 314–41

[16]   L. Betelli *et al.*, *Particle Physics with Bubble Chamber Photographs*, CERN/INFN-Preprint (1993)

[17]   P. Galison, *Bubbles, Sparks and the Postwar Laboratory*, Proc. Batavia Conf. 1985, *Pions to Quarks* (1989) 213–51

[18]   CERN Photo Archive

[19]   K. Kleinknecht, *Detectors for Particle Radiation*, 2nd edition, Cambridge University Press, Cambridge (1998); *Detektoren für Teilchenstrahlung*, Teubner, Wiesbaden (2005)

[20]   W.J. Bolte *et al.*, A Bubble Chamber for Dark Matter Detection (the COUPP project status), *J. Phys. Conf. Ser.* **39** (2006) 126–8

[21]   Y.L. Ju, J.R. Dodd, W.J. Willis (Nevis Labs, Columbia U.) & L.X. Jia (Brookhaven), Cryogenic Design and Operation of Liquid Helium in Electron Bubble Chamber, *AIP Conf. Proc.* **823** (2006) 433–40

[22]   H. Bingham *et al.*, *Holography of Particle Tracks in the Fermilab 15-Foot Bubble Chamber*, E-632 Collaboration, CERN-EF-90-3 (1990); *Nucl. Instr. Meth.* **A297** (1990) 364–89

[23]   W. Kittel, *Bubble Chambers in High Energy Hadron Collisions*, Nijmegen Preprint HEN-365 (1993)

[24]   P. Rice-Evans, *Spark, Streamer, Proportional and Drift Chambers*, Richelieu Press, London (1974)

[25]   V. Eckardt, *Die Speicherung von Teilchenspuren in einer Streamerkammer*, Ph.D. Thesis, University of Hamburg (1971)

[26]   F. Bulos *et al.*, *Streamer Chamber Development*, SLAC-Technical-Report, SLAC-R-74, UC-28 (1967)

[27]   F. Rohrbach, *Streamer Chambers at CERN During the Past Decade and Visual Techniques of the Future*, CERN-EF-88-17 (1988)

[28]   G. Charpak, Principes et Essais Préliminaires D'un Nouveau Détecteur Permettant De Photographier la Trajectoire des Particules Ionisantes Dans un Gas, *J. Phys. Rad.* **18** (1957) 539–47

[29]   M. Conversi, *The Development of the Flash and Spark Chambers in the 1950's*, CERN-EP-82-167 (1982)

[30] M. Conversi & A. Gozzini, The 'Hodoscope Chamber': A New Instrument for Nuclear Research, *Nuovo Cim.* **2** (1955) 189–95

[31] M. Conversi *et al.*, A New Type of Hodoscope of High Spatial Resolution, *Nuovo Cim. Suppl.* **4** (1956) 234–9

[32] M. Conversi & L. Frederici, Flash Chambers of Plastic Material, *Nucl. Instr. Meth.* **151** (1978) 93–106

[33] C.A. Ayre & M.G. Thompson, Digitization of Neon Flash Tubes, *Nucl. Instr. Meth.* **69** (1969) 106–110

[34] C.G. Dalton & G.J. Krausse, Digital Readout for Flash Chambers, *Nucl. Instr. Meth.* **158** (1979) 289–97

[35] F. Ashton & J. King, The Electric Charge of Interacting Cosmic Ray Particles at Sea Level, *J. Phys.* **A4** (1971) L31–3

[36] F. Ashton, private communication (1991)

[37] J. Sonnemeyer, Staatsexamensarbeit, Universität Siegen (1979)

[38] J. Trümper, E. Böhm & M. Samorski, private communication (1969)

[39] J.W. Keuffel, Parallel Plate Counters, *Rev. Sci. Instr.* **20** (1949) 202–11

[40] S. Fukui & S. Miyamoto, A New Type of Particle Detector: The Discharge Chamber, *Nuovo Cim.* **11** (1959) 113–15

[41] O.C. Allkofer *et al.*, Die Ortsbestimmung geladener Teilchen mit Hilfe von Funkenzählern und ihre Anwendung auf die Messung der Vielfachstreuung von Mesonen in Blei, *Phys. Verh.* **6** (1955) 166–71; P.G. Henning, *Die Ortsbestimmung geladener Teilchen mit Hilfe von Funkenzählern*, Ph.D. Thesis, University of Hamburg (1955); *Atomkernenergie* **2** (1957) 81–9

[42] F. Bella, C. Franzinetti & D.W. Lee, On Spark Counters, *Nuovo Cim.* **10** (1953) 1338–40; F. Bella & C. Franzinetti, Spark Counters, *Nuovo Cim.* **10** (1953) 1461–79

[43] T.E. Cranshaw & J.F. De Beer, A Triggered Spark Counter, *Nuovo Cim.* **5** (1957) 1107–16

[44] V.S. Kaftanov & V.A. Liubimov, Spark Chamber Use in High Energy Physics, *Nucl. Instr. Meth.* **20** (1963) 195–202

[45] S. Attenberger, Spark Chamber with Multi-Track Capability, *Nucl. Instr. Meth.* **107** (1973) 605–10

[46] R. Kajikawa, Direct Measurement of Shower Electrons with 'Glass-Metal' Spark Chambers, *J. Phys. Soc. Jpn* **18** (1963) 1365–73

[47] A.S. Gavrilov *et al.*, Spark Chambers with the Recording of Information by Means of Magnetostrictive Lines, *Instr. Exp. Techn.* **6** (1966) 1355–63

[48] S. Kinoshita, Photographic Action of the α-Particles Emitted from Radio-Active Substances, *Proc. R. Soc. Lond.* **83** (1910) 432–53

[49] M.M. Shapiro, Nuclear Emulsions, in S. Flügge (ed.), *Handbuch der Physik*, Band **XLV**, Springer, Berlin (1958) 342–436

[50] R. Reinganum, Streuung und Photographische Wirkung der α-Strahlen, *Z. Phys.* **12** (1911) 1076–81

[51] D.H. Perkins, *Cosmic Ray Work with Emulsions in the 40's and 50's*, Oxford University Preprint OUNP 36/85 (1985)

[52] C.F. Powell, P.H. Fowler & D.H. Perkins, *The Study of Elementary Particles by the Photographic Method*, Pergamon Press, London (1959)

[53] S. Aoki *et al.*, Fully Automated Emulsion Analysis System, *Nucl. Instr. Meth.* **B51** (1990) 466–73

[54] M. Simon, Lawrence Berkeley Lab. XBL-829-11834, private communication (1992)

[55] G.D. Rochester, Atomic Nuclei From Outer Space, *British Association for the Advancement of Science* (December 1970) 183–94; G.D. Rochester & K.E. Turver, Cosmic Rays of Ultra-high Energy, *Contemp. Phys.* **22** (1981) 425–50; G.D. Rochester & A.W. Wolfendale, Cosmic Rays at Manchester and Durham, *Acta Phys. Hung.* **32** (1972) 99–114

[56] J. Sacton, *The Emulsion Technique and its Continued Use*, University of Brussels, Preprint IISN 0379-301X/IIHE-93.06 (1993)

[57] Th. Wendnagel, University of Frankfurt am Main, private communication (1991)

[58] C. Childs & L. Slifkin, Room Temperature Dislocation Decoration Inside Large Crystals, *Phys. Rev. Lett.* **5(11)** (1960) 502–3; A New Technique for Recording Heavy Primary Cosmic Radiation and Nuclear Processes in Silver Chloride Single Crystals, *IEEE Trans. Nucl. Sci.* **NS-9 (3)** (1962) 413–15

[59] Th. Wendnagel *et al.*, Properties and Technology of Monocrystalline AgCl-Detectors; 1. Aspects of Solid State Physics *and* Properties and Technology of AgCl-Detectors; 2. Experiments and Technological Performance, in S. Francois (ed.), *Proc. 10th Int. Conf. on SSNTD, Lyon 1979*, Pergamon Press, London (1980)

[60] A. Noll, *Methoden zur Automatischen Auswertung von Kernwechselwirkungen in Kernemulsionen und AgCl-Kristallen*, Ph.D. Thesis, University of Siegen (1990)

[61] C.M.G. Lattes, Y. Fujimoto & S. Hasegawa, *Hadronic Interactions of High Energy Cosmic Rays Observed by Emulsion Chambers*, ICR-Report-81-80-3, University of Tokyo (1980)

[62] Mt. Fuji Collaboration (M. Akashi *et al.*), *Energy Spectra of Atmospheric Cosmic Rays Observed with Emulsion Chambers*, ICR-Report-89-81-5, University of Tokyo (1981)

[63] J. Nishimura *et al.*, Emulsion Chamber Observations of Primary Cosmic Ray Electrons in the Energy Range 30–1000 GeV, *Astrophys. J.* **238** (1980) 394–409

[64] I. Ohta *et al.*, *Characteristics of X-Ray Films Used in Emulsion Chambers and Energy Determination of Cascade Showers by Photometric Methods*, 14th Int. Cosmic Ray Conf. München, Vol. 9, München (1975) 3154–9

[65] A.F. McKinley, *Thermoluminescence Dosimetry*, Adam Hilger Ltd, Bristol (1981)

[66] M. Oberhofer & A. Scharmann (eds.), *Applied Thermoluminescence Dosimetry*, Adam Hilger Ltd, Bristol (1981)

[67] Y.S. Horowitz, *Thermoluminescence and Thermoluminescent Dosimetry*, CRC Press (1984)

[68]   E. Sauter, *Grundlagen des Strahlenschutzes*, Siemens AG, Berlin/München (1971); *Grundlagen des Strahlenschutzes*, Thiemig, München (1982)

[69]   Y. Okamoto *et al.*, *Thermoluminescent Sheet to Detect the High Energy Electromagnetic Cascades*, 18th Int. Cosmic Ray Conf., Bangalore, Vol. 8 (1983) 161–5

[70]   R.L. Fleischer, P.B. Price & R.M. Walker, *Nuclear Tracks in Solids; Principles and Application*, University of California Press, Berkeley (1975)

[71]   P.H. Fowler & V.M. Clapham (eds.), *Solid State Nuclear Track Detectors*, Pergamon Press, Oxford (1982)

[72]   F. Granzer, H. Paretzke & E. Schopper (eds.), *Solid State Nuclear Track Detectors*, Vols. 1 & 2, Pergamon Press, Oxford (1978)

[73]   W. Enge, Introduction to Plastic Nuclear Track Detectors, *Nucl. Tracks* **4** (1980) 283–308

[74]   W. Heinrich *et al.*, Application of Plastic Nuclear Track Detectors in Heavy Ion Physics, *Nucl. Tracks Rad. Measurements* **15(1–4)** (1988) 393–400

[75]   Afzal Ahmed & Julie Oliveaux, Life Science Data Archive, Johnson Space Center, Houston (2005). web-page: http://lsda.jsc.nasa.gov/scripts/photoGallery/detail_result.cfm?image_id=1664

[76]   M. Lang, U.A. Glasmacher, R. Neumann & G.A. Wagner, Etching Behaviour of Alpha-Recoil Tracks in Natural Dark Mica Studied via Artificial Ion Tracks, *Nucl. Instr. Meth.* **B209** (2003) 357–61

[77]   J.A. Miller, J.A.C. Horsfall, N. Petford & R.H. Tizard, Counting Fission Tracks in Mica External Detectors, *Pure and Applied Geophysics*, **140(4)** (December 1993)

[78]   T. Xiaowei *et al.*, A Nuclear Detector with Super-High Spatial Resolution, *Nucl. Instr. Meth.* **A320** (1992) 396–7

7

Track detectors

Some scientists find, or so it seems, that they get their best ideas when smoking; others by drinking coffee or whiskey. Thus there is no reason why I should not admit that some may get their ideas by observing or by repeating observations.

Karl R. Popper

The measurement of particle trajectories is a very important issue for any experiment in high energy physics. This provides information about the interaction point, the decay path of unstable particles, angular distributions and, when the particle travels in a magnetic field, its momentum. Track detectors, used intensively in particle physics up to the early seventies, have been described in Chap. 6.

A new epoch was opened by the invention of the multiwire proportional chamber [1, 2]. Now gaseous wire chambers and micropattern detectors almost play a dominant rôle in the class of track detectors.

The fast progress of semiconductor detectors has resulted in a growth of the number of high energy physics experiments using tracking systems based on semiconductor microstrip or pixel detectors, especially in areas where extremely high spatial accuracy is required.

7.1 Multiwire proportional chambers

A *multiwire proportional chamber* (MWPC) [1–4] is essentially a planar layer of proportional counters without separating walls (Fig. 7.1). The shape of the electric field is somewhat modified compared to the pure cylindrical arrangement in proportional counters (Fig. 7.2) [5, 6].

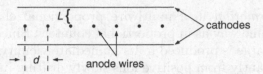

Fig. 7.1. Schematic layout of the construction of a multiwire proportional chamber.

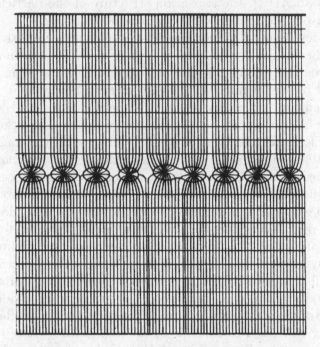

Fig. 7.2. Field and equipotential lines in a multiwire proportional chamber. The effect of a minor displacement of one anode wire on the field quality is clearly visible [5, 6].

When the coordinates of the wires are $y = 0, x = 0, \pm d, \pm 2d, \dots$ the potential distribution is approximated by an analytical form [6]:

$$U(x, y) = \frac{CV}{4\pi\varepsilon_0} \left\{ \frac{2\pi L}{d} - \ln\left[4\left(\sin^2\frac{\pi x}{d} + \sinh^2\frac{\pi y}{d}\right)\right] \right\} , \qquad (7.1)$$

where L and d are defined in Fig. 7.1, V is the anode voltage, ε_0 the permittivity of free space ($\varepsilon_0 = 8.854 \cdot 10^{-12}\,\mathrm{F/m}$), and C the capacitance per unit length given by the formula

$$C = \frac{4\pi\varepsilon_0}{2\left(\frac{\pi L}{d} - \ln\frac{2\pi r_i}{d}\right)} , \qquad (7.2)$$

where r_i is the anode-wire radius.

Avalanche formation in a multiwire proportional chamber proceeds exactly in the same way as in proportional counters. Since for each anode wire the bulk charge is produced in its immediate vicinity, the signal originates predominantly from positive ions slowly drifting in the direction of the cathode, see Eq. (5.41) and Fig. 5.8. If the anode signal is read out with a high-time-resolution oscilloscope or with a fast analogue-to-digital converter (flash ADC), the ionisation structure of the particle track can also be resolved in the multiwire proportional chamber.

The time development of the avalanche formation in a multiwire proportional chamber can be detailed as follows (Fig. 7.3). A primary electron drifts towards the anode (a), the electron is accelerated in the strong electric field in the vicinity of the wire in such a way that it can gain a sufficient amount of energy on its path between two collisions so that it can ionise further gas atoms. At this moment the avalanche formation starts (b). Electrons and positive ions are created in the ionisation processes essentially in the same place. The multiplication of charge carriers comes to an end when the space charge of positive ions reduces the external electric field below a critical value. After the production of charge carriers, the electron and ion clouds drift apart (c). The electron cloud drifts in the direction of the wire and broadens slightly due to lateral diffusion. Depending on the direction of incidence of the primary electron, a slightly asymmetric density distribution of secondary electrons around the wire will be formed. This asymmetry is even more pronounced in streamer tubes. In this case, because of the use of thick anode wires and also because of the strong absorption of photons, the avalanche formation is completely restricted to the side of the anode wire where the electron was incident (see also Figs. 5.7 and 5.13) (d). In a last step the ion cloud recedes radially and slowly drifts to the cathode (e).

In most cases gold-plated tungsten wires with diameters between $10\,\mu\text{m}$ and $30\,\mu\text{m}$ are used as anodes. A typical anode-wire distance is $2\,\text{mm}$. The distance between the anode wire and the cathode is on the order of $10\,\text{mm}$. The individual anode wires act as independent detectors. The cathodes can be made from metal foils or also as layers of stretched wires.

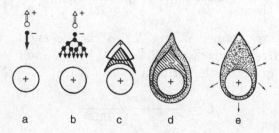

Fig. 7.3. Temporal and spatial development of an electron avalanche.

As *counting gases* all gases and gas mixtures, which also are standard for the operation of proportional counters, namely, noble gases like Ar, Xe with admixtures of CO_2, CH_4, isobutane, and other hydrocarbons can be used [7–9]. Typical gas amplifications of 10^5 are achieved in multiwire proportional chambers. To obtain fast signals, gases with high electron mobility are used. For example, in the work of [10] a time resolution of 4.1 ns was achieved with a proportional chamber using a $CF_4 + 10\%$ i-C_4H_{10} filling.

In most chambers the possibility to process the analogue information on the wires is not taken advantage of. Instead, only thresholds for the incoming signals are set. In this mode of operation the multiwire proportional chamber is only used as a track detector. For an anode-wire distance of $d = 2$ mm the root-mean-square deviation of the spatial resolution is given by, see Eq. (2.6),

$$\sigma(x) = \frac{d}{\sqrt{12}} = 577 \, \mu m \ . \tag{7.3}$$

The fundamental reason that limits a reduction of the wire spacing d is the *electrostatic repulsion* between long anode wires. This effect should be taken into account for MWPC construction. The central wire position is stable only if the *wire tension T* satisfies the relation

$$V \leq \frac{d}{lC} \sqrt{4\pi\varepsilon_0 T} \ , \tag{7.4}$$

where V is the anode voltage, d the wire spacing, l the wire length and C the capacitance per unit length of the detector [11, 12], Formula (7.2) (see Fig. 7.1). Using this equation, the required wire tension for stable wires can be calculated taking into account Eq. (7.2),

$$T \geq \left(\frac{V \cdot l \cdot C}{d}\right)^2 \cdot \frac{1}{4\pi\varepsilon_0} \tag{7.5}$$

$$\geq \left(\frac{V \cdot l}{d}\right)^2 \cdot 4\pi\varepsilon_0 \left[\frac{1}{2\left(\frac{\pi L}{d} - \ln\frac{2\pi r_i}{d}\right)}\right]^2 \ . \tag{7.6}$$

For a wire length $l = 1$ m, an anode voltage $V = 5$ kV, an anode–cathode distance of $L = 10$ mm, an anode-wire spacing of $d = 2$ mm and an anode-wire diameter of $2r_i = 30 \, \mu m$, Eq. (7.6) yields a minimum mechanical wire tension of 0.49 N corresponding to a stretching of the wire with a mass of about 50 g.

Longer wires must be stretched with larger forces or, if they cannot withstand higher tensions, they must be supported at fixed distances. This will, however, lead to locally inefficient zones.

For a reliable operation of MWPCs it is also important that the wires do not sag too much gravitationally due to their own mass [13]. A sag of the anode wire would reduce the distance from anode to cathode, thereby reducing the homogeneity of the electric field.

A horizontally aligned wire of length l stretched with a tension T would exhibit a *sag* due to the pull of gravity of [14] (see also Problem 7.5)

$$f = \frac{\pi r_i^2}{8} \cdot \varrho \cdot g \frac{l^2}{T} = \frac{mlg}{8T} \qquad (7.7)$$

(m, l, ϱ, r_i – mass, length, density and radius of the unsupported wire, g – acceleration due to gravity, and T – wire tension [in N]).

Taking our example from above, a gold-plated tungsten wire ($r_i = 15\,\mu m$; $\varrho_W = 19.3\,g/cm^3$) would develop a sag in the middle of the wire of

$$f = 34\,\mu m \ , \qquad (7.8)$$

which would be acceptable if the anode–cathode distance is on the order of 10 mm.

Multiwire proportional chambers provide a relatively poor spatial resolution which is on the order of $\approx 600\,\mu m$. They also give only the coordinate perpendicular to the wires and not along the wires. An improvement in the performance can be obtained by a segmentation of the cathode and a measurement of the induced signals on the cathode segments. The cathode, for example, can be constructed of parallel strips, rectangular pads ('mosaic counter') or of a layer of wires (Fig. 7.4).

In addition to the anode signals, the induced signals on the cathode strips are now also recorded. The coordinate along the wire is given by the centre of gravity of the charges, which is derived from the signals induced on the cathode strips. Depending on the subdivision of the cathode, spatial resolutions along the wires of $\approx 50\,\mu m$ can be achieved, using

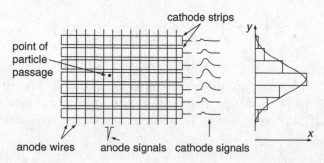

Fig. 7.4.   Illustration of the cathode readout in a multiwire proportional chamber.

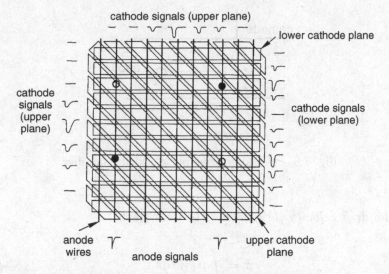

cathode signals (upper plane)

lower cathode plane

cathode signals (upper plane)

cathode signals (lower plane)

anode wires

anode signals

upper cathode plane

Fig. 7.5. Illustration of the resolution of ambiguities for two-particle detection in a multiwire proportional chamber.

this procedure. In case of multiple tracks also the second cathode must be segmented to exclude ambiguities.

Figure 7.5 sketches the passage of two particles through a multiwire proportional chamber. If only one cathode were segmented, the information from the anode wires and cathode strips would allow the reconstruction of four possible track coordinates, two of which, however, would be 'ghost coordinates'. They can be excluded with the help of signals from a second segmented cathode plane. A larger number of simultaneous particle tracks can be successfully reconstructed if cathode pads instead of cathode strips are used. Naturally, this results also in an increased number of electronic channels.

Further progress in the position resolution of MWPCs as well as in the *rate capability* has been achieved with the development of gaseous micropattern chambers. These detectors are discussed in Sect. 7.4.

7.2 Planar drift chambers

The principle of a *drift chamber* is illustrated by Fig. 7.6. The time Δt between the moment of the particle passage through the chamber and the arrival time of the charge cloud at the anode wire depends on the point of passage of the particle through the chamber. If v^- is the constant drift velocity of the electrons, the following linear relation holds:

$$x = v^- \cdot \Delta t \tag{7.9}$$

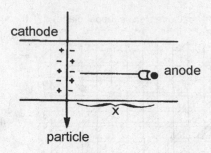

Fig. 7.6. Working principle of a drift chamber.

or, if the *drift velocity* varies along the drift path,

$$x = \int v^-(t)\,\mathrm{d}t \ . \tag{7.10}$$

In order to produce a suitable drift field, potential wires are introduced between neighbouring anode wires.

The measurement of the drift time allows the number of anode wires in a drift chamber to be reduced considerably in comparison to an MWPC or, by using small anode-wire spacings, to improve significantly the *spatial resolution*. Normally, both advantages can be achieved at the same time [15]. Taking a drift velocity of $v^- = 5\,\mathrm{cm/\mu s}$ and a time resolution of the electronics of $\sigma_t = 1\,\mathrm{ns}$, spatial resolutions of $\sigma_x = v^-\sigma_t = 50\,\mathrm{\mu m}$ can be achieved. However, the spatial resolution has contributions not only from the time resolution of the electronics, but also from the diffusion of the drifting electrons and the fluctuations of the statistics of primary ionisation processes. The latter are most important in the vicinity of the anode wire (Fig. 7.7 [5, 16]).

For a particle trajectory perpendicular to the chamber, the statistical production of electron–ion pairs along the particle track becomes important. The electron–ion pair closest to the anode wire is not necessarily produced on the connecting line between anode and potential wire. Spatial fluctuations of charge-carrier production result in large drift-path differences for particle trajectories close to the anode wire while they have only a minor effect for distant particle tracks (Fig. 7.8).

Naturally, the time measurement cannot discriminate between particles having passed the anode wire on the right- or on the left-hand side. A double layer of drift cells where the layers are staggered by half a cell width can resolve this *left–right ambiguity* (Fig. 7.9).

Drift chambers can be made very large [17–19]. For larger drift volumes the potential between the anode-wire position and the negative potential

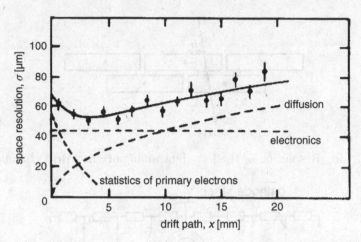

Fig. 7.7. Spatial resolution in a drift chamber as a function of the drift path [5, 16].

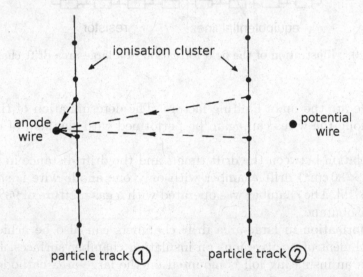

Fig. 7.8. Illustration of different drift paths for 'near' and 'distant' particle tracks to explain the dependence of the spatial resolution on the primary ionisation statistics.

on the chamber ends is divided linearly by using cathode strips connected to a chain of resistors (Fig. 7.10).

The maximum achievable spatial resolution for large-area drift chambers is limited primarily by mechanical tolerances. For large chambers typical values of $200\,\mu m$ are obtained. In small chambers ($10 \times 10\,cm^2$) spatial resolutions of $20\,\mu m$ have been achieved. In the latter case the time resolution of the electronics and the diffusion of electrons on their way to

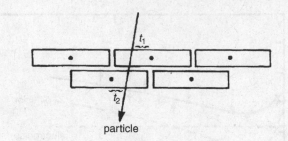

Fig. 7.9. Resolution of the left–right ambiguity in a drift chamber.

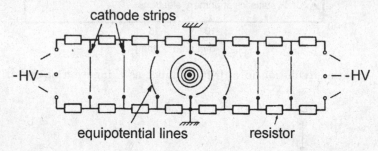

Fig. 7.10. Illustration of the field formation in a large-area drift chamber.

the anode are the main limiting factors. The determination of the coordinate along the wires can again be performed with the help of cathode pads.

The relation between the drift time t and the drift distance in a large-area $(80 \times 80\,\mathrm{cm}^2)$ drift chamber with only one anode wire is shown in Fig. 7.11 [19]. The chamber was operated with a gas mixture of 93% argon and 7% isobutane.

Field formation in large-area drift chambers can also be achieved by the attachment of positive ions on insulating chamber surfaces. In these chambers an insulating foil is mounted on the large-area cathode facing the drift space (Fig. 7.12). In the time shortly after the positive high voltage on the anode wire has been switched on, the field quality is insufficient to expect a reasonable electron drift with good spatial resolution over the whole chamber volume (Fig. 7.13a). Positive ions which have been produced by the penetrating particle now start to drift along the field lines to the electrodes. The electrons will be drained by the anode wire, but the positive ions will get stuck on the inner side of the insulator on the cathode thereby forcing the field lines out of this region. After a certain while ('charging-up time') no field lines will end on the covers of the chamber and an ideal drift-field configuration will have been formed (Fig. 7.13b, [20, 21]). If the chamber walls are not completely insulating, i.e., their volume or surface resistance is finite, some field lines will still

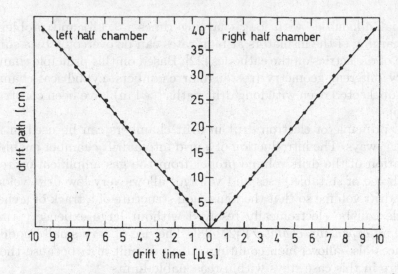

Fig. 7.11. Drift-time–space relation in a large drift chamber ($80 \times 80\,\text{cm}^2$) with only one anode wire [19].

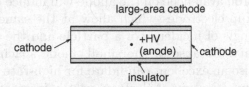

Fig. 7.12. Principle of construction of an *electrodeless drift chamber*.

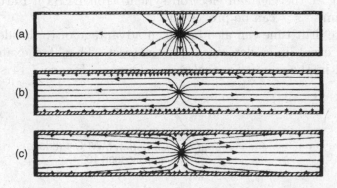

Fig. 7.13. Field formation in an electrodeless drift chamber by ion attachment [20, 21].

end on the chamber covers (Fig. 7.13c). Although, in this case, no ideal field quality is achieved, an overcharging of the cathodes is avoided since the chamber walls have a certain conductivity or transparency to allow for a removal of surplus surface charges.

Initial difficulties with long charging times ($\approx 1\,h$) and problems of overcharging of the insulators at high rates can be overcome by a suitable choice of dielectrics on the cathodes [22]. Based on this principle chambers of very different geometry (rectangular chambers, cylindrical chambers, drift tubes, etc.) even with long drift paths ($> 1\,m$) have been constructed [23–26].

The principle of electron drift in drift chambers can be used in many different ways. The introduction of a grid into a drift chamber enables the separation of the drift volume proper from the gas amplification region. The choice of suitable gases and voltages allows very low drift velocities in the drift volume so that the ionisation structure of a track of a charged particle can be electronically resolved without large expense (principle of a *time expansion chamber*) [27, 28]. The use of very small anode-wire distances also allows high counting rates per unit area because the rate per wire in this case stays within reasonable limits.

The *induction drift chamber* [29–31] also allows high spatial resolutions by using anode and potential wires with small relative distances. The formation of an electron avalanche on the anode will induce charge signals on neighbouring pickup electrodes which allow at the same time the determination of the angle of incidence of a particle and the resolution of the right–left ambiguity. Because of the small anode spacing the induction drift chamber is also an excellent candidate for high-rate experiments, for example, for the investigation of electron–proton interactions in a storage ring at high repetition frequencies (e.g. in HERA, the hadron–electron storage ring at the German electron synchrotron DESY). Particle rates up to $10^6\,\mathrm{mm^{-2}\,s^{-1}}$ can be processed.

The finite drift time can also be taken advantage of to decide whether or not an event in a detector is of interest. This, for example, can be realised in the *multistep avalanche chamber*. Figure 7.14 shows the

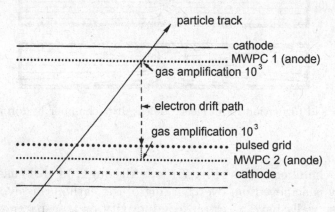

Fig. 7.14.   Principle of operation of a multistep avalanche chamber [32].

principle of operation [32]. The detector consists of two multiwire proportional chambers (MWPC 1 and 2), whose gas amplifications are arranged to be relatively small ($\approx 10^3$). All particles penetrating the detector will produce relatively weak signals in both proportional chambers. Electrons from the avalanche in MWPC 1 can be transferred with a certain probability into the drift region situated between the two chambers. Depending on the width of the drift space these electrons require several hundred nanoseconds to arrive at the second multiwire proportional chamber. The end of the drift space is formed by a wire grid which is only opened by a voltage signal if some external logic signals an interesting event. In this case the drifting electrons are again multiplied by a gas amplification factor 10^3 so that a gas amplification of $10^6 \cdot \varepsilon$ in MWPC 2 is obtained, where ε is the mean transfer probability of an electron produced in chamber 1 into the drift space. If ε is sufficiently large (e.g. > 0.1), the signal in chamber 2 will be large enough to trigger the conventional readout electronics of this chamber. These 'gas delays', however, are nowadays mainly realised by purely electronic delay circuits.

Experiments at electron–positron storage rings and at future proton–proton colliders require large-area chambers for muon detection. There are many candidates for muon chambers, such as layers of streamer tubes. For the accurate reconstruction of decay products of the searched-for Higgs particles, for example, excellent spatial resolutions over very large areas are mandatory. These conditions can be met with modular drift chambers [33, 34].

7.3 Cylindrical wire chambers

For storage-ring experiments cylindrical detectors have been developed which fulfill the requirement of a maximum solid-angle coverage, i.e. *hermeticity*. In the very first experiments cylindrical multigap spark chambers (see Chap. 6) and multiwire proportional chambers were used, however, at present drift chambers have been almost exclusively adopted for the measurement of particle trajectories and the determination of the specific ionisation of charged particles.

There are several types of such detectors: cylindrical drift chambers whose wire layers form cylindrical surfaces; jet chambers, where the drift spaces are segmented in azimuthal direction; and time-projection chambers, which are in the sensitive volume free of any material (apart from the counting gas), and where the information on particle trajectories is drifted to circular end-plate detectors.

Cylindrical drift chambers operated in a magnetic field allow the determination of the momenta of charged particles. The transverse momentum

p of charged particles is calculated from the axial magnetic field and the bending radius of the track, ρ, to be (see Chap. 11)

$$p\,[\mathrm{GeV}/c] = 0.3\ B\,[\mathrm{T}] \cdot \rho\,[\mathrm{m}]\ . \tag{7.11}$$

7.3.1   Cylindrical proportional and drift chambers

Figure 7.15 shows the principle of construction of a *cylindrical drift chamber*. All wires are stretched in an axial direction (in the z direction, the direction of the magnetic field). For cylindrical drift chambers a potential wire is stretched between two anode wires. Two neighbouring readout layers are separated by a cylindrical layer of potential wires. In the most simple configuration the individual drift cells are trapezoidal where the boundaries are formed by eight potential wires. Figure 7.15 shows a projection in the $r\varphi$ plane, where r is the distance from the centre of the chamber and φ is the azimuthal angle. Apart from this trapezoidal drift cell other drift-cell geometries are also in use [35].

In the so-called *open trapezoidal cells* every second potential wire on the potential-wire planes is left out (Fig. 7.16).

The field quality can be improved by using closed cells (Fig. 7.17) at the expense of a larger number of wires. The compromise between the aforementioned drift-cell configurations is a hexagonal structure of the cells (Fig. 7.18). In all these configurations the potential wires are of larger diameter ($\varnothing \approx 100\,\mu\mathrm{m}$) compared to the anode wires ($\varnothing \approx 30\,\mu\mathrm{m}$).

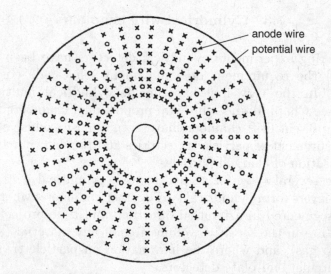

Fig. 7.15.   Schematic layout of a cylindrical drift chamber. The figure shows a view of the chamber along the wires.

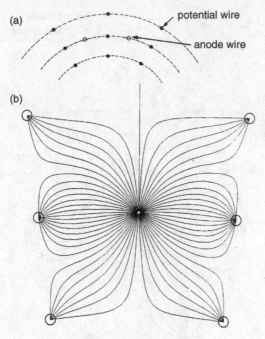

Fig. 7.16. (a) Illustration of an open drift-cell geometry. (b) Field lines in an open drift cell [36].

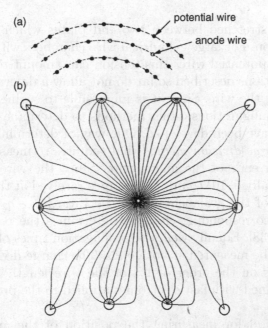

Fig. 7.17. (a) Illustration of a closed drift-cell geometry. (b) Field lines in a closed drift cell [36].

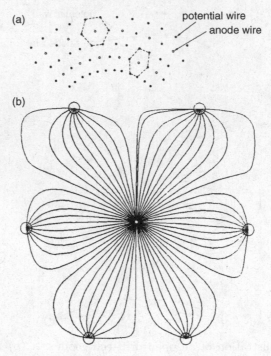

Fig. 7.18.   (a) Hexagonal drift-cell geometry. (b) Field lines in a hexagonal drift cell [36].

All wires are stretched between two end plates which must take the whole wire tension. For large cylindrical wire chambers with several thousand anode and potential wires this tension can amount to several tons.

The configurations described so far do not allow a determination of the coordinate along the wire. Since it is impossible to segment the cathode wires in these configurations, other methods to determine the coordinate along the wire have been developed. One way of determining the z coordinate is the *charge-division method* that requires to measure the signals arriving at both ends of the anode wire. Since the wire has a certain resistivity (typically 5–10 Ω/cm), the charges received at the ends depend on the position of the avalanche. Then the ratio $(q_1 - q_2)/(q_1 + q_2)$ (q_1 and q_2 are the corresponding charges) determines the point of particle intersection [37, 38]. Equally well, the propagation times of signals on the anode wires can be measured at both ends. The charge-division technique allows accuracies on the order of 1% of the wire length. This precision can also be obtained with fast electronics applied to the propagation-time technique.

Another method for measuring the position of the avalanche along the sense wire uses *spiral-wire delay lines*, of diameter smaller than 2 mm, stretched parallel to the sense wire [39]. This technique, which is mechanically somewhat complicated for large detector systems, allows accuracies on the order of 0.1% along the wires. If the delay line is placed between two closely spaced wires, it also resolves the left–right ambiguity. More sophisticated delay-line readouts allow even higher spatial resolutions [40, 41].

However, there is also a fourth possibility by which one can determine the z coordinate along the wire. In this case, some anode wires are stretched not exactly parallel to the cylinder axis, but are tilted by a small angle with respect to this axis (*stereo wires*). The spatial resolution $\sigma_{r,\varphi}$ measured perpendicular to the anode wires is then translated into a resolution σ_z along the wire according to

$$\sigma_z = \frac{\sigma_{r,\varphi}}{\sin\gamma} \; , \tag{7.12}$$

if γ is the 'stereo angle' (Fig. 7.19). For typical $r\varphi$ resolutions of $200\,\mu m$ z resolutions on the order of $\sigma_z = 3\,mm$ are obtained, if the stereo angle is $\gamma \approx 4°$. In this case, the z resolution does not depend on the wire length. The magnitude of the stereo angle is limited by the maximum allowed transverse cell size. Cylindrical drift chambers with stereo wires are also known as *hyperbolic chambers*, because the tilted stereo wires appear to sag hyperbolically with respect to the axial anode wires.

In all these types of chambers, where the drift field is perpendicular to the magnetic field, special attention must be paid to the Lorentz angle (see Sect. 1.4).

Figure 7.20 shows the drift trajectories of electrons in an open rectangular drift cell with and without an axial magnetic field [42, 43].

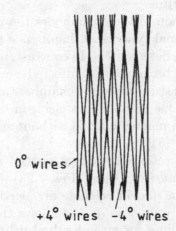

$0°$ wires

$+4°$ wires $-4°$ wires

Fig. 7.19.   Illustration of the determination of the coordinate along the anode wire by use of stereo wires.

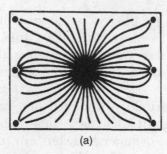 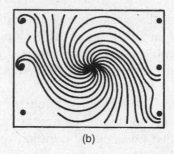

(a)   (b)

Fig. 7.20. Drift trajectories of electrons in an open rectangular drift cell (a) without and (b) with magnetic field [42, 43].

Figure 7.21 shows the $r\varphi$ projections of reconstructed particle tracks from an electron–positron interaction (PLUTO) in a cylindrical multiwire proportional chamber [44]. Figure 7.21a shows a clear two-jet structure which originated from the process $e^+e^- \rightarrow q\bar{q}$ (production of a quark–antiquark pair). Part (b) of this figure exhibits a particularly interesting event of an electron–positron annihilation from the aesthetic point of view. The track reconstruction in this case was performed using only the fired anode wires without making use of drift-time information (see Sect. 7.1). The spatial resolutions obtained in this way, of course, cannot compete with those that can be reached in drift chambers.

Cylindrical multiwire proportional chambers can also be constructed from layers of so-called *straw chambers* (Fig. 7.22) [45–49]. Such straw-tube chambers are frequently used as vertex detectors in storage-ring experiments [50, 51]. These straw chambers are made from thin aluminised mylar foils. The straw tubes have diameters of between 5 mm and 10 mm and are frequently operated at overpressure. These detectors allow for spatial resolutions of 30 μm.

Due to the construction of these chambers the risk of broken wires is minimised. In conventional cylindrical chambers a single broken wire can disable large regions of a detector [52]. In contrast, in straw-tube chambers only the straw with the broken wire is affected.

Because of their small size straw-tube chambers are candidates for high-rate experiments [53]. Due to the short electron drift distance they can also be operated in high magnetic fields without significant deterioration of the spatial resolution [54].

Very compact configurations with high spatial resolution can also be obtained with *multiwire drift modules* (Fig. 7.23) [51, 55, 56].

In the example shown, 70 drift cells are arranged in a hexagonal structure of 30 mm diameter only. Figure 7.24 shows the structure of electric field and equipotential lines for an individual drift cell [55]. Figure 7.25 shows a single particle track through such a multiwire drift module [55].

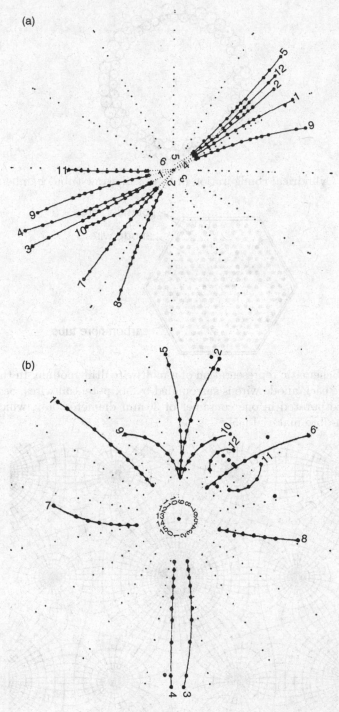

Fig. 7.21. Multitrack events of electron–positron interactions measured in the PLUTO central detector [44].

7 Track detectors

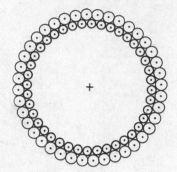

Fig. 7.22.   Cylindrical configuration of thin-wall straw-tube chambers [45, 47].

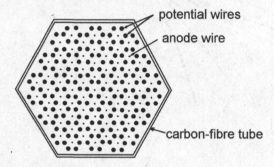

Fig. 7.23.   Schematic representation of a multiwire drift module. In this hexagonal structure each anode wire is surrounded by six potential wires. Seventy drift cells are incorporated in one container of 30 mm diameter only, which is made from carbon-fibre material [55].

Fig. 7.24.   Calculated electric field and equipotential lines in one individual hexagonal drift cell of the multiwire drift module [55].

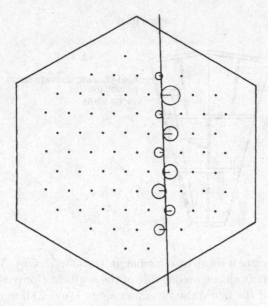

Fig. 7.25. Example of a single particle passage through a multiwire drift module. The circles indicate the measured drift times of the fired anode wires. The particle track is a tangent to all drift circles [55].

7.3.2 Jet drift chambers

Cylindrical drift chambers used now at collider experiments have up to 50 layers of anode wires. These are often used for multiple dE/dx measurements, e.g. to discriminate charged pions against kaons.

In *jet drift chambers*, especially suited for these tasks, an accurate measurement of the energy loss by ionisation is performed by determining the specific ionisation on as large a number of anode wires as possible. The central detector of the JADE experiment [57, 58] at PETRA determined the energy loss of charged particles on 48 wires, which are stretched parallel to the magnetic field. The cylindrical volume of the drift chamber is subdivided into 24 radial segments. Figure 7.26 sketches the principle of the arrangement of one of these sectors, which is itself again subdivided into smaller drift regions of 16 anode wires each.

The field formation is made by potential strips at the boundaries between two sectors. The electric field is perpendicular to the counting-wire planes and also perpendicular to the direction of the magnetic field. For this reason the electron drift follows the *Lorentz angle* which is determined from the electric and magnetic field strengths and the drift velocity. For the solenoidal $\vec{B}$ field of 0.45 T in JADE a Lorentz angle of $\alpha = 18.5°$ is obtained. To reach a maximum accuracy for an individual energy-loss measurement the chamber is operated under a pressure of 4 atm. This

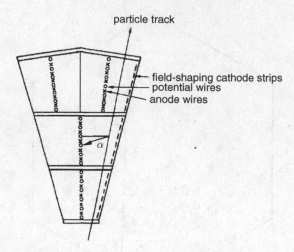

Fig. 7.26.   Segment of a jet drift chamber (after [35, 57–59]). The field-forming cathode strips are only shown on one side of the segment (for reasons of simplicity and not to overload the figure the two inner rings 1 and 2 show only five and the outer ring 3 only six anode wires).

overpressure also suppresses the influence of primary ion statistics on the spatial resolution. However, it is important not to increase the pressure to too high a value since the logarithmic rise of the energy loss, which is the basis for particle separation, may be reduced by the onset of the density effect.

The determination of the coordinate along the wire is done by using the charge-division method.

The $r\varphi$ projection of trajectories of particles from an electron–positron interaction in the JADE drift chamber is shown in Fig. 7.27 [57, 58]. The 48 coordinates along each track originating from the interaction vertex can clearly be recognised. The left–right ambiguity in this chamber is resolved by staggering the anode wires (see also Fig. 7.28). An even larger jet drift chamber was mounted in the OPAL detector at the Large Electron–Positron collider LEP at CERN [60].

The structure of the MARK II jet chamber (Fig. 7.28 [61, 62]) is very similar to that of the JADE chamber. The ionisation produced by particle tracks in this detector is collected on the anode wires. Potential wires between the anodes and layers of field-forming wires produce the drift field. The field quality at the ends of the drift cell is improved by additional potential wires. The drift trajectories in this jet chamber in the presence of the magnetic field are shown in Fig. 7.29 [61, 62].

JADE

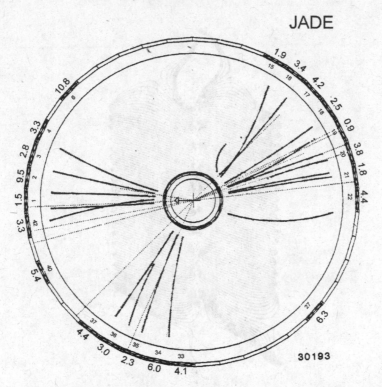

Fig. 7.27. The $r\varphi$ projection of interaction products from an electron–positron collision (gluon production: $e^+ + e^- \rightarrow q + \bar{q} + g$, producing three jets) in the JADE central detector [57, 58]. The bent tracks correspond to charged particles and the dotted tracks to neutral particles which are not affected by the magnetic field (and are not registered in the chamber).

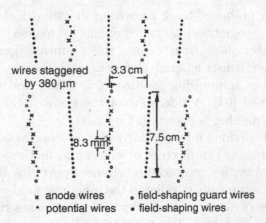

Fig. 7.28. Drift-cell geometry of the MARK II jet drift chamber [61, 62].

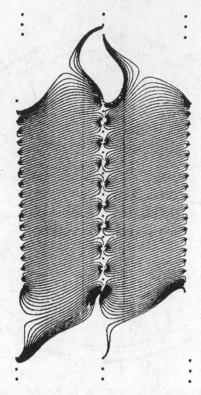

Fig. 7.29. Calculated drift trajectories in a jet-chamber drift cell in the presence of a magnetic field [61, 62].

7.3.3   Time-projection chambers (TPCs)

The *crème de la crème* of track recording in cylindrical detectors (also suited for other geometries) at the moment is realised with the *time-projection chamber* [63]. Apart from the counting gas this detector contains no other constructional elements and thereby represents the optimum as far as minimising multiple scattering and photon conversions are concerned [64]. A side view of the construction principle of a time-projection chamber is shown in Fig. 7.30.

The chamber is divided into two halves by means of a central electrode. A typical counting gas is a mixture of argon and methane (90:10).

The primary ionisation produced by charged particles drifts in the electric field – which is typically parallel to the magnetic field – in the direction of the end plates of the chamber, which in most cases consist of multiwire proportional detectors. The magnetic field suppresses the diffusion perpendicular to the field. This is achieved by the action of the magnetic forces on the drifting electrons which, as a consequence, spiral around the

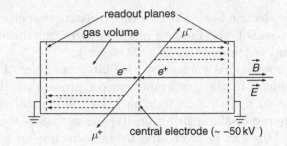

Fig. 7.30. Working principle of a time-projection chamber (TPC) [63] for a collider experiment. For simplicity the beam pipe is not shown.

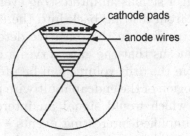

Fig. 7.31. Principle of operation of a pad readout in an endcap multiwire proportional chamber. The anode wires and some cathode pads are shown for one sector.

direction of the magnetic field. For typical values of electric and magnetic field strengths, Larmor radii below $1\,\mu$m are obtained. The arrival time of primary electrons at the end plates supplies the z coordinate along the cylinder axis. The layout of one end plate is sketched in Fig. 7.31.

The gas amplification of the primary ionisation takes place at the anode wires, which are stretched in azimuthal direction. The radial coordinate r can in principle be obtained from the fired wire (for short wires). To obtain three-dimensional coordinates the cathodes of endcap-multiwire-proportional-chamber segments are usually structured as pads. Therefore the radial coordinate is also provided by reading the position of the fired pad. In addition, the pads supply the coordinate along the anode wire resulting in a determination of the azimuthal angle φ. Therefore, the time-projection chamber allows the determination of the coordinates r, φ and z, i.e. a three-dimensional space point, for each cluster of primary electrons produced in the ionisation process.

The analogue signals on the anode wires provide information on the specific energy loss and can consequently be used for particle identification. Typical values of the magnetic field are around $1.5\,$T, and around

20 kV/m for the electric field. Since in this construction electric and magnetic field are parallel, the Lorentz angle is zero and the electrons drift parallel to $\vec{E}$ and $\vec{B}$ (there is no '$\vec{E} \times \vec{B}$ effect').

A problem, however, is caused by the large number of positive ions which are produced in the gas amplification process at the end plates and which have to drift a long way back to the central electrode. The strong space charge of the drifting positive ions causes the field quality to deteriorate. This can be overcome by introducing an additional grid ('gate') between the drift volume and the endcap multiwire proportional chamber (Fig. 7.32).

The gate is normally closed. It is only opened for a short period of time if an external trigger signals an interesting event. In the closed state the gate prevents ions from drifting back into the drift volume. Thereby the quality of the electric field in the sensitive detector volume remains unchanged [35]. This means that the gate serves a dual purpose. On the one hand, electrons from the drift volume can be prevented from entering the gas amplification region of the endcap multiwire proportional chamber if there is no trigger which would signal an interesting event. On the other hand – for gas-amplified interesting events – the positive ions are prevented from drifting back into the detector volume. Figure 7.33 shows the operation principle of the gate in the ALEPH TPC [65].

Time-projection chambers can be made very large (diameter ≥ 3 m, length ≥ 5 m). They contain a large number of analogue readout channels (number of anode wires ≈ 5000 and cathode pads $\approx 50\,000$). Several hundred samples can be obtained per track, which ensure an excellent determination of the radius of curvature and allow an accurate measurement of the energy loss, which is essential for particle identification [65–67]. The drawback of the time-projection chamber is the fact that high particle rates cannot be handled, because the drift time of the electrons in the detector volume amounts to $40\,\mu$s (for a drift path of 2 m) and the readout of the analogue information also requires several microseconds.

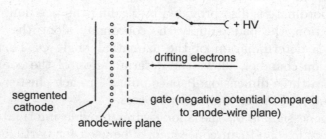

Fig. 7.32.   *Gating principle* in a time-projection chamber.

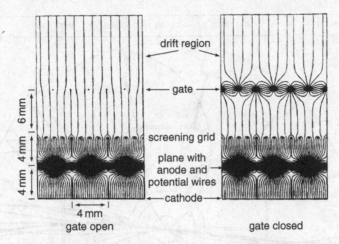

Fig. 7.33. Working principle of the gate in the ALEPH TPC [65]. For an open gate the ionisation electrons are not prevented from entering the gas amplification region. The closed gate, however, confines the positive ions to the gas amplification region. A closed gate also stops electrons in the drift volume from entering the gas-amplification region. For an event of interest the gate is first opened to allow the primary electrons to enter the gas amplification region and then it is closed to prevent the positive ions produced in the avalanche process from drifting back into the detector volume.

In large time-projection chambers typical spatial resolutions of $\sigma_z = 1\,\mathrm{mm}$ and $\sigma_{r,\varphi} = 160\,\mu\mathrm{m}$ are obtained. In particular, the resolution of the z coordinate requires an accurate knowledge of the drift velocity. This, however, can be calibrated and monitored by UV-laser-generated ionisation tracks.

Figure 7.34 shows the $r\varphi$ projection of an electron–positron annihilation in the ALEPH time-projection chamber [65, 66].

Time-projection chambers can also be operated with liquid noble gases. Such *liquid-argon time-projection chambers* represent an electronic replacement for bubble chambers with the possibility of three-dimensional event reconstruction. In addition, they can serve simultaneously as a calorimetric detector (see Chap. 8), are permanently sensitive, and can intrinsically supply a trigger signal by means of the scintillation light produced in the liquid noble gas (see Sect. 5.4) [68–73]. The electronic resolution of the bubble-chamber-like pictures is on the order of $100\,\mu\mathrm{m}$. The operation of large liquid-argon TPCs, however, requires ultrapure argon (contaminants $< 0.1\,\mathrm{ppb}$ ($1\,\mathrm{ppb} \equiv 10^{-9}$)) and high-performance low-noise preamplifiers since no gas amplification occurs in the counting medium. Multi-kiloton liquid-argon TPCs appear to be good candidates to study rare phenomena in underground experiments ranging from the search for nucleon decay to solar-neutrino observations [74, 75].

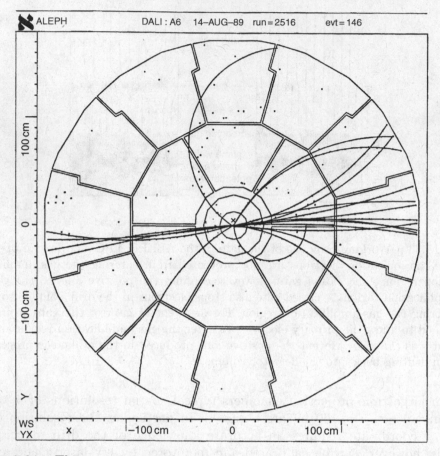

Fig. 7.34.   The $r\varphi$ projection of an electron–positron annihilation in the ALEPH
time-projection chamber [65, 66]. The end-plate detector is structured into two
rings which are made from six (inner ring) and twelve (outer ring) multiwire-
proportional-chamber segments.

Self-triggering time-projection chambers have also been operated successfully with liquid xenon [76, 77].

7.4   Micropattern gaseous detectors

The construction of multiwire proportional chambers would be simplified
and their stability and flexibility would be greatly enhanced if anodes
were made in the form of strips or dots on insulating or semiconducting
surfaces instead of stretching anode wires in the counter volume. The rate
capability improves by more than one order of magnitude for these devices

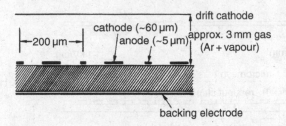

Fig. 7.35. Schematic arrangement of a microstrip gas detector.

[78, 79]. At present the class of *micropattern gaseous detectors* is already rather wide and many new promising devices are under study [80–82].

These microstrip gaseous chambers (MSGCs) are miniaturised multiwire proportional chambers, in which the dimensions are reduced by about a factor of 10 in comparison to conventional chambers (Fig. 7.35). The typical pitch is 100–200 µm and the gas gap varies between 2–10 mm. This has been made possible because the electrode structures can be reduced with the help of electron lithography. The wires are replaced by strips which are evaporated onto a thin substrate. Cathode strips arranged between the anode strips allow for an improved field quality and a fast removal of positive ions. The segmentation of the otherwise planar cathodes in the form of strips or pixels [83, 84] also permits two-dimensional readout. Instead of mounting the electrode structures on ceramic substrates, they can also be arranged on thin plastic foils. In this way, even light, flexible detectors can be constructed which exhibit a high spatial resolution. Possible disadvantages lie in the electrostatic charging-up of the insulating plastic structures which can lead to time-dependent amplification properties because of the modified electric fields [85–90].

The gain of an MSGC can be up to 10^4. The spatial resolution of this device for point-like ionisation, measured with soft X rays, reaches 20–30 µm rms. For minimum-ionising charged particles crossing the gap, the resolution depends on the angle of incidence. It is dominated by primary ionisation statistics [91].

The obvious advantages of these *microstrip detectors* – apart from their excellent spatial resolution – are the low dead time (the positive ions being produced in the avalanche will drift a very short distance to the cathode strips in the vicinity of the anodes), the reduced radiation damage (because of the smaller sensitive area per readout element) and the high-rate capability.

Microstrip proportional chambers can also be operated in the drift mode (see Sect. 7.2).

However, the MSGC appeared to be prone to ageing and discharge damages [92]. To avoid these problems many different designs of micropattern detectors were suggested. Here we consider two of them, Micromegas [93]

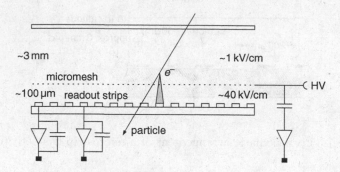

Fig. 7.36.   The layout of the Micromegas detector [11, 95].

and GEM [94] detectors, widely used now by many groups. Both of them demonstrate good performance.

The *Micromegas* design is shown in Fig. 7.36. Electrons released by charged particles in the conversion gap of 2–5 mm width drift to the multiplication gap. This gap of 50–100 μm width is bordered by a fine cathode mesh and an anode readout strip or pad structure. A constant distance between cathode and anode is kept with dielectric pillars with a pitch of ≈ 1 mm.

A high electric field in the multiplication gap (30–80 kV/cm) provides a gain up to 10^5. Since most of the ions produced in the avalanche are collected by the nearby cathode, this device has excellent timing properties [96] and a high-rate capability [97].

Another structure providing charge multiplication is the *Gas Electron Multiplier* (GEM). This is a thin (≈ 50 μm) insulating kapton foil coated with a metal film on both sides. It contains chemically produced holes of 50–100 μm in diameter with 100–200 μm pitch. The metal films have different potential to allow gas multiplication in the holes. A GEM schematic view and the electric field distribution is presented in Figs. 7.37 and 7.38. A GEM-based detector contains a drift cathode separated from one or several GEM layers and an anode readout structure as shown in Fig. 7.37.

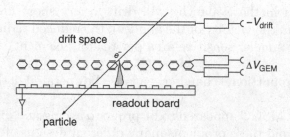

Fig. 7.37.   Detailed layout of a GEM detector.

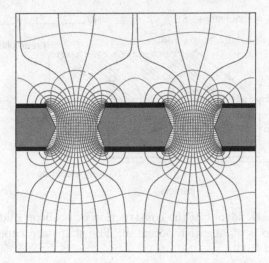

Fig. 7.38.   Electric field distribution in a GEM detector [11, 94].

The electrons are guided by the electric drift field to the GEM where they experience a high electric field in the GEM channels thereby starting avalanche formation in them. Most of these secondary electrons will drift to the anode while the majority of ions is collected by the GEM electrodes. One GEM only can provide a gain of up to several thousand which is sufficient to detect minimum-ionising particles in the thin gaseous layer. By using two or three GEM detectors on top of each other, one can obtain a substantial total gain while a moderately low gain at each stage provides better stability and a higher discharge threshold [95, 98, 99].

7.5 Semiconductor track detectors

Basically, the *semiconductor track detector* is a set of semiconductor diodes described in Sect. 5.3. The main features of detectors of this family are discussed in various reviews [100–102].

The electrodes of the solid-state track detectors are segmented in the form of strips or pads. Figure 7.39 shows the operation principle of a silicon microstrip detector with sequential cathode readout [103].

A minimum-ionising particle crossing the depletion gap produces on average 90 electron–hole pairs per $1\,\mu$m of its path. For a typical detector of $300\,\mu$m thickness this resulted in a total collected charge well above the noise level of available electronics. The optimal pitch is determined by the carrier diffusion and by the spread of δ electrons which is typically $25\,\mu$m.

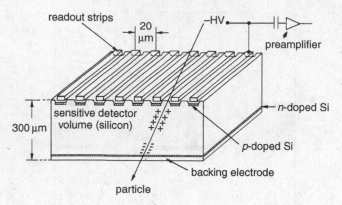

Fig. 7.39. Schematic layout of the construction of a silicon microstrip detector. Each readout strip is at negative potential. The strips are capacitively coupled (not to scale, from [103]).

The charge distribution on the readout strips allows a spatial resolution on the order of $10\,\mu m$ or even better [104, 105]. To reduce the number of electronics channels only every second or third strip may be read out. Due to the large capacity between neighbouring strips, the so-called *floating strips* contribute to the centre-of-gravity value. Such silicon microstrip counters are frequently used in storage-ring experiments as vertex detectors, in particular, to determine the lifetimes of unstable hadrons in the picosecond range and to tag short-lived mesons in complicated final states. This technique of using silicon microstrip detectors in the vicinity of interaction points mimics the ability of high-resolution bubble chambers or nuclear emulsions (see Chap. 6) but uses a purely electronic readout. Because of the high spatial resolution of microstrip detectors, secondary vertices can be reconstructed and separated from the primary interaction relatively easily.

To measure the second coordinate, the n^+ side can also be divided into orthogonal strips. This encounters some technical difficulties which, however, can be overcome with a more complex structure [101, 102]. Readout electronics for microstrip detectors includes specifically developed chips bonded to the sensor plate. Each chip contains a set of preamplifiers and a circuit which sends signals to a digitiser using multiplexing techniques.

If a silicon chip is subdivided in a matrix-like fashion into many pads that are electronically shielded by potential wells with respect to one another, the energy depositions produced by complex events which are stored in the cathode pads can be read out pixel by pixel. The readout time is rather long because of the sequential data processing. It supplies, however, two-dimensional images in a plane perpendicular to the beam direction. For a pixel size of $20 \times 20\,\mu m^2$ spatial resolutions of $5\,\mu m$ can be obtained. Because of the charge coupling of the pads, this type of silicon detector is also called a *charge-coupled device*. Commercially available *CCD detectors* with external dimensions of $1 \times 1\,cm^2$ have about 10^5 pixels [35, 106–108]. Modern devices in commercial cameras have up to 10 megapixels.

However, CCD detectors have some limitations. The depletion area is typically thin, 20–$40\,\mu m$, which implies a rather small number of electron–hole pairs per minimum-ionising particle and, hence, requiring the necessity of CCD cooling to keep the dark current on an acceptable low level. Another drawback of CCDs is their slow data readout, a fraction of millisecond, that renders the use of this device at high-luminosity colliders difficult.

To avoid these limitations one has to return to the design as shown in Fig. 7.39 but with a segmentation of the p^+ side to pixels and connecting each pixel to individual preamplifiers. This technology has also been developed for the LHC experiments [109–111]. At present, the hybrid pixel technology is rather well established. Such a detector consists of the sensor and an integrated circuit board containing front-end electronics (Fig. 7.40). The connection of these two elements is made with the help of either solder (PbSn) or indium bump bonds.

Detectors of this type are being constructed now for LHC experiments [110, 111] as well as for X-ray counting systems [112, 113]. For medical imaging sensors with high-Z semiconductors, Cd(Zn)Te or GaAs, have also been made [114].

A pixel size of $\approx 50 \times 50\,\mu m^2$ is about the limit for hybrid pixel detectors. More promising is the technology of monolithic pixel detectors where both sensor and front-end electronics are integrated onto one silicon crystal. At present this technology is in the research and development stage [115, 116]. It is worth to mention that pixel detectors basically have a low capacity per pixel. According to Eq. (5.69) this results in low electronics noise.

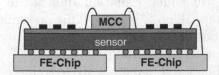

Fig. 7.40. An example of a hybrid pixel detector layout developed for the ATLAS detector [109]. The sensor plate containing pixels of $50\,\mu m \times 400\,\mu m$ is bonded to front-end (FE) chips via bump connection. The flex hybrid kapton layer atop the sensor carries additional electrical components and the module control chip (MCC).

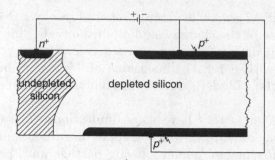

Fig. 7.41.   Fabrication principle of a silicon drift chamber by sideward depletion [100, 119].

Silicon microstrip detectors can also operate as *solid-state drift chambers*. A review of this type of detectors can be found in [100, 117–119]. The working principle of a silicon drift chamber can be understood from Fig. 7.41.

The ohmic n^+ contact, normally extending over the whole surface of the detector, is now concentrated at one distinct position, which can be placed anywhere on the undepleted conducting bulk. Then diodes of p^+ layers can be put on both sides of the silicon wafer. At sufficiently high voltage between the n^+ contact and the p^+ layer the conductive bulk will be depleted and will retract into the vicinity of the n^+ electrode. In this way a potential valley is made in which electrons can move by, e.g. diffusion towards the n^+ readout contact. The produced holes will drift to the nearby p^+ contact.

If now an electric field with a field component parallel to the detector surface is added to such a structure, one arrives at a silicon drift chamber, where now the electrons produced by a photon or a charged particle in the depletion layer are guided by the drift field down the potential valley to the n^+ anode. Such a graded potential can be achieved by dividing the p^+ electrode into strips of suitably different potential (Fig. 7.42) [100, 119].

Silicon detectors will suffer radiation damage in a harsh radiation environment (see Chap. 12 on 'Ageing'). This presents a problem, in particular, at high-luminosity colliders where silicon detectors with excellent radiation hardness are required [100].

Silicon strip, pixel or voxel detectors are extremely useful in many different fields. Their small size, their high granularity, low intrinsic noise and the possibility to make them largely radiation resistant under special treatment render them a high-resolution detector of exceptional versatility. They are already now used at the heart of many particle physics experiments as vertex detectors or as focal-plane detectors in

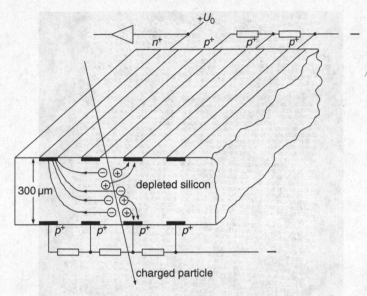

Fig. 7.42. Silicon drift chamber with graded potential [100, 117–119].

very successful satellite missions (Chandra, XMM Newton). Important applications in many other fields, like medicine (Compton cameras), art, material science and engineering (see Chap. 16) complete the large-scope applications.

7.6 Scintillating fibre trackers

A separate readout of individual *scintillating fibres* also provides an excellent spatial resolution which can even exceed the spatial resolution of drift chambers [120–122]. Similarly, thin capillaries (*macaronis*) filled with liquid scintillator can be used for tracking charged particles [123, 124]. In this respect, scintillating fibre calorimeters or, more generally, light-fibre systems can also be considered as tracking detectors. In addition, they represent, because of the short decay time of the scintillation flash, a genuine alternative to gas-discharge detectors, which are rather slow because of the intrinsically slow electron drift. Figure 7.43 shows the track of a charged particle in a stack of scintillating fibres. The fibre diameter in this case amounts to 1 mm [125].

Scintillating fibres, however, can also be produced with much smaller diameters. Figure 7.44 shows a microphotograph of a bundle consisting of scintillating fibres with 60 μm diameter. Only the central fibre is illuminated. A very small fraction of the light is scattered into the neighbouring fibres [126, 127]. The fibres are separated by a very thin

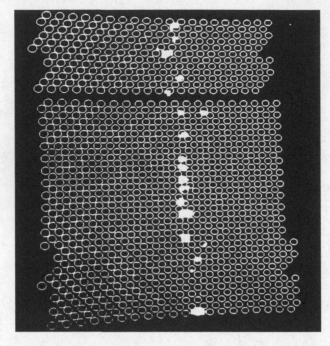

Fig. 7.43.   Particle track in a stack of scintillating fibres; fibre diameter ∅ = 1 mm [125].

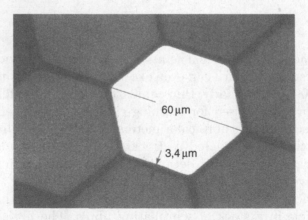

Fig. 7.44.   Microphotograph of a bundle consisting of seven scintillating fibres. The fibres have a diameter of 60 μm. Only the central fibre is illuminated. The fibres are optically separated by a very thin cladding (3.4 μm) of lower refractive index to trap the light by total internal reflection. Only a small amount of light is scattered into the neighbouring fibres [126, 127].

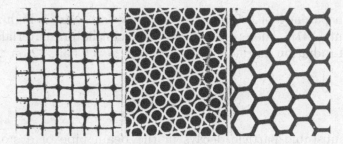

Fig. 7.45. Bundles of scintillating fibres from different companies (left: 20 μm; Schott (Mainz); centre: 20 μm; US Schott (Mainz); right: 30 μm plastic fibres; Kyowa Gas (Japan)) [129].

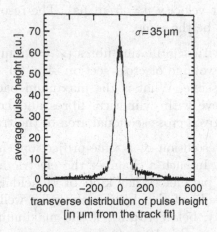

Fig. 7.46. Transverse pulse-height distribution of charged particles in a stack of 8000 scintillating fibres with 30 μm diameter [130].

cladding (3.4 μm). The optical readout system for such light-fibre systems, however, has to be made in such a way that the granular structure of the light fibres is resolved with sufficient accuracy, e.g. with optical pixel systems [128].

Arrangements of such fibre bundles are excellent candidates for tracking detectors in experiments with high particle rates requiring high time and spatial resolution. Figure 7.45 shows different patterns of bundles of scintillating fibres from different companies [129]. Figure 7.46 shows the spatial resolution for charged particles obtained in a stack consisting of 8000 scintillating fibres (30 μm diameter). A single-track resolution of 35 μm and a two-track resolution of 83 μm have been achieved [130].

The transparency of scintillators can deteriorate in a high-radiation environment [131]. There exist, however, scintillator materials with a substantial radiation hardness [132, 133] (see also Chap. 12 on 'Ageing').

7.7 Problems

7.1 An unstable particle decays in the beam pipe of a storage-ring experiment into two charged particles which are measured in a three-layer drift chamber giving the drift times T_1^i, T_2^i, T_3^i ($T_1^1 = 300\,\text{ns}, T_2^1 = 195\,\text{ns}, T_3^1 = 100\,\text{ns}, T_1^2 = 400\,\text{ns}, T_2^2 = 295\,\text{ns}, T_3^2 = 200\,\text{ns}$). The two tracks make an angle of $\alpha = 60°$. Estimate the uncertainty with which the vertex of the two tracks can be reconstructed (drift velocity $v = 5\,\text{cm}/\text{µs}$). The resolution for all wires is assumed to be the same.

7.2 In a tracker with scintillating fibres ($\varnothing = 1\,\text{mm}$) one would like to instrument a volume of cross section $A = 20 \times 20\,\text{cm}^2$ as closely packed as possible. What is the maximum packing fraction that one can achieve with cylindrical fibres and how many does one need for the given cross-sectional area of the tracker?

7.3 The spatial resolution of a time-projection chamber is assumed to be $100\,\text{µm}$. In such a chamber the electric and magnetic fields are normally parallel. What kind of B field is required to keep the Larmor radii of the drifting electrons well below the spatial resolution (say, below $10\,\text{µm}$), if the maximum electron velocity perpendicular to B is $10\,\text{cm}/\text{µs}$?

7.4 $60\,\text{keV}$ X rays from a ^{241}Am source are to be measured in a proportional counter. The counter has a capacity of $180\,\text{pF}$. What kind of gas gain is required if the amplifier connected to the anode wire requires $10\,\text{mV}$ at the input for a good signal-to-noise performance? The average energy to produce an electron–ion pair in the argon-filled counter is $W = 26\,\text{eV}$. What is the intrinsic energy resolution of the $60\,\text{keV}$ line (the Fano factor for argon is $F = 0.17$)?

7.5 The wires in a multiwire proportional chamber (length $\ell = 1\,\text{m}$, diameter $30\,\text{µm}$, made of gold-plated tungsten) are strung with a weight of $50\,\text{g}$. Work out the sag of the wires under the influence of gravity!

The solution of this problem is normally given using variational methods [134–140]. Deviating from that solution the key assumption here is that the local horizontal and vertical forces in the wire define its local slope, where the wire itself is inelastic.

References

[1] G. Charpak, *Electronic Imaging of Ionizing Radiation with Limited Avalanches in Gases*, Nobel-Lecture 1992, CERN-PPE-93-25 (1993); *Rev. Mod. Phys.* **65** (1993) 591–8

[2] G. Charpak *et al.*, The Use of Multiwire Proportional Chambers to Select and Localise Charged Particles, *Nucl. Instr. Meth.* **62** (1968) 202–68

[3] W. Bartl, G. Neuhofer, M. Regler & A. Taurok (eds.), Proc. 6th Int. Conf. on Wire Chambers, Vienna 1992, *Nucl. Instr. Meth.* **A323** (1992)

[4] G. Charpak *et al.*, Some Developments in the Operation of Multiwire Proportional Chambers, *Nucl. Instr. Meth.* **80** (1970) 13–34

[5] F. Sauli, *Principles of Operation of Multiwire Proportional and Drift Chambers*, CERN-77-09 (1977) and references therein

[6] G.A. Erskine, Electrostatic Problems in Multiwire Proportional Chambers, *Nucl. Instr. Meth.* **105** (1972) 565–72

[7] J. Va'vra, *Wire Chamber Gases*, SLAC-Pub-5793 (1992)

[8] J. Va'vra, Wire Chamber Gases, *Nucl. Instr. Meth.* **A323** (1992) 34–47

[9] Y.H. Chang, Gases for Drift Chambers in SSC/LHC Environments, *Nucl. Instr. Meth.* **A315** (1992) 14–20

[10] E.V. Anashkin *et al.*, Z-Chamber and Trigger of the CMD-2 Detector, *Nucl. Instr. Meth.* **A323** (1992) 178–83

[11] Particle Data Group, Review of Particle Physics, S. Eidelman *et al.*, *Phys. Lett.* **B592 Vol. 1–4** (2004) 1–1109; W.-M. Yao *et al.*, *J. Phys.* **G33** (2006) 1–1232; http://pdg.lbl.gov

[12] T. Trippe, *Minimum Tension Requirement for Charpak-Chamber Wires*, CERN NP Internal Report 69-18 (1969); P. Schilly *et al.*, Construction and Performance of Large Multiwire Proportional Chambers, *Nucl. Instr. Meth.* **91** (1971) 221–30

[13] M. Chew *et al.*, Gravitational Wire Sag in Non-Rigid Drift Chamber Structures, *Nucl. Instr. Meth.* **A323** (1992) 345–9

[14] H. Netz, *Formeln der Technik*, Hanser, München/Wien (1983)

[15] A.H. Walenta *et al.*, The Multiwire Drift Chamber: A New Type of Proportional Wire Chambers, *Nucl. Instr. Meth.* **92** (1971) 373–80

[16] A. Filatova *et al.*, Study of a Drift Chamber System for a K-e Scattering Experiment at the Fermi National Accelerator Lab., *Nucl. Instr. Meth.* **143** (1977) 17–35

[17] U. Becker *et al.*, A Comparison of Drift Chambers, *Nucl. Instr. Meth.* **128** (1975) 593–5

[18] M. Rahman *et al.*, A Multitrack Drift Chamber with 60 cm Drift Space, *Nucl. Instr. Meth.* **188** (1981) 159–63

[19] K. Mathis, *Test einer großflächigen Driftkammer*, Thesis, University of Siegen (1979)

[20] J. Allison, C.K. Bowdery & P.G. Rowe, *An Electrodeless Drift Chamber*, Int. Report, Univ. Manchester MC 81/33 (1981)

[21] J. Allison *et al.*, An Electrodeless Drift Chamber, *Nucl. Instr. Meth.* **201** (1982) 341–57

[22] Yu.A. Budagov *et al.*, How to Use Electrodeless Drift Chambers in Experiments at Accelerators, *Nucl. Instr. Meth.* **A255** (1987) 493–500

[23] A. Franz & C. Grupen, Characteristics of a Circular Electrodeless Drift Chamber, *Nucl. Instr. Meth.* **200** (1982) 331–4

[24] Ch. Becker *et al.*, Wireless Drift Tubes, Electrodeless Drift Chambers and Applications, *Nucl. Instr. Meth.* **200** (1982) 335–43

[25] G. Zech, Electrodeless Drift Chambers, *Nucl. Instr. Meth.* **217** (1983) 209–12

[26] R. Dörr, C. Grupen & A. Noll, Characteristics of a Multiwire Circular Electrodeless Drift Chamber, *Nucl. Instr. Meth.* **A238** (1985) 238–44

[27] A.H. Walenta & J. Paradiso, *The Time Expansion Chamber as High Precision Drift Chamber*, Proc. Int. Conf. on Instrumentation for Colliding Beam Physics; Stanford; SLAC-Report SLAC-R-250 UC-34d (1982) and SI-82-07 (1982) 1–29

[28] H. Anderhub *et al.*, Operating Experience with the Mark J Time Expansion Chamber, *Nucl. Instr. Meth.* **A265** (1988) 50–9

[29] E. Roderburg *et al.*, The Induction Drift Chamber, *Nucl. Instr. Meth.* **A252** (1986) 285–91

[30] A.H. Walenta *et al.*, Study of the Induction Drift Chamber as a High Rate Vertex Detector for the ZEUS Experiment, *Nucl. Instr. Meth.* **A265** (1988) 69–77

[31] E. Roderburg *et al.*, Measurement of the Spatial Resolution and Rate Capability of an Induction Drift Chamber, *Nucl. Instr. Meth.* **A323** (1992) 140–9

[32] D.C. Imrie, *Multiwire Proportional and Drift Chambers: From First Principles to Future Prospects*, Lecture delivered at the School for Young High Energy Physicists, Rutherford Lab. (September 1979)

[33] V.D. Peshekhonov, Wire Chambers for Muon Detectors on Supercolliders, *Nucl. Instr. Meth.* **A323** (1992) 12–21

[34] H. Faissner *et al.*, Modular Wall-less Drift Chambers for Muon Detection at LHC, *Nucl. Instr. Meth.* **A330** (1993) 76–82

[35] K. Kleinknecht, *Detektoren für Teilchenstrahlung*, Teubner, Stuttgart (1984, 1987, 1992); *Detectors for Particle Radiation*, Cambridge University Press, Cambridge (1986)

[36] S. Schmidt, private communication (1992)

[37] W.R. Kuhlmann *et al.*, Ortsempfindliche Zählrohre, *Nucl. Instr. Meth.* **40** (1966) 118–20

[38] H. Foeth, R. Hammarström & C. Rubbia, On the Localization of the Position of the Particle Along the Wire of a Multiwire Proportional Chamber, *Nucl. Instr. Meth.* **109** (1973) 521–4

[39] A. Breskin *et al.*, Two-Dimensional Drift Chambers, *Nucl. Instr. Meth.* **119** (1974) 1–8

[40] E.J. De Graaf *et al.*, Construction and Application of a Delay Line for Position Readout of Wire Chambers, *Nucl. Instr. Meth.* **166** (1979) 139–49

[41] L.G. Atencio *et al.*, Delay-Line Readout Drift Chamber, *Nucl. Instr. Meth.* **187** (1981) 381–6

[42] J.A. Jaros, *Drift and Proportional Tracking Chambers*, SLAC-Pub-2647 (1980)

[43] W. de Boer *et al.*, *Behaviour of Large Cylindrical Drift Chambers in a Superconducting Solenoid*, Proc. Wire Chamber Conf., Vienna (1980); *Nucl. Instr. Meth.* **176** (1980) 167–80

[44] PLUTO Collaboration, L. Criegee & G. Knies, e^+e^--Physics with the PLUTO Detector, *Phys. Rep.* **83** (1982) 151–280

[45] W.H. Toki, *Review of Straw Chambers*, SLAC-Pub-5232 (1990)

[46] V.M. Aulchenko *et al.*, Vertex Chamber for the KEDR Detector, *Nucl. Instr. Meth.* **A283** (1989) 528–31

[47] C. Biino *et al.*, A Very Light Proportional Chamber Constructed with Aluminised Mylar Tubes for Drift Time and Charge Division Readouts, *IEEE Trans. Nucl. Sci.* **36** (1989) 98–100

[48] G.D. Alekseev *et al.*, Operating Properties of Straw Tubes, *JINR-Rapid Communications*, No. 2 [41] (1990) 27–32

[49] V.N. Bychkov *et al.*, A High Precision Straw Tube Chamber with Cathode Readout, *Nucl. Instr. Meth.* **A325** (1993) 158–60

[50] F. Villa (ed.), *Vertex Detectors*, Plenum Press, New York (1988)

[51] D.H. Saxon, Multicell Drift Chambers, *Nucl. Instr. Meth.* **A265** (1988) 20–32

[52] E. Roderburg & S. Walsh, Mechanism of Wire Breaking Due to Sparks in Proportional or Drift Chambers, *Nucl. Instr. Meth.* **A333** (1993) 316–19

[53] J.A. Kadyk, J. Va'vra & J. Wise, Use of Straw Tubes in High Radiation Environments, *Nucl. Instr. Meth.* **A300** (1991) 511–17

[54] U.J. Becker *et al.*, Fast Gaseous Detectors in High Magnetic Fields, *Nucl. Instr. Meth.* **A335** (1993) 439–42

[55] R. Bouclier *et al.*, Fast Tracking Detector Using Multidrift Tubes, *Nucl. Instr. Meth.* **A265** (1988) 78–84

[56] Yu.P. Gouz *et al.*, Multi-Drift Module Simulation, *Nucl. Instr. Meth.* **A323** (1992) 315–21

[57] W. Bartel *et al.*, Total Cross-Section for Hadron Production by e^+e^- Annihilation at PETRA Energies, *Phys. Lett.* **B88** (1979) 171–7

[58] H. Drumm *et al.*, Experience with the JET-Chamber of the JADE Detector at PETRA, *Nucl. Instr. Meth.* **176** (1980) 333–4

[59] A. Wagner, Central Detectors, *Phys. Scripta* **23** (1981) 446–58

[60] O. Biebel *et al.*, *Performance of the OPAL Jet Chamber*, CERN-PPE-92-55 (1992); *Nucl. Instr. Meth.* **A323** (1992) 169–77

[61] F. Sauli, *Experimental Techniques*, CERN-EP-86-143 (1986)

[62] J. Bartelt, *The New Central Drift Chamber for the Mark II Detector at SLC*, Contribution to the 23rd Proc. Int. Conf. on High Energy Physics, Berkeley, Vol. 2 (1986) 1467–9

[63] D.R. Nygren, Future Prospects of the Time Projection Chamber Idea, *Phys. Scripta* **23** (1981) 584–96

[64] T. Lohse & W. Witzeling, The Time-Projection Chamber, in F. Sauli (ed.), *Instrumentation in High Energy Physics*, World Scientific, Singapore

(1992); The Time-Projection Chamber, *Adv. Ser. Direct. High Energy Phys.* **9** (1992) 81–155

[65]   ALEPH Collaboration, D. Decamp *et al.*, ALEPH: A Detector for Electron–Positron Annihilations at LEP, *Nucl. Instr. Meth.* **A294** (1990) 121–78

[66]   W.B. Atwood *et al.*, Performance of the ALEPH Time Projection Chamber, *Nucl. Instr. Meth.* **A306** (1991) 446–58

[67]   Y. Sacquin, The DELPHI Time Projection Chamber, *Nucl. Instr. Meth.* **A323** (1992) 209–12

[68]   C. Rubbia, *The Liquid Argon Time Projection Chamber: A New Concept for Neutrino Detectors*, CERN-EP-77-08 (1977)

[69]   P. Benetti *et al.*, *The ICARUS Liquid Argon Time Projection Chamber: A New Detector for ν_τ-Search*, CERN-PPE-92-004 (1992)

[70]   A. Bettini *et al.*, The ICARUS Liquid Argon TPC: A Complete Imaging Device for Particle Physics, *Nucl. Instr. Meth.* **A315** (1992) 223–8

[71]   F. Pietropaolo *et al.*, *The ICARUS Liquid Argon Time Projection Chamber: A Full Imaging Device for Low Energy e^+e^- Colliders?*, Frascati INFN-LNF 91-036 (R) (1991)

[72]   G. Buehler, *The Liquid Argon Time Projection Chamber*, Proc. Opportunities for Neutrino Physics at BNL, Brookhaven (1987) 161–8

[73]   J. Seguinot *et al.*, Liquid Xenon Ionization and Scintillation: Studies for a Totally Active-Vector Electromagnetic Calorimeter, *Nucl. Instr. Meth.* **A323** (1992) 583–600

[74]   P. Benetti *et al.*, *A Three Ton Liquid Argon Time Projection Chamber*, INFN-Report DFPD 93/EP/05, University of Padua (1993); *Nucl. Instr. Meth.* **A332** (1993) 395–412

[75]   C. Rubbia, *The Renaissance of Experimental Neutrino Physics*, CERN-PPE-93-08 (1993)

[76]   G. Carugno *et al.*, A Self Triggered Liquid Xenon Time Projection Chamber, *Nucl. Instr. Meth.* **A311** (1992) 628–34

[77]   E. Aprile *et al.*, Test of a Two-Dimensional Liquid Xenon Time Projection Chamber, *Nucl. Instr. Meth.* **A316** (1992) 29–37

[78]   A. Oed, Position Sensitive Detector with Microstrip Anode for Electron Multiplication with Gases, *Nucl. Instr. Meth.* **A263** (1988) 351–9

[79]   P.M. McIntyre *et al.*, Gas Microstrip Chambers, *Nucl. Instr. Meth.* **A315** (1992) 170–6

[80]   F. Sauli, *New Developments in Gaseous Detectors*, CERN-EP-2000-108 (2000)

[81]   L. Shekhtman, Micro-pattern Gaseous Detectors, *Nucl. Instr. Meth.* **A494** (2002) 128–41

[82]   M. Hoch, Trends and New Developments in Gaseous Detectors, *Nucl. Instr. Meth.* **A535** (2004) 1–15

[83]   F. Angelini *et al.*, *A Microstrip Gas Chamber with True Two-dimensional and Pixel Readout*, INFN-PI/AE 92/01 (1992); *Nucl. Instr. Meth.* **A323** (1992) 229–35

[84] F. Angelini, A Thin, Large Area Microstrip Gas Chamber with Strip and Pad Readout, *Nucl. Instr. Meth.* **A336** (1993) 106–15

[85] F. Angelini *et al.*, *A Microstrip Gas Chamber on a Silicon Substrate*, INFN, Pisa PI/AE 91/10 (1991); *Nucl. Instr. Meth.* **A314** (1992) 450–4

[86] F. Angelini *et al.*, *Results from the First Use of Microstrip Gas Chambers in a High Energy Physics Experiment*, CERN-PPE-91-122 (1991)

[87] J. Schmitz, *The Micro Trench Gas Counter: A Novel Approach to High Luminosity Tracking in HEP*, NIKHEF-H/91-14 (1991)

[88] R. Bouclier *et al.*, *Microstrip Gas Chambers on Thin Plastic Supports*, CERN-PPE-91-227 (1991)

[89] R. Bouclier *et al.*, *Development of Microstrip Gas Chambers on Thin Plastic Supports*, CERN-PPE-91-108 (1991)

[90] R. Bouclier *et al.*, *High Flux Operation of Microstrip Gas Chambers on Glass and Plastic Supports*, CERN-PPE-92-53 (1992)

[91] F. van den Berg *et al.*, Study of Inclined Particle Tracks in Micro Strip Gas Counters, *Nucl. Instr. Meth.* **A349** (1994) 438–46

[92] B. Schmidt, Microstrip Gas Chambers: Recent Developments, Radiation Damage and Long-Term Behavior, *Nucl. Instr. Meth.* **A419** (1998) 230–8

[93] Y. Giomataris *et al.*, MICROMEGAS: A High-Granularity Position-Sensitive Gaseous Detector for High Particle-Flux Environments, *Nucl. Instr. Meth.* **A376** (1996) 29–35

[94] F. Sauli, GEM: A New Concept for Electron Amplification in Gas Detectors, *Nucl. Instr. Meth.* **A386** (1997) 531–4

[95] F. Sauli, Progress with the Gas Electron Multiplier, *Nucl. Instr. Meth.* **A522** (2004) 93–8 (Most recent review)

[96] B. Peyaud *et al.*, KABES: A Novel Beam Spectrometer for NA48, *Nucl. Instr. Meth.* **A535** (2004) 247–52

[97] D. Thers *et al.*, Micromegas as a Large Microstrip Detector for the COMPASS Experiment, *Nucl. Instr. Meth.* **A469** (2001) 133–46

[98] A. Buzulutskov, A. Breskin, R. Chechik *et al.*, The GEM Photomultiplier Operated with Noble Gas Mixtures, *Nucl. Instr. Meth.* **A443** (2000) 164–80. (Triple-GEM introduction. High-gain operation of the triple-GEM in pure noble gases and their mixtures. GEM-based photomultiplier with CsI photocathode.)

[99] A. Bondar, A. Buzulutskov, A. Grebenuk, *et al.* Two-Phase Argon and Xenon Avalanche Detectors Based on Gas Electron Multipliers, *Nucl. Instr. Meth.* **A556** (2006) 273–80 (GEM operation at cryogenic temperatures, including in the two-phase mode)

[100] G. Lutz, *Semiconductor Radiation Detectors*, Springer, Berlin (1999)

[101] H. Dijkstra & J. Libby, Overview of Silicon Detectors, *Nucl. Instr. Meth.* **A494** (2002) 86–93

[102] M. Turala, Silicon Tracking Detectors – Historical Overview, *Nucl. Instr. Meth.* **A541** (2005) 1–14

[103] R. Horisberger, *Solid State Detectors*, Lectures given at the III ICFA School on Instrumentation in Elementary Particles Physics, Rio de Janeiro, July 1990, and PSI-PR-91-38 (1991)

[104] J. Straver *et al.*, *One Micron Spatial Resolution with Silicon Strip Detectors* CERN-PPE-94-26 (1994)

[105] R. Turchetta, Spatial Resolution of Silicon Microstrip Detectors, *Nucl. Instr. Meth.* **A335** (1993) 44–58

[106] P.N. Burrows, A CCD Vertex Detector for a High-Energy Linear e^+e^- Collider, *Nucl. Instr. Meth.* **A447** (2000) 194–201

[107] M. Cargnelli *et al.*, Performance of CCD X-ray Detectors in Exotic Atom Experiments, *Nucl. Instr. Meth.* **A535** (2004) 389–93

[108] M. Kuster *et al.*, *PN-CCDs in a Low-Background Environment: Detector Background of the Cast X-Ray Telescope*, Conference on UV, X-ray and Gamma-ray Space Instrumentation for Astronomy, San Diego, California, 1–3 August 2005, e-Print Archive: physics/0508064

[109] N. Wermes, Pixel Detector for Particle Physics and Imaging Applications, *Nucl. Instr. Meth.* **A512** (2003) 277–88

[110] W. Erdmann, The CMS Pixel Detector, *Nucl. Instr. Meth.* **A447** (2000) 178–83

[111] F. Antinori, A Pixel Detector System for ALICE, *Nucl. Instr. Meth.* **A395** (1997) 404–9

[112] S.R. Amendolia *et al.*, Spectroscopic and Imaging Capabilities of a Pixellated Photon Counting System, *Nucl. Instr. Meth.* **A466** (2001) 74–8

[113] P. Fischer *et al.*, A Counting Pixel Readout Chip for Imaging Applications, *Nucl. Instr. Meth.* **A405** (1998) 53–9

[114] M. Lindner *et al.*, Comparison of Hybrid Pixel Detectors with Si and GaAs Sensors, *Nucl. Instr. Meth.* **A466** (2001) 63–73

[115] P. Holl *et al.*, Active Pixel Matrix for X-ray Satellite Mission, *IEEE Trans. Nucl. Sci.* **NS-47**, No. 4 (2000) 1421–5

[116] P. Klein *et al.*, A DEPFET Pixel Bioscope for the Use in Autoradiography, *Nucl. Instr. Meth.* **A454** (2000) 152–7

[117] E. Gatti *et al.* (ed.), Proc. Sixth European Symp. on Semiconductor Detectors, New Developments in Radiation Detectors, *Nucl. Instr. Meth.* **A326** (1993)

[118] E. Gatti & P. Rehak, Semiconductor Drift Chamber – An Application of a Novel Charge Transport Scheme, *Nucl. Instr. Meth.* **A225** (1984) 608–14

[119] L. Strüder, High-Resolution Imaging X-ray Spectrometers, in H.J. Besch, C. Grupen, N. Pavel, A.H. Walenta (eds.), *Proceedings of the 1st International Symposium on Applications of Particle Detectors in Medicine, Biology and Astrophysics, SAMBA '99* (1999) 73–113

[120] A. Simon, *Scintillating Fiber Detectors in Particle Physics*, CERN-PPE-92-095 (1992)

[121] M. Adinolfi *et al.*, *Application of a Scintillating Fiber Detector for the Study of Short-Lived Particles*, CERN-PPE-91-66 (1991); *Nucl. Instr. Meth.* **A310** (1991) 485–9

[122] D. Autiero *et al.*, Study of a Possible Scintillating Fiber Tracker at the LHC and Tests of Scintillating Fibers, *Nucl. Instr. Meth.* **A336** (1993) 521–32

[123]   J. Bähr *et al.*, *Liquid Scintillator Filled Capillary Arrays for Parti-cle Tracking*, CERN-PPE-91-46 (1991); *Nucl. Instr. Meth.* **A306** (1991) 169–76

[124]   N.I. Bozhko *et al.*, *A Tracking Detector Based on Capillaries Filled with Liquid Scintillator*, Serpukhov Inst., High Energy Phys. 91-045 (1991); *Nucl. Instr. Meth.* **A317** (1992) 97–100

[125]   *CERN: Tracking by Fibers*, CERN-Courier, **27(5)** (1987) 9–11

[126]   *Working with High Collision Rates*, CERN-Courier, **29(10)** (1989) 9–11

[127]   C. D'Ambrosio *et al.*, Reflection Losses in Polystyrene Fibers, *Nucl. Instr. Meth.* **A306** (1991) 549–56; private communication by C. D'Ambrosio (1994)

[128]   M. Salomon, New Measurements of Scintillating Fibers Coupled to Multianode Photomultipliers, *IEEE Trans. Nucl. Sci.* **39** (1992) 671–3

[129]   *Scintillating Fibers*, CERN-Courier, **30(8)** (1990) 23–5

[130]   D. Acosta *et al.*, Advances in Technology for High Energy Subnuclear Physics. Contribution of the LAA Project, *Riv. del Nuovo Cim.* **13 (10–11)** (1990) 1–228; G. Anzivino *et al.*, The LAA Project, *Riv. del Nuovo Cim.* **13(5)** (1990) 1–131

[131]   G. Marini *et al.*, *Radiation Damage to Organic Scintillation Materials*, CERN-85-08 (1985)

[132]   J. Proudfoot, *Conference Summary: Radiation tolerant Scintillators and Detectors*, Argonne Nat. Lab. ANL-HEP-CP-92-046 (1992)

[133]   G.I. Britvich *et al.*, Investigation of Radiation Resistance of Polystyrene-Based Scintillators, *Instr. Exp. Techn.* **36** (1993) 74–80

[134]   J.J. O'Connor & E.F. Robertson, Catenary, www-gap.dcs.st-and.ac.uk/ ~history/Curves/Catenary.html (1997)

[135]   Catenary: http://mathworld.wolfram.com/Catenary.html (2003)

[136]   The Schiller Institute, Two Papers on the Catenary Curve and Logarith-mic Curve (Acta Eruditorium, 1691, G.W. Leibniz, translated by Pierre Beaudry), www.schillerinstitute.org/fid_97-01/011_catenary.html (2001)

[137]   Hanging chain www.math.udel.edu/MECLAB/UndergraduateResearch/ Chain/Main_Page.html

[138]   Th. Peters, *Die Kettenlinie*, www.mathe-seiten.de (2004)

[139]   http://mathsrv.ku-eichstaett.de/MGF/homes/grothmann/Projekte/ Kettenlinie/

[140]   Die Mathe-Redaktion, *Die Kettenlinie als Minimalproblem*, http://mathe planet.com/default3.html?article=506 (2003)

8

Calorimetry

Most particles end their journey in calorimeters.

Anonymous

Methods of particle energy measurement in modern high energy physics have to cover a large dynamical range of more than 20 orders of magnitude in energy. Detection of extremely small energies (milli-electron-volts) is of great importance in astrophysics if one searches for the remnants of the Big Bang. At the other end of the spectrum, one measures cosmic-ray particles with energies of up to 10^{20} eV, which are presumably of extragalactic origin.

Calorimetric methods imply total absorption of the particle energy in a bulk of material followed by the measurement of the deposited energy. Let us take as an example a 10 GeV muon. Passing through material this particle loses its energy mainly by the ionisation of atoms while other contributions are negligible. To absorb all the energy of the muon one needs about 9 m of iron or about 8 m of lead. It is quite a big bulk of material!

On the other hand, high-energy photons, electrons and hadrons can interact with media producing secondary particles which leads to a shower development. Then the particle energy is deposited in the material much more efficiently. Thus calorimeters are most widely used in high energy physics to detect the electromagnetic and hadronic showers. Accordingly, such detector systems are referred to as *electromagnetic* and *hadron* calorimeters.

At very high energies ($\geq 1\,$TeV), however, also muon calorimetry becomes possible because TeV muons in iron and lead undergo mainly interaction processes where the energy loss is proportional to the muon energy (see Chap. 1), thus allowing muon calorimetry. This technique will become relevant for very high-energy colliders ($\geq 1\,$TeV muon energy).

230

8.1 Electromagnetic calorimeters

8.1.1 Electron–photon cascades

The dominating interaction processes for spectroscopy in the MeV energy range are the photoelectric and Compton effect for photons and ionisation and excitation for charged particles. At high energies (higher than 100 MeV) electrons lose their energy almost exclusively by bremsstrahlung while photons lose their energy by electron–positron pair production [1] (see Sect. 1.2).

The radiation losses of electrons with energy E can be described by the simplified formula:

$$-\left(\frac{\mathrm{d}E}{\mathrm{d}x}\right)_{\mathrm{rad}} = \frac{E}{X_0} \; , \tag{8.1}$$

where X_0 is the radiation length. The probability of electron–positron pair production by photons can be expressed as

$$\frac{\mathrm{d}w}{\mathrm{d}x} = \frac{1}{\lambda_{\mathrm{prod}}}\, \mathrm{e}^{-x/\lambda_{\mathrm{prod}}} \; , \quad \lambda_{\mathrm{prod}} = \frac{9}{7}X_0 \; . \tag{8.2}$$

A convenient measure to consider shower development is the distance normalised in radiation lengths, $t = x/X_0$.

The most important properties of electron cascades can be understood in a very simplified model [2, 3]. Let E_0 be the energy of a photon incident on a bulk of material (Fig. 8.1).

After one radiation length the photon produces an e^+e^- pair; electrons and positrons emit after another radiation length one bremsstrahlung photon each, which again are transformed into electron–positron pairs. Let us assume that the energy is symmetrically shared between the particles at

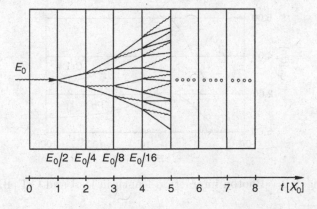

Fig. 8.1. Sketch of a simple model for shower parametrisation.

each step of the multiplication. The number of shower particles (electrons, positrons and photons together) at depth t is

$$N(t) = 2^t \ , \tag{8.3}$$

where the energy of the individual particles in generation t is given by

$$E(t) = E_0 \cdot 2^{-t} \ . \tag{8.4}$$

The multiplication of the shower particles continues as long as $E_0/N > E_c$. When the particle energy falls below the critical value E_c, absorption processes like ionisation for electrons and Compton and photoelectric effects for photons start to dominate. The position of the shower maximum is reached at this step of multiplication, i.e. when

$$E_c = E_0 \cdot 2^{-t_{\max}} \ . \tag{8.5}$$

This leads to

$$t_{\max} = \frac{\ln(E_0/E_c)}{\ln 2} \propto \ln(E_0/E_c) \ . \tag{8.6}$$

Let us take as an example the shower in a CsI crystal detector initiated by a 1 GeV photon. Using the value $E_c \approx 10\,\mathrm{MeV}$ we obtain for the number of particles in the shower maximum $N_{\max} = E_0/E_c = 100$ and for the depth of the shower maximum to be $\approx 6.6\,X_0$.

After the shower maximum electrons and positrons* having an energy below the critical value E_c will stop in a layer of $1\,X_0$. Photons of the same energy can penetrate a much longer distance. Figure 8.2 presents the energy dependence of the photon interaction length in CsI and lead.

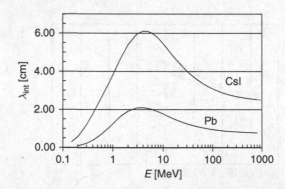

Fig. 8.2.   Photon interaction length in lead and CsI [4].

---

* Throughout this chapter both electrons and positrons are referred to as electrons.

As we can see, this function has a broad maximum between $1\,\mathrm{MeV}$ and $10\,\mathrm{MeV}$ where it amounts to about $3\,X_0$. The energy of photons in the shower maximum close to E_c is just in this range. Thus, to absorb 95% of photons produced in the shower maximum, an additional 7–$9\,X_0$ of material is necessary which implies that the thickness of a calorimeter with high shower containment should be at least 14–$16\,X_0$. The energy deposition in an absorber is a result of the ionisation losses of electrons and positrons. Since the $(\mathrm{d}E/\mathrm{d}x)_{\mathrm{ion}}$ value for relativistic electrons is almost energy-independent, the amount of energy deposited in a thin layer of absorber is proportional to the number of electrons and positrons crossing this layer.

This very simple model already correctly describes the most important qualitative characteristics of *electromagnetic cascades*.

(i) To absorb most of the energy of the incident photon the total calorimeter thickness should be more than 10–$15\,X_0$.

(ii) The position of the shower maximum increases slowly with energy. Hence, the thickness of the calorimeter should increase as the logarithm of the energy but not proportionally as for muons.

(iii) The energy leakage is caused mostly by soft photons escaping the calorimeter at the sides (lateral leakage) or at the back (rear leakage).

In reality the shower development is much more complicated. This is sketched in Fig. 8.3. An accurate description of the shower development is

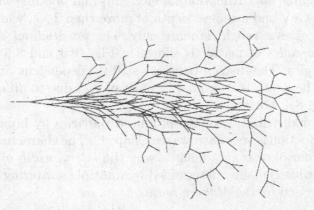

Fig. 8.3. Schematic representation of an electromagnetic cascade. The wavy lines are photons and the solid lines electrons or positrons.

a difficult task. Earlier, large efforts were undertaken to develop an analytical approach [5]. At present, due to the increase of the computer capacity, an accurate description is obtained from Monte Carlo simulations.

The longitudinal distribution of the energy deposition in electromagnetic cascades is reasonably described by an approximation based on the Monte Carlo programme EGS [6, 7],

$$\frac{\mathrm{d}E}{\mathrm{d}t} = E_0 b \frac{(bt)^{a-1} \, \mathrm{e}^{-bt}}{\Gamma(a)} \ , \tag{8.7}$$

where $\Gamma(a)$ is Euler's Γ function, defined by

$$\Gamma(g) = \int_0^\infty \mathrm{e}^{-x} x^{g-1} \, \mathrm{d}x \ . \tag{8.8}$$

The gamma function has the property

$$\Gamma(g+1) = g \, \Gamma(g) \ . \tag{8.9}$$

Here a and b are model parameters and E_0 is the energy of the incident particle. In this approximation the maximum of the shower development is reached at

$$t_{\max} = \frac{a-1}{b} = \ln\left(\frac{E_0}{E_c}\right) + C_{\gamma e} \ , \tag{8.10}$$

where $C_{\gamma e} = 0.5$ for a gamma-induced shower and $C_{\gamma e} = -0.5$ for an incident electron. The parameter b as obtained from simulation results is $b \approx 0.5$ for heavy absorbers from iron to lead. Then the energy-dependent parameter a can be derived from Eq. (8.10).

The experimentally measured distributions [8–10] are well described by a Monte Carlo simulation with the code EGS4 [1, 6]. Formula (8.7) provides a reasonable approximation for electrons and photons with energies larger than 1 GeV and a shower depth of more than $2\,X_0$, while for other conditions it gives a rough estimate only. The *longitudinal development of electron cascades* in matter is shown in Figs. 8.4 and 8.5 for various incident energies. The distributions are slightly dependent on the material (even if the depth is measured in units of X_0) due to different E_c, as shown in Fig. 8.4, bottom.

The angular distribution of the produced particles by bremsstrahlung and pair production is very narrow (see Chap. 1). The characteristic angles are on the order of $m_e c^2 / E_\gamma$. That is why the *lateral width of an electromagnetic cascade* is mainly determined by multiple scattering and can be best characterised by the *Molière radius*

$$R_{\mathrm{M}} = \frac{21\,\mathrm{MeV}}{E_c} X_0 \ \{\mathrm{g/cm}^2\} \ . \tag{8.11}$$

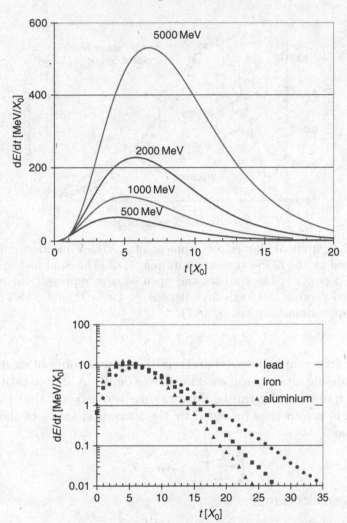

Fig. 8.4. Longitudinal shower development of electromagnetic cascades. Top: approximation by Formula (8.7). Bottom: Monte Carlo simulation with EGS4 for 10 GeV electron showers in aluminium, iron and lead [11].

Figure 8.6 shows the longitudinal and lateral development of a 6 GeV electron cascade in a lead calorimeter (based on [12, 13]). The lateral width of an electromagnetic shower increases with increasing longitudinal shower depth. The largest part of the energy is deposited in a relatively narrow shower core. Generally speaking, about 95% of the shower energy is contained in a cylinder around the shower axis whose radius is $R(95\%) = 2R_M$ almost independently of the energy of the incident particle. The dependence of the containment radius on the material is taken into account by the critical energy and radiation length appearing in Eq. (8.11).

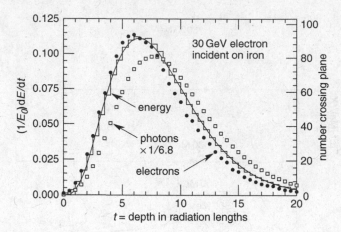

Fig. 8.5. Longitudinal shower development of a 30 GeV electron-induced cascade obtained by the EGS4 simulation in iron [1, 6]. The solid histogram shows the energy deposition; black circles and open squares represent the number of electrons and photons, respectively, with energy larger than 1.5 MeV; the solid line is the approximation given by (8.7).

Another important shower characteristic is the number of electrons and photons crossing a plane at a certain shower depth. A simple estimation of the electron number N_e can be done taking into account that the energy deposition in a shower is provided by the ionisation losses of the charged particle and

$$\left(\frac{\mathrm{d}E}{\mathrm{d}x}\right)_{\mathrm{ion}} \cdot X_0 = E_{\mathrm{c}} \ . \tag{8.12}$$

Then one can estimate

$$N_e(t) = \frac{1}{E_{\mathrm{c}}}\frac{\mathrm{d}E}{\mathrm{d}t} \ . \tag{8.13}$$

However, a considerable part of the shower particles is soft. Since only electrons above a certain threshold are detected, the effective number of shower particles becomes much smaller. Figure 8.5 shows the numbers of electrons and photons with energy above 1.5 MeV as well as $\mathrm{d}E/\mathrm{d}t$ values for a 30 GeV shower in iron [1]. We can see that N_e in this case is about a factor of two lower than given by Formula (8.13).

At very high energies the development of electromagnetic cascades in dense media is influenced by the *Landau–Pomeranchuk–Migdal* (LPM) *effect* [14, 15]. This effect predicts that the production of low-energy photons by high-energy electrons is suppressed in dense media. When an electron interacts with a nucleus producing a bremsstrahlung photon the longitudinal momentum transfer between the electron and nucleus is very

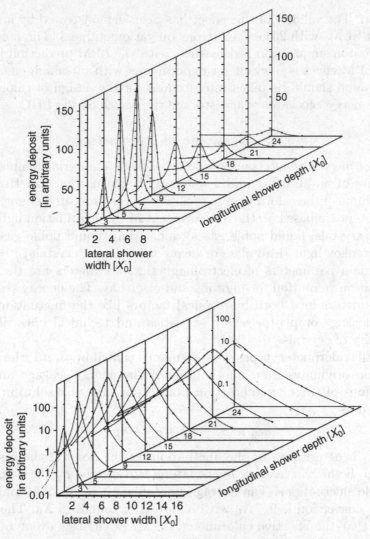

Fig. 8.6. Longitudinal and lateral development of an electron shower (6 GeV) in lead shown in linear and logarithmic scales (based on [12, 13]).

small. Heisenberg's uncertainty principle therefore requires that the interaction must take place over a long distance, which is called the *formation zone*. If the electron is disturbed while travelling this distance, the photon emission can be disrupted. This can occur for very dense media, where the distance between scattering centres is small compared to the spatial extent of the wave function. The Landau–Pomeranchuk–Migdal effect predicts that in dense media multiple scattering of electrons is sufficient to suppress photon production at the low-energy end of the bremsstrahlung spectrum. The validity of this effect has been demonstrated by an experiment at SLAC with 25 GeV electrons on various targets. The magnitude of the photon suppression is consistent with the LPM prediction [16, 17].

The LPM effect is relevant for experiments with ultrahigh-energy cosmic rays and should be taken into account for the design of calorimeters at high-energy accelerators and storage rings such as the LHC.

8.1.2 Homogeneous calorimeters

Homogeneous calorimeters are constructed from a material combining the properties of an absorber and a detector. It means that practically the total volume of the calorimeter is sensitive to the deposited energy. These calorimeters are based on the measurement of the scintillation light (scintillation crystals, liquid noble gases), ionisation (liquid noble gases) and the Cherenkov light (lead glass or heavy transparent crystals).

The main parameters of electromagnetic calorimeters are the energy and position resolution for photons and electrons. The energy resolution σ_E/E is determined both by physical factors like the fluctuation of the energy leakage or photoelectron statistics and technical ones like non-uniformity of crystals.

For all calorimeter types the common contribution to the energy resolution originates from fluctuations of the energy leakage and from fluctuations of the first interaction point. The energy resolution can be expressed as

$$\sigma_{\text{int}}^2 = \sigma_1^2 + \sigma_r^2 + \sigma_l^2 + \sigma_b^2 \ , \qquad (8.14)$$

where σ_1 is determined by the fluctuations of the point of the first interaction, σ_r is the rear leakage, σ_l the lateral leakage and σ_b the leakage due to albedo fluctuations. The average photon path in the material before the first conversion is $9/7\,X_0$ with a spread of roughly $1\,X_0$. The spread implies that the effective calorimeter thickness changes event by event. Looking at the transition curve of Fig. 8.6 we can estimate σ_1 as

$$\sigma_1 \approx \left(\frac{\mathrm{d}E}{\mathrm{d}t}\right)_{t=t_{\text{cal}}} X_0 \ , \qquad (8.15)$$

where t_{cal} is the total calorimeter thickness. It is clear that σ_1 is getting larger with increasing energy.

As discussed earlier, the energy leakage is mostly due to low-energy (1–10 MeV) photons. The albedo is usually quite small ($< 1\%$ of the initial energy) and the induced contribution to the energy resolution is negligible. At first glance the lateral size of the calorimeter should be chosen as large as necessary to make the lateral energy leakage negligible. But in a real experiment, where an event contains several or many particles, a lateral size of the area assigned to a certain particle should be limited by a few R_M. The fraction of lateral energy leakage is practically independent of the photon energy. Even though the number of escaping photons increases with the photon energy, the relative fluctuations σ_l/E_0 should go down.

The value of σ_r/E_0 has a slow energy dependence. Often the terms σ_l and σ_r are considered combined. A detailed review of the physics of shower development and fluctuations can be found in the book by Wigmans [11].

Crystal calorimeters are based on heavy scintillation crystals (see Sect. 5.4, Table 5.2). These detectors are usually built as hodoscopes with a transverse size of elements of order one to two R_M. Then the shower energy is deposited in a group of crystals usually referred to as *cluster*. The light readout is performed by photomultiplier tubes, vacuum phototriodes or silicon photodiodes (see Sect. 5.5). One of the calorimeters of this kind is described in Chap. 13. At present the best energy resolutions are obtained with calorimeters of this type [18–22].

A typical energy spectrum measured in a calorimeter is shown in Fig. 8.7 [23]. For a high-resolution detector system it is usually asymmetric, with a rather long 'tail' to lower energies, and the energy resolution is conventionally parametrised as

$$\sigma_E = \frac{\text{FWHM}}{2.35} . \tag{8.16}$$

This asymmetric distribution can be approximated, for example, by the logarithmic Gaussian shape

$$\mathrm{d}W = \exp\left\{ -\frac{\ln^2[1 - \eta(E - E_\mathrm{p})/\sigma]}{2s_0^2} - \frac{s_0^2}{2} \right\} \frac{\eta \, \mathrm{d}E}{\sqrt{2\pi}\sigma s_0} , \tag{8.17}$$

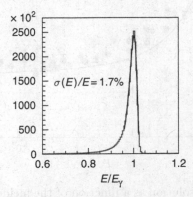

Fig. 8.7. Typical energy spectrum measured in a calorimeter [23] for photons of 4–7 GeV. The solid line is the approximation by Formula (8.17).

where E_{p} is the energy corresponding to the peak, $\sigma = \mathrm{FWHM}/2.35$, η the asymmetry parameter and s_0 can be written as

$$s_0 = \frac{2}{\xi}\,\mathrm{arsinh}\left(\frac{\eta\xi}{2}\right) \ , \quad \xi = 2.35 \ . \tag{8.18}$$

When $\eta \to 0$, the distribution becomes Gaussian.

Various approximations were used to describe the energy dependence of the resolution of calorimeters. Figure 8.8 shows the energy resolution of a calorimeter made of $16\,X_0$ CsI crystals for photons in the range from $20\,\mathrm{MeV}$ to $5.4\,\mathrm{GeV}$ [24]. The light readout was done with two $2\,\mathrm{cm}^2$ photodiodes per crystal. The energy resolution was approximated as

$$\frac{\sigma_E}{E} = \sqrt{\left(\frac{0.066\%}{E_{\mathrm{n}}}\right)^2 + \left(\frac{0.81\%}{\sqrt[4]{E_{\mathrm{n}}}}\right)^2 + (1.34\%)^2} \ , \quad E_{\mathrm{n}} = E/\mathrm{GeV} \ , \tag{8.19}$$

where the term proportional to $1/E$ stands for the electronics-noise contribution.

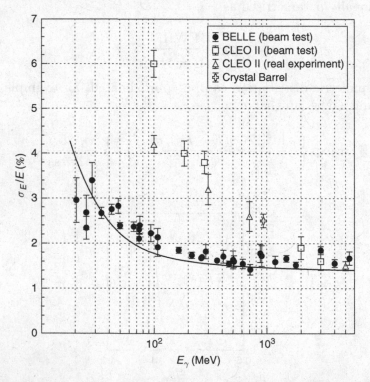

Fig. 8.8. The energy resolution as a function of the incident-photon energy [24]. The solid line is the result of an MC simulation. For the Belle data a cluster of 5×5 crystals at a threshold of $0.5\,\mathrm{MeV}$ was used.

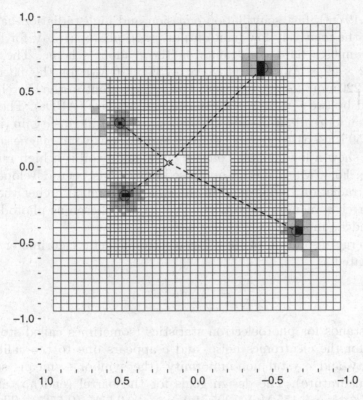

Fig. 8.9. A view of the clusters in the KTEV calorimeter for a typical event of a $K_L \to \pi^0\pi^0$ decay [22]. The calorimeter modules have a cross section of 5×5 cm^2 (2.5×2.5 cm^2) in the outer (inner) part. The hit crystals are shaded.

In discussing crystal calorimeters for high energies we have to mention the one for the KTEV experiment that was based on about 3200 pure CsI crystals of 50 cm ($27\,X_0$) length [22]. This device was intended for the detection of photons with energies up to 80 GeV, and an impressive energy resolution σ_E/E better than 1% for energies larger than 5 GeV was achieved. Figure 8.9 presents a view of the energy clusters in this calorimeter for a typical event of a $K_L \to \pi^0\pi^0$ decay. All photons are clearly separated.

At present the most sophisticated project of a calorimeter of this type is under development for the Compact Muon Solenoid (CMS) detector [25] at the CERN LHC proton–proton collider. The CMS electromagnetic calorimeter [26] incorporates 80 000 lead-tungstate (PbWO$_4$ or PWO) crystals mounted with other CMS subdetectors including the hadron calorimeter inside the superconducting solenoid, which produces a 4 T magnetic field. These crystals (see Table 5.2) have been chosen as a detector medium due to their short radiation length (0.89 cm), small Molière radius (2.19 cm), fast scintillation emission and high radiation hardness. However, the relatively low light output, ≈ 50 photons/MeV for full-size crystals, imposes hard constraints on the readout scheme. The crystal size is $22 \times 22 \times 230 \, \text{mm}^3$ ($1 \, R_M \times 1 \, R_M \times 26 \, X_0$) for the barrel and $30 \times 30 \times 220 \, \text{mm}^3$ for the endcaps. The light readout in the barrel part is performed by two $5 \times 5 \, \text{mm}^2$ avalanche photodiodes (APDs). The APDs were chosen for readout because in addition to their intrinsic gain (in CMS a gain of 50 is used) APDs are compact and insensitive to magnetic fields; they also show a low nuclear counter effect and exhibit a high radiation resistance. For CMS a special optimised device has been developed [27]. Since the radiation background in the endcaps is much higher than that in the barrel, a vacuum phototriode (VPT) was chosen as photodetector for the endcap modules.

The energy resolution of the CMS electromagnetic calorimeter can be approximated as

$$\frac{\sigma_E}{E} = \frac{a}{\sqrt{E}} \oplus \frac{b}{E} \oplus c \, , \qquad (8.20)$$

where a stands for photoelectron statistics (sometimes called stochastic term), b for the electronics noise, and c appears due to the calibration uncertainty and crystal non-uniformity (the symbol $\oplus$ means summation in quadrature). The design goals for the barrel (endcap) are $a = 2.7\%$ (5.7%), $b = 155 \, \text{MeV}$ (205 MeV), $c = 0.55\%$ (0.55%). This was confirmed by tests with a prototype [28].

A disadvantage of crystal calorimeters is the high cost of the scintillation crystals and limitations in the production of large volumes of this material. To circumvent these constraints, *lead-glass blocks* can be used in homogeneous calorimeters instead of crystals. The properties of typical lead glass (Schott SF-5 or Corning CEREN 25) are: density of about $4 \, \text{g/cm}^3$, radiation length of $X_0 \approx 2.5$ cm and refractive index of $n \approx 1.7$. The Cherenkov-radiation threshold energy for electrons in this glass is quite low, $T_{\text{ct}}^e \approx 120 \, \text{keV}$ implying that the total number of Cherenkov photons is proportional to the total track length of all charged particles in a shower developing in the lead-glass absorber. Since the energy deposition in the electron–photon shower is provided by the ionisation losses of electrons, which is also proportional to the total track length, one can assume that the total number of Cherenkov photons is proportional to the deposited energy.

However, the amount of Cherenkov light is much less (by, roughly, a factor of 1000) compared to that of conventional scintillators. This results in a large contribution of photoelectron statistics to the energy resolution of *lead-glass calorimeters*. The OPAL experiment at CERN [29], which used lead glass for the endcap calorimeter, reported an energy resolution of

$$\frac{\sigma_E}{E} = \frac{5\%}{\sqrt{E\,[\text{GeV}]}} \ , \qquad\qquad (8.21)$$

dominated by the stochastic term. Recently, the SELEX experiment at Fermilab demonstrated a high performance of its lead-glass calorimeter [30]. However, it should be noted that at present homogeneous Cherenkov calorimeters are becoming quite rare. The main reason probably is the progress in sampling calorimeters (discussed later), which achieve now the same range of energy resolution.

Homogeneous *ionisation calorimeters* can be built as an array of ionisation chambers immersed into liquid xenon [31, 32] or liquid krypton [33, 34] (see also Sect. 5.2). The energy resolution achieved with calorimeters of this type is close to that for crystal detectors. The NA48 experiment approximates the energy resolution of its LKr calorimeter [33] by Formula (8.20) with a set of the following parameters:

$$a = 3.2\% \ , \quad b = 9\% \ , \quad c = 0.42\% \ . \qquad\qquad (8.22)$$

This device is intended as a photon detector in the 10–100 GeV energy range. One more example is the LKr calorimeter of the KEDR detector [32]. The energy resolution obtained with a prototype is described by the same formula with $a = 0.3\%$, $b = 1.6\%$, $c = 1.6\%$ [35].

The initial layers of the LXe or LKr calorimeters can be designed as a series of fine-grained strips or wire ionisation chambers. Then the lateral position of the photon conversion point can be measured with high accuracy. For example, in [35] the photon spatial resolution was measured to be about $\sigma_{\rm r} \approx 1$ mm, almost independent of the photon energy.

In calorimeters without longitudinal segmentation the photon angles (or coordinates) are measured usually as corrected centre of gravity of the energy deposition,

$$\theta_\gamma = \frac{\sum \theta_i E_i}{\sum E_i} F_\theta(\varphi, \theta, E) \ , \quad \varphi_\gamma = \frac{\sum \varphi_i E_i}{\sum E_i} F_\varphi(\varphi, \theta, E) \ , \qquad (8.23)$$

where E_i, θ_i, φ_i are, respectively, the energy deposited in the ith calorimeter element with the angular coordinates θ_i and φ_i. The correction functions (F) can be usually written as a product of functions containing only one of the angles and the energy. The angular resolution depends on the energy and the calorimeter granularity. A general limitation is due to the finite number of particles in a shower. Since the shower cross section is almost energy-independent, the uncertainty in the lateral shower position can be roughly estimated as

$$\sigma_{\rm lp} = \frac{R_{\rm M}}{\sqrt{N_{\rm tot}}} = \frac{R_{\rm M}}{\sqrt{E/E_{\rm c}}} \ , \qquad\qquad (8.24)$$

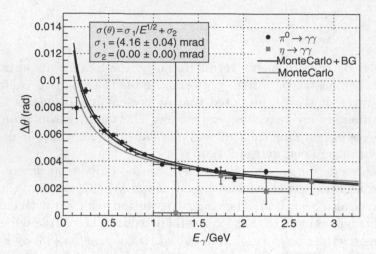

Fig. 8.10. Angular resolution of the calorimeter of the BaBar detector. The lower curve is a Monte Carlo simulation, and the upper one includes background (BG). The central line is a fit to the data, where the fit parameters are given in the inset [20].

where E_c is the critical energy. This leads to $\sigma_{lp} \approx 4 \, \text{mm}$ for $E_\gamma = 1 \, \text{GeV}$ and a CsI crystal. This is in surprisingly good agreement with the experimental results. A typical energy dependence of the angular resolution (obtained by the BaBar detector [20]) is presented in Fig. 8.10. The energy dependence is parametrised by

$$\sigma(\theta) = \frac{4.2 \, \text{mrad}}{\sqrt{E \, [\text{GeV}]}} \; . \tag{8.25}$$

8.1.3 Sampling calorimeters

There is a simpler and more economical way to measure the photon energy if the ultimate energy resolution is not crucial. Let us look again at the simplest shower model and place a thin flat counter behind a thick layer of an absorber corresponding to the depth of the shower maximum. In this naïve model the number of electrons crossing the counter, see Formulae (8.5) and (8.6), is just 2/3 of $N_{\max} = E_\gamma/E_c$, because $N_{\max}$ is equally shared between electrons, positrons and photons. The amplitude of the counter signal is normally proportional to the number of charged particles. For a lead absorber ($E_c = 7.4 \, \text{MeV}$) and $E_\gamma = 1 \, \text{GeV}$, one gets $N_e \approx 90$. The relative fluctuation of this value is

$$\frac{\sigma(N_e)}{N_e} = \frac{1}{\sqrt{N_e}} \approx 10\% \; ; \tag{8.26}$$

that provides not so bad an energy resolution! Of course, the real pattern of the shower development is much more complicated (see Figs. 8.3 and 8.4). In a realistic model the number of electrons crossing the plane at a certain depth is much smaller than that expected from Formula (8.26).

To take advantage of the discussed idea one normally designs a calorimeter as an array of thin counters separated by layers of absorbers. These types of calorimeters are referred to as *sampling calorimeters* since only a sample of the energy deposition is measured. In addition to the general energy-leakage fluctuation the energy resolution of these calorimeters is affected by sampling fluctuations.

If the energy is determined by detectors in which only track segments of shower particles are registered, the number of intersection points with the detector layers is given by

$$N_{\text{tot}} = \frac{T}{d} , \tag{8.27}$$

where T is the total track length and d is the thickness of one sampling layer (absorber plus detector). The value of T can be estimated just as $T = (E_\gamma/E_c) \cdot X_0$, see Eq. (8.12). For the example considered above and $d = 1\,X_0$ we get $N_{\text{tot}} \approx 135$ and the sampling fluctuations are $1/\sqrt{N_{\text{tot}}} \approx 8.6\%$.

Actually, as discussed earlier, the number of detected particles is strongly dependent on the detection threshold. The measurable track length can be parametrised by [36]

$$T_{\text{m}} = F(\xi) \cdot \frac{E_\gamma}{E_c} \cdot X_0 \ \{\text{g/cm}^2\} , \tag{8.28}$$

where $T_{\text{m}} \leq T$ and the parameter ξ is a function of the detection energy threshold ϵ_{th}. However, the $\xi(\epsilon_{\text{th}})$ dependence is not very pronounced if ϵ_{th} is chosen to be sufficiently small ($\approx$ MeV). The function $F(\xi)$ takes into account the effect of the cutoff parameter on the total measurable track length for completely contained electromagnetic cascades in a calorimeter. $F(\xi)$ can be parametrised as [36]

$$F(\xi) = [1 + \xi \ln(\xi/1.53)]\,e^\xi , \tag{8.29}$$

where

$$\xi = 2.29 \cdot \frac{\epsilon_{\text{th}}}{E_c} . \tag{8.30}$$

Using the measurable track length defined by Eq. (8.28), the number of track segments is then

$$N = F(\xi) \cdot \frac{E_\gamma}{E_c} \cdot \frac{X_0}{d} . \tag{8.31}$$

Here we neglected the fact that, because of multiple scattering, the shower particles have a certain angle θ with respect to the shower axis. The effective sampling thickness is therefore not d, but rather $d/\cos\theta$. However, the average value $\langle 1/\cos\theta \rangle$ is not large; it is in the range between 1 and 1.3 depending on the energy E_γ.

Using Poisson statistics the sampling fluctuations limit the energy resolution to

$$\left[\frac{\sigma(E_\gamma)}{E_\gamma} \right]_{\mathrm{samp}} = \sqrt{\frac{E_{\mathrm{c}} \cdot d}{F(\xi) \cdot E_\gamma \cdot X_0 \cdot \cos\theta}} \ . \tag{8.32}$$

As can be seen from Eq. (8.32), the energy resolution of a sampling calorimeter for a fixed given material improves with $\sqrt{d/E_\gamma}$. However, Formula (8.32) does not take into account the correlations which are induced by electrons penetrating through two or several counter planes. These correlations become quite important when $d \ll 1\,X_0$ and limit the improvement of the resolution at small d.

A more accurate and simpler expression is suggested in [11] for the *sampling fluctuations* of calorimeters with counters based on condensed material:

$$\frac{\sigma_{\mathrm{samp}}}{E} = \frac{2.7\%}{\sqrt{E\,[\mathrm{GeV}]}} \sqrt{\frac{s\,[\mathrm{mm}]}{f_{\mathrm{samp}}}} \ . \tag{8.33}$$

Here s is the thickness of the sensitive layer and f_{samp} is the so-called *sampling fraction*, which is the ratio of ionisation losses of minimum-ionising particles in the sensitive layer to the sum of the losses in the sensitive layer and absorber. Figure 8.11 presents the energy resolution of some calorimeters versus the value $\sqrt{s/f_{\mathrm{samp}}}$ [11]. Anyway, these empirical formulae are only used for a preliminary estimate and general understanding of the sampling-calorimeter characteristics, while the final parameters are evaluated by a Monte Carlo simulation.

As sensitive elements of sampling calorimeters, gas-filled chambers, liquid-argon ionisation detectors, '*warm' liquids* (e.g. TMS) and scintillators are used. Energy depositions from large energy transfers in ionisation processes can further deteriorate the energy resolution. These Landau fluctuations are of particular importance for thin detector layers. If δ is the average energy loss per detector layer, the Landau fluctuations of the ionisation loss yield a contribution to the energy resolution of [36, 37]

$$\left[\frac{\sigma(E)}{E} \right]_{\mathrm{Landau\ fluctuations}} \propto \frac{1}{\sqrt{N}\ln(k \cdot \delta)} \ , \tag{8.34}$$

where k is a constant and δ is proportional to the matter density per detector layer.

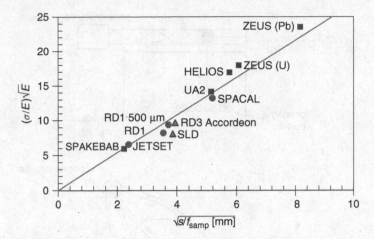

Fig. 8.11. The energy resolution of some sampling calorimeters. The solid line is approximation (8.33) [11]. (The energy is measured in GeV and the ordinate values are given in per cent.)

Since fluctuations of the ionisation losses are much higher in gases than in dense materials, the energy resolution for calorimeters with gaseous counters ($\sigma_E/E \approx 5\%$–20% at 1 GeV) is worse compared to that for liquid argon or scintillator sampling.

In *streamer-tube calorimeters* tracks are essentially counted, at least as long as the particles are not incident under too large an angle with respect to the shower axis, which is assumed to be perpendicular to the detector planes. For each ionisation track exactly one streamer is formed – independent of the ionisation produced along the track. For this reason Landau fluctuations have practically no effect on the energy resolution for this type of detector [9].

In general, the energy resolution of scintillator or liquid-argon sampling calorimeters is superior to that achievable with gaseous detectors. The layers in the liquid-argon sampling calorimeters can be arranged as planar chambers or they can have a more complex shape (*accordion type*). The achieved energy resolution with LAr calorimeters is 8%–10% at 1 GeV [38, 39].

If, as is the case in calorimeters, a sufficient amount of light is available, the light emerging from the end face of a scintillator plate can be absorbed in an external wavelength-shifter rod. This wavelength shifter re-emits the absorbed light isotropically at a larger wavelength and guides it to a photosensitive device (Fig. 8.12).

It is very important that a small air gap remains between the scintillator face and the wavelength-shifter rod. Otherwise, the frequency-shifted, isotropically re-emitted light would not be contained in the wavelength-shifter rod by internal reflection. This method of light transfer normally

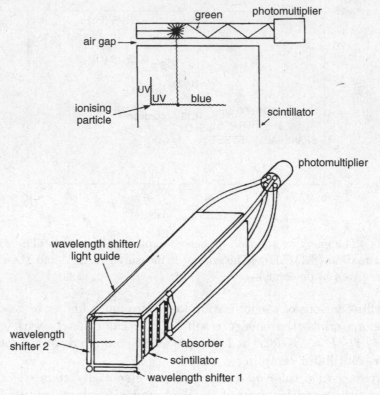

Fig. 8.12. *Wavelength-shifter readout* of a scintillator and two-step wavelength-shifter readout of a calorimeter.

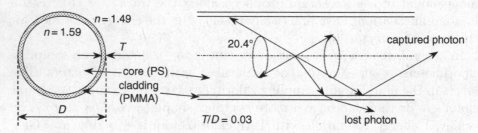

Fig. 8.13. Structure and principle of operation of the scintillation and light-guiding fibres [40, 41].

entails an appreciable loss of light; typical conversion values are around 1% to 5%. However, now single- and multicladding scintillation and light-guide fibres are available. The structure and operation principle of such fibres are explained in Fig. 8.13 [40, 41]. The fibres of this type allow light transfer over long distances at small light losses. The fraction of the captured light is typically 3% for single-cladding fibres and up to 6% for multicladding ones. The cladding-fibre light guides can be glued to the scintillator without any air gap.

A normal sampling calorimeter of absorber plates and scintillator sheets can also be read out by wavelength-shifter rods or fibres running through the scintillator plates perpendicularly [42–44]. The technique of wavelength-shifter readout allows to build rather compact calorimeters.

The scintillation counters used in calorimeters must not necessarily have the form of plates alternating with absorber layers. They can also be embedded as scintillating fibres, for example, in a lead matrix [45, 46]. In this case the readout is greatly simplified because the scintillating fibres can be bent rather strongly without loss of internal reflection. Scintillating fibres can either be read out directly or via light-guide fibres by photomultipliers (*spaghetti calorimeter*). The energy resolution of the scintillation-fibre-based calorimeter of the KLOE detector achieved a value of $\sigma_E/E = 5.7\%/\sqrt{E \, [\text{GeV}]}$. In addition to high energy resolution, this calorimeter provides precise timing for photons ($\sigma_t \approx 50 \, \text{ps}/\sqrt{E \, [\text{GeV}]}$) due to the short decay time of the light flash of the plastic scintillator [46]. Recently, even a better energy resolution, $4\%/\sqrt{E \, [\text{GeV}]}$, was reported for a '*shashlik*'-*type sampling calorimeter* developed for the KOPIO experiment [43].

The scintillator readout can also be accomplished by inserting wavelength-shifting fibres into grooves milled into planar scintillator sheets (*tile calorimeter*) [47–49].

8.2 Hadron calorimeters

In principle, *hadron calorimeters* work along the same lines as electron–photon calorimeters, the main difference being that for hadron calorimeters the longitudinal development is determined by the average nuclear interaction length λ_I, which can be roughly estimated as [1]

$$\lambda_I \approx 35 \, \text{g/cm}^2 A^{1/3} \; . \tag{8.35}$$

In most detector materials this is much larger than the radiation length X_0, which describes the behaviour of electron–photon cascades. This is the reason why hadron calorimeters have to be much larger than electromagnetic shower counters.

Frequently, electron and hadron calorimeters are integrated in a single detector. For example, Fig. 8.14 [50] shows an iron–scintillator calorimeter with separate wavelength-shifter readout for electrons and hadrons. The electron part has a depth of 14 radiation lengths, and the hadron section corresponds to 3.2 interaction lengths.

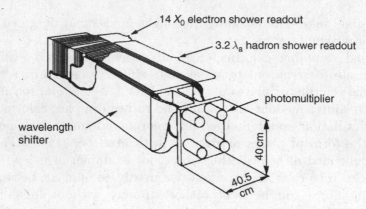

Fig. 8.14.   Typical set-up of an iron–scintillator calorimeter with wavelength-shifter readout [50].

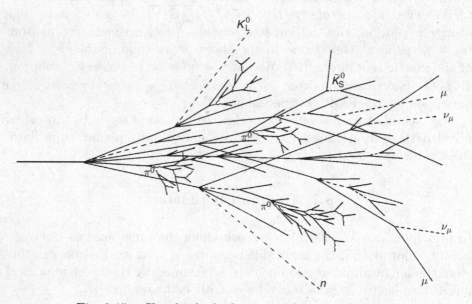

Fig. 8.15.   Sketch of a hadron cascade in an absorber.

Apart from the larger *longitudinal development* of hadron cascades, their *lateral width* is also sizably increased compared to electron cascades. While the lateral structure of electron showers is mainly determined by multiple scattering, in hadron cascades it is caused by large transverse momentum transfers in nuclear interactions. Typical processes in a hadron cascade are shown in Fig. 8.15.

Different structures of 250 GeV photon- and proton-induced cascades in the Earth's atmosphere are clearly visible from Fig. 8.16 [51]. The results shown in this case were obtained from a Monte Carlo simulation.

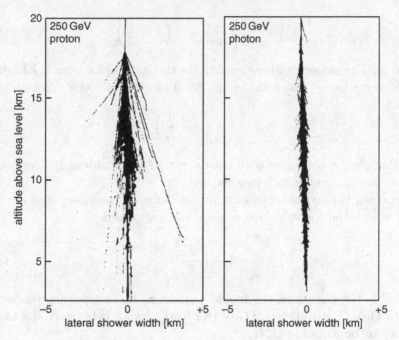

Fig. 8.16.   Monte Carlo simulations of the different development of hadronic and electromagnetic cascades in the Earth's atmosphere, induced by 250 GeV protons and photons [51].

The production of secondary particles in a hadron cascade is caused by inelastic hadronic processes. Mainly charged and neutral pions, but, with lower multiplicities, also kaons, nucleons and other hadrons are produced. The average particle multiplicity per interaction varies only weakly with energy ($\propto \ln E$). The average transverse momentum of secondary particles can be characterised by

$$\langle p_{\mathrm{T}} \rangle \approx 0.35 \, \mathrm{GeV}/c \ . \tag{8.36}$$

The average inelasticity, that is, the fraction of energy which is transferred to secondary particles in the interaction, is around 50%.

A large component of the secondary particles in hadron cascades are neutral pions, which represent approximately one third of the pions produced in each inelastic collision. Neutral pions decay rather quickly ($\approx 10^{-16}\,\mathrm{s}$) into two energetic photons, thereby initiating electromagnetic subcascades in a hadron shower. Therefore, after the first collision 1/3 of the energy is deposited in the form of an electromagnetic shower, at the second stage of multiplication the total fraction of this energy, f_{em}, will be

$$\frac{1}{3} + \left(1 - \frac{1}{3}\right)\frac{1}{3} = 1 - \left(1 - \frac{1}{3}\right)^2 , \tag{8.37}$$

and so on. The same argument applies for the leaving hadron. If a hadronic shower comprises n generations, the total electromagnetic fraction is

$$f_{\mathrm{em}} = 1 - \left(1 - \frac{1}{3}\right)^n . \tag{8.38}$$

Assuming that n increases with the energy of the incident hadron we can see that the f_{em} value increases as well.

Of course, this consideration is rather naïve. This effect was analysed in [52] where the following expression was suggested:

$$f_{\mathrm{em}} = 1 - \left(\frac{E}{E_0}\right)^{k-1} , \tag{8.39}$$

where E is the energy of the incident hadron, E_0 is a parameter varying from $0.7\,\mathrm{GeV}$ (for iron) to $1.3\,\mathrm{GeV}$ (for lead), and k is between 0.8 to 0.85. Details can be found in [11].

π^0 production, however, is subject to large fluctuations, which are determined essentially by the properties of the first inelastic interaction.

Some part of the energy in the hadronic shower is deposited via ionisation losses of the charged hadrons (f_{ion}).

In contrast to electrons and photons, whose electromagnetic energy is almost completely recorded in the detector, a substantial fraction of the energy in hadron cascades remains *invisible* (f_{inv}). This is related to the fact that some part of the hadron energy is used to break up nuclear bonds. This *nuclear binding energy* is provided by the primary and secondary hadrons and does not contribute to the visible energy.

Furthermore, extremely short-range nuclear fragments are produced in the break-up of nuclear bonds. In sampling calorimeters, these fragments do not contribute to the signal since they are absorbed before reaching the detection layers. In addition, long-lived or stable neutral particles like neutrons, K_{L}^0, or neutrinos can escape from the calorimeter, thereby reducing the visible energy. Muons created as decay products of pions and kaons deposit in most cases only a very small fraction of their energy in the calorimeter (see the example at the beginning of this chapter). As a result of all these effects, the energy resolution for hadrons is significantly inferior to that of electrons because of the different interaction and particle-production properties. The total *invisible energy fraction* of a hadronic cascade can be estimated as $f_{\mathrm{inv}} \approx 30\% - 40\%$ [11].

It is important to remember that only the electromagnetic energy and the energy loss of charged particles can be recorded in a calorimeter.

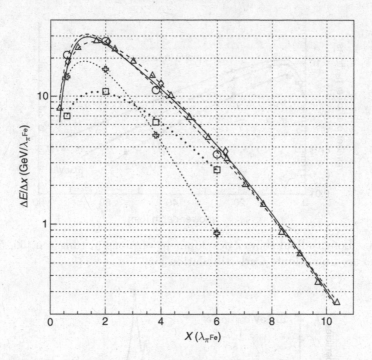

Fig. 8.17. The longitudinal energy distribution in a hadronic shower in iron induced by 100 GeV pions. The depth X is measured in units of the interaction length λ_I. Open circles and triangles are experimental data, diamonds are predictions of a simulation. The dash-dotted line is a simple fit by Formula (8.7) with optimal a and b, the other lines are more sophisticated approximations. Crosses and squares are contributions of electromagnetic showers and the non-electromagnetic part, respectively [53].

Consequently, a hadron signal for the same particle energy is normally smaller than an electron signal.

Figure 8.17 shows the measured longitudinal shower development of 100 GeV pions in iron [53] in comparison to Monte Carlo calculations and empirical approximations. The energy-deposition distributions for a tungsten calorimeter obtained for different pion energies are presented in Fig. 8.18 [54–58]. The lateral shower profiles of 10 GeV/c pions in iron are shown in Fig. 8.19.

The so-called length of a hadron cascade depends on exactly how this is defined. Regardless of the definition, the length increases with the energy of the incident particle. Figure 8.20 shows the shower lengths and centre of gravity of hadronic cascades for various definitions [55]. One possible definition is given by the requirement that the shower length is reached if, on average, only one particle or less is registered at the depth t. According to this definition a 50 GeV-pion shower in an iron–scintillator calorimeter

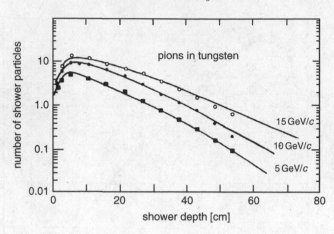

Fig. 8.18.   Longitudinal shower development of pions in tungsten [56, 57]. The solid lines are from Monte Carlo simulation [58].

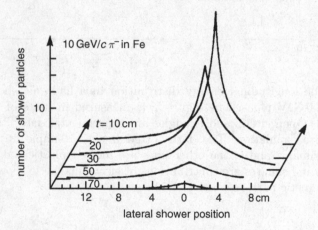

Fig. 8.19.   Lateral shower profile of $10\,\text{GeV}/c$ pions in iron [59].

is approximately 120 cm Fe 'long'. An alternative definition is given by the depth before which a certain fraction of the primary energy (e.g. 95%) is contained. A 95% energy containment would lead to a length of 70 cm iron for a 50 GeV-pion shower. The longitudinal centre of gravity of the shower only increases logarithmically with the energy. The position of the centre of gravity of the shower is also shown in Fig. 8.20.

The 95%-longitudinal-containment length in iron can be approximated by [2]

$$L\,(95\%) = (9.4\,\ln(E/\text{GeV}) + 39)\,\text{cm}\ . \tag{8.40}$$

This estimation scaled by the interaction length λ_{I} characterises the hadronic showers in other materials as well.

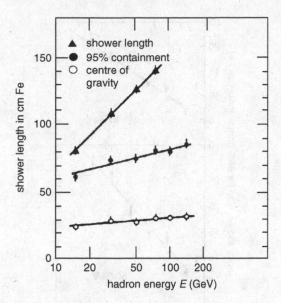

Fig. 8.20. Shower lengths and centre of gravity of hadron cascades for various definitions [55].

Similarly, the *lateral distribution* of cascades can be characterised by a radial width. The lateral distribution of a hadron shower is initially very narrow but becomes wider with increasing calorimeter depth (see Fig. 8.19). The required lateral calorimeter radius for a 95% containment as a function of the longitudinal shower depth is shown in Fig. 8.21 for pions of two different energies in iron [55].

The energy resolution for hadrons is significantly worse compared to electrons because of the large fluctuations in the hadron-shower development. A large contribution to this fact is caused by the difference in the calorimeter response to electrons and hadrons. Due to this difference the fluctuations in the number of neutral pions produced in the hadronic shower create a sizable effect for the energy resolution.

It is, however, possible to regain some part of the 'invisible' energy in hadron cascades, thereby equalising the response to electrons and hadrons. This hadron-calorimeter *compensation* is based on the following physical principles [11, 60, 61].

If uranium is used as an absorber material, neutrons will also be produced in nuclear interactions. These neutrons may induce fission of other target nuclei producing more neutrons as well and energetic γ rays as a consequence of nuclear transitions. These neutrons and γ rays can enhance the amplitude of the hadron-shower signal if their energy is recorded. Also for absorber materials other than uranium where fission processes are

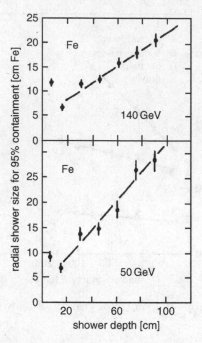

Fig. 8.21. Radius of hadronic showers for 95% containment as a function of the depth in iron [55]. The corresponding total width of the hadron shower is twice the radius.

endotherm, neutrons and γ rays may be produced. The γ rays can contribute to the visible energy by a suitable choice of sampling detectors, and neutrons can produce low-energy recoil protons in (n, p) reactions in detector layers containing hydrogen. These recoil protons also increase the hadron signal.

For energies below 1 GeV even in uranium sampling calorimeters, the lost energy in hadron cascades cannot be regained. By suitable combination (uranium/liquid argon, uranium/copper/scintillator) compensation can be achieved for energies exceeding several GeV. For very high energies (≥ 100 GeV) even *overcompensation* can occur. Such overcompensation can be avoided by limiting the sampling time. Overcompensation can also be caused by a reduction of the electron signal due to saturation effects in the detector layers. Because of the different lateral structure of electron and hadron cascades, saturation effects affect the electron and hadron signals differently.

The best hadron sampling calorimeters (e.g. uranium/scintillator, uranium/liquid argon) reach an energy resolution of [62]

$$\frac{\sigma(E)}{E} = \frac{35\%}{\sqrt{E \text{ [GeV]}}} \ . \tag{8.41}$$

However, hadron calorimeters recently developed for the detection of high-energy hadrons at LHC achieved a rather good energy resolution even without compensation. For example, for the ATLAS detector a resolution of about $42\%/\sqrt{E}$ [GeV] was obtained for pions at a total calorimeter thickness of about $8.2\,\lambda_I$, and an e/h ratio[†] of about 1.37 was measured [63]. A possible constant term in the parametrisation of the energy resolution usually can safely be neglected for hadronic cascades because the large sampling fluctuations dominate the energy resolution. Only for extremely high energies ($\approx 1000\,\text{GeV}$) a constant term will limit the energy resolution.

The energy resolution attainable in hadron calorimeters varies with the number of detector layers (sampling planes) similarly to electromagnetic calorimeters. Experimentally one finds that absorber thicknesses $d < 2\,\text{cm}$ of iron do not lead to an improvement of the energy resolution [2]. Depending on the application as well as on the available financial resources, a large variety of sampling detectors can be considered. Possible candidates for sampling elements in calorimeters are scintillators, liquid-argon or liquid-xenon layers, multiwire proportional chambers, layers of proportional tubes, flash chambers, streamer tubes, Geiger–Müller tubes (with local limitation of the discharge – 'limited Geiger mode'), parallel-plate chambers and layers of 'warm' (i.e. room temperature) liquids (see Chap. 5). Ionisation chambers under high pressure can also be used [64]. For absorber materials, uranium, copper, tungsten and iron are most commonly used, although aluminium and marble calorimeters have also been constructed and operated.

A prominent feature of calorimeters is that their energy resolution $\sigma(E)/E$ improves with increasing energy like $1/\sqrt{E}$, quite in contrast to momentum spectrometers, whose resolution σ_p/p deteriorates linearly with increasing momentum. In addition, calorimeters are rather compact even for high energies, because the shower length only increases logarithmically with the particle energy.

In cosmic-ray experiments involving the energy determination of protons, heavy nuclei and photons of the primary cosmic radiation in the energy range $> 10^{14}\,\text{eV}$, various calorimetric measurement methods are needed to account for the low particle intensities. Cosmic-ray particles initiate in the Earth's atmosphere hadronic or electromagnetic cascades (see Fig. 8.16) which can be detected by quite different techniques. The energy of extensive air showers is traditionally determined by sampling their lateral distribution at sea level. This classical method quite obviously suffers from a relatively inaccurate energy determination [65]. Better results are

---

[†] The e/h ratio is the ratio of energy deposits of an electron-initiated shower compared to that of a hadron-initiated shower for the same initial energy of electrons and hadrons.

obtained if the scintillation or Cherenkov light of the shower particles produced in the atmosphere is recorded (compare Sect. 16.12). To observe the very rare highest-energy cosmic rays, an as large as possible detection volume is necessary. In this case the scintillation of nitrogen produced by the cosmic-ray shower can be detected [66, 67]. Both of these techniques, Cherenkov and air scintillation, however, require – because of the low light yield – clear and moonless nights.

A possible way out or an alternative method is the detection of *geosynchrotron radiation* in the radio band (40–80 MHz) of extensive air showers, which is generated by the deflection of the large number of shower particles in the Earth's magnetic field [68, 69]. It is also conceivable to measure high-energy extensive air showers by *acoustic detection techniques* [70].

An alternative method can be considered for the energy determination of high-energy cosmic neutrinos or muons. These particles easily penetrate the Earth's atmosphere, so that one can also take advantage of the clear and highly transparent water of the ocean, deep lakes or even polar ice as a Cherenkov medium. Muons undergo energy losses at high energies ($> 1\,\mathrm{TeV}$) mainly by bremsstrahlung and direct electron-pair production (see Fig. 1.6). These two energy-loss processes are both proportional to the muon energy. A measurement of the energy loss using a three-dimensional matrix of photomultipliers in deep water, shielded from sunlight, allows a determination of the muon energy. Similarly, the energy of electron or muon neutrinos can be roughly determined, if these particles produce electrons or muons in inelastic interactions in water, that, for the case of electrons, induce electromagnetic cascades, and, for the case of muons, they produce a signal proportional to the energy loss. The deep ocean, lake water or polar ice in this case are both interaction targets and detectors for the Cherenkov light produced by the interaction products. Electrons or muons produced in neutrino interactions closely keep the direction of incidence of the neutrinos. Therefore, these deep-water neutrino detectors are at the same time *neutrino telescopes* allowing one to enter the domain of neutrino astronomy in the TeV energy range [71–73].

8.3 Calibration and monitoring of calorimeters

In the modern experiments on particle physics the information is collected as digitised data (see Chap. 15). The pulse height A_i measured in an event from a certain (ith) element of the calorimeter is related to the energy E_i deposited in this element by

$$E_i = \alpha_i(A_i - P_i) \ , \tag{8.42}$$

where P_i is the pedestal, i.e. the origin of the scale, and α_i is the calibration coefficient. Thus, to keep a good performance of the calorimeter, the following procedures are usually carried out:

- *Pedestal determination* by providing a trigger from a pulser without any signal at the input of the ADC ('random trigger events').

- Electronics channel control by test pulses applied to the input of the electronics chain.

- *Monitoring* of the stability of the calibration coefficients α_i.

- Absolute *energy calibration*, i.e. determination of the α_i values.

In general, the dependence (8.42) can be non-linear. In this case more calibration coefficients are needed to describe the E/A relation.

Prior to real physics experiments a study of the parameters of individual calorimeter elements and modules is usually done in accelerator-test beams which supply identified particles of known momenta. By varying the beam energy the linearity of the calorimeter can be tested and characteristic shower parameters can be recorded. For the calibration of calorimeters designed for low energies, e.g. semiconductor detectors, radioactive sources are normally used. Preferentially used are K-line emitters, like ^{207}Bi, with well-defined monoenergetic electrons or gamma-ray lines, which allow a calibration via the total-absorption peaks.

In addition to energy calibration, the dependence of the calorimeter signal on the point of particle impact, the angle of incidence and the behaviour in magnetic fields is of great importance. In particular, for calorimeters with gas sampling, magnetic-field effects can cause spiralling electrons, which can significantly modify the calibration. In gas sampling calorimeters the particle rate can have influence on the signal amplitude because of dead-time or recovery-time effects. A thorough calibration of a calorimeter therefore requires an extensive knowledge of the various parameter-dependent characteristics.

Big experiments can contain a large number of calorimeter modules, not all of which can be calibrated in test beams. If some of the modules are calibrated in a test beam, the rest can be adjusted relative to them. This relative calibration can be done by using minimum-ionising muons that penetrate many calorimeter modules. In uranium calorimeters, the constant noise caused by the natural radioactivity of the uranium can be used for a relative calibration. If one uses non-radioactive absorber materials in gas sampling calorimeters, a test and relative calibration can also be performed with radioactive noble gases like ^{85}Kr.

Scintillator calorimeters can best be calibrated by feeding defined light signals, e.g. via light-emitting diodes (LEDs), into the detector layers and recording the output signals from the photomultipliers. To avoid variations in the injected light intensity, which may be caused by different light yields of the individual light diodes, a single light source can be used (e.g. a laser), which distributes its light via a manifold of light fibres to the scintillation counters [2].

Once a complex calorimeter system has been calibrated, one has to ensure that the calibration constants do not vary or, if they do, the drift of the calibration parameters must be monitored. The time stability of the calibration can be checked with, e.g., cosmic-ray muons. In some cases some calorimeter modules may be positioned unfavourably so that the rate of cosmic-ray muons is insufficient for accurate stability control. Therefore, reference measurements have to be performed periodically by injecting calibrated reference signals into the various detector layers or into the inputs of the readout electronics. The calibration and monitoring of scintillation crystal calorimeters can be performed using cosmic-ray muons as it was demonstrated in [74, 75].

In gas sampling calorimeters the output signal can in principle only vary because of a change of gas parameters and high voltage. In this case, a test chamber supplied with the detector gas can be used for monitoring. To do that, the current, the pulse rate or the spectrum under the exposition to characteristic X rays of a radioactive source should be continuously measured. A change in the measured X-ray energy in this test chamber indicates a time-dependent calibration which can be compensated by an adjustment of the high voltage.

In some experiments there are always particles that can be used for calibration and monitoring. For example, elastic Bhabha scattering $(e^+e^- \to e^+e^-)$ can be used to calibrate the electromagnetic calorimeters in an e^+e^- scattering experiment, since the final-state particles – if one neglects radiative effects – have known beam energy. In the same way, the reaction $e^+e^- \to q\bar{q}$ (e.g. going through a resonance of known mass, like m_Z, if one wants to be independent of initial-state radiation) with subsequent hadronisation of the quarks can be used to check the performance of a hadron calorimeter. Finally, muon-pair production $(e^+e^- \to \mu^+\mu^-)$ supplies final-state muons with known momentum (= beam momentum at high energies), which can reach all detector modules because of their nearly flat angular distribution $(\mathrm{d}\sigma/\mathrm{d}\Omega \propto 1 + \cos^2\theta$, where θ is the angle between e^- and μ^-).

It should be noted that the energy of an electron or hadron absorbed in the calorimeter is distributed over a cluster of crystals. The total deposited energy can be expressed as a sum

$$E = \sum_{i=1}^{M} \alpha_i A_i , \qquad (8.43)$$

where pedestals are assumed to be already subtracted. Then, the calibration coefficients are determined by minimisation of the functional

$$L = \sum_{k=1}^{N} \left(\sum_{i=1}^{M} \alpha_i A_{ik} - E_{0k} \right)^2 , \qquad (8.44)$$

where the first summation is performed over all N events selected for calibration, A_{ik} is the response of the ith calorimeter element in the kth event and E_{0k} is the known incident particle energy in the kth event. Requiring for all α_j

$$\frac{\partial L}{\partial \alpha_j} = 0 , \qquad (8.45)$$

we obtain a linear equation system for the determination of the calibration constants,

$$\sum_{i=1}^{M} \alpha_i \left(\sum_{k=1}^{N} A_{jk} A_{ik} \right) = \sum_{k=1}^{N} E_{0k} A_{jk} . \qquad (8.46)$$

8.4 Cryogenic calorimeters

The calorimeters described so far can be used for the spectroscopy of particles from the MeV range up to the highest energies. For many investigations the detection of particles of extremely low energy in the range between 1 eV and 1000 eV is of great interest. Calorimeters for such low-energy particles are used for the detection of and search for low-energy cosmic neutrinos, weakly interacting massive particles (WIMPs) or other candidates of dark, non-luminous matter, X-ray spectroscopy for astrophysics and material science, single-optical-photon spectroscopy and in other experiments [76–79]. In the past 20 years this field of experimental particle physics has developed intensively and by now it comprises dozens of projects [80, 81].

To reduce the detection threshold and improve at the same time the calorimeter energy resolution, it is only natural to replace the ionisation or electron–hole pair production by *quantum transitions* requiring lower energies (see Sect. 5.3).

Phonons in solid-state materials have energies around 10^{-5} eV for temperatures around 100 mK. The other types of quasiparticles at low temperature are *Cooper pairs* in a superconductor which are bound states of two electrons with opposite spin that behave like bosons and will form at sufficiently low temperatures a Bose condensate. Cooper pairs in super-conductors have binding energies in the range between $4 \cdot 10^{-5}$ eV (Ir) and $3 \cdot 10^{-3}$ eV (Nb). Thus, even extremely low energy depositions would produce a large number of phonons or break up Cooper pairs. To avoid thermal excitations of these quantum processes, such calorimeters, how-ever, would have to be operated at extremely low temperatures, typically in the milli-Kelvin range. For this reason, such calorimeters are called *cryogenic detectors*. Cryogenic calorimeters can be subdivided in two main categories: first, detectors for quasiparticles in superconducting materials or suitable crystals, and secondly, phonon detectors in insulators.

One detection method is based on the fact that the superconductivity of a substance is destroyed by energy deposition if the detector element is sufficiently small. This is the working principle of superheated supercon-ducting granules [82]. In this case the cryogenic calorimeter is made of a large number of superconducting spheres with diameters in the microme-tre range. If these granules are embedded in a magnetic field, and the energy deposition of a low-energy particle transfers one particular granule from the superconducting to the normal-conducting state, this transition can be detected by the suppression of the *Meissner effect*. This is where the magnetic field, which does not enter the granule in the supercon-ducting state, now again passes through the normal-conducting granule. The transition from the superconducting to the normal-conducting state can be detected by pickup coils coupled to very sensitive preamplifiers or by SQUIDs (*Superconducting Quantum Interference Devices*) [83]. These *quantum interferometers* are extremely sensitive detection devices for magnetic effects. The operation principle of a SQUID is based on the *Josephson effect*, which represents a tunnel effect operating between two superconductors separated by thin insulating layers. In contrast to the normal one-particle tunnel effect, known, e.g. from α decay, the Joseph-son effect involves the tunnelling of Cooper pairs. In Josephson junctions, interference effects of the tunnel current occur which can be influenced by magnetic fields. The structure of these interference effects is related to the size of the magnetic flux quanta [84–86].

An alternative method to detect *quasiparticles* is to let them directly tunnel through an insulating foil between two superconductors (SIS – Superconducting–Insulating–Superconducting transition) [87]. In this case the problem arises of keeping undesired leakage currents at an extremely low level.

In contrast to Cooper pairs, phonons, which can be excited by energy depositions in insulators, can be detected with methods of classical calorimetry. If ΔE is the absorbed energy, this results in a temperature rise of

$$\Delta T = \Delta E/mc \; , \qquad\qquad (8.47)$$

where c is the specific heat capacity and m the mass of the calorimeter. If these calorimetric measurements are performed at very low temperatures, where c can be very small (the lattice contribution to the specific heat is proportional to T^3 at low temperatures), this method is also used to detect individual particles. In a real experiment, the temperature change is recorded with a thermistor, which is basically an NTC resistor (negative temperature coefficient), embedded into or fixed to an ultrapure crystal. The crystal represents the absorber, i.e. the detector for the radiation that is to be measured. Because of the discrete energy of phonons, one would expect discontinuous thermal energy fluctuations which can be detected with electronic filter techniques.

In Fig. 8.22 the principle of such a calorimeter is sketched [88].

In this way α particles and γ rays have been detected in a large TeO_2 crystal at $15\,\text{mK}$ in a purely thermal detector with thermistor readout with an energy resolution of $4.2\,\text{keV}$ FWHM for $5.4\,\text{MeV}$ α particles [89]. Special bolometers have also been developed in which heat and ionisation signals are measured simultaneously [90, 91].

Thermal detectors provide promise for improvements of energy resolutions. For example, a $1\,\text{mm}$ cubic crystal of silicon kept at $20\,\text{mK}$ would have a heat capacity of $5 \cdot 10^{-15}\,\text{J/K}$ and a FWHM energy resolution of $0.1\,\text{eV}$ (corresponding to $\sigma = 42\,\text{meV}$) [92].

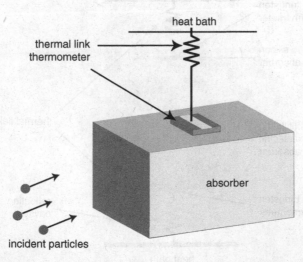

Fig. 8.22.   Schematic of a cryogenic calorimeter. The basic components are the absorber for incident particles, a thermometer for detecting the heat signal and a thermal link to the heat bath [88].

Joint efforts in the fields of cryogenics, particle physics and astrophysics are required, which may lead to exciting and unexpected results. One interesting goal would be to detect relic neutrinos of the Big Bang with energies around $200\,\mu eV$ [92].

At present cryogenic calorimeters are most frequently used in the search for *weakly interacting massive particles* (WIMPs). The interaction cross section for WIMP interactions is extremely small, so that possible backgrounds have to be reduced to a very low level. Unfortunately, also the energy transfer of a WIMP to a target nucleus in a cryogenic detector is only in the range of $\approx 10\,keV$. An excellent method to discriminate a WIMP signal against the background caused, e.g., by local radioactivity is to use scintillating crystals like $CaWO_4$, $CdWO_4$ or $ZnWO_4$. These scintillators allow to measure the light yield at low temperatures and the phonon production by WIMP interactions at the same time. *Nuclear recoils* due to WIMP–nucleon scattering produce mainly phonons and very little scintillation light, while in *electron recoils* also a substantial amount of scintillation light is created. A schematic view of such a cryogenic detector system is shown in Fig. 8.23 [88].

Particles are absorbed in a scintillating dielectric crystal. The scintillation light is detected in a silicon wafer while the phonons are measured in two tungsten thermometers, one of which can be coupled to the silicon detector to increase the sensitivity of the detector. The whole detector

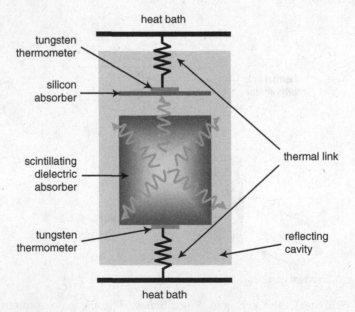

Fig. 8.23. Schematic view of a cryogenic detector with coincident phonon and light detection [88].

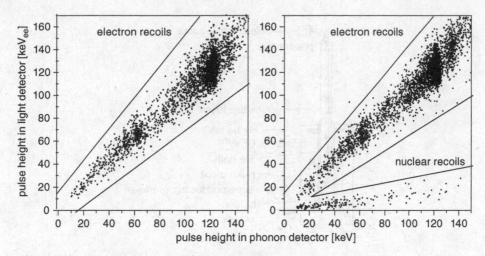

Fig. 8.24. Scatter plot of the pulse height in the light detector from photons of the $CaWO_4$ crystal versus the pulse height from phonons from the same crystal. The left-hand part of the figure shows the response of the detector to photons and electrons only, while in the right-hand part also neutron interactions are included. The purpose of the lines is just to guide the eye [88, 93].

setup is enclosed in a reflecting cavity and operated at milli-Kelvin temperatures.

The response of a $CaWO_4$ cryogenic calorimeter to electron recoils and nuclear recoils is shown in Fig. 8.24 [88, 93].

Electron recoils were created by irradiating the crystal with 122 keV and 136 keV photons from a ^{57}Co source and electrons from a ^{90}Sr β source (left panel). To simulate also WIMP interactions the detector was bombarded with neutrons from an americium–beryllium source leading to phonon and scintillation-light yields as shown in the right-hand plot of Fig. 8.24. The light output due to electron recoils caused by photons or electrons (which constitute the main background for WIMP searches) is quite high, whereas nuclear recoils created by neutrons provide a strong phonon signal with only low light yield. It is conjectured that WIMP interactions will look similar to neutron scattering, thus allowing a substantial background rejection if appropriate cuts in the scatter diagram of light versus phonon yield are applied. However, the figure also shows that the suppression of electron recoils at energies below 20 keV becomes rather difficult.

The set-up of a cryogenic detector, based on the energy absorption in *superheated superconducting granules*, is shown in Fig. 8.25 [94]. The system of granules and pickup coil was rotatable by 360° around an axis perpendicular to the magnetic field. This was used to investigate the dependence of the critical field strength for reaching the superconducting

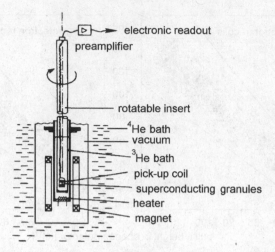

Fig. 8.25. Experimental set-up of a cryogenic detector based on superheated superconducting granules (SSG) [94].

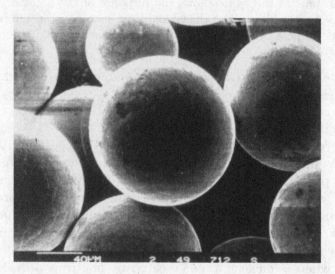

Fig. 8.26. Tin granules (diameter $= 130\,\mu m$) as a cryogenic calorimeter. A small energy absorption can warm the granules by an amount sufficient to cause a change from the superconducting state to the normal-conducting state, thereby providing a detectable signal [82].

state on the orientation of the granules with respect to the magnetic field. This system succeeded in detecting quantum transitions in tin, zinc and aluminium granules at ^{4}He and ^{3}He temperatures. Figure 8.26 shows a microphotograph of tin granules [82, 95]. At present it is already possible to manufacture tin granules with diameters as small as $5\,\mu m$.

With a detector consisting of superheated superconducting granules, it has already been shown that one can detect minimum-ionising particles unambiguously [95].

The detection of transitions from the superconducting into the normal-conducting state with signal amplitudes of about $100\,\mu V$ and recovery times of $10\,ns$ to $50\,ns$ already indicates that *superconducting strip counters* are possible candidates for microvertex detectors for future generations of particle physics experiments [96].

8.5 Problems

8.1 In an experiment an η meson with total energy $E_0 = 2000\,MeV$ is produced in the laboratory frame. Estimate the width of the η mass peak measured in a calorimeter which has an energy and angular resolution of $\sigma_E/E = 5\%$ and $\sigma_\theta = 0.05$ radian, respectively ($m_\eta = 547.51\,MeV$).

8.2 Photons of $1\,GeV$ ($100\,MeV$) energy are detected in a NaI(Tl) calorimeter which has an energy resolution $\sigma_E/E = 1.5\%/(E\,[GeV])^{1/4}$. Determine how the pulse-height distribution would change if an aluminium sheet of $L = 0.5\,X_0$ thickness would be placed in front of the calorimeter. Estimate the resulting decrease of the energy resolution.

8.3 Estimate the quality of a pion/electron separation for a total particle energy of $E = 500\,MeV$ using the energy deposition in a calorimeter based on NaI(Tl) crystals of $15\,X_0$ length. For the estimation assume that the main mixing effect consists of pion charge exchange on nuclei which occurs with 50% probability when the pion interacts with nuclei. In this charge-exchange reaction the charged pion is transformed into a neutral pion which initiates an electromagnetic cascade.

References

[1] Particle Data Group, Review of Particle Physics, S. Eidelman *et al.*, *Phys. Lett.* **B592 Vol. 1–4** (2004) 1–1109; W.-M. Yao *et al.*, *J. Phys.* **G33** (2006) 1–1232; http://pdg.lbl.gov

[2] K. Kleinknecht, *Detektoren für Teilchenstrahlung*, Teubner, Stuttgart (1984, 1987, 1992); *Detectors for Particle Radiation*, Cambridge University Press, Cambridge (1986)

[3]	O.C. Allkofer & C. Grupen, Introduction II: Cosmic Rays in the Atmosphere, in A. Bruzek & H. Pilkuhn (eds.), *Lectures on Space Physics 1*, Bertelsmann Universitätsverlag, Gütersloh (1973) 35–54

[4]	O.F. Nemets & Yu.V. Gofman, *Spravochnik po yadernoi fizike*, Naukova dumka, Kiev (1975)

[5]	B. Rossi, *High Energy Particles*, Prentice-Hall, Englewood Cliffs (1952)

[6]	W.R. Nelson *et al.*, *The EGS4 Code System*, SLAC-R-265 (1985)

[7]	E. Longo & I. Sestili, Monte Carlo Calculation of Photon-Initiated Electromagnetic Showers in Lead Glass, *Nucl. Instr. Meth.* **128** (1975) 283–307; Erratum, *ibid.*, **135** (1976) 587–90

[8]	R. Baumgart, *Messung und Modellierung von Elektron- und Hadron-Kaskaden in Streamerrohrkalorimetern*, Ph.D. Thesis, University of Siegen (1987)

[9]	R. Baumgart *et al.*, Performance Characteristics of an Electromagnetic Streamer Tube Calorimeter, *Nucl. Instr. Meth.* **A256** (1987) 254–60

[10]	N. Akchurin *et al.*, Electromagnetic Shower Profile Measurements in Iron with 500 MeV Electrons, *Nucl. Instr. Meth.* **A471** (2001) 303–13

[11]	R. Wigmans, *Calorimetry: Energy Measurement in Particle Physics*, Clarendon Press, Oxford (2000)

[12]	S. Iwata, *Calorimeters*, Nagoya University Preprint DPNU-13-80 (1980)

[13]	S. Iwata, *Calorimeters (Total Absorption Detectors) for High Energy Experiments at Accelerators*, Nagoya University Preprint DPNU-3-79 (1979)

[14]	L.D. Landau, *The Collected Papers of L.D. Landau*, Pergamon Press, London (1965); A.B. Migdal, Bremsstrahlung and Pair Production in Condensed Media at High Energies, *Phys. Rev.* **103** (1956) 1811–20

[15]	E. Konishi *et al.*, *Three Dimensional Cascade Showers in Lead Taking Account of the Landau-Pomeranchuk-Migdal Effect*, Inst. for Cosmic Rays, Tokyo, ICR Report 36-76-3 (1976)

[16]	*Photon Theory Verified after 40 Years*, CERN-Courier **34(1)** (1994) 12–13

[17]	R. Becker-Szendy *et al.* (SLAC-E-146 Collaboration), *Quantummechanical Suppression of Bremsstrahlung*, SLAC-Pub-6400 (1993)

[18]	M. Oreglia *et al.*, Study of the Reaction $\psi' \to \gamma\gamma J/\psi$, *Phys. Rev.* **D25** (1982) 2259–77

[19]	E. Blucher *et al.*, Tests of Cesium Iodide Crystals for Electromagnetic Calorimeter, *Nucl. Instr. Meth.* **A249** (1986) 201–27

[20]	B. Aubert, The BaBar Detector, *Nucl. Instr. Meth.* **A479** (2002) 1–116; B. Lewandowski, The BaBar Electromagnetic Calorimeter, *Nucl. Instr. Meth.* **A494** (2002) 303–7

[21]	A. Abashian *et al.* (Belle collaboration), The Belle Detector, *Nucl. Instr. Meth.* **A479** (2002) 117–232

[22]	P.N. Shanahan, *The Performance of a New CsI Calorimeter for the KTeV Experiment at Fermilab*, Proc. of the Sixth Int. Conf. on Calorimetry in High Energy Physics, Frascati, 8–14 June 1996, printed in Frascati Physics Series, Vol. VI, Frascati (1996); V. Prasad, Performance of the Cesium Iodide Calorimeter at the KTeV Experiment at Fermilab, *Nucl. Instr. Meth.* **A461** (2001) 341–3

[23] B.A. Shwartz, *Performance and Upgrade Plans of the Belle Calorimeter*, Proc. of the 10th Int. Conf. on Calorimetry in Particle Physics, Pasadena, 25–29 March 2002, 182–6

[24] H. Ikeda *et al.*, A Detailed Test of the CsI(Tl) Calorimeter for Belle with Photon Beams of Energy between 20 MeV and 5.4 GeV, *Nucl. Instr. Meth.* **A441** (2000) 401–26

[25] *The Compact Muon Solenoid Technical Proposal*, CERN/LHCC **94-38** (1994)

[26] *9. CMS ECAL Technical Design Report*, CERN/LHCC **97-33** (1997)

[27] K. Deiters *et al.*, Properties of the Avalanche Photodiodes for the CMS Electromagnetic Calorimeter, *Nucl. Instr. Meth.* **A453** (2000) 223–6

[28] A. Ghezzi, Recent Testbeam Results of the CMS Electromagnetic Calorimeter, *Nucl. Phys. B Proc. Suppl.* **B150** (2006) 93–7

[29] http://opal.web.cern.ch/Opal/

[30] SELEX collaboration (M.Y. Balatz *et al.*), *The Lead-Glass Electromagnetic Calorimeter for the SELEX Experiment*, FERMILAB-TM-2252 (July 2004) 42pp.; *Nucl. Instr. Meth.* **A545** (2005) 114–38

[31] T. Doke, A Historical View on the R&D for Liquid Rare Gas Detectors, *Nucl. Instr. Meth.* **A327** (1993) 113–18

[32] A.A. Grebenuk, Liquid Noble Gas Calorimeters for KEDR and CMD-2M Detectors, *Nucl. Instr. Meth.* **A453** (2000) 199–204

[33] M. Jeitler, The NA48 Liquid-Krypton Calorimeter, *Nucl. Instr. Meth.* **A494** (2002) 373–7

[34] V.M. Aulchenko *et al.*, Investigation of Electromagnetic Calorimeter Based on Liquid Krypton, *Nucl. Instr. Meth.* **A289** (1990) 468–74

[35] V.M. Aulchenko *et al.*, The Test of the LKR Calorimeter Prototype at the Tagged Photon Beam, *Nucl. Instr. Meth.* **A394** (1997) 35–45

[36] U. Amaldi, Fluctuations in Calorimetric Measurements, *Phys. Scripta* **23** (1981) 409–24

[37] C.W. Fabjan, Calorimetry in High Energy Physics, in T. Ferbel (ed.), *Proceedings on Techniques and Concepts of High Energy Physics*, Plenum, New York (1985) 281; CERN-EP-85-54 (1985)

[38] D. Axen *et al.*, The Lead-Liquid Argon Sampling Calorimeter of the SLD Detector, *Nucl. Instr. Meth.* **A328** (1993) 472–94

[39] B. Aubert *et al.*, Construction, Assembly and Tests of the ATLAS Electromagnetic Barrel Calorimeter, *Nucl. Instr. Meth.* **A558** (2006) 388–93

[40] Scintillation materials, Catalogue, Kuraray Co. Ltd. (2000)

[41] Scintillation products, Scintillating optical fibers, Saint-Gobain brochure, Saint-gobain ceramics & plastics Inc. (2005)

[42] F. Barreiro *et al.*, An Electromagnetic Calorimeter with Scintillator Strips and Wavelength Shifter Read Out, *Nucl. Instr. Meth.* **A257** (1987) 145–54

[43] G.S. Atoian *et al.*, Development of Shashlyk Calorimeter for KOPIO, *Nucl. Instr. Meth.* **A531** (2004) 467–80

[44] The LHCb Collaboration, LHCb Calorimeters (Technical Design Report), CERN/LHCC 2000-0036, LHCb TDR 2 (2000) CERN, Geneve

[45] D. Acosta *et al.*, Lateral Shower Profiles in a Lead Scintillating-Fiber Calorimeter, *Nucl. Instr. Meth.* **A316** (1992) 184–201

[46] M. Adinolfi *et al.*, The KLOE Electromagnetic Calorimeter, *Nucl. Instr. Meth.* **A482** (2002) 364–86

[47] Y. Fujii, *Design Optimisation, Simulation, and Bench Test of a Fine-Granularity Tile/Fiber EM Calorimeter Test Module*, International Workshop on Linear Colliders (LCWS 2002), Jeju Island, Korea, 26–30 August 2002; published in 'Seogwipo 2002, Linear colliders' 588–91

[48] K. Hara *et al.*, Design of a 2 × 2 Scintillating Tile Calorimeter Package for the SDC Barrel Electromagnetic Tile/Fiber Calorimeter, *Nucl. Instr. Meth.* **A373** (1996) 347–57

[49] S. Aota *et al.*, A Scintillating Tile/Fiber System for the CDF Upgrade em Calorimeter, *Nucl. Instr. Meth.* **A352** (1995) 557–68

[50] B.M. Bleichert *et al.*, The Response of a Simple Modular Electron/Hadron Calorimeter to Electrons, *Nucl. Instr. Meth.* **199** (1982) 461–4

[51] T.C. Weekes, Very High Energy Gamma-Ray Astronomy, *Phys. Rep.* **160** (1988) 1–121

[52] T.A. Gabriel *et al.*, Energy Dependence of Hadronic Activity, *Nucl. Instr. Meth.* **A338** (1994) 336–47

[53] P. Amaral *et al.*, Hadronic Shower Development in Iron-Scintillator Tile Calorimetry, *Nucl. Instr. Meth.* **A443** (2000) 51–70

[54] M. Holder *et al.*, A Detector for High Energy Neutrino Interactions, *Nucl. Instr. Meth.* **148** (1978) 235–49

[55] M. Holder *et al.*, Performance of a Magnetized Total Absorption Calorimeter Between 15 GeV and 140 GeV, *Nucl. Instr. Meth.* **151** (1978) 69–80

[56] D.L. Cheshire *et al.*, Measurements on the Development of Cascades in a Tungsten-Scintillator Ionization Spectrometer, *Nucl. Instr. Meth.* **126** (1975) 253–62

[57] D.L. Cheshire *et al.*, Inelastic Interaction Mean Free Path of Negative Pions in Tungsten, *Phys. Rev.* **D12** (1975) 2587–93

[58] A. Grant, A Monte Carlo Calculation of High Energy Hadronic Cascades in Iron, *Nucl. Instr. Meth.* **131** (1975) 167–72

[59] B. Friend *et al.*, Measurements of Energy Flow Distributions of 10 GeV/c Hadronic Showers in Iron and Aluminium, *Nucl. Instr. Meth.* **136** (1976) 505–10

[60] R. Wigmans, *Advances in Hadron Calorimetry*, CERN-PPE-91-39 (1991)

[61] C. Leroy & P. Rancoita, Physics of Cascading Shower Generation and Propagation in Matter: Principles of High-Energy, Ultrahigh-Energy and Compensating Calorimetry, *Rep. Prog. Phys.* **63** (2000) 505–606

[62] A. Bernstein *et al.*, Beam Tests of the ZEUS Barrel Calorimeter, *Nucl. Instr. Meth.* **336** (1993) 23–52

[63] P. Puso *et al.*, ATLAS Calorimetry, *Nucl. Instr. Meth.* **A494** (2002) 340–5

[64] S. Denisov *et al.*, A Fine Grain Gas Ionization Calorimeter, *Nucl. Instr. Meth.* **A335** (1993) 106–12

[65] P. Baillon, *Detection of Atmospheric Cascades at Ground Level*, CERN-PPE-91-012 (1991)

[66] M. Kleifges and the Auger Collaboration, Status of the Southern Pierre Auger Observatory, *Nucl. Phys. B Proc. Suppl.* **150** (2006) 181–5

[67] High Resolution Fly's Eye: http://hires.physics.utah.edu/; www.telesco pearray.org/

[68] H. Falcke *et al.*, LOPES Collaboration, Detection and Imaging of Atmospheric Radio Flashes from Cosmic Ray Air Showers, *Nature* **435** (2005) 313–16

[69] C. Grupen *et al.*, Radio Detection of Cosmic Rays with LOPES, *Braz. J. Phys.* **36(4A)** (2006) 1157–64

[70] X. Gao, Y. Liu & S. Du, *Acoustic Detection of Air Shower Cores*, 19th Intern. Cosmic Ray Conf., Vol. 8 (1985) 333–6

[71] S. Barwick *et al.*, Neutrino Astronomy on the $1\,\mathrm{km}^2$ Scale, *J. Phys.* **G18** (1992) 225–47

[72] Y. Totsuka, Neutrino Astronomy, *Rep. Progr. Phys.* **55(3)** (1992) 377–430

[73] Chr. Spiering, Neutrinoastronomie mit Unterwasserteleskopen, *Phys. Bl.* **49(10)** (1993) 871–5

[74] M.N. Achasov *et al.*, Energy Calibration of the NaI(Tl) Calorimeter of the SND Detector Using Cosmic Muons, *Nucl. Instr. Meth.* **A401** (1997) 179–86

[75] E. Erlez *et al.*, Cosmic Muon Tomography of Pure Cesium Iodide Calorimeter Crystals, *Nucl. Instr. Meth.* **A440** (2000) 57–85

[76] K. Pretzl, Cryogenic Calorimeters in Astro and Particle Physics, *Nucl. Instr. Meth.* **A454** (2000) 114–27

[77] E. Previtali, 20 years of Cryogenic Particle Detectors: Past, Present and Future, *Nucl. Phys. B Proc. Suppl.* **A150** (2006) 3–8

[78] E. Fiorini, Introduction or 'Low-Temperature Detectors: Yesterday, Today and Tomorrow', *Nucl. Instr. Meth.* **A520** (2004) 1–3

[79] A. Nucciotti, Application of Cryogenic Detectors in Subnuclear and Astroparticle Physics, *Nucl. Instr. Meth.* **A559** (2006) 334–6

[80] T.O. Niinikoski, Early Developments and Future Directions in LTDs, *Nucl. Instr. Meth.* **A559** (2006) 330–3

[81] G. Waysand, A Modest Prehistory of Low-Temperature Detectors, *Nucl. Instr. Meth.* **520** (2004) 4–10

[82] K.P. Pretzl, Superconducting Granule Detectors, *Particle World* **1** (1990) 153–62

[83] V.N. Trofimov, *SQUIDs in Thermal Detectors of Weakly Interacting Particles*, Dubna-Preprint E8-91-67 (1991)

[84] C. Kittel, *Introduction to Solid State Physics*, 8th edition, Wiley Interscience, New York (2005); *Einführung in die Festkörperphysik*, Oldenbourg, München/Wien (1980)

[85] N.W. Ashcroft & N.D. Mermin, *Solid State Physics*, Holt-Saunders, New York (1976)

[86] K.H. Hellwege, *Einführung in die Festkörperphysik*, Springer, Berlin/Heidelberg/New York (1976)

[87] J.R. Primack, D. Seckel & B. Sadoulet, Detection of Cosmic Dark Matter, *Ann. Rev. Nucl. Part. Sci.* **38** (1988) 751–807

[88] I. Bavykina, *Investigation of ZnWO$_4$ Crystals as an Absorber in the CRESST Dark Matter Search*, Master Thesis, University of Siegen (March 2006)

[89] A. Alessandrello, A Massive Thermal Detector for Alpha and Gamma Spectroscopy, *Nucl. Instr. Meth.* **A440** (2000) 397–402

[90] D. Yvon et al., *Bolometer Development, with Simultaneous Measurement of Heat and Ionization Signals*, Saclay-Preprint CEN-DAPNIA-SPP 93-11 (1993)

[91] F. Petricca et al., CRESST: First Results with the Phonon-Light Technique, *Nucl. Instr. Meth.* **A559** (2006) 375–7

[92] E. Fiorini, Underground Cryogenic Detectors, *Europhys. News* **23** (1992) 207–9

[93] P. Meunier et al., Discrimination between Nuclear Recoils and Electron Recoils by Simultaneous Detection of Phonons and Scintillation Light, *Appl. Phys. Lett.* **75(9)** (1999) 1335–7

[94] M. Frank et al., Study of Single Superconducting Grains for a Neutrino and Dark Matter Detector, *Nucl. Instr. Meth.* **A287** (1990) 583–94

[95] S. Janós et al., The Bern Cryogenic Detector System for Dark Matter Search, *Nucl. Instr. Meth.* **A547** (2005) 359–67

[96] A. Gabutti et al., A Fast, Self-Recovering Superconducting Strip Particle Detector Made with Granular Tungsten, *Nucl. Instr. Meth.* **A312** (1992) 475–82

9

Particle identification

It is impossible to trap modern physics into predicting anything with perfect determinism because it deals with probabilities from the outset.

Sir Arthur Eddington

One of the tasks of particle detectors is, apart from measuring characteristic values like momentum and energy, to determine the identity of particles. This implies the determination of the mass and charge of a particle. In general, this is achieved by combining information from several detectors.

For example, the radius of curvature ρ of a charged particle of mass m_0 in a magnetic field supplies information on the momentum p and the charge z via the relation

$$\rho \propto \frac{p}{z} = \frac{\gamma m_0 \beta c}{z} \; . \tag{9.1}$$

The velocity $\beta = v/c$ can be obtained by time-of-flight measurements using

$$\tau \propto \frac{1}{\beta} \; . \tag{9.2}$$

The determination of the energy loss by ionisation and excitation can approximately be described by, see Chap. 1,

$$-\frac{\mathrm{d}E}{\mathrm{d}x} \propto \frac{z^2}{\beta^2} \ln(a\gamma\beta) \; , \tag{9.3}$$

where a is a material-dependent constant. An energy measurement yields

$$E_{\mathrm{kin}} = (\gamma - 1)m_0 c^2 \; , \tag{9.4}$$

273

since normally only the kinetic energy rather than the total energy is measured.

Equations (9.1) to (9.4) contain three unknown quantities, namely m_0, β and z; the Lorentz factor γ is related to the velocity β according to $\gamma = 1/\sqrt{1 - \beta^2}$. Three of the above-mentioned four measurements are in principle sufficient to positively identify a particle. In the field of elementary particle physics one mostly deals with singly charged particles ($z = 1$). In this case, two different measurements are sufficient to determine the particle's identity. For particles of high energy, however, the determination of the velocity does not provide sufficient information, since for all relativistic particles, independent of their mass, β is very close to 1 and therefore cannot discriminate between particles of different mass.

In large experiments all systems of a general-purpose detector contribute to *particle identification* by providing relevant parameters which are combined to joint likelihood functions (see Chap. 15). These functions are used as criteria to identify and distinguish different particles. In practice, the identification is never perfect. Let us assume that particles of type I should be selected in the presence of high background of particles of type II (pion versus kaon, electron versus hadron, etc.). Then any selection criterion is characterised by the identification efficiency $\varepsilon_{\mathrm{id}}$ for type I at certain probability p_{mis} to misidentify the particle of type II as type I.

9.1   Charged-particle identification

A typical task of experimental particle physics is to identify a charged particle when its momentum is measured by a magnetic spectrometer.

9.1.1   Time-of-flight counters

A direct way to determine the particle velocity is to measure its time of flight (TOF) between two points separated by a distance L. These two points can be defined by two counters providing 'start' and 'stop' signals or by the moment of particle production and a stop counter. In the latter case the 'start' signal synchronised with the beam–beam or beam–target collision can be produced by the accelerator system. A more detailed review of TOF detectors in high-energy particle experiments can be found in [1, 2].

Two particles of mass m_1 and m_2 have for the same momentum and flight distance L the time-of-flight difference

$$\Delta t = L \left(\frac{1}{v_1} - \frac{1}{v_2} \right) = \frac{L}{c} \left(\frac{1}{\beta_1} - \frac{1}{\beta_2} \right) \ . \tag{9.5}$$

Using $pc = \beta E$ we obtain

$$\Delta t = \frac{L}{pc^2}(E_1 - E_2) = \frac{L}{pc^2}\left(\sqrt{p^2c^2 + m_1^2c^4} - \sqrt{p^2c^2 + m_2^2c^4}\right) . \quad (9.6)$$

Since in this case $p^2c^2 \gg m_{1,2}^2c^4$, the expansion of the square roots leads to

$$\Delta t = \frac{Lc}{2p^2}(m_1^2 - m_2^2) . \quad (9.7)$$

Suppose that for a mass separation a significance of $\Delta t = 4\sigma_t$ is demanded. That is, a time-of-flight difference four times the time resolution is required. In this case a pion/kaon separation can be achieved up to momenta of $1\,\mathrm{GeV}/c$ for a flight distance of $1\,\mathrm{m}$ and a time resolution of $\sigma_t = 100\,\mathrm{ps}$, which can be obtained with, e.g., scintillation counters [1, 2]. For higher momenta the time-of-flight systems become increasingly long since $\Delta t \propto 1/p^2$.

At present the most developed and widely used technique for *TOF measurements* in high energy physics is based on plastic scintillation counters with PM-tube readout (see Sect. 5.4). A typical layout is shown in Fig. 9.1. The beam-crossing signal related to the interaction point starts the TDC (time-to-digital converter). The signal from the PM anode, which reads out the 'stop' counter, is fed to a discriminator, a device which generates a standard (logic) output pulse when the input pulse exceeds a certain threshold. The discriminator output is connected to the 'stop' input of the TDC. The signal magnitude is measured by an ADC (amplitude-to-digital converter). Since the moment of threshold crossing usually depends on the pulse height, a measurement of this value helps to make corrections in the off-line data processing.

The time resolution can be approximated by the formula

$$\sigma_t = \sqrt{\frac{\sigma_{\mathrm{sc}}^2 + \sigma_{\mathrm{l}}^2 + \sigma_{\mathrm{PM}}^2}{N_{\mathrm{eff}}} + \sigma_{\mathrm{el}}^2} , \quad (9.8)$$

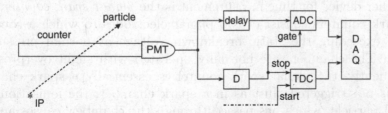

Fig. 9.1. The principle of time-of-flight measurements: IP – interaction point, D – discriminator, TDC – time-to-digital converter, ADC – amplitude-to-digital converter, DAQ – data-acquisition system.

where σ_{sc} is the contribution of the light-flash duration, σ_l the variation of the travel time due to different particle impact points and various
emission angles of scintillation photons, σ_{PM} the photoelectrons' transit-
time spread, N_{eff} the effective number of photoelectrons produced at the
PM photocathode, and σ_{el} is the electronics contribution to the time resolution. The quantity N_{eff} is usually smaller than the total number of
photoelectrons since some of them will arrive too late at the first dynode of the PM tube due to large emission angles to be useful for signal
generation. The total photoelectron number is given by Eq. (5.58), where
the deposited energy E_{dep} is proportional to the scintillator thickness. For
large-size counters the light attenuation length becomes crucial to obtain
a large N_{eff}.

For long counters the measured time depends on the point x where the
particle crosses the counter,

$$t_m = t_0 + \frac{x}{v_{eff}} \ , \qquad (9.9)$$

where v_{eff} is the effective light speed in the scintillator. To compensate
this dependence the scintillation bar is viewed from both edges. Then
the average of two measured times, $(t_1 + t_2)/2$, is – at least partially –
compensated. Further corrections can be applied taking into account the
impact coordinates provided by the tracking system.

The time resolution achieved for counters of 2–3 m length and $(5–10) \times$
$(2–5)\,cm^2$ cross section is about 100 ps [3–6]. Even better resolutions, 40–
60 ps, were reported for the TOF counters of the GlueX experiment [7].

Very promising results were reported recently for TOF counters based
on Cherenkov-light detection [8, 9]. The light flash in this case is extremely
short. Moreover, the variations in the photon path length can be kept
small in comparison to the scintillation light as all Cherenkov photons
are emitted at the same angle to the particle trajectory. To demonstrate
this, $4 \times 4 \times 1\,cm^3$ glass plates viewed by a microchannel plate (MCP)
PM tube (see Sect. 5.5) were used in [10]. A time resolution of about 6 ps
was achieved.

Another device for time measurement is the *planar spark counter*. Planar spark counters consist of two planar electrodes to which a constant
voltage exceeding the static breakdown voltage at normal pressure is
applied. The chambers are normally operated with slight overpressure.
Consequently, the planar spark counter is essentially a spark chamber
which is not triggered. Just as in a spark chamber, the ionisation of a
charged particle, which has passed through the chamber, causes an avalanche, which develops into a conducting plasma channel connecting the
electrodes. The rapidly increasing anode current can be used to generate
a voltage signal of very short rise time via a resistor. This voltage pulse

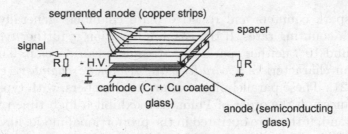

Fig. 9.2. Working principle of a planar spark counter [11, 12]. In many cases the anode is either coated with a semiconducting material or with a material of high specific bulk resistivity.

can serve as a very precise timing signal for the arrival time of a charged particle in the spark counter.

Figure 9.2 shows the working principle of a planar spark counter [1, 11, 12]. If metallic electrodes are used, the total capacitance of the chamber will be discharged in one spark. This may lead to damages of the metallic surface and also causes a low multitrack efficiency. If, however, the electrodes are made from a material with high specific bulk resistivity [13, 14], only a small part of the electrode area will be discharged via the sparks (Pestov counters). These do not cause surface damage because of the reduced current in the spark. A high multitrack efficiency is also guaranteed in this way. In addition to determining the arrival time of charged particles, the chamber also allows a spatial resolution if the anode is segmented. Noble gases with quenchers which suppress secondary spark formation are commonly used as gas filling.

Planar spark counters provide excellent time resolution ($\sigma_t \leq 30\,\mathrm{ps}$) if properly constructed [15]. This, however, requires narrow electrode gaps on the order of $100\,\mu\mathrm{m}$. The production of large-area spark counters, therefore, requires very precise machining to guarantee parallel electrodes with high surface quality.

Planar spark counters can also be operated at lower gas amplifications, and are then referred to as *resistive plate chambers* (RPCs), if, for example, instead of semiconducting electrode materials, graphite-covered glass plates are used. These chambers are most commonly operated in the streamer or in the avalanche mode [1, 16, 17]. Instead of graphite-covered glass plates other materials with suitable surface resistivity like Bakelite, a synthetic resin, can also be used. These resistive plate chambers also supply very fast signals and can – just as scintillation counters – be used for triggering with high time resolution. If the electrodes of the resistive plate chambers are segmented, they may also provide an excellent position resolution.

Planar spark counters and resistive plate chambers generally do not permit high counting rates. If the gas amplification is further reduced to values around 10^5, neither sparks nor streamers can develop. This mode of operation characterises a *parallel-plate avalanche chamber* (PPAC or PPC) [18–21]. These parallel-plate avalanche chambers, with typical electrode distances on the order of 1 mm, also exhibit a high time resolution ($\approx 500\,\mathrm{ps}$) and, if they are operated in the proportional mode, have as well an excellent energy resolution [22]. An additional advantage of parallel-plate avalanche chambers, compared to spark counters and resistive plate chambers, is that they can be operated at high counting rates because of the low gas amplification involved.

All these chamber types have in common that they provide excellent timing resolution due to the small electrode gaps. The present status of the counters with localised discharge and its applications is reviewed in [23].

9.1.2 Identification by ionisation losses

Since the specific ionisation energy loss depends on the particle energy, this can be used for identification (see Chap. 1). The average energy loss of electrons, muons, pions, kaons and protons in the momentum range between $0.1\,\mathrm{GeV}/c$ and $100\,\mathrm{GeV}/c$ in a 1 cm layer of argon–methane (80%:20%) is shown in Fig. 9.3 [24, 25]. It is immediately clear that a muon/pion separation on the basis of an energy-loss measurement is practically impossible, because they are too close in mass. However, a $\pi/K/p$ separation should be achievable. The logarithmic rise of the energy loss in gases ($\propto \ln\gamma$, see Eq. (1.11)) amounts to 50% up to 60% compared to the energy loss of minimum-ionising particles at a pressure of 1 atm [25, 26].

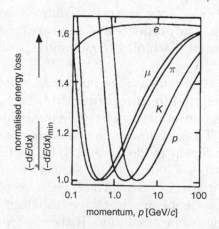

Fig. 9.3. Average energy loss of electrons, muons, pions, kaons and protons, normalised to the minimum-ionising value [24].

It should be noted that the relativistic rise of dE/dx is almost completely suppressed in solid-state materials by the density effect. Thus, solid-state detectors, like semiconductors or scintillators, can be used for dE/dx particle identification only in the low β range.

The key problem of *particle identification* by dE/dx is the fluctuation of the ionisation losses (see Chap. 1). A typical energy-loss distribution of $50\,\mathrm{GeV}/c$ pions and kaons in a layer of 1 cm argon–methane mixture (80%:20%) is sketched in Fig. 9.4 (left). The width of this distribution (FWHM) for gaseous media is in the range of 40%–100%. A real distribution measured for 3 GeV electrons in a thin-gap multiwire chamber is shown in Fig. 9.4 (right) [27]. To improve the resolution, multiple dE/dx measurements for the particles are used.

However, asymmetric energy-loss distributions with extended high-energy-loss tails render the direct averaging of the measured values inefficient. The origin of such long tails is caused by single δ electrons which can take away an energy ϵ_δ that is much larger than the average ionisation loss. The widely used '*truncated mean*' *method* implies an exclusion of a certain part (usually 30%–60%) of all individual energy-loss measurements with the largest values and averaging

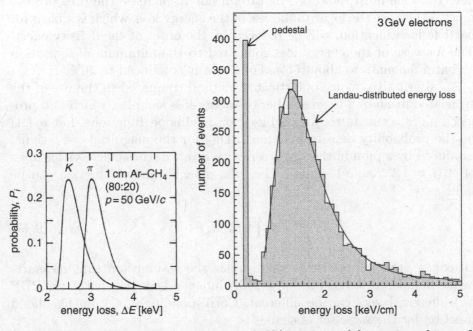

Fig. 9.4. Energy-loss distribution of $50\,\mathrm{GeV}/c$ pions and kaons in a layer of 1 cm argon and methane (left). The distribution measured for 3 GeV electrons in a thin-gap multiwire chamber (right) [27]. See also the discussion of this figure in Sect. 1.1.

over the remaining ones. This method excludes high energy transfers caused by the occasional production of energetic δ' electrons. Sometimes also the lowest $\mathrm{d}E/\mathrm{d}x$ measurements are discarded (typically 10%) to aim for a Gaussian-like energy-loss distribution. With about 100 $\mathrm{d}E/\mathrm{d}x$ measurements energy-loss resolutions of

$$\frac{\sigma(\mathrm{d}E/\mathrm{d}x)}{(\mathrm{d}E/\mathrm{d}x)} \approx (2\text{--}3)\% \tag{9.10}$$

for pions, kaons and protons of 50 GeV can be achieved [28].

The resolution can be improved by increasing the number N of individual measurements according to $1/\sqrt{N}$, which means, to improve the $\mathrm{d}E/\mathrm{d}x$ resolution by a factor of two, one has to take four times as many $\mathrm{d}E/\mathrm{d}x$ measurements. For a fixed total length of a detector, however, there exists an optimum number of measurements. If the detector is subdivided in too many $\mathrm{d}E/\mathrm{d}x$ layers, the energy loss per layer will eventually become too small, thereby increasing its fluctuation. Typically, the $\mathrm{d}E/\mathrm{d}x$ resolution for the drift chambers used in high energy physics experiments is in the range from 3% to 10% [2, 26].

The resolution should also improve with increasing gas pressure in the detector. One must, however, be careful not to increase the pressure too much, otherwise the logarithmic rise of the energy loss, which is a basis for particle identification, will be reduced by the onset of the density effect. The increase of the energy loss compared to the minimum of ionisation at 1 atm amounts to about 55%. For 7 atm it is reduced to 30%.

An alternative, more sophisticated method compared to the use of the truncated mean of a large number of energy-loss samples, which also provides more accurate results, is based on likelihood functions. Let $p_\pi(A)$ be the probability density function (PDF) for the magnitude of a signal produced by a pion in the single sensitive layer. Each particle yields a set of $A_i (i = 1, 2, \ldots, N)$ signals. Then the pion likelihood function can be built as

$$L_\pi = \prod_{i=1}^{N} p_\pi(A_i) . \tag{9.11}$$

Of course, this expression is valid under the assumption that measurements in different layers are statistically independent. In general, the PDF for different layers can be different. Correspondingly, a kaon likelihood function for the same set of signals is

$$L_K = \prod_{i=1}^{N} p_K(A_i) . \tag{9.12}$$

Then the most efficient parameter to choose one of the two alternative hypotheses on the type of the particle (pion or kaon) is a likelihood ratio as it was suggested by Neyman and Pearson (see, e.g., [29] for details):

$$R_L = \frac{L_\pi}{L_\pi + L_K} \ . \tag{9.13}$$

The likelihood-ratio method is rather time consuming, but it uses all available information and provides better results compared to the truncated-mean method.

Figure 9.5 shows the results of energy-loss measurements in a mixed particle beam [24]. This figure very clearly shows that the method of particle separation by dE/dx sampling only works either below the minimum of ionisation ($p < 1\,\mathrm{GeV}/c$) or in the relativistic-rise region. The identification by the 'truncated mean' method in various momentum ranges is illustrated by Fig. 9.6. These results were obtained with the ALEPH TPC which produced up to 344 measurements per track. A '60% truncated mean' was used providing about 4% for the resolution $\sigma(dE/dx)/(dE/dx)$ [30].

9.1.3 Identification using Cherenkov radiation

The main principles of Cherenkov counters are described in Sect. 5.6. This kind of device is widely used for particle identification in high energy physics experiments. The gaseous threshold counters are often exploited

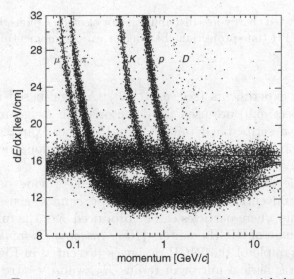

Fig. 9.5. Energy-loss measurements in a mixed particle beam [24].

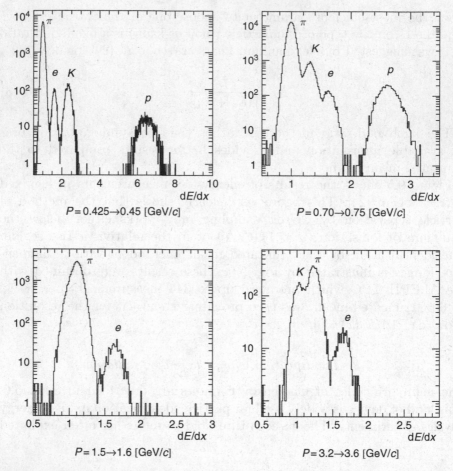

Fig. 9.6.   Truncated energy-loss distributions for electrons, pions, kaons and protons in the ALEPH time-projection chamber in various momentum ranges [31].

in fixed-target experiments (see, e.g., [32]). Aerogel-based multielement systems (see Sect. 5.6) are used for detectors with 4π geometry. One of such systems is described in detail in Chap. 13. Other examples are considered in [33–35]. Counters of this type can provide a pion/kaon separation up to 2.5–3 GeV.

Although the differential Cherenkov counters provide better particle identification, conventional differential counters cannot be used in storage-ring experiments where particles can be produced over the full solid angle. This is the domain of RICH counters (*Ring Imaging Cherenkov counters*) [36, 37]. An example of the RICH design is presented in Fig. 9.7 [38]. In this example a spherical mirror of radius R_S, whose centre of curvature coincides with the interaction point, projects the cone of Cherenkov light

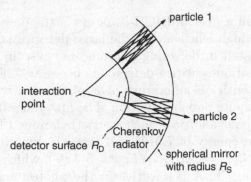

Fig. 9.7.   Working principle of a RICH counter [38].

produced in the radiator onto a ring on the surface of a spherical detector of radius R_D (see Fig. 9.7).

The radiator fills the volume between the spherical surfaces with radii R_S and R_D. In general, one takes $R_D = R_S/2$, since the focal length f of a spherical mirror is $R_S/2$. Because all Cherenkov photons are emitted at the same angle θ_c with respect to the particle trajectory pointing away from the sphere centre, all of them will be focussed to the thin detector ring on the inner sphere. One can easily calculate the radius of the Cherenkov ring on the detector surface,

$$r = f \cdot \theta_c = \frac{R_S}{2} \cdot \theta_c \ . \tag{9.14}$$

The measurement of r allows one to determine the particle velocity via

$$\cos\theta_c = \frac{1}{n\beta} \ \rightarrow \ \beta = \frac{1}{n \cos\left(\frac{2r}{R_S}\right)} \ . \tag{9.15}$$

It should be noted that many other designs exist, for example [39–43]. As Cherenkov radiators, heavy gases, like freons, or UV-transparent crystals, for example CaF_2 or LiF, are typically used.

If the momentum of the charged particle is already known, e.g. by magnetic deflection, then the particle can be identified (i.e. its mass m_0 determined) from the size of the Cherenkov ring, r. From Eq. (9.15) the measurement of r yields the particle velocity β, and by use of the relation

$$p = \gamma m_0 \beta c = \frac{m_0 c \beta}{\sqrt{1 - \beta^2}} \tag{9.16}$$

the mass m_0 can be determined.

The most crucial aspect of RICH counters is the detection of Cherenkov photons with high efficiency on the large detector surface. Since one is not only interested in detecting photons, but also in measuring their coordinates, a position-sensitive detector is necessary. Multiwire proportional chambers, with an admixture of a photosensitive vapour in the counter gas, are a quite popular solution. The first generation of the RICH detectors used vapour additions such as triethylamine (TEA: $(C_2H_5)_3N$) with an ionisation energy of 7.5 eV and tetrakis-dimethylaminoethylene (TMAE: $[(CH_3)_2N]_2C = C_5H_{12}N_2$; $E_{ion} = 5.4$ eV), which yields 5–10 photoelectrons per ring. TEA is sensitive in the photon energy range from 7.5 eV to 9 eV which requires a crystal window like CaF_2 or LiF, while TMAE photo-ionisation occurs by photons of 5.5 eV to 7.5 eV allowing the work with quartz windows. Figure 9.8 shows the pion/kaon separation in a RICH counter at 200 GeV/c. For the same momentum kaons are slower

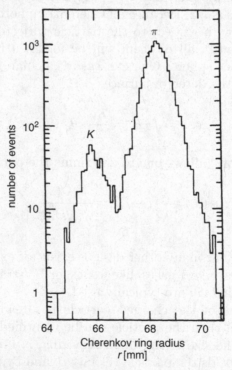

Fig. 9.8. Distribution of Cherenkov ring radii in a pion–kaon beam at 200 GeV/c. The Cherenkov photons have been detected in a multiwire proportional chamber filled with helium (83%), methane (14%) and TEA (3%). Calcium-fluoride crystals (CaF_2 crystal), having a high transparency in the ultraviolet region, were used for the entrance window [44].

compared to pions, and consequently produce, see Eqs. (9.14) and (9.15), Cherenkov rings with smaller radii [44].

Better *Cherenkov rings* are obtained from fast heavy ions, because the number of produced photons is proportional to the square of the projectile charge. Figure 9.9 [45] shows an early measurement of a Cherenkov ring produced by a relativistic heavy ion. The centre of the ring is also visible since the ionisation loss in the photon detector leads to a high energy deposit at the centre of the ring (see Fig. 9.7). Spurious signals, normally not lying on the Cherenkov ring, are caused by δ rays, which are produced in interactions of heavy ions with the chamber gas.

Figure 9.10 [46] shows an example of Cherenkov rings obtained by superimposing 100 collinear events from a monoenergetic collinear particle beam. The four square contours show the size of the calcium-fluoride

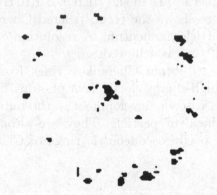

Fig. 9.9. Cherenkov ring of a relativistic heavy ion in a RICH counter [45].

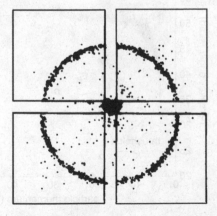

Fig. 9.10. Superposition of Cherenkov rings of 100 collinear events in a RICH counter. The square contours indicate the calcium-fluoride entrance windows of the photon detector [46].

crystals ($10 \times 10 \, \text{cm}^2$ each), which served as the entrance window for the photon detector. The ionisation loss of the particles is also seen at the centre of the Cherenkov rings.

At present TEA and TMAE are still used widely as photo-converters, but solid CsI photocathodes become popular in RICH detectors. In addition to the gaseous or crystal radiators recently aerogel came in use as a Cherenkov medium. In modern RICH projects single- and multi-anode conventional PM tubes as well as hybrid PM tubes are often used (see reviews [25, 43, 47, 48] and references therein). Micropattern gaseous detectors (see Sect. 7.4) with a CsI photocathode are also good candidates as photon sensors for RICH systems. Modern RICH detectors are characterised by a number of photoelectrons in the range of 10–30 per ring and a resolution on the Cherenkov angle of $\sigma_{\theta_c} \approx 3$–5 mrad [40, 49]. Figure 9.11 (left) exhibits two intersecting Cherenkov rings measured by a system of multichannel PMTs in the HERA-B RICH detector [50]. The right part of this figure shows the reconstructed Cherenkov angle in its dependence on the particle momentum. A resolution of $\sigma_{\theta_c} \approx 1$ mrad for momenta exceeding 10 GeV is achieved.

It is even possible to obtain Cherenkov rings from electromagnetic cascades initiated by high-energy electrons or photons. The secondary particles produced during cascade development in the radiator follow closely the direction of the incident particle. They are altogether highly relativistic and therefore produce concentric rings of Cherenkov light with

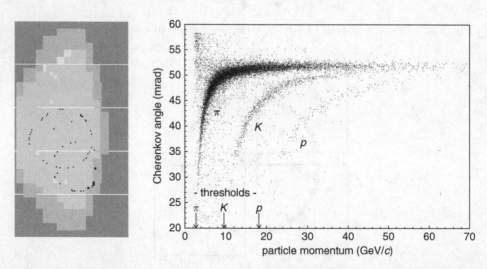

Fig. 9.11.   Left: two intersecting Cherenkov rings measured by the system of multichannel PMTs in the HERA-B RICH detector. Right: reconstructed Cherenkov angle in the same detector [50].

Fig. 9.12. Cherenkov ring produced by a high-energy (5 GeV) electron [51].

equal radii lying on top of one another. Figure 9.12 shows a distinct Cherenkov ring produced by a 5 GeV electron [51]. The large number of produced Cherenkov photons can be detected via the photoelectric effect in a position-sensitive detector.

The shape and position of such Cherenkov rings (elliptically distorted for inclined angles of incidence) can be used to determine the direction of incidence of high-energy gamma rays in the field of gamma-ray astronomy [52], where high-energy photons from cosmic-ray sources induce electromagnetic cascades in the Earth's atmosphere. Another example of particle identification by Cherenkov rings comes from neutrino physics. An important aspect of atmospheric neutrino studies is the correct identification of neutrino-induced muons and electrons. Figures 9.13 and 9.14 show a neutrino-induced event ($\nu_\mu + N \rightarrow \mu^- + X$) with subsequent (0.9 μs later) decay $\mu^- \rightarrow e^- + \bar{\nu}_e + \nu_\mu$ in the SNO experiment which contains a spherical vessel with 1000 tons of heavy water viewed by 10 000 PMTs [53]. The particle-identification capability of large-volume neutrino detectors is clearly seen.

A new generation of Cherenkov detectors uses the internal reflection in the radiator along with a PM-tube readout of the photons. The idea of the DIRC (Detector of Internally Reflected Cherenkov light), which was developed for particle identification in the BaBar detector [54], is illustrated in Fig. 9.15. The radiators of this detector are quartz bars of rectangular cross section. Most of the Cherenkov light generated by the particle is captured inside the bar due to internal reflection. The photon angle does not change during its travel with multiple reflections to the edge of the bar. After leaving the bar the photon is detected by PM tubes placed at some distance from the bar edge. Of course, quartz bars for this system should have the highest possible surface quality as well as a high accuracy of fabrication. The photon arrival time is measured as well – that helps to reject background hits.

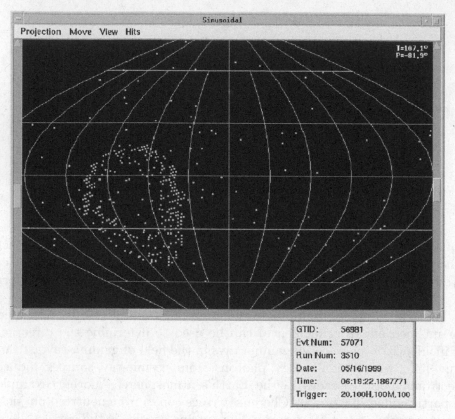

Fig. 9.13. Neutrino-induced muon in the SNO experiment [53].

The DIRC system of the BaBar detector contains 144 quartz bars that are 17 mm thick, 35 mm wide and 4.9 m long, which are viewed by the 896 PM tubes. The number of detected Cherenkov photons varies from 20 to 50 depending on the track's polar angle. This allows a reliable pion/kaon separation from 1 GeV to 4 GeV as shown in Fig. 9.16 [54, 55].

As discussed above, the basic DIRC idea is to measure two coordinates of the photons leaving the quartz bar, one of which is given by the end face of the quartz bar and the other by the impact position of the photon on the photon detector. However, to determine the Cherenkov angle, also two different variables, one spatial coordinate and the photon's time of propagation, are sufficient since one knows the particle track position and direction from the tracking system. This is the main idea of further developments of the Cherenkov-ring technique called the time-of-propagation (TOP) counter [56]. This device is quite promising and much more compact than a DIRC, however, it requires ultimate time resolution for single photons, better than 50 ps.

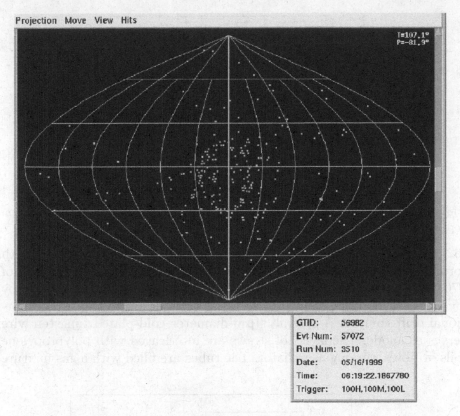

GTID: 56982
Evt Num: 57072
Run Num: 3510
Date: 05/16/1999
Time: 06:19:22.1867780
Trigger: 100H,100M,100L

Fig. 9.14. Cherenkov ring produced by an electron from muon decay, where the muon was created by a muon neutrino [53].

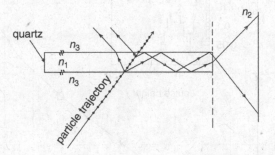

Fig. 9.15. The working principle of the DIRC counter [54].

9.1.4 Transition-radiation detectors

The effect of *transition radiation* [57] is used for high-energy particle identification in many current and planned experiments [58–62].

Let us consider as an example the ATLAS transition-radiation tracker (TRT). This sophisticated system is the largest present-day TRD detector

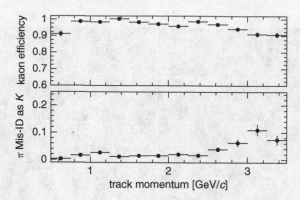

Fig. 9.16. Efficiency (top) and misidentification probability for pion/kaon separation (bottom) [55].

[62]. The TRT is part of the ATLAS inner detector and it is used both for charged-particle tracking and electron/pion separation. It consists of 370 000 cylindrical drift tubes (straws). Made from kapton and covered by a conductive film, the straw tube serves as cathode of a cylindrical proportional drift counter. A central 30 μm-diameter gold-plated tungsten wire serves as anode. The layers of straws are interleaved with polypropylene foils or fibres working as radiator. The tubes are filled with a gas mixture

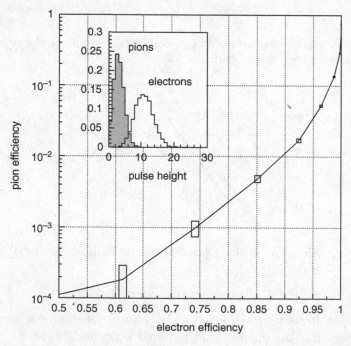

Fig. 9.17. Electron/pion separation capability measured with a prototype. The insert shows the energy depositions in a single straw for pions and electrons [63].

70% Xe + 27% CO_2 + 3% O_2, which provides a high X-ray absorption and proper counting characteristics.

The coordinate determination is performed by a drift-time measurement resulting in a spatial resolution of about 130 μm. The electron/pion separation is based on the energy deposition. A typical energy of the transition-radiation photon in the TRT is 8–10 keV, while a minimum-ionising particle deposits in one straw about 2 keV on average (see Fig. 9.17, left). As separation parameter the number of straws along the particle track having an energy deposition exceeding a certain threshold can be defined. Figure 9.18 shows a simulated event with a decay

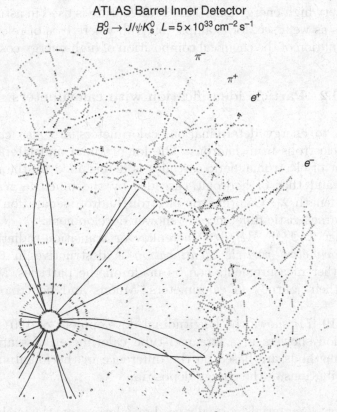

ATLAS Barrel Inner Detector

$B_d^0 \rightarrow J/\psi K_s^0 \quad L = 5 \times 10^{33}\,\mathrm{cm^{-2}\,s^{-1}}$

Fig. 9.18. A simulated event with a decay $B_d^0 \rightarrow J/\psi\,K_s$, where $J/\psi \rightarrow e^+ e^-$ and $K_s \rightarrow \pi^+ \pi^-$. Solid lines are reconstructed tracks beyond the TRT. Pion tracks are characterised by low energy depositions while electron tracks exhibit many straws with high energy deposition (black points > 6 keV, transition-radiation hits) [62]. It is also visible that low-energy δ electrons produced in ionisation processes with large energy transfers create high energy deposits because of the $1/\beta^2$ dependence of the ionisation energy loss. These unwanted 'transition-radiation hits' will complicate the pattern and particle identification.

$B_d^0 \to J/\psi\, K_s$, where $J/\psi \to e^+ e^-$ and $K_s \to \pi^+ \pi^-$. It is observed that the number of high-energy hits along electron tracks is larger than for pion tracks. The separation efficiency measured with a prototype straw chamber is presented in Fig. 9.17. For 90% electron efficiency, the probability of pion misidentification as electrons was measured to be 1.2% [63, 64].

TRD detectors are used rather widely for cosmic-ray experiments especially for the measurements above the Earth's atmosphere. For these experiments devices with a large sensitive area and low weight are required which are well met by TRDs [65]. Examples of such TRDs used or planned in the experiments HEAT, PAMELA and AMS can be found in [66–68]. It should be noted that the number of transition-radiation photons increases with z^2 of the particle, which makes it useful for the detection and identification of very high-energy ions. This TRD feature is used in astroparticle experiments as well; see, for example [69, 70], where it is of relevance for the determination of the chemical composition of high-energy cosmic rays.

9.2   Particle identification with calorimeters

In addition to energy determination, calorimeters are also capable of separating electrons from hadrons. The longitudinal and lateral shower development of electromagnetic cascades is determined by the radiation length X_0, and that of hadronic cascades by the much larger nuclear interaction length λ_I. Calorimetric electron/hadron separation is based on these characteristic differences of shower development.

In contrast to TOF, $\mathrm{d}E/\mathrm{d}x$, Cherenkov or transition-radiation techniques, *calorimetric particle identification* is destructive in the sense that no further measurements can be made on the particles. Most particles end their journey in calorimeters. Muons and neutrinos are an exception.

Figure 9.19 [71] shows the longitudinal development of 100 GeV electron and pion showers in a streamer-tube calorimeter. Essentially, the separation methods are based on the difference in the longitudinal and lateral distributions of the energy deposition.

- Since for all materials normally used in calorimeters the nuclear interaction length λ_I is much larger than the radiation length X_0, electrons interact earlier in the calorimeter compared to hadrons. Thus, electrons deposit the largest fraction of their energy in the front part of a calorimeter. Usually, the electromagnetic and hadron calorimeters are separated and the ratio of the energy deposited in the electromagnetic calorimeter to the particle momentum serves as an electron/hadron separation parameter. In case of a longitudinal segmentation of the

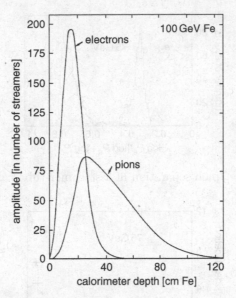

Fig. 9.19.   Comparison of the longitudinal development of 100 GeV pions and electrons in a streamer-tube calorimeter [71].

calorimeter, the starting point of the shower development can be used as an additional separation criterion.

- Hadronic cascades are much wider compared to electromagnetic showers (see Figs. 8.6 and 8.19). In a compact iron calorimeter 95% of the electromagnetic energy is contained in a cylinder of 3.5 cm radius. For hadronic cascades the 95%-lateral-containment radius is larger by a factor of about five, depending on the energy. From the different lateral behaviour of electromagnetic and hadronic cascades a typical characteristic compactness parameter can be derived.
- Finally, the longitudinal centre of gravity of the shower can also be used as an electron/hadron separation criterion.

Each separation parameter can be used to define a likelihood function corresponding to the electron or pion hypothesis in an unseparated electron–pion beam. The combined likelihood function including functions for all separation parameters allows to obtain much better electron/pion separation in calorimeters. One must take into account, however, that the separation criteria may be strongly correlated. Figure 9.20 [72, 73] shows such combined parameter distributions exhibiting only a small overlap between the electron and pion hypothesis. The resulting e/π *misidentification* probability for a given electron efficiency is shown in Fig. 9.21 [72, 73]. For a 95% electron acceptance one obtains in this example a 1% pion contamination for a particle energy of 75 GeV. With more sophisticated

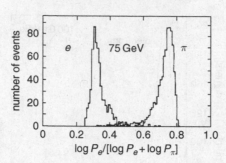

Fig. 9.20. Electron/pion separation in a streamer-tube calorimeter [72, 73].

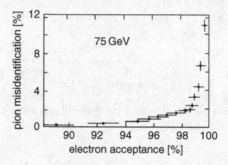

Fig. 9.21. Electron/pion misidentification probability in a streamer-tube calorimeter [72, 73]. The electron acceptance is the fraction of electrons accepted by a cut in the probability distribution. Correspondingly, the pion misidentification represents the fraction of accepted electron candidates that are really pions.

calorimeters a pion contamination as low as 0.1% can be reached with calorimetric methods.

Figure 9.22 demonstrates the separation capability of a crystal calorimeter for low-energy particles. The data were taken with the CMD-2 detector, in which the processes $e^+e^- \to e^+e^-, \mu^+\mu^-, \pi^+\pi^-$ were studied at a centre-of-mass energy of about 0.8 GeV. The two-dimensional plot presents the energy for final-state particles measured in the CsI-crystal electromagnetic calorimeter [74]. e^+e^- events concentrate in the upper right-hand corner while minimum-ionising particles, $\mu^+\mu^-$, $\pi^+\pi^-$, and a small admixture of the cosmic-ray background populate the lower left-hand area. The $\pi^+\pi^-$ distribution has long tails to higher energies due to pion nuclear interactions. The electrons are well separated from other particles. One can note that the event-separation quality strongly improves when one has two particles of the same type.

High-energy muons can be distinguished not only from pions but also from electrons by their low energy deposition in calorimeters and by their longer range. Figure 9.23 [71] shows the amplitude distributions of 50 GeV

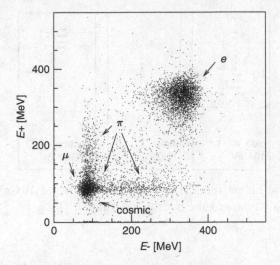

Fig. 9.22. The energies of final particles in $e^+e^- \rightarrow e^+e^-, \mu^+\mu^-, \pi^+\pi^-$ processes, measured by the CsI-crystal calorimeter of the CMD-2 detector [74]. The centre-of-mass energy of the experiment is about 0.8 GeV.

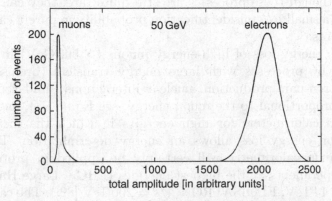

Fig. 9.23. Amplitude distribution of 50 GeV electrons and muons in a streamer-tube calorimeter [71].

electrons and muons. The possibility of an excellent electron/muon separation is already evident from this diagram.

The digital hit pattern of a 10 GeV pion, muon and electron in a streamer-tube calorimeter is shown in Fig. 9.24 [75]. Detectors operating at energies below 10–20 GeV are often equipped with a muon-range system instead of a hadron calorimeter. This system usually consists of absorber plates alternating with sensitive layers (see, e.g., Chap. 13). Then the particle of known momentum is identified by comparing the measured

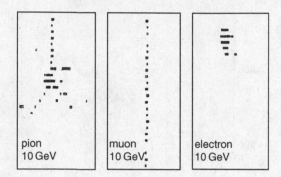

Fig. 9.24. Digital hit patterns (i.e. dot = fired tube) of 10 GeV pions, muons and electrons in a streamer-tube hadron calorimeter [75].

range with the one expected for a muon as well as by the lateral hit pattern.

For higher beam energies the interaction probability of muons for processes with higher energy transfers, e.g. by *muon bremsstrahlung*, increases [76–81]. Although these processes are still quite rare, they can nevertheless lead to a small μ/e misidentification probability in purely calorimetric measurements.

Since the energy loss of high-energy muons ($> 500\,\mathrm{GeV}$) in matter is dominated by processes with large energy transfers (bremsstrahlung, direct electron-pair production, nuclear interactions), and these energy losses are proportional to the muon energy, see Eq. (1.74), one can even build muon calorimeters for high energies in which the measurement of the muon energy loss allows an energy determination. This possibility of muon calorimetry will certainly be applied in proton–proton collision experiments at the highest energies (LHC – Large Hadron Collider, $\sqrt{s} = 14\,\mathrm{TeV}$; ELOISATRON, $\sqrt{s} = 200\,\mathrm{TeV}$ [82]). The calorimetric method of muon energy determination can also be employed in deep-water and ice experiments used as neutrino telescopes.

9.3   Neutron detection

Depending on the energy of the neutrons, different detection techniques must be employed. Common to all methods is that charged particles have to be produced in neutron interactions, which then are seen by the detector via 'normal' interaction processes like, e.g. ionisation or the production of light in scintillators [83–85].

For low-energy neutrons ($E_n^{\text{kin}} < 20\,\text{MeV}$) the following conversion reactions can be used:

$$n + {}^6\text{Li} \rightarrow \alpha + {}^3\text{H}\ , \tag{9.17}$$

$$n + {}^{10}\text{B} \rightarrow \alpha + {}^7\text{Li}\ , \tag{9.18}$$

$$n + {}^3\text{He} \rightarrow p + {}^3\text{H}\ , \tag{9.19}$$

$$n + p \rightarrow n + p\ . \tag{9.20}$$

The cross sections for these reactions depend strongly on the neutron energy. They are plotted in Fig. 9.25 [85].

For energies between $20\,\text{MeV} \leq E_n \leq 1\,\text{GeV}$ the production of recoil protons via the elastic (n, p) scattering can be used for neutron detection, Eq. (9.20). Neutrons of high energy ($E_n > 1\,\text{GeV}$) produce hadron cascades in inelastic interactions which are easy to identify.

To be able to distinguish neutrons from other particles, a *neutron counter* basically always consists of an anti-coincidence counter, which vetoes charged particles, and the actual neutron detector.

Thermal neutrons ($E_n \approx \frac{1}{40}\,\text{eV}$) are easily detected with ionisation chambers or proportional counters, filled with boron-trifluoride gas (BF_3). To be able to detect higher-energy neutrons also in these counters, the neutrons first have to be moderated, since otherwise the neutron interaction cross section would be too small (see Fig. 9.25). The moderation of non-thermal neutrons can best be done with substances containing many protons, because neutrons can transfer a large amount of energy to collision partners of the same mass. In collisions with heavy nuclei essentially only elastic scattering with small energy transfers occurs. Paraffin or water are preferred moderators. Neutron counters for non-thermal neutrons are

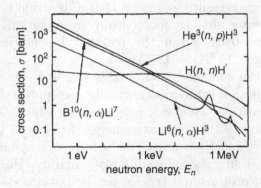

Fig. 9.25. Cross sections for neutron-induced reactions as a function of the neutron energy (1 barn $= 10^{-24}\,\text{cm}^2$) [85].

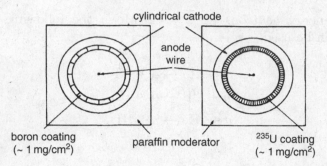

Fig. 9.26.　Neutron detection with proportional counters [83].

therefore covered with these substances. With BF_3 counters, neutron detection efficiencies on the order of 1% can be achieved.

Thermal neutrons can also be detected via a fission reaction (n, f) (f = fission). Figure 9.26 shows two special proportional counters which are covered on the inside with either a thin boron or uranium coating to induce the neutrons to undergo either (n, α) or (n, f) reactions [83]. To moderate fast neutrons these counters are mounted inside a paraffin barrel.

Thermal or *quasi-thermal neutrons* can also be detected with solid-state detectors. For this purpose, a lithium-fluoride (^{6}LiF) coating is evaporated onto the surface of a semiconductor counter in which, according to Eq. (9.17), α particles and tritons are produced. These can easily be detected by the solid-state detector.

Equally well europium-doped lithium-iodide scintillation counters LiI(Eu) can be used for neutron detection since α particles and tritons produced according to Eq. (9.17) can be measured via their scintillation light. Slow neutrons or neutrons with energies in the MeV range can be detected in multiwire proportional chambers filled with a gas mixture of ^{3}He and Kr at high pressure by means of the Reaction (9.19).

For *slow neutrons*, due to momentum conservation, ^{3}H and p are produced back to back. From the reaction kinematics one can find $E_p = 0.57\,\mathrm{MeV}$ and $E(^3\mathrm{H}) = 0.19\,\mathrm{MeV}$.

A typical neutron counter based on this reaction commonly employed in the field of radiation protection normally uses polyethylene spheres as moderator along with a ^{3}He-recoil proportional detector. Since the cross section for Reaction (9.19) is strongly energy-dependent, the performance and the sensitivity of such a counter can be improved by neutron absorbers in the moderator. Using special gas fillings – mainly ^{3}He/CH$_4$ are used – the yield of recoil protons and tritons can be optimised. The moderator parameters can be determined by appropriate simulation programs for neutron transport [86].

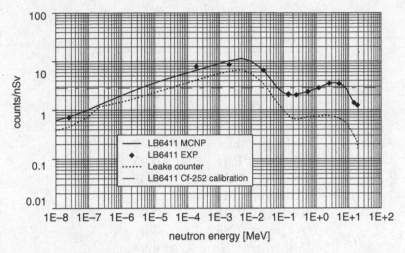

Fig. 9.27. Energy dependence of the neutron detection sensitivity of ^{3}He proportional counters with polyethylene shielding (from [87, 88]).

Typical sensitivities of several counts per nano-Sievert can be achieved with a scatter of ±30% for neutron energies between 50 keV and 10 MeV. For lower energies (10 meV to 100 eV) larger variations in sensitivity are unavoidable (Fig. 9.27, [87, 88]).

Due to the massive moderator the sensitivity for α, β or γ radiation is extremely small making such a ^{3}He counter ideally suited for reliable neutron measurements even in an environment of other radiation fields.

Possible applications are neutron dosimeters in nuclear power plants or hospitals where separate neutron-dose measurements are required because the *relative biological effectiveness* of neutrons is rather high compared to β and γ rays. It is also conceivable to search for illegal trafficking of radioactive neutron-emitting sources (such as weapon-grade plutonium) or for hidden sources which are otherwise difficult to detect, because α, β or γ rays can easily be shielded while neutrons cannot, providing a possibility to trace radioactive material [87, 88].

The elastic recoil reaction (9.20) can also be used in multiwire proportional chambers containing hydrogen-rich components (e.g. CH_4 + Ar). The size of a neutron counter should be large compared to the maximum range of the recoil protons: 10 cm in typical gases [89]. In solids the range of protons is reduced approximately in reverse proportion to the density (see Sect. 1.1.9).

Neutrons in the energy range 1–100 MeV can also be detected in organic scintillation counters via the production of recoil protons (i.e. via the H (n, n') H$'$ reaction) according to Eq. (9.20). However, the cross section

Table 9.1. *Threshold reactions for neutron energy measurements [83]*

Reaction	Threshold energy [MeV]
Fission of ^{234}U	0.3
Fission of ^{236}U	0.7
^{31}P (n,p) ^{31}Si	0.72
^{32}S (n,p) ^{32}P	0.95
Fission of ^{238}U	1.3
^{27}Al (n,p) ^{27}Mg	1.9
^{56}Fe (n,p) ^{56}Mn	3.0
^{27}Al (n,α) ^{24}Na	3.3
^{24}Mg (n,p) ^{24}Na	4.9
^{65}Cu $(n,2n)$ ^{64}Cu	10.1
^{58}Ni $(n,2n)$ ^{57}Ni	12.0

for this reaction decreases rapidly with increasing neutron energy (see Fig. 9.25) so that the neutron-detection efficiency is reduced. For neutrons of 10 MeV the np scattering cross section is about 1 barn. Then, for an organic scintillator of 1 cm thickness (density $\varrho = 1.2\,\text{g/cm}^3$ assumed) with a 30% molar fraction of free protons, a neutron-detection efficiency of about 2.5% is obtained.

In some applications – e.g. in the field of radiation protection – the measurement of the *neutron energy* is of great importance because the relative biological effectiveness of neutrons is strongly energy-dependent. The measurement of the neutron energy is frequently carried out with threshold detectors. Such a detector consists of a carrier foil covered with an isotope that only reacts with neutrons above a certain threshold energy. The particles or charged nuclei liberated in these reactions can be detected, e.g. in plastic detectors (cellulose-nitrate or cellulose-acetate foils) by an etching technique, and evaluated under a microscope or with automatic pattern-recognition methods (compare Sect. 6.11). Table 9.1 lists several threshold reactions used for neutron detection.

To cover different energy ranges of neutrons in a single exposure, one uses stacks of plastic foils coated with different isotopes. From the counting rates in the individual carrier foils with different energy thresholds, a rough determination of the neutron energy spectrum can be performed [83].

9.4 Problems

9.1 What are the Cherenkov angles for 3, 4 and $5\,\mathrm{GeV}/c$ pions in Lucite, silica aerogel, Pyrex and lead glass?

	index of refraction
Lucite	1.49
silica aerogel	1.025–1.075
Pyrex	1.47
lead glass	1.92

9.2 Calculate the Cherenkov energy emitted in water in the visible range (400–700 nm) per cm by a $2.2\,\mathrm{GeV}/c$ kaon!

9.3 How would you design a water Cherenkov detector that gives about 12 collected photoelectrons for $5\,\mathrm{GeV}/c$ protons?

Assume that the quantum efficiency of the used photomultiplier is 20%, the light collection efficiency to be 25% and the transfer probability from the photocathode to the first dynode to be 80%.

9.4 A $3\,\mathrm{GeV}/c$ proton is passing through Lucite. Estimate the number of visible photons emitted by δ rays using an approximation for $\mathrm{d}E/\mathrm{d}x$ over the relevant energy range? Assume a radiator thickness of $x = 10\,\mathrm{g/cm^2}$ ($\hat{=} 6.71\,\mathrm{cm}$).

9.5 The Cherenkov angle of relativistic particles in air ($n = 1.000295$) is $1.4°$. Still, in experiments with Imaging Air Cherenkov telescopes typical Cherenkov angles around $1°$ are reported. What is the reason for that?

9.6 In an experiment for particle identification the energy loss $\mathrm{d}E/\mathrm{d}x$ is measured in a $300\,\mu\mathrm{m}$ silicon counter and the energy is obtained from a total-absorption calorimeter. In a mixed beam of muons and pions of 10 MeV kinetic energy a product $\Delta E \cdot E_{\mathrm{kin}} = 5.7\,\mathrm{MeV^2}$ is obtained. Was this due to a muon or a pion?

($\varrho_{\mathrm{Si}} = 2.33\,\mathrm{g/cm^3}$, $Z_{\mathrm{Si}} = 14$, $A_{\mathrm{Si}} = 28$, $I_{\mathrm{Si}} \approx 140\,\mathrm{eV}$.)

The same setup is used to separate the beryllium isotopes $^7\mathrm{Be}$ and $^9\mathrm{Be}$ of 100 MeV kinetic energy with the result $\Delta E \cdot E_{\mathrm{kin}} = 3750\,\mathrm{MeV^2}$. Identify the beryllium isotope that produced this result. Why did $^8\mathrm{Be}$ not show up in this beam of beryllium isotopes?

($m(^7\mathrm{Be}) = 6.55\,\mathrm{GeV}/c^2$, $m(^9\mathrm{Be}) = 8.42\,\mathrm{GeV}/c^2$.)

References

[1]   W. Klempt, Review of Particle Identification by Time of Flight Techniques, *Nucl. Instr. Meth.* **A433** (1999) 542–53

[2]   M. Bonesini, *A Review of Recent Techniques of TOF Detectors*, Proc. 8th Int. Conf. on Advanced Technology and Particle Physics (ICATPP 2003): Astroparticle, Particle, Space Physics, Detectors and Medical Physics Applications, Como, Italy, 6–10 October 2003, Como 2003, Astroparticle, Particles and Space Physics, Detectors and Medical Physics Applications, 455–61

[3]   Y. Kubota *et al.*, The CLEO II Detector, *Nucl. Instr. Meth.* **A320** (1992) 66–113

[4]   H. Kichimi *et al.*, The BELLE TOF System, *Nucl. Instr. Meth.* **A453** (2000) 315–20

[5]   Ch. Paus *et al.*, Design and Performance Tests of the CDF Time-of-Flight System, *Nucl. Instr. Meth.* **A461** (2001) 579–81; S. Cabrera *et al.*, The CDF Time of Flight Detector, FERMILAB-CONF-03-404-E (2004)

[6]   G. Osteria *et al.*, The Time-of-Flight System of the PAMELA Experiment on Satellite, *Nucl. Instr. Meth.* **A535** (2004) 152–7

[7]   S. Denisov *et al.*, Characteristics of the TOF Counters for GlueX Experiment, *Nucl. Instr. Meth.* **A494** (2002) 495–9

[8]   M. Akatsu *et al.*, MCP-PMT Timing Property for Single Photons, *Nucl. Instr. Meth.* **A528** (2004) 763–75

[9]   Y. Enari *et al.*, Cross-talk of a Multi-anode PMT and Attainment of a sigma Approx. 10-ps TOF Counter, *Nucl. Instr. Meth.* **A547** (2005) 490–503

[10]   K. Inami *et al.*, A 5 ps TOF-Counter with an MCP-PMT, *Nucl. Instr. Meth.* **A560** (2006) 303–8

[11]   K. Kleinknecht, *Detektoren für Teilchenstrahlung*, Teubner, Stuttgart (1984, 1987, 1992); *Detectors for Particle Radiation*, Cambridge University Press, Cambridge (1986)

[12]   W. Braunschweig, Spark Gaps and Secondary Emission Counters for Time of Flight Measurement, *Phys. Scripta* **23** (1981) 384–92

[13]   M.V. Babykin *et al.*, Plane-Parallel Spark Counters for the Measurement of Small Times; Resolving Time of Spark Counters, *Sov. J. Atomic Energy* **IV** (1956) 627–34; Atomic Energy, Engineering, Physics and Astronomy and Russian Library of Science, Springer, **1(4)** (1956) 487–94

[14]   V.V. Parkhomchuck, Yu.N. Pestov & N.V. Petrovykh, A Spark Counter with Large Area, *Nucl. Instr. Meth.* **A93** (1971) 269–70

[15]   E. Badura *et al.*, Status of the Pestov Spark Counter Development for the ALICE Experiment, *Nucl. Instr. Meth.* **A379** (1996) 468–71

[16]   I. Crotty *et al.*, Investigation of Resistive Plate Chambers, *Nucl. Instr. Meth.* **A329** (1993) 133–9

[17]   E. Cerron-Zeballos *et al.*, A Very Large Multigap Resistive Plate Chamber, *Nucl. Instr. Meth.* **A434** (1999) 362–72

[18] A. Peisert, G. Charpak, F. Sauli & G. Viezzoli, Development of a Multistep Parallel Plate Chamber as Time Projection Chamber Endcap or Vertex Detector, *IEEE Trans. Nucl. Sci.* **31** (1984) 125–9

[19] P. Astier *et al.*, Development and Applications of the Imaging Chamber, *IEEE Trans. Nucl. Sci.* **NS-36** (1989) 300–4

[20] V. Peskov *et al.*, Organometallic Photocathodes for Parallel-Plate and Wire Chambers, *Nucl. Instr. Meth.* **A283** (1989) 786–91

[21] M. Izycki *et al.*, *A Large Multistep Avalanche Chamber: Description and Performance*, in Proc. 2nd Conf. on Position Sensitive Detectors, London (4–7 September 1990), *Nucl. Instr. Meth.* **A310** (1991) 98–102

[22] G. Charpak *et al.*, *Investigation of Operation of a Parallel Plate Avalanche Chamber with a CsI Photocathode Under High Gain Conditions*, CERN-PPE-91-47 (1991); *Nucl. Instr. Meth.* **A307** (1991) 63–8

[23] Yu.N. Pestov, Review on Counters with Localised Discharge, *Nucl. Instr. Meth.* **A494** (2002) 447–54

[24] J.N. Marx & D.R. Nygren, The Time Projection Chamber, *Physics Today* (October 1978) 46–53

[25] J. Va'vra, Particle Identification Methods in High-Energy Physics, *Nucl. Instr. Meth.* **A453** (2000) 262–78

[26] M. Hauschild, Progress in dE/dx Techniques Used for Particle Identification, *Nucl. Instr. Meth.* **A379** (1996) 436–41

[27] K. Affholderbach *et al.*, Performance of the New Small Angle Monitor for BAckground (SAMBA) in the ALEPH Experiment at CERN, *Nucl. Instr. Meth.* **A410** (1998) 166–75

[28] I. Lehraus *et al.*, Performance of a Large Scale Multilayer Ionization Detector and its Use for Measurements of the Relativistic Rise in the Momentum Range of 20–110 GeV/c, *Nucl. Instr. Meth.* **153** (1978) 347–55

[29] W.T. Eadie *et al.*, *Statistical Methods in Experimental Physics*, Elsevier–North Holland, Amsterdam/London (1971)

[30] W.B. Atwood *et al.*, Performance of the ALEPH Time Projection Chamber, *Nucl. Instr. Meth.* **A306** (1991) 446–58

[31] A. Ngac, Diploma Thesis, University of Siegen, 2000

[32] J.M. Link *et al.*, Cherenkov Particle Identification in FOCUS, *Nucl. Instr. Meth.* **A484** (2002) 270–86

[33] A.Yu. Barnyakov *et al.*, Development of Aerogel Cherenkov Detectors at Novosibirsk, *Nucl. Instr. Meth.* **A553** (2005) 125–9

[34] B. Tonguc *et al.*, The BLAST Cherenkov Detectors, *Nucl. Instr. Meth.* **A553** (2005) 364–9

[35] E. Cuautle *et al.*, Aerogel Cherenkov Counters for High Momentum Proton Identification, *Nucl. Instr. Meth.* **A553** (2005) 25–9

[36] J. Seguinot & T. Ypsilantis, Photo-Ionization and Cherenkov Ring Imaging, *Nucl. Instr. Meth.* **142** (1977) 377–91

[37] E. Nappi & T. Ypsilantis (eds.), *Experimental Techniques of Cherenkov Light Imaging*, Proc. of the First Workshop on Ring Imaging Cherenkov Detectors, Bari, Italy 1993; *Nucl. Instr. Meth.* **A343** (1994) 1–326

[38] C.W. Fabjan & H.G. Fischer, *Particle Detectors*, CERN-EP-80-27 (1980)

[39] K. Abe *et al.*, Operational Experience with SLD CRID at the SLC, *Nucl. Instr. Meth.* **A379** (1996) 442–3

[40] M. Artuso *et al.*, Construction, Pattern Recognition and Performance of the CLEO III LiF-TEA RICH Detector, *Nucl. Instr. Meth.* **A502** (2003) 91–100

[41] C. Matteuzzi, Particle Identification in LHCb, *Nucl. Instr. Meth.* **A494** (2002) 409–15

[42] H.E. Jackson, The HERMES Dual-Radiator RICH Detector, *Nucl. Instr. Meth.* **A502** (2003) 36–40

[43] J. Va'vra, *Cherenkov Imaging Techniques for the Future High Luminosity Machines*, SLAC-Pub-11019 (September 2003)

[44] T. Ekelöf, *The Use and Development of Ring Imaging Cherenkov Counters*, CERN-PPE-91-23 (1991)

[45] R. Stock, NA35-Collaboration, private communication (1990)

[46] F. Sauli, *Gas Detectors: Recent Developments and Applications*, CERN-EP-89-74 (1989); and Le Camere Proporzionali Multifili: Un Potente Instrumento Per la Ricera Applicata, *Il Nuovo Saggiatore* **2** (1986) 2/26

[47] J. Seguinot & T. Ypsilantis, Evolution of the RICH Technique, *Nucl. Instr. Meth.* **A433** (1999) 1–16

[48] F. Piuz, Ring Imaging CHerenkov Systems Based on Gaseous Photo-Detectors: Trends and Limits around Particle Accelerators, *Nucl. Instr. Meth.* **A502** (2003) 76–90

[49] F. Piuz *et al.*, Final Tests of the CsI-based Ring Imaging Detector for the ALICE Experiment, *Nucl. Instr. Meth.* **A433** (1999) 178–89

[50] I. Arino *et al.*, The HERA-B Ring Imaging Cherenkov Counter, *Nucl. Instr. Meth.* **A516** (2004) 445–61

[51] *Cherenkov Telescopes for Gamma-Rays*, CERN-Courier **28(10)** (1988) 18–20

[52] E. Lorenz, Air Shower Cherenkov Detectors, *Nucl. Instr. Meth.* **A433** (1999) 24–33

[53] Sudbury Neutrino Observatory home page: www.sno.phy.queensu.ca/sno/events/

[54] I. Adam *et al.*, The DIRC Particle Identification System for the BABAR Experiment, *Nucl. Instr. Meth.* **A538** (2005) 281–357

[55] J. Schwiening, The DIRC Detector at the SLAC B-factory PEP-II: Operational Experience and Performance for Physics Applications, *Nucl. Instr. Meth.* **A502** (2003) 67–75

[56] Y. Enari *et al.*, Progress Report on Time-of-Propagation Counter – A New Type of Ring Imaging Cherenkov Detector, *Nucl. Instr. Meth.* **A494** (2002) 430–5

[57] V.L. Ginzburg & I.M. Frank, Radiation of a Uniformly Moving Electron due to its Transitions from One Medium into Another, *JETP* **16** (1946) 15–29

[58] G. Bassompierre *et al.*, A Large Area Transition Radiation Detector for the NOMAD Experiment, *Nucl. Instr. Meth.* **A403** (1998) 363–82

[59] K.N. Barish *et al.*, TEC/TRD for the PHENIX Experiment, *Nucl. Instr. Meth.* **A522** (2004) 56–61

[60] www-alice.gsi.de/trd/

[61] J.D. Jackson, *Classical Electrodynamics*, 3rd edition, John Wiley & Sons, New York (1998); L.C.L. Yuan & C.S. Wu (eds.), *Methods of Experimental Physics*, Vol. 5A, Academic Press, New York (1961) 163; W.W.M. Allison & P.R.S. Wright, The Physics of Charged Particle Identification: dE/dx, Cherenkov Radiation and Transition Radiation, in T. Ferbel (ed.), *Experimental Techniques in High Energy Physics*, Addison-Wesley, Menlo Park, CA (1987) 371

[62] ATLAS Inner Detector Community, Inner detector. Technical design report, Vol. II, ATLAS TDR 5, CERN/LHCC 97-17 (1997); Fido Dittus, private communication (2006)

[63] V.A. Mitsou for the ATLAS collaboration, *The ATLAS Transition Radiation Tracker*, Proc. of 8th Int. Conf. on Advanced Technology and Particle Physics (ICATPP 2003), Astroparticle, Particle, Space Physics, Detectors and Medical Physics Applications, Como, Italy (6–10 October 2003) 497–501

[64] T. Akesson *et al.*, ATLAS Transition Radiation Tracker Test-Beam Results, *Nucl. Instr. Meth.* **A522** (2004) 50–5

[65] D. Müller, Transition Radiation Detectors in Particle Astrophysics, *Nucl. Instr. Meth.* **A522** (2004) 9–15

[66] S.W. Barwick *et al.*, The High-Energy Antimatter Telescope (HEAT): An Instrument for the Study of Cosmic-Ray Positrons, *Nucl. Instr. Meth.* **A400** (2004) 34–52

[67] M. Ambriola and PAMELA Collaboration, Performance of the Transition Radiation Detector of the PAMELA Space Mission, *Nucl. Phys. B Proc. Suppl.* **113** (2002) 322–8

[68] F.R. Spada, The AMS Transition Radiation Detector, *Int. J. Mod. Phys.* **A20** (2005) 6742–4

[69] J. L'Heureux *et al.*, A Detector for Cosmic-Ray Nuclei at Very High Energies, *Nucl. Instr. Meth.* **A295** (1990) 246–60

[70] F. Gahbauer *et al.*, A New Measurement of the Intensities of the Heavy Primary Cosmic-Ray Nuclei around 1 TeV amu, *Astrophys. J.* **607** (2004) 333–41

[71] R. Baumgart, *Messung und Modellierung von Elektron- und Hadron-Kaskaden in Streamerrohrkalorimetern*, Ph.D. Thesis, University of Siegen (1987)

[72] U. Schäfer, *Untersuchungen zur Magnetfeldabhängigkeit und Pion/Elektron Unterscheidung in Elektron-Hadron Kalorimetern*, Ph.D. Thesis, University of Siegen (1987)

[73] R. Baumgart *et al.*, Electron–Pion Discrimination in an Iron/Streamer Tube Calorimeter up to 100 GeV, *Nucl. Instr. Meth.* **272** (1988) 722–6

[74] R.R. Akhmetshin *et al.*, Measurement of $e^+e^- \rightarrow \pi^+\pi^-$ Cross-section with CMD-2 around ρ-meson, *Phys. Lett.* **B527** (2002) 161–72

[75] ALEPH Collaboration, D. Decamp *et al.*, ALEPH: A Detector for Electron–Positron Annihilations at LEP, *Nucl. Instr. Meth.* **A294** (1990) 121–78

[76] C. Grupen, Electromagnetic Interactions of High Energy Cosmic Ray Muons, *Fortschr. der Physik* **23** (1976) 127–209

[77] W. Lohmann, R. Kopp & R. Voss, *Energy Loss of Muons in the Energy Range 1 – 10.000 GeV*, CERN-85-03 (1985)

[78] W.K. Sakumoto *et al.*, *Measurement of TeV Muon Energy Loss in Iron*, University of Rochester UR-1209 (1991); *Phys. Rev.* **D45** (1992) 3042–50

[79] R. Baumgart *et al.*, Interaction of 200 GeV Muons in an Electromagnetic Streamer Tube Calorimeter, *Nucl. Instr. Meth.* **A258** (1987) 51–7

[80] C. Zupancic, *Physical and Statistical Foundations of TeV Muon Spectroscopy*, CERN-EP-85-144 (1985)

[81] M.J. Tannenbaum, *Comparison of Two Formulas for Muon Bremsstrahlung*, CERN-PPE-91-134 (1991)

[82] E. Nappi & J. Seguinot (eds.), *INFN Eloisatron Project: 42nd Workshop On Innovative Detectors For Supercolliders*, World Scientific, Singapore (2004)

[83] E. Sauter, *Grundlagen des Strahlenschutzes*, Siemens AG, Berlin/München (1971); *Grundlagen des Strahlenschutzes*, Thiemig, München (1982)

[84] W. Schneider, *Neutronenmeßtechnik*, Walter de Gruyter, Berlin/New York (1973)

[85] H. Neuert, *Kernphysikalische Meßverfahren*, G. Braun, Karlsruhe (1966)

[86] R.A. Forster & T.N.K. Godfrey, *MCNP – A General Monte Carlo Code for Neutron and Photon Transport. Version 3A*, ed. J.F. Briesmeister, Los Alamos, LA 739-6M, Rev. 2 (1986); J.S. Hendricks & J.F. Briesmeister, Recent MCNP Developments, *IEEE Trans. Nucl. Sci.* **39** (1992) 1035–40

[87] A. Klett, Plutonium Detection with a New Fission Neutron Survey Meter, *IEEE Trans. Nucl. Sci.* **46** (1999) 877–9

[88] A. Klett *et al.*, *Berthold Technologies, 3rd International Workshop on Radiation Safety of Synchrotron Radiation Sources*; RadSynch'04 SPring-8, Mikazuki, Hyogo, Japan, 17–19 November 2004

[89] P. Marmier, *Kernphysik I*, Verlag der Fachvereine, Zürich (1977)

10

Neutrino detectors

Neutrino physics is largely an art of learning a great deal by observing nothing.

Haim Harari

10.1 Neutrino sources

The detection of neutrinos is a challenge. Due to the smallness of neutrino interaction cross sections, neutrino detectors are required to be very massive to provide measurable rates. The cross section for neutrino–nucleon scattering of $10\,\mathrm{GeV}$ neutrinos is on the order of $7 \cdot 10^{-38}\,\mathrm{cm}^2/\mathrm{nucleon}$. Thus, for a target of $10\,\mathrm{m}$ of solid iron the interaction probability

$$R = \sigma \cdot N_\mathrm{A}\,[\mathrm{mol}^{-1}]/\mathrm{g} \cdot d \cdot \varrho \qquad (10.1)$$

(σ – nuclear cross section, N_A – Avogadro's number, d – target thickness, ϱ – density) is only

$$R = 7 \cdot 10^{-38}\,\mathrm{cm}^2 \cdot 6.023 \cdot 10^{23}\,\mathrm{g}^{-1} \cdot 10^3\,\mathrm{cm} \cdot 7.6\,\frac{\mathrm{g}}{\mathrm{cm}^3} = 3.2 \cdot 10^{-10} \ . \quad (10.2)$$

Therefore, it is very unlikely that neutrinos interact even in a massive detector. The situation is even worse for low-energy neutrinos. Solar neutrinos of $100\,\mathrm{keV}$ are associated with a cross section for neutrino–nucleon scattering of

$$\sigma(\nu_e N) \approx 10^{-45}\,\mathrm{cm}^2/\mathrm{nucleon} \ . \qquad (10.3)$$

The interaction probability of these neutrinos with our planet Earth at a central collision is only $\approx 4 \cdot 10^{-12}$. In addition to low cross sections threshold effects also play an important rôle. Energies in the several $100\,\mathrm{keV}$ range are below the threshold for inverse β decay ($\bar{\nu}_e + p \to n + e^+$), where a minimum antineutrino energy of 1.8 MeV is required to induce this reaction. To get measurable interaction rates therefore high neutrino fluxes and not too low energies are necessary.

Neutrinos are generated in *weak interactions* and decays [1–3]. Reactor neutrinos originate from nuclear β decay,

$$n \rightarrow p + e^- + \bar{\nu}_e \quad (\beta^- \text{ decay}) , \tag{10.4}$$

$$p \rightarrow n + e^+ + \nu_e \quad (\beta^+ \text{ decay}) , \tag{10.5}$$

$$p + e^- \rightarrow n + \nu_e \quad (\text{electron capture}) . \tag{10.6}$$

These neutrinos have typically MeV energies.

Stars produce energy by nuclear fusion, creating only electron-type neutrinos, mostly via proton–proton fusion

$$p + p \rightarrow d + e^+ + \nu_e , \tag{10.7}$$

but also in electron-capture reactions with ^{7}Be,

$$^7\text{Be} + e^- \rightarrow ^7\text{Li} + \nu_e \tag{10.8}$$

and decays from boron,

$$^8\text{B} \rightarrow ^8\text{Be} + e^+ + \nu_e . \tag{10.9}$$

Solar neutrinos range from the keV region to about 15 MeV.

Neutrinos are also copiously produced in the atmosphere in air showers initiated by primary cosmic-ray nuclei where mainly pions and kaons are the parents of muon neutrinos,

$$\pi^+ \rightarrow \mu^+ + \nu_\mu , \tag{10.10}$$

$$\pi^- \rightarrow \mu^- + \bar{\nu}_\mu , \tag{10.11}$$

$$K^+ \rightarrow \mu^+ + \nu_\mu , \tag{10.12}$$

$$K^- \rightarrow \mu^- + \bar{\nu}_\mu . \tag{10.13}$$

The decay of muons yields also electron-type neutrinos,

$$\mu^+ \rightarrow e^+ + \nu_e + \bar{\nu}_\mu , \tag{10.14}$$

$$\mu^- \rightarrow e^- + \bar{\nu}_e + \nu_\mu . \tag{10.15}$$

Atmospheric neutrinos can be very energetic ($\geq$ GeV).

Supernova explosions are a strong source of neutrinos. These neutrinos can originate from the deleptonisation phase where protons and electrons are merged,

$$p + e^- \rightarrow n + \nu_e , \tag{10.16}$$

producing only electron neutrinos, while all neutrino flavours are democratically generated via weak decays of virtual Zs made in e^+e^- interactions,

$$e^+e^- \rightarrow Z \rightarrow \nu_\alpha + \bar{\nu}_\alpha \quad (\alpha = e, \mu, \tau) \ . \tag{10.17}$$

High-energy neutrinos can be obtained from earthbound or cosmic accelerators in *beam-dump experiments*, where they are created in weak decays of short-lived hadrons. The total cross section for high-energy neutrinos rises linearly with energy until propagator effects from Z or W exchange saturate the cross section.

Finally, the Big Bang was a rich source of neutrinos which have cooled down during the expansion of the universe to a present temperature of $1.9\,\mathrm{K}$, corresponding to $\approx 0.16\,\mathrm{MeV}$ [4, 5].

10.2 Neutrino reactions

Neutrinos can be detected in weak interactions with nucleons. There are characteristic *charged-current interactions* for the different neutrino flavours,

$$\nu_e + n \rightarrow e^- + p \ , \tag{10.18}$$

$$\bar{\nu}_e + p \rightarrow e^+ + n \ , \tag{10.19}$$

$$\nu_\mu + n \rightarrow \mu^- + p \ , \tag{10.20}$$

$$\bar{\nu}_\mu + p \rightarrow \mu^+ + n \ , \tag{10.21}$$

$$\nu_\tau + n \rightarrow \tau^- + p \ , \tag{10.22}$$

$$\bar{\nu}_\tau + p \rightarrow \tau^+ + n \ . \tag{10.23}$$

The corresponding *neutral-current interactions* are not very helpful for neutrino detection since a large fraction of the energy is carried away by the final-state neutrino.

However, neutrinos can also be detected in neutral-current interactions with atomic electrons,

$$\nu_\alpha + e^- \rightarrow \nu_\alpha + e^- \quad (\alpha = e, \mu, \tau) \ , \tag{10.24}$$

where a fraction of the neutrino energy is transferred to the final-state electron, which can be measured. For antineutrinos such reactions are also possible, with the speciality that in $\bar{\nu}_e e^-$ scattering – if the centre-of-mass energy is sufficiently high – also muons and taus can be produced,

$$\bar{\nu}_e + e^- \rightarrow \mu^- + \bar{\nu}_\mu \ , \tag{10.25}$$

$$\bar{\nu}_e + e^- \rightarrow \tau^- + \bar{\nu}_\tau \ . \tag{10.26}$$

As already indicated, the cross sections for the different neutrino interactions are very small, in particular, for low-energy neutrinos. To circumvent this problem, the missing-energy or missing-momentum technique has been invented in elementary particle physics experiments, where the neutrino flavour and its four-momentum are inferred from the particles seen in the detector. This procedure requires the knowledge of the available energy for the reaction. If, e.g., a W pair is produced in e^+e^- collisions,

$$e^+ + e^- \rightarrow W^+ + W^- \; , \tag{10.27}$$

where one W decays hadronically $(W^- \rightarrow u\bar{d})$ and the other leptonically $(W^+ \rightarrow \mu^+ + \bar{\nu}_\mu)$, the energy and momentum of the $\bar{\nu}_\mu$ can be inferred from the four-momenta of all visible particles if the centre-of-mass energy of the collision is known. The flavour of the neutrino is obvious from the generated muon.

10.3   Some historical remarks on neutrino detection

Cowan and Reines [6] discovered the $\bar{\nu}_e$ in the 1950s via the reaction

$$\bar{\nu}_e + p \rightarrow e^+ + n \; , \tag{10.28}$$

where the positron was identified in the annihilation process

$$e^+ + e^- \rightarrow \gamma + \gamma \tag{10.29}$$

yielding two back-to-back photons of 511 keV each in delayed coincidence with photons originating from the γ decay of an excited nucleus after neutron capture. The neutrino detector used in this 'Poltergeist' experiment consisted of a large liquid-scintillation counter.

Muon neutrinos as distinct from electron neutrinos were first seen in the famous 'two-neutrino experiment' by Lederman, Schwartz, and Steinberger [7] in 1962 through the reaction

$$\nu_\mu + n \rightarrow \mu^- + p \; , \tag{10.30}$$

where the large spark-chamber detector could easily tell muons from electrons, because muons only produce straight tracks, while electrons – if they would have been generated in such a reaction – would have initiated electromagnetic cascades (see also Chap. 16), which exhibit a distinctly different pattern in the spark-chamber stack.

The existence of tau neutrinos was indirectly inferred from μ, e events observed in e^+e^- interactions [8],

$$e^+ + e^- \to \tau^+ + \tau^-$$
$$\hookrightarrow e^- + \bar{\nu}_e + \nu_\tau \qquad (10.31)$$
$$\hookrightarrow \mu^+ + \nu_\mu + \bar{\nu}_\tau \ .$$

The direct observation of τ neutrinos by the DONUT experiment in the year 2000,

$$\nu_\tau + N \to \tau^- + X \ , \qquad (10.32)$$

with subsequent decay of the τ^- completed the third family of leptons [9]. Due to the short lifetime of the tau this experiment required a massive but fine-grained detector of very high spatial resolution. This was achieved by a large-volume nuclear-emulsion detector which, however, required tedious scanning to find the τ-decay vertices.

10.4 Neutrino detectors

Neutrino detection is always very indirect. Neutrinos are caused to undergo interactions in which charged particles, excited nuclei or excited atoms are produced which then can be detected with standard measurement techniques. The simplest form is neutrino counting. This is the basis of radiochemical experiments in which solar neutrinos have first been detected in the chlorine experiment in the Homestake Mine [10, 11].

If the neutrino exceeds a certain threshold energy, the following reaction can occur:

$$\nu_e + {}^{37}\text{Cl} \to {}^{37}\text{Ar} + e^- \ , \qquad (10.33)$$

where a neutron in the chlorine nucleus is transformed into a proton. The argon isotope is purged out of the detector volume and counted in a proportional counter, which has an extremely low background rate. The argon isotope undergoes electron capture,

$$^{37}\text{Ar} + e^- \to {}^{37}\text{Cl} + \nu_e \ , \qquad (10.34)$$

which leaves the chlorine in an excited atomic state. This in turn transforms into the ground state by the emission of characteristic X rays or Auger electrons which are eventually the evidence that a neutrino detection has occurred. The gallium experiments for solar-neutrino detection work along similar lines.

Figure 10.1 shows the proportional-tube detector for the GALLEX experiment [12], which has measured the radioactive decay of the

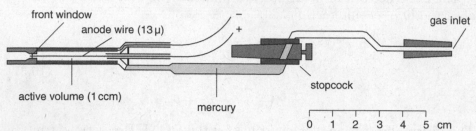

Fig. 10.1. Proportional-tube detector for the GALLEX experiment [12]; the radioactive decay of the produced ^{71}Ge is measured in a similar way as in the chlorine experiment by Ray Davis. The produced ^{71}Ge in the form of germane (GeH$_4$) is pressed into the proportional tube with the help of mercury. The germane (70%) is mixed with xenon (30%) to increase the photo absorption of the characteristic X rays [13].

neutrino-induced ^{71}Ge in a similar way as the ^{37}Ar was counted in the chlorine experiment by Ray Davis.

Calorimetric neutrino detectors for high-energy neutrinos are based on the measurement of the total energy of the final-state hadrons produced in a neutrino–nucleon interaction. These calorimeters are mostly sandwiches consisting of alternating passive targets and active detectors (e.g. scintillators) like those being used for hadron calorimeters. Of course, also total-absorption calorimeters can be built, where the target at the same time must be an active detector element. The large neutrino detectors at CERN (CDHS and Charm) were sampling detectors while KARMEN and SuperKamiokande are large-volume total-absorption devices exploiting Cherenkov- and scintillation-light detection. Depending on which neutrino flavour is being detected the calorimeter must be sensitive not only to hadrons but also to electrons or muons.

A photo of the sampling calorimeter used by the CDHS collaboration is shown in Fig. 10.2 [14]. The track-reconstruction capability of this experiment is demonstrated by the di-muon event as shown in Fig. 10.3 [14]. The KARMEN experiment (Fig. 10.4) is an example of a total-absorption scintillation calorimeter. The central part of the detector consists of a stainless-steel tank filled with 65 000 l of liquid scintillator. Details of the photomultiplier readout of the end faces are explained in Fig. 10.4 [15].

If the active elements in a total-absorption or sampling calorimetric system provide some spatial information, one is able to distinguish different final-state products in such neutrino tracking detectors. In reactions like

$$\nu_\mu + N \rightarrow \mu^- + \text{hadrons} \tag{10.35}$$

Fig. 10.2. Photo of the CDHS experiment [14].

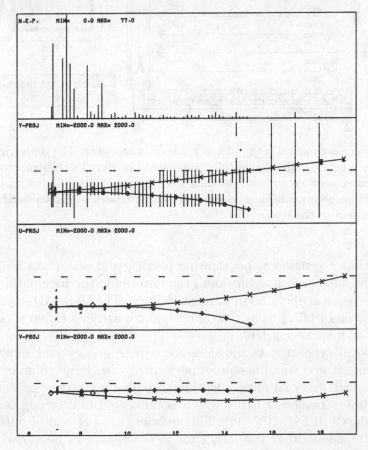

Fig. 10.3. Event display of a di-muon event in the CDHS detector [14]. In this event reconstruction the energy deposits and the muon tracks in different projections are shown.

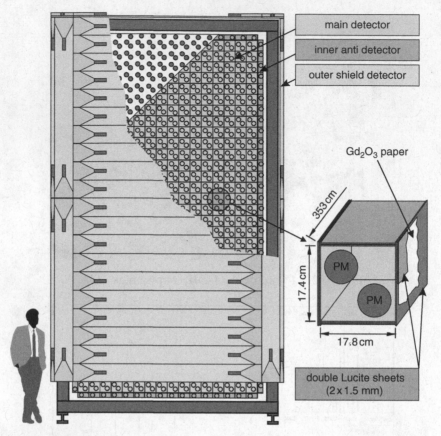

main detector

inner anti detector

outer shield detector

Gd_2O_3 paper

353 cm

17.4 cm

PM

PM

17.8 cm

double Lucite sheets
(2 × 1.5 mm)

Fig. 10.4. Experimental setup of the KARMEN detector. The detector is constructed as a large-volume liquid-scintillation calorimeter. The central part of the detector consists of a stainless-steel tank filled with 65 000 l of liquid scintillator. Details of the photomultiplier readout of the end faces is shown in the lower right-hand part of the figure [15].

one identifies the muon as penetrating particle and one might be able to resolve the final-state hadrons and even determine the momentum if the tracking system is operated in a magnetic field. The NOMAD experiment (Figs. 10.5 and 10.6), just as the CDHS and Charm experiments, provides this kind of information [16].

A classical type of neutrino detector, where energy and momentum measurements are simultaneously performed, are large-volume bubble chambers with external muon identifiers. A photograph of the Big European Bubble Chamber BEBC, which is immersed in a strong magnetic field, is shown in Fig. 10.7 [17]. This bubble chamber can be filled with different liquids so that the target for interactions can be varied to the needs of the experimenter.

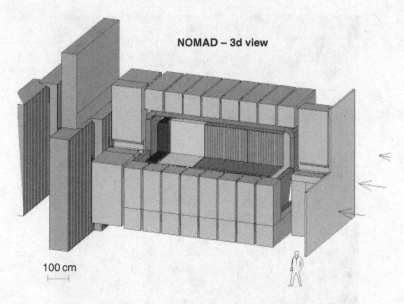

Fig. 10.5.   Sketch of the NOMAD experiment [16].

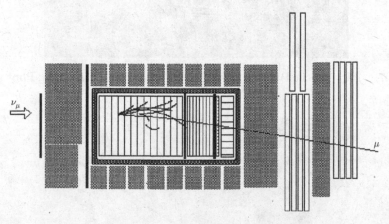

Fig. 10.6. Muon-neutrino-induced inelastic interaction in the NOMAD experiment [16].

Bubble-chamber pictures allow to obtain very detailed information on the final-state products of the neutrino interaction at the expense of a tedious scanning. Figure 10.8 shows the world's first neutrino observation in a 12-foot hydrogen bubble chamber at Argonne [18]. The invisible neutrino strikes a proton where three particle tracks originate. The neutrino turns into a muon, visible by the long track. The short track is the proton.

Figure 10.9 is an example of the rich information a bubble chamber can provide. It shows a charged-current inelastic interaction of a muon

Fig. 10.7.   Photo of the Big European Bubble Chamber, BEBC [17]; Photo credit CERN.

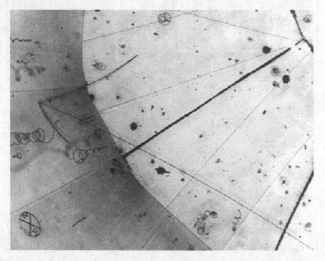

Fig. 10.8.   The world's first neutrino observation in a 12-foot hydrogen bubble chamber at Argonne. The invisible neutrino strikes a proton where three particle tracks originate. The neutrino turns into a muon, the long track. The short track is the proton. The third track is a pion created by the collision; $\nu_\mu + p \to \mu^- + p + \pi^+$; Photo credit Argonne National Laboratory [18].

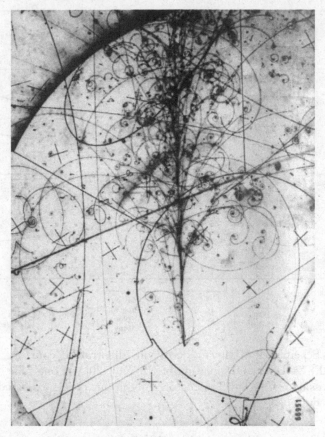

Fig. 10.9.   Charged-current inelastic interaction of a muon neutrino in the BEBC bubble chamber filled with a liquid neon–hydrogen mixture. Along with the muon in the final state also hadrons have been formed, among them also neutral pions, whose decay products – two energetic photons – initiate electromagnetic cascades. The charged-particle tracks are bent in a 3.5 T magnetic field oriented perpendicular to the plane shown [19]; Photo credit CERN.

neutrino in the BEBC bubble chamber filled with a liquid neon–hydrogen mixture [19]. Along with the muon in the final state also hadrons have been formed, among them also neutral pions, whose decay products – two energetic photons – initiate electromagnetic cascades. The charged-particle tracks are bent in a 3.5 T magnetic field oriented perpendicular to the plane shown.

In recent times nuclear emulsions have been rejuvenated in the field of neutrino interactions. Nuclear emulsions provide µm resolution which is required for measurement and identification of ν_τ interactions. τ leptons with a $c\tau_\tau = 87\,\mu\text{m}$ require excellent spatial resolution for their identification, because not only the production vertex, but also the decay

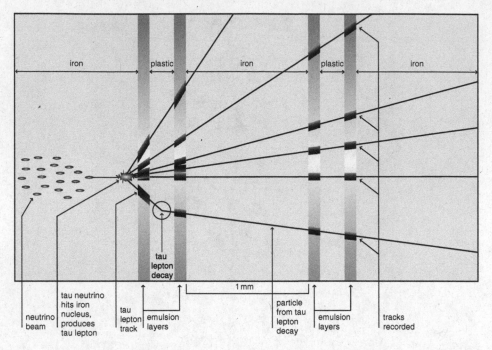

Fig. 10.10. Schematics of the ν_τ detection; illustration courtesy of Fermilab Experiment E872, Direct Observation of Nu Tau [20]. Of one billion*(10^{12}) tau neutrinos crossing the DONUT detector, only about one is expected to interact with an iron nucleus.

vertex must be unambiguously identified. Figures 10.10 and 10.11 show the schematics of the ν_τ detection and a real event in the DONUT experiment [20, 21]. The reconstructed event display shows different projections of the ν_τ interaction.

The detection of solar neutrinos or neutrinos from supernova explosions has been performed in large water Cherenkov detectors. In these detectors only ν_e or $\bar{\nu}_e$ can be measured because solar or SN neutrinos have insufficient energy to create other lepton flavours. Typical reactions are

$$\nu_e + e^- \rightarrow \nu_e + e^- \tag{10.36}$$

or

$$\bar{\nu}_e + p \rightarrow e^+ + n \ , \tag{10.37}$$

where the final-state electron or positron is detected.

If higher-energy neutrinos are available, such as from interactions of primary cosmic rays in the atmosphere and the decay of pions and kaons, one

---

* Beware: the European 'billion' is 10^{12} in contrast to an American 'billion', which is just 10^9.

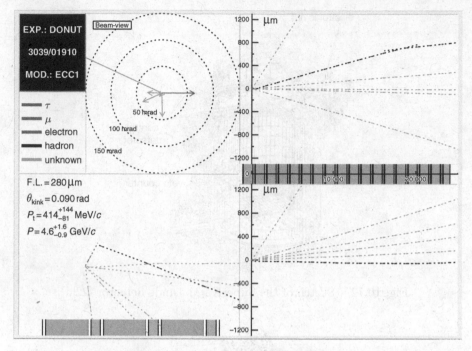

Fig. 10.11. Detection of the τ neutrino in the DONUT experiment; illustration courtesy of Fermilab Experiment E872, Direct Observation of Nu Tau [21].

can produce energetic electrons and muons which can easily be identified in large-volume water Cherenkov counters. The Cherenkov light produced by the charged final-state particles is measured by a large assembly of photomultipliers, where a substantial coverage of the detector end faces eases energy measurements and spatial reconstruction. Figure 10.12 shows the large-volume water Cherenkov counter SuperKamiokande with its central detector surrounded by an anticoincidence shield which vetoes remnant charged particles produced in the atmosphere [22].

Figures 10.13 and 10.14 show signatures of an electron and a muon produced in an electron-neutrino and muon-neutrino interaction, respectively [22]. The barrel part of the SuperKamiokande detector has been unwrapped, and the lower and the upper circular part of the cylindrically shaped detector has been added in the appropriate place. Electrons initiate electromagnetic showers in the water. Therefore their Cherenkov pattern shows some fuzziness at the edges of the Cherenkov ring, while muons do not start showers leading to a clear, distinct Cherenkov pattern.

Because of the large background from atmospheric neutrinos, the detection of neutrinos from cosmic-ray sources in our galaxy and beyond is limited to the TeV region. The expected flux of such neutrinos is low, so

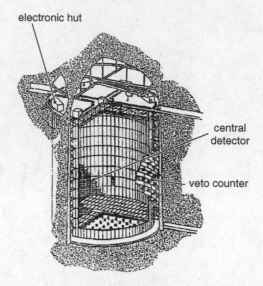

Fig. 10.12.   Sketch of the SuperKamiokande detector [22].

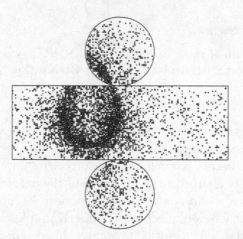

Fig. 10.13.   Signature of an electron in the SuperKamiokande experiment [22].

huge detectors must be provided. Large-volume water or ice Cherenkov detectors (Baikal, AMANDA, IceCube, ANTARES, NESTOR or NEMO) are being prepared or are taking data already. The measurement principle is the detection of energetic muons produced in $\nu_\mu N$ interactions, where in the TeV region the energy loss of muons by bremsstrahlung and direct electron-pair production – being proportional to the muon energy – provides calorimetric information. Figure 10.15 shows the layout of the AMANDA-II detector in the antarctic ice at the South Pole [23], and Fig. 10.16 is an example of an upgoing muon produced by a

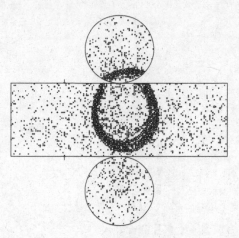

Fig. 10.14. Signature of a muon in the SuperKamiokande experiment [22].

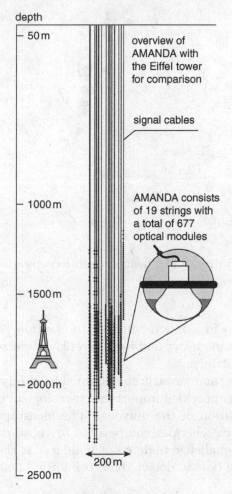

Fig. 10.15. Sketch of the AMANDA-II array at the South Pole [23].

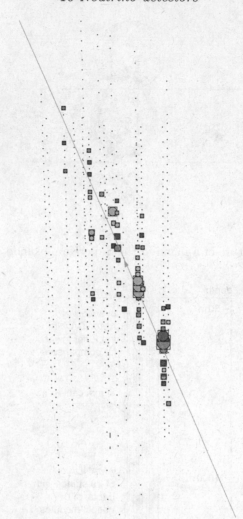

Fig. 10.16. Event display of a neutrino-induced upward-going muon [24].

cosmic-ray neutrino in a muon-neutrino interaction [24]. Until now the operating large-volume water and ice Cherenkov detectors have only seen atmospheric ν_μ neutrinos.

The observation and measurement of blackbody photons in the microwave band has provided important cosmological information on the structure and evolution of the universe. The measurement of Big Bang neutrinos which have energies comparable to those of 2.7 K microwave photons presents a challenge to detector builders. At the moment no technique is conceivable how to detect these primordial neutrinos in the MeV range.

Neutrino physics is mostly an issue at accelerators, in cosmic rays or with reactor experiments. Now that neutrino propagation in the framework of neutrino oscillations is understood [25], neutrinos can also be used as probes, e.g., to search for oil or hydrocarbons in the Earth's crust, to investigate the inner part of the Earth, or to help to clarify the question of the radiogenic contribution to the terrestrial heat production by measuring geoneutrinos from the decay of the naturally occurring radioisotopes uranium (^{238}U), thorium (^{232}Th) and potassium (^{40}K).

10.5 Problems

10.1 The Sun converts protons into helium according to the reaction

$$4p \rightarrow {}^4\text{He} + 2e^+ + 2\nu_e \ .$$

The solar constant describing the power of the Sun at Earth is $P \approx 1400\,\text{W/m}^2$. The energy gain for this fusion reaction of 26.1 MeV differs slightly from the binding energy of helium ($E_\text{B}({}^4\text{He}) = 28.3\,\text{MeV}$), because the neutrinos produced in this chain also take some energy. How many solar neutrinos arrive at Earth?

10.2 If solar electron neutrinos oscillate into muon or tau neutrinos they could in principle be detected via the reactions

$$\nu_\mu + e^- \rightarrow \mu^- + \nu_e \ , \quad \nu_\tau + e^- \rightarrow \tau^- + \nu_e \ .$$

Work out the threshold energy for these reactions to occur. (Assume that the target electrons are at rest.)

10.3 Radiation exposure due to solar neutrinos. Use

$$\sigma(\nu_e N) \approx 10^{-45}\,\text{cm}^2/\text{nucleon}$$

to work out the number of interactions of solar neutrinos in the human body (tissue density $\varrho \approx 1\,\text{g/cm}^3$). Neutrinos interact in the human body by

$$\nu_e + N \rightarrow e^- + N' \ ,$$

where the radiation damage is caused by the electrons. Estimate the annual dose for a human under the assumption that on average 50% of the neutrino energy is transferred to the electron. The equivalent dose is defined as

$$H = (\Delta E/m)\,w_R$$

(m is the mass of the human body, w_R the radiation weighting factor ($= 1$ for electrons), $[H] = 1\,\mathrm{Sv} = 1\,w_R\ \mathrm{J/kg}$, and ΔE the energy deposit in the human body). Work out the annual equivalent dose due to solar neutrinos and compare it with the normal dose due to natural radiation from the environment of $H_0 \approx 2\,\mathrm{mSv/a}$.

10.4 Work out the muon and neutrino energies in pion and leptonic kaon decay at rest ($\pi^+ \rightarrow \mu^+ + \nu_\mu$, $K^+ \rightarrow \mu^+ + \nu_\mu$).

10.5 Two electron neutrinos with energies E_1 and E_2 and an assumed rest mass m_0 are emitted from the supernova 1987A at exactly the same time. What is their arrival-time difference at Earth (distance to SN 1987A is r), and how can their mass be inferred from such a time-difference measurement if $m_0 c^2 \ll E$ is assumed for the neutrino?

10.6 It is considered realistic that a point source in our galaxy produces a neutrino spectrum according to

$$\frac{\mathrm{d}N}{\mathrm{d}E_\nu} = 2 \cdot 10^{-11}\,\frac{100}{E_\nu^2\,[\mathrm{TeV}^2]}\ \mathrm{cm}^{-2}\,\mathrm{s}^{-1}\,\mathrm{TeV}^{-1}\ . \tag{10.38}$$

This leads to an integral flux of neutrinos of

$$\Phi_\nu(E_\nu > 100\,\mathrm{TeV}) = 2 \cdot 10^{-11}\ \mathrm{cm}^{-2}\,\mathrm{s}^{-1}\ . \tag{10.39}$$

Work out the annual interaction rate of $> 100\,\mathrm{TeV}$ neutrinos in IceCube ($d = 1\,\mathrm{km} = 10^5\,\mathrm{cm}$, $\varrho(\mathrm{ice}) \approx 1\,\mathrm{g/cm}^3$, $A_{\mathrm{eff}} = 1\,\mathrm{km}^2$).

References

[1] John N. Bahcall, *Neutrino Astrophysics*, Cambridge University Press, Cambridge (1989)

[2] H.V. Klapdor-Kleingrothaus & K. Zuber, *Particle Astrophysics*, Institute of Physics Publishing, Bristol (2000)

[3] H.V. Klapdor-Kleingrothaus & K. Zuber, *Teilchenphysik ohne Beschleuniger*, Teubner, Stuttgart (1995)

[4] J.A. Peacock, *Cosmological Physics*, Cambridge University Press, Cambridge (1999)

[5] P. Adhya, D.R. Chaudhuri & S. Hannestad, Late-Time Entropy Production from Scalar Decay and the Relic Neutrino Temperature, *Phys. Rev.* **D68** (2003) 083519, 1–6

[6] F. Reines, C.L. Cowan, F.B. Harrison, A.D. McGuire & H.W. Kruse (Los Alamos), Detection of the Free Antineutrino, *Phys. Rev.* **117** (1960) 159–73; C.L. Cowan, F. Reines, F.B. Harrison, H.W. Kruse & A.D. McGuire

(Los Alamos), Detection of the Free Neutrino: A Confirmation, *Science* **124** (1956) 103–4

[7] G. Danby, J.M. Gaillard, K. Goulianos, L.M. Lederman, N. Mistry, M. Schwartz & J. Steinberger (Columbia U. and Brookhaven) 1962, Observation of High Energy Neutrino Reactions and the Existence of Two Kinds of Neutrinos, *Phys. Rev. Lett.* **9** (1962) 36–44

[8] M. Perl *et al.*, Evidence for Anomalous Lepton Production in $e^+ - e^-$ Annihilation, *Phys. Rev. Lett.* **35** (1975) 1489–92

[9] K. Kodama *et al.*, Detection and Analysis of Tau-Neutrino Interactions in DONUT Emulsion Target, *Nucl. Instr. Meth.* **A493** (2002) 45–66

[10] J.N. Bahcall & R. Davis Jr, *Essays in Nuclear Astrophysics*, Cambridge University Press, Cambridge (1982)

[11] R. Davis Jr, *A Half Century with Solar Neutrinos, Nobel Lecture, from 'Les Prix Nobel'*, ed. T. Frängsmyr, Stockholm (2003)

[12] R. Wink *et al.*, The Miniaturised Proportional Counter HD-2(Fe)/(Si) for the GALLEX Solar Neutrino Experiment, *Nucl. Instr. Meth.* **A329** (1993) 541–50; www.mpi-hd.mpg.de/nuastro/gallex/counter.gif

[13] W. Hampel, private communication (2006)

[14] CERN–Dortmund–Heidelberg–Saclay Collaboration (J. Knobloch *et al.*) 1981, In Wailea 1981, Proceedings, Neutrino '81, Vol. 1, 421–8, http:// knobloch.home.cern.ch/knobloch/cdhs/cdhs.html; and private communication by J. Knobloch

[15] KARMEN Collaboration (W. Kretschmer for the collaboration), Neutrino Physics with KARMEN, *Acta Phys. Polon.* **B33** (2002) 1775–90; www-ik1. fzk.de/www/karmen/karmen_e.html

[16] NOMAD Collaboration (A. Cardini for the collaboration), The NOMAD Experiment: A Status Report, Prepared for 4th International Workshop on Tau Lepton Physics (TAU 96), Estes Park, Colorado, 16–9 September 1996, *Nucl. Phys. B Proc. Suppl.* **55** (1997) 425–32; http://nomadinfo.cern.ch/

[17] H.P. Reinhard (CERN), First Operation of Bebc, in Frascati 1973, Proceedings, High Energy Instrumentation Conference, Frascati 1973, 87–96; and *Status and Problems of Large Bubble Chambers*, Frascati 1973, 3–12; www.bo.infn.it/antares/bolle_proc/foto.html; http:// doc.cern.ch/archive/electronic/cern/others/PHO/photo-hi/7701602.jpeg; Photo credit CERN

[18] www.anl.gov/OPA/news96arch/news961113.html

[19] W.S.C. Williams, *Nuclear and Particle Physics*, Clarendon Press, Oxford (1991)

[20] www.fnal.gov/pub/inquiring/physics/neutrino/discovery/photos/signal_ low.jpg

[21] www-donut.fnal.gov/web_pages/

[22] T. Kajita & Y. Totsuka, Observation of Atmospheric Neutrinos, *Rev. Mod. Phys.* **73** (2001) 85–118

[23] AMANDA Collaboration (K. Hanson *et al.*), July 2002, 3pp., Prepared for 31st International Conference on High Energy Physics (ICHEP 2002), Amsterdam, The Netherlands, 24–31 July 2002, Amsterdam 2002, ICHEP

126–8; http://amanda.uci.edu/; and private communication by Chr. Spiering (2003), Chr. Walck & P.O. Hulth (2007)

[24] AMANDA Collaboration, Christian Spiering, private communications (2004)

[25] S.M. Bilenky & B. Pontecorvo, Lepton Mixing and Neutrino Oscillations, *Phys. Rep.* **41** (1978) 225–61

11

Momentum measurement and muon detection

I think that a particle must have a separate reality independent of the measurements. That is, an electron has spin, location, and so forth even when it is not being measured. I like to think that the moon is there even if I am not looking at it.

Albert Einstein

Momentum measurement and, in particular, muon detection is an important aspect of any experiment of particle physics, astronomy or astrophysics. Ultra-high-energy cosmic rays are currently at the forefront of astroparticle physics searching for the accelerators in the sky. These questions can be studied by the detection of extensive air showers at ground level by measuring secondary electrons, muons and hadrons produced by primary cosmic rays which initiate hadronic cascades in the Earth's atmosphere. The detectors have to operate for many years in order to map the galactic sources of high-energy cosmic rays which may be visible at the experimental sites. There are several experiments dedicated to studying these air showers that employ large detector arrays for electron and muon detection. Apart from water Cherenkov and scintillation counters, typical detectors such as limited streamer tubes [1] and resistive-plate chambers are also used [2].

In the field of high energy physics, over the last several decades many outstanding discoveries have been made from the studies of muons along with other precision measurements of leptons and hadrons. Notable are the determination of the number of neutrino generations by the LEP detectors, charm production (J/ψ), the observation of the electroweak bosons ($W^{\pm}, Z$), and the top quark (t). Although these particles have higher branching ratios for their hadronic decay channels, it is difficult to measure and isolate hadrons. At the time of the writing of this book – the Large Hadron Collider is under construction (at CERN,

Geneva) – advances in the understanding of the Standard Model of particle physics have led to believe that the physics of Higgs particles as well as new phenomena, like supersymmetry, should show up at a mass scale of approximately a few TeV. ATLAS, CMS, ALICE and LHCb are experiments having large, i.e. several thousand square metres of muon-detection surfaces to signal the presence of new physics. Also precision experiments like Belle and BaBar looking into *B* physics have to rely on efficient muon identification and accurate momentum measurement. Muons can be identified by the large penetrating power and the relevant parameters to be measured very precisely are energy and momentum. Energies of muons beyond the TeV range can be measured with calorimetric techniques, because the energy loss at high energies is dominated by bremsstrahlung and direct electron-pair production, both of which processes are proportional to the muon energy.

The momenta of muons, just as for all charged particles, are usually determined in magnetic spectrometers. The Lorentz force causes the particles to follow circular or helical trajectories around the direction of the magnetic field. The bending radius of particle tracks is related to the magnetic field strength and the momentum component of the particle perpendicular to the magnetic field. Depending on the experimental situation, different magnetic spectrometers are used.

11.1   Magnetic spectrometers for fixed-target experiments

The basic set-up of a *magnetic spectrometer* for fixed-target experiments (in contrast to storage-ring experiments) is sketched in Fig. 11.1. Particles

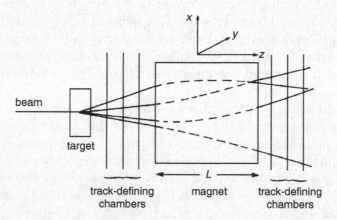

Fig. 11.1.   Schematic representation of a magnetic spectrometer in a fixed-target experiment with a stationary target.

of known identity and also, in general, of known energy are incident on a target thereby producing secondary particles in an interaction. The purpose of the spectrometer is to measure the momenta of the charged secondary particles.

Let the magnetic field B be oriented along the y axis, $\vec{B} = (0, B_y, 0)$, whereas the direction of incidence of the primary particles is taken to be parallel to the z axis. In hadronic interactions typical *transverse momenta* of

$$p_\mathrm{T} \approx 350 \,\mathrm{MeV}/c \qquad (11.1)$$

are transferred to secondary particles, where

$$p_\mathrm{T} = \sqrt{p_x^2 + p_y^2} \;. \qquad (11.2)$$

Normally, $p_x, p_y \ll p_z$, where the momenta of outgoing particles are described by $\vec{p} = (p_x, p_y, p_z)$. The trajectories of particles incident into the spectrometer are determined in the most simple case by track detectors before they enter and after they have left the magnet. Since the magnetic field is oriented along the y axis, the deflection of charged particles is in the xz plane. Figure 11.2 sketches the track of a charged particle in this plane.

The Lorentz force provides a centripetal acceleration v^2/ρ directed along the bending radius. We choose our coordinate system in such a way that the particles incident into the spectrometer are parallel to the z axis, i.e. $|\vec{p}| = p_z = p$, where $\vec{p}$ is the momentum of the particle to be

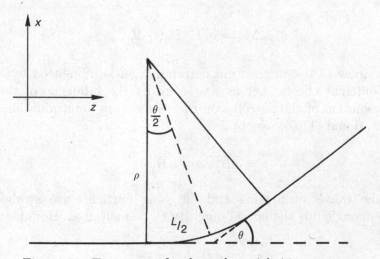

Fig. 11.2.   Trajectory of a charged particle in a magnet.

measured. One then has (for $\vec{p} \perp \vec{B}$, where m – mass, v – velocity and ρ – bending radius of the track in the magnetic field):

$$\frac{mv^2}{\rho} = e\,v\,B_y \; . \tag{11.3}$$

The *bending radius* ρ itself is obtained from Eq. (11.3) by

$$\rho = \frac{p}{eB_y} \; . \tag{11.4}$$

With standard units, which are common in particle and astroparticle physics, this formula leads to

$$\rho\,[\mathrm{m}] = \frac{p\,[\mathrm{GeV}/c]}{0.3\,B\,[\mathrm{T}]} \; . \tag{11.5}$$

The particles pass through the magnet following a circular trajectory, where the bending radius ρ, however, is normally very large compared to the magnet length L. Therefore, the deflection angle θ can be approximated by

$$\theta = \frac{L}{\rho} = \frac{L}{p}eB_y \; . \tag{11.6}$$

Because of the magnetic deflection, the charged particles obtain an additional transverse momentum of

$$\Delta p_x = p \cdot \sin\theta \approx p \cdot \theta = L\,e\,B_y \; . \tag{11.7}$$

If the magnetic field varies along L, Eq. (11.7) is generalised to

$$\Delta p_x = e \int_0^L B_y(l)\,\mathrm{d}l \; . \tag{11.8}$$

The accuracy of the momentum determination is influenced by a number of different effects. Let us first consider the influence of the finite track resolution of the detector on the momentum determination. Using Eqs. (11.4) and (11.6), we obtain

$$p = e\,B_y \cdot \rho = e\,B_y \cdot \frac{L}{\theta} \; . \tag{11.9}$$

Since the tracks of ingoing and outgoing particles are straight, the deflection angle θ is the actual quantity to be measured. Because of

$$\left| \frac{\mathrm{d}p}{\mathrm{d}\theta} \right| = e\,B_y\,L \cdot \frac{1}{\theta^2} = \frac{p}{\theta} \; , \tag{11.10}$$

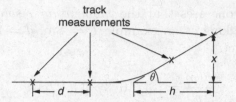

Fig. 11.3.   Sketch illustrating the determination of the track measurement error.

one has

$$\frac{\mathrm{d}p}{p} = \frac{\mathrm{d}\theta}{\theta} \tag{11.11}$$

and

$$\frac{\sigma(p)}{p} = \frac{\sigma(\theta)}{\theta} \ . \tag{11.12}$$

Let us assume that to determine the deflection angle, θ_{def}, four track coordinates are measured, i.e. two in front of and two behind the magnet (although for a circular orbit three coordinates would in principle be sufficient). If the distance between the sensors in each pair is d (Fig. 11.3), then the input, output and deflection angles are expressed as:

$$\vartheta_{\mathrm{in}} \approx \frac{x_2 - x_1}{d} \ , \quad \vartheta_{\mathrm{out}} \approx \frac{x_4 - x_3}{d} \ , \tag{11.13}$$

$$\theta_{\mathrm{def}} = \vartheta_{\mathrm{out}} - \vartheta_{\mathrm{in}} \approx \frac{x_2 - x_1 - x_4 + x_3}{d} \ . \tag{11.14}$$

If all track measurements have the same measurement error $\sigma(x)$, the variance of the deflection angle is obtained to be

$$\sigma^2(\theta) \propto \sum_{i=1}^{4} \sigma_i^2(x) = 4\sigma^2(x) \ , \tag{11.15}$$

and

$$\sigma(\theta) = \frac{2\sigma(x)}{d} \ . \tag{11.16}$$

Using Eq. (11.12), this leads to

$$\frac{\sigma(p)}{p} = \frac{2\sigma(x)}{d} \frac{p}{LeB_y} = \frac{p \ [\mathrm{GeV}/c]}{0.3 \, L \ [\mathrm{m}]} B \ [\mathrm{T}] \cdot \frac{2\sigma(x)}{d} \ . \tag{11.17}$$

From Eq. (11.17) one sees that the *momentum resolution* $\sigma(p)$ is proportional to p^2. Taking as an example $L = 1\,\text{m}$, $d = 1\,\text{m}$, $B = 1\,\text{T}$ and $\sigma_x = 0.2\,\text{mm}$, we get

$$\frac{\sigma(p)}{p} = 1.3 \cdot 10^{-3}\, p \,[\text{GeV}/c] \; . \tag{11.18}$$

Depending on the quality of the track detectors, one may obtain

$$\frac{\sigma(p)}{p} = (10^{-3} \text{ to } 10^{-4}) \cdot p \,[\text{GeV}/c] \; . \tag{11.19}$$

In cosmic-ray experiments, it has become usual practice to define a *maximum detectable momentum* (mdm). This is defined by

$$\frac{\sigma(p_{\text{mdm}})}{p_{\text{mdm}}} = 1 \; . \tag{11.20}$$

For a magnetic spectrometer with a momentum resolution given by Eq. (11.19), the maximum detectable momentum would be

$$p_{\text{mdm}} = 1\,\text{TeV}/c \text{ to } 10\,\text{TeV}/c \; . \tag{11.21}$$

The momentum measurement is normally performed in an *air-gap magnet*. The effect of multiple scattering is low in this case and influences the measurement accuracy only at low momenta. Because of the high penetrating power of muons, their momenta can also be analysed in *solid-iron magnets*. For this kind of application, however, the influence of multiple scattering cannot be neglected.

A muon penetrating a solid-iron magnet of thickness L obtains a transverse momentum $\Delta p_{\text{T}}^{\text{MS}}$ due to multiple scattering according to

$$\Delta p_{\text{T}}^{\text{MS}} = p \cdot \sin\theta_{\text{rms}} \approx p \cdot \theta_{\text{rms}} = 19.2 \sqrt{\frac{L}{X_0}} \,\text{MeV}/c \tag{11.22}$$

(Fig. 11.4 and Eq. (1.53) for $p \gg m_0 c$ and $\beta \approx 1$).

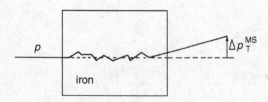

Fig. 11.4.   Illustration of the multiple-scattering error.

Since the magnetic deflection is in the x direction, only the multiple-scattering error projected onto this direction is of importance:

$$\Delta p_x^{\text{MS}} = \frac{19.2}{\sqrt{2}} \sqrt{\frac{L}{X_0}} \; \text{MeV}/c = 13.6 \sqrt{\frac{L}{X_0}} \; \text{MeV}/c \; . \tag{11.23}$$

The momentum resolution limited by the effect of multiple scattering is given by the ratio of the deflection by multiple scattering to the magnetic deflection according to [3]

$$\frac{\sigma(p)}{p} \bigg|^{\text{MS}} = \frac{\Delta p_x^{\text{MS}}}{\Delta p_x^{\text{magn}}} = \frac{13.6 \sqrt{L/X_0} \; \text{MeV}/c}{e \int_0^L B_y(l) \, dl} \; . \tag{11.24}$$

Both the deflection angle θ caused by the Lorentz force and the multiple-scattering angle are inversely proportional to the momentum. Therefore, the momentum resolution in this case does not depend on the momentum of the particle.

For solid-iron magnetic spectrometers ($X_0 = 1.76$ cm) typical values of $B = 1.8$ T are used, leading to a momentum resolution of, see Eq. (11.24),

$$\frac{\sigma(p)}{p} \bigg|^{\text{MS}} = 0.19 \cdot \frac{1}{\sqrt{L \, [\text{m}]}} \; . \tag{11.25}$$

This gives for $L = 3$ m

$$\frac{\sigma(p)}{p} \bigg|^{\text{MS}} = 11\% \; . \tag{11.26}$$

This equation only contains the effect of *multiple scattering* on the momentum resolution. In addition, one has to consider the momentum-measurement error from the uncertainty of the position measurement. This error can be obtained from Eq. (11.17) or from the determination of the *sagitta* (Fig. 11.5) [4]. The sagitta s is related to the magnetic bending radius ρ and the magnetic deflection angle θ by

$$s = \rho - \rho \cos \frac{\theta}{2} = \rho \left(1 - \cos \frac{\theta}{2} \right) \; . \tag{11.27}$$

Because of $1 - \cos \frac{\theta}{2} = 2 \sin^2 \frac{\theta}{4}$, one obtains

$$s = 2\rho \sin^2 \frac{\theta}{4} \; . \tag{11.28}$$

Since $\theta \ll 1$, the sagitta can be approximated by (θ in radians)

$$s = \frac{\rho \theta^2}{8} \; . \tag{11.29}$$

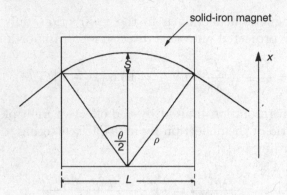

Fig. 11.5.   Illustration of the sagitta method for momentum determination [4].

In the following we will replace B_y by B for simplicity. Using Eqs. (11.9) and (11.4) for θ and ρ the sagitta can be expressed by

$$s = \frac{\rho}{8} \cdot \left(\frac{eBL}{p}\right)^2 = \frac{eBL^2}{8p} \ . \tag{11.30}$$

For fixed units one gets

$$s \ [\mathrm{m}] = 0.3 \, B \ [\mathrm{T}] \, (L \ [\mathrm{m}])^2/(8p \ [\mathrm{GeV}/c]) \ . \tag{11.31}$$

The determination of the sagitta requires at least 3 position measurements x_i ($i = 1, 2, 3$). These can be obtained from 3 tracking detectors positioned at the entrance (x_1) and at the exit (x_3) of the magnet, while one chamber could be placed in the centre of the magnet (x_2). Because of

$$s = x_2 - \frac{x_1 + x_3}{2} \tag{11.32}$$

and under the assumption that the *track measurement errors* $\sigma(x)$ are the same for all chambers, it follows that

$$\sigma(s) = \sqrt{\frac{3}{2}} \, \sigma(x) \ . \tag{11.33}$$

This leads to a momentum resolution from track measurement errors of

$$\left.\frac{\sigma(p)}{p}\right|^{\text{track error}} = \frac{\sigma(s)}{s} = \frac{\sqrt{\frac{3}{2}}\sigma(x) \ [\mathrm{m}] \cdot 8p \ [\mathrm{GeV}/c]}{0.3 \, B \ [\mathrm{T}] \, (L \ [\mathrm{m}])^2} \ . \tag{11.34}$$

If the track is measured not only at 3 but at N points equally distributed over the magnet length L, it can be shown that the momentum resolution due to the finite track measurement error is given by [5]

$$\left.\frac{\sigma(p)}{p}\right|^{\text{track error}} = \frac{\sigma(x)\ [\text{m}]}{0.3\ B\ [\text{T}]\ (L\ [\text{m}])^2}\sqrt{720/(N+4)} \cdot p\ [\text{GeV}/c]\ . \quad (11.35)$$

For $B = 1.8\,\text{T}$, $L = 3\,\text{m}$, $N = 4$ and $\sigma(x) = 0.5\,\text{mm}$ Eq. (11.35) leads to

$$\left.\frac{\sigma(p)}{p}\right|^{\text{track error}} \approx 10^{-3} \cdot p\ [\text{GeV}/c]\ . \quad (11.36)$$

If the N measurements are distributed over L in k constant intervals, one has

$$L = k \cdot N \quad (11.37)$$

and thereby (if $N \gg 4$):

$$\left.\frac{\sigma(p)}{p}\right|^{\text{track error}} \propto (L\ [\text{m}])^{-5/2} \cdot (B\ [\text{T}])^{-1} \cdot p\ [\text{GeV}/c]\ . \quad (11.38)$$

To obtain the total error on the momentum determination, the multiple-scattering and track-resolution error have to be combined. Both contributions according to Eqs. (11.26) and (11.36) are plotted in Fig. 11.6 for the

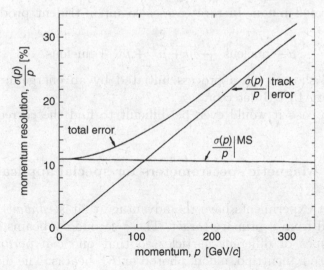

Fig. 11.6. Contributions to the momentum resolution for a solid-iron magnetic spectrometer.

aforementioned parameters of a solid-iron magnetic spectrometer. At low momenta multiple scattering dominates the error and at high momenta it is limited by the track measurement error.

For an air-gap magnet the error contribution due to multiple scattering is naturally much smaller. If Eq. (11.24) is applied to an air-gap magnet, $(X_0 = 304\,\mathrm{m})$, one obtains

$$\left.\frac{\sigma(p)}{p}\right|^{\mathrm{MS}} = 1.4 \cdot 10^{-3}/\sqrt{L\,[\mathrm{m}]} \ , \tag{11.39}$$

which means for $L = 3\,\mathrm{m}$:

$$\left.\frac{\sigma(p)}{p}\right|^{\mathrm{MS}} = 0.08\% \ . \tag{11.40}$$

For a realistic experiment one has to consider another effect that will degrade the momentum resolution of muons. In particular, at high energies muons will undergo electromagnetic interactions, sometimes with large energy transfers, in the solid-iron magnet, like bremsstrahlung and direct electron-pair production. In addition, muons can undergo photonuclear interactions. A monoenergetic muon beam will develop a 'radiative tail' due to bremsstrahlung and pair-production losses. The probability for an energy transfer of more than $10\,\mathrm{GeV}$ for a $200\,\mathrm{GeV}$ muon in a $2\,\mathrm{m}$ long iron magnet is already 3% [6]. This increases to 12% for a $1\,\mathrm{TeV}$ muon in $2\,\mathrm{m}$ of iron [7]. The secondaries produced by a muon might also emerge from the solid-iron magnet, thereby complicating the track reconstruction of the deflected muon. In rare cases also muon-trident production can occur, i.e.

$$\mu + \text{nucleus} \ \to \ \mu + \mu^+ + \mu^- + \text{nucleus}' \ . \tag{11.41}$$

Figure 11.7 shows such a process initiated by an energetic cosmic-ray muon in the ALEPH detector.

In such a case it would even be difficult to find the correct outgoing muon.

11.2   Magnetic spectrometers for special applications

Fixed-target experiments have the advantage that *secondary beams* can be produced from a primary target. These secondary beams can consist of many types of different particles so that one can perform experiments with, e.g. neutrino, muon, photon or K_L^0 beams. The disadvantage with fixed-target experiments, however, is that the available centre-of-mass energy is relatively small. Therefore, investigations in the field of

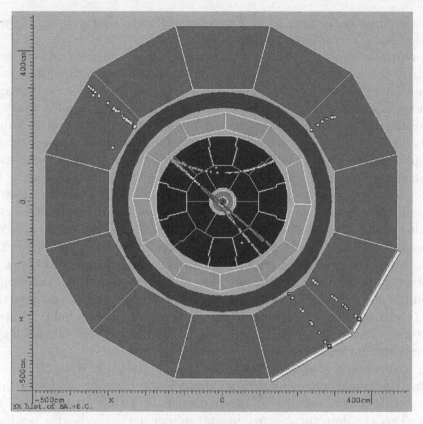

Fig. 11.7. Cosmic-ray muon undergoing a muon-trident production in the ALEPH detector. The muon pair is created in the flux return of the solenoidal magnetic field. The bending of one of the secondary muons in the iron is seen to be opposite to the bending in the central detector [8].

high energy physics are frequently done at storage rings. In storage-ring experiments, the centre-of-mass system is identical with the laboratory system (for a crossing angle of zero), if the colliding beams have the same energy and are opposite in momentum. The event rates are in general rather low because the target density – one beam represents the target for the other and vice versa – is low compared to fixed-target experiments. There are, however, important differences between collider and fixed-target experiments: in the first case the interaction products are emitted into the full solid angle, while in the latter case the products are released within a narrow cone around the incident direction. Therefore – in contrast to fixed-target experiments – storage-ring detectors normally have to cover the full solid angle of 4π surrounding the interaction point. Such a *hermeticity* allows a complete reconstruction of individual events.

Depending on the type of storage ring, different magnetic-field configurations can be considered.

For proton–proton (or $p\bar{p}$) storage rings *dipole magnets* can be used, where the magnetic field is perpendicular to the beam direction. Since such a dipole also bends the stored beam, its influence must be corrected by compensation coils. The compensation coils are also dipoles, but with opposite field gradient, so that there is no net effect on the stored beams. Such a configuration is rarely used for electron–positron storage rings – except at relatively low energies [9] – because the strong dipole field would cause the emission of intense synchrotron radiation, which cannot be tolerated for the storage-ring operation and the safe running of the detectors.

A dipole magnet can be made self-compensating if two dipoles with opposite field gradient on both sides of the interaction point are used instead of only one dipole. Compensation is automatically fulfilled in this case, but at the expense of strongly inhomogeneous magnetic fields at the interaction point which complicate the track reconstruction considerably. If, on the other hand, *toroidal magnets* are employed, one can achieve that the beams traverse the spectrometer in a region of zero field. Multiple scattering, however, on the inner cylinder of the toroidal magnet limits the momentum resolution.

In most cases a *solenoidal magnetic field* is chosen, in which the stored beams run essentially – apart from small beam crossing angles or betatron oscillations – parallel to the magnetic field (like in Fig. 11.7 [8]). Therefore, the detector magnet has no influence on the beams, and also no or very little synchrotron radiation is produced. In either case one has to consider that any magnetic spectrometer used in the detector becomes an integral element of the accelerator and should be properly accounted for and compensated.

The track detectors are mounted inside the magnetic coil and are therefore also cylindrical. The longitudinal magnetic field acts only on the transverse momentum component of the produced particles and leads to a momentum resolution given by Eq. (11.35), where $\sigma(x)$ is the coordinate resolution in the plane perpendicular to the beam axis. Figure 11.8 shows schematically two tracks originating from the interaction point in a projection perpendicular to the beam ('$r\varphi$ plane') and parallel to the beam ('rz plane'). The characteristic track parameters are given by the polar angle θ, the azimuthal angle φ and the radial coordinate r, i.e. the distance from the interaction point. A sketch of a simulated muon-track reconstruction in the Compact Muon Solenoid (CMS) at CERN is shown in Fig. 11.9 [10]. A simulated event of the production of supersymmetric particles in the ATLAS experiment with two muons escaping to the left can be seen in Fig. 11.10 [11]..

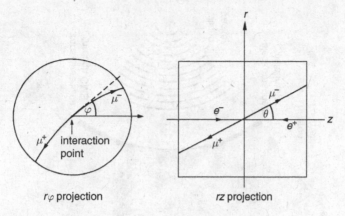

$r\varphi$ projection rz projection

Fig. 11.8. Track reconstruction in a solenoid detector (shown for an event $e^+ e^- \to \mu^+ \mu^-$).

If N coordinates are measured along a track of total length L with an accuracy of $\sigma_{r\varphi}$ in a magnetic field B, the transverse momentum resolution caused by the track measurement error is found to be [5], see Eq. (11.35),

$$\left. \frac{\sigma(p)}{p_T} \right|^{\text{track error}} = \frac{\sigma_{r\varphi} \; [\text{m}]}{0.3 \; B \; [\text{m}] \; (L \; [\text{m}])^2} \sqrt{\frac{720}{N+4}} \cdot p_T \; [\text{GeV}/c] \; . \qquad (11.42)$$

In addition to the track error one has to consider the multiple-scattering error. This is obtained from Eq. (11.24) for the general case of also non-relativistic velocities β as

$$\left. \frac{\sigma(p)}{p_T} \right|^{\text{MS}} = 0.045 \; \frac{1}{\beta} \frac{1}{B \; [\text{T}] \sqrt{L \; [\text{m}] \; X_0 \; [\text{m}]}} \; , \qquad (11.43)$$

where X_0 is the average radiation length of the material traversed by the particle.

The total momentum of the particle is obtained from p_T and the polar angle θ to be

$$p = \frac{p_T}{\sin \theta} \; . \qquad (11.44)$$

As in the transverse plane, the measurement of the polar angle contains a track error and multiple-scattering error.

If the z coordinate in the track detector is determined with an accuracy $\sigma(z)$, the error on the measurement of the polar angle can be derived from a simple geometrical consideration to be

$$\sigma(\theta) = \sin^2(\theta) \frac{\sigma(z)}{r} = \sin(2\theta) \frac{\sigma(z)}{2z} \; . \qquad (11.45)$$

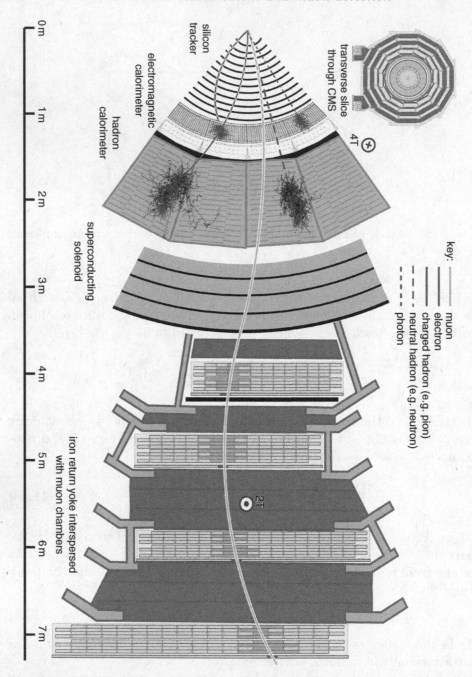

Fig. 11.9.   Sketch demonstrating the particle-identification possibilities in the CMS experiment at CERN. A muon originating from the vertex is deflected in the central solenoidal magnet. The backbending of the muon is clearly visible in the outer magnetic spectrometer [10].

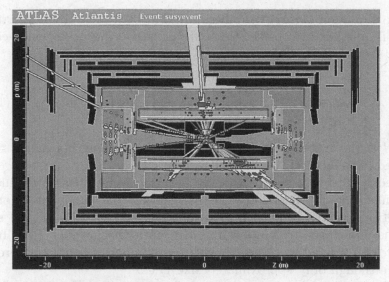

Fig. 11.10. Event simulation for the production of supersymmetric particles in ATLAS and reconstruction of the tracks in the various subdetector components with two energetic muons escaping to the left [11]. The central part of ATLAS incorporates a solenoidal field, while the outer section of the experiment uses toroidal magnets.

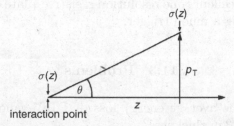

Fig. 11.11. Illustration of the polar-angle measurement error for the case of only two coordinates, defining a track. p_T is the transverse momentum to the beam.

(For high-energy particles the particle track in the rz plane is a straight line, see Fig. 11.11.) If the particle track is measured in N equidistant steps each with an error $\sigma(z)$ along the track length L, the angular uncertainty is obtained to be [4, 5]

$$\sigma(\theta)|^{\text{track error}} = \frac{\sigma(z)}{L}\sqrt{\frac{12(N-1)}{N(N+1)}} \ . \tag{11.46}$$

In this formula z is the projected track length in the z direction which is normally on the same order of magnitude as the transverse length of a track. Equation (11.46) describes only the track measurement error. In addition, one has to consider the multiple-scattering error which can be derived from Eq. (1.50) to be

$$\sigma(\theta)|^{\mathrm{MS}} = \frac{0.0136}{\sqrt{3}} \cdot \frac{1}{p\,[\mathrm{GeV}/c]} \cdot \sqrt{\frac{l}{X_0}} \,, \qquad (11.47)$$

where l is the track length (in units of radiation lengths) and $\beta = 1$ is assumed. The factor $1/\sqrt{3}$ is motivated in [12, 13].

Gaseous detectors with extremely low transverse mass are generally used in solenoids. Therefore, the momentum measurement error due to multiple scattering plays only a minor rôle. Equation (11.42) shows that the momentum resolution improves with the product BL^2. It also improves for a fixed track length with the number of track measurement points although only approximately like $1/\sqrt{N}$.

In the past multiwire proportional chambers or drift chambers have been used as particle trackers in muon spectrometers. To cover large areas, streamer tubes with digital readout, muon drift tubes or resistive-plate chambers – often inserted into slots in the magnetised iron – can be used. For experiments at the Large Hadron Collider at CERN momentum resolutions of $\Delta p/p < 10^{-4} \times p/(\mathrm{GeV}/c)$ for $p > 300\,\mathrm{GeV}/c$ are envisaged. Because of their excellent time resolution resistive-plate chambers can also be used for deriving a muon trigger.

11.3 Problems

11.1 What is the average energy loss of 1 TeV muons in a solid-iron magnet of 3 m thickness?

11.2 In gaseous detectors track reconstruction is often hindered by δ electrons which spiral in the magnetic field thus producing many hits. For LHC experiments track multiplicities of 100 charged particles per beam crossing are not uncommon. Low-momentum electrons are not so serious because their helices occupy only a small volume. High-momentum electrons are only slightly deflected. It is the δ rays with bending radii between 5 cm and 20 cm that represent problems.

Estimate the number of δ rays with bending radii between 5 cm and 20 cm in a 3 m-diameter argon-filled track detector at atmospheric pressure for a magnetic field of 2 T. Assume that the charged particles that create δ rays are very energetic ($\gg 10\,\mathrm{GeV}$).

11.3 High-resolution β-ray spectroscopy can be accomplished with a *double-focussing semicircular magnetic spectrometer* [14–16]. The magnetic field in this spectrometer is axially symmetric but inhomogeneous in the radial direction like

$$B(\rho) = B(\rho_0) \left(\frac{\rho_0}{\rho} \right)^n , \quad 0 < n < 1 ,$$

where ρ_0 is the bending radius of the central orbit. Focussing in radial direction is achieved after an angle of

$$\Theta_\rho = \frac{\pi}{\sqrt{1-n}}$$

and in axial direction after [16]

$$\Theta_\varphi = \frac{\pi}{\sqrt{n}} .$$

(a) Work out the radial dependence of the guiding field and determine the angle at which double focussing is achieved.
(b) What kind of average energy loss will a 10 keV electron experience in such a spectrometer ($\rho_0 = 50\,\text{cm}$, $\frac{dE}{dx}(10\,\text{keV}) = 27\,\text{keV/cm}$, pressure $p = 10^{-3}\,\text{Torr}$)? How many ionisation processes would this correspond to?

11.4 Most colliders use magnetic quadrupoles to focus the beam into the interaction point, because a beam of small transverse dimensions ensures high luminosity. The magnetic bending power must be proportional to the distance of the charged particle from the ideal orbit, i.e., particles far away from the central orbit must experience a stronger deflection than those that are already close to the desired orbit.

It has been shown that the bending angle θ depends on the length of the magnetic field, ℓ, and the bending radius ρ like, see Eq. (11.6),

$$\theta = \frac{\ell}{\rho} = \frac{\ell}{p} e B_y \propto x , \quad \text{i.e.} \quad B_y \cdot \ell \propto x ,$$

where B_y is the magnetic field strength that causes the focussing in the x direction. In the direction perpendicular to x a bending field B_x is required with the corresponding property

$$\theta \propto B_x \ell \propto y .$$

For practical reasons the length ℓ of the quadrupole is fixed. How has the shape of the iron yoke of the quadrupole to look like so that it produces a magnetic field with the desired properties?

References

[1] KASCADE-Grande, T. Antoni, A. Bercuci, *et al.*, A Large Area Limited Streamer Tube Detector for the Air Shower Experiment, KASCADE-Grande, *Nucl. Instr. Meth.* **533** (2003) 387–403

[2] C. Bacci *et al.*, Performance of the RPC's for the ARGO Detector Operated at the YANGBAJING laboratory (4300 m a.s.l.), Prepared for 6th Workshop on Resistive Plate Chambers and Related Detectors (RPC 2001), Coimbra, Portugal, 26–27 November 2001, *Nucl. Instr. Meth.* **A508** (2003) 110–15

[3] K. Kleinknecht, *Detectors for Particle Radiation*, 2nd edition, Cambridge University Press, Cambridge (1998); *Detektoren für Teilchenstrahlung*, Teubner, Wiesbaden (2005)

[4] K. Kleinknecht, *Detektoren für Teilchenstrahlung*, Teubner, Stuttgart (1984, 1987, 1992); *Detectors for Particle Radiation*, Cambridge University Press, Cambridge (1986)

[5] R.L. Glückstern, Uncertainties in Track Momentum and Direction due to Multiple Scattering and Measurement Errors, *Nucl. Instr. Meth.* **24** (1963) 381–9

[6] R. Baumgart *et al.*, Interaction of 200 GeV Muons in an Electromagnetic Streamer Tube Calorimeter, *Nucl. Instr. Meth.* **A258** (1987) 51–7

[7] C. Grupen, Electromagnetic Interactions of High Energy Cosmic Ray Muons, *Fortschr. der Physik* **23** (1976) 127–209

[8] CosmoALEPH Collaboration, F. Maciuc *et al.*, Muon-Pair Production by Atmospheric Muons in CosmoALEPH, *Phys. Rev. Lett.* **96** (2006) 021801, 1–4

[9] S.E. Baru *et al.*, Experiments with the MD-1 Detector at the e^+e^- Collider VEPP-4 in the Energy Region of Upsilon Mesons, *Phys. Rep.* **267** (1996) 71–159

[10] CMS Collaboration, http://cmsinfo.cern.ch/Welcome.html/; http://cmsinfo.cern.ch/Welcome.html/CMSdocuments/DetectorDrawings/Slice/CMS_Slice.gif

[11] ATLAS Collaboration: http://atlantis.web.cern.ch/atlantis/

[12] Particle Data Group, Review of Particle Properties, *Phys. Lett.* **239** (1990) 1–516

[13] Particle Data Group, Review of Particle Properties, *Phys. Rev.* **D45** (1992) 1–574; Particle Data Group, *Phys. Rev.* **D46** (1992) 5210 (Errata)

[14] K. Siegbahn (ed.), *Alpha, Beta and Gamma-Ray Spectroscopy*, Vols. 1 and 2, Elsevier–North Holland, Amsterdam (1968)

[15] G. Hertz, *Lehrbuch der Kernphysik*, Bd. 1, Teubner, Leipzig (1966)

[16]   N. Svartholm & K. Siegbahn, An Inhomogeneous Ring-Shaped Magnetic Field for Two-Directional Focusing of Electrons and Its Applications to β-Spectroscopy, *Ark. Mat. Astron. Fys. Ser.* **A33**, Nr. 21 (1946) 1–28; N. Svartholm, The Resolving Power of a Ring-Shaped Inhomogeneous Magnetic Field for Two-Directional Focusing of Charged Particles, *Ark. Mat. Astron. Fys. Ser.* **A33**, Nr. 24 (1946) 1–10; K. Siegbahn & N. Svartholm, Focusing of Electrons in Two Dimensions by an Inhomogeneous Magnetic Field, *Nature* **157** (1946) 872–3; N. Svartholm, Velocity and Two-Directional Focusing of Charged Particles in Crossed Electric and Magnetic Fields, *Phys. Rev.* **74** (1948) 108–9

12

Ageing and radiation effects

Life would be infinitely happier if we could only be born at the age of
eighty and gradually approach eighteen.

Mark Twain

12.1   Ageing effects in gaseous detectors

Ageing processes in gaseous detectors are about as complicated and
unpredictable as in humans.

Avalanche formation in multiwire proportional or drift chambers can
be considered as a microplasma discharge. In the plasma of an electron
avalanche, chamber gases, vapour additions and possible contaminants
are partially decomposed, with the consequence that aggressive radicals
may be formed (molecule fragments). These *free radicals* can then form
long chains of molecules, i.e., *polymerisation* can set in. These polymers
may be attached to the electrodes of the wire chamber, thereby reducing
the gas amplification for a fixed applied voltage: the chamber ages. After
a certain amount of charge deposited on the anodes or cathodes, the
chamber properties deteriorate so much that the detector can no longer
be used for accurate measurements (e.g. energy-loss measurements for
particle identification).

Ageing phenomena represent serious problems for the uses of gaseous
detectors especially in *harsh radiation environments*, such as at future
high-intensity experiments at the Large Hadron Collider at CERN. It is
not only that gas mixtures for detectors have to be properly chosen, also
all other components and construction materials of the detector systems
have to be selected for extraordinary radiation hardness.

346

Which processes are of importance now for the premature ageing of gaseous detectors, and which steps can be taken to increase the lifetime of the chambers?

Ageing processes are very complex. Different experimental results concerning the question of ageing are extremely difficult to compare, since ageing phenomena depend on a large number of parameters and each experiment usually has different sets of parameters. Nevertheless, some clear conclusions can be drawn even though a detailed understanding of ageing processes has yet to come. The main parameters which are related to wire-chamber ageing are characterised below [1–13].

A multiwire proportional chamber, drift chamber or more general gaseous detector is typically filled with a mixture of a noble gas and one or several vapour additions. *Contaminants*, which are present in the chamber gas or enter it by outgassing of detector components, cannot be completely avoided. The electron avalanche, which forms in such a gas environment in the immediate vicinity of the anode structure, produces a large number of molecules. The energy required for the break-up of covalent molecule bonds is typically a factor of three lower than the ionisation potential. If electrons or photons from the avalanche break up a gas molecule bond, radicals that normally have quite a large dipole moment are formed. Because of the large electric field strength in the vicinity of the electrodes, these radicals are attracted mainly by the anode and may form in the course of time a poorly or non-conducting *anode coating*, which can cause the electrodes to be noisy. Conducting anode deposits increase the anode diameter, thereby reducing the gas amplification. Because of the relatively large chemical activity of radicals, different compounds can be produced on the anode in this way. The rate of polymerisation is expected to be proportional to the density of radicals which in itself is proportional to the electron density in the avalanche. Polymerisation effects, therefore, will increase with increasing charge deposition on the anode. However, not only the anode is affected. In the course of polymer formation (e.g. positive) polymers may be formed which migrate slowly to the cathode. This is confirmed by patterns of 'wire shadows' which can be formed by deposits on planar cathodes [1, 2].

Typical deposits consist of carbon, thin oxide layers or silicon compounds. Thin metal oxide layers are extremely photosensitive. If such layers are formed on cathodes, even low-energy photons can free electrons from the cathodes via the photoelectric effect. These photoelectrons are gas amplified thus increasing the charge deposition on the anode, thereby accelerating the ageing process. Deposits on the electrodes can even be caused during the construction of the chamber, e.g. by finger prints. Also the gases which are used, even at high purity, can be contaminated in the course of the manufacturing process by very small oil droplets or silicon dust (SiO_2). Such contaminants at the level of several ppm can cause significant ageing effects.

Once a coating on the electrodes has been formed by deposition, high electric fields between the deposit layer and the electrode can be produced by secondary electron emission from the electrode coating (*Malter effect* [14]). As a consequence of this, these strong electric fields may cause field-electron emission from the electrodes, thereby reducing the lifetime of the chamber.

Which are now the most sensitive parameters that cause ageing or accelerate ageing, and which precautions have to be considered for chamber construction? In addition, it is an interesting question whether there are means to clean up (*rejuvenate*) aged wires.

Generally, it can be assumed that pure gases free of any contaminants will delay ageing effects. The gases should be as resistant as possible to polymerisation. It only makes sense, however, to use ultrapure gases if it can be guaranteed that contamination by outgassing of chamber materials or gas pipes into the detector volume can be prevented.

These precautions are particularly important for the harsh environments at high-intensity colliders where long-term operation with limited access to the detectors is foreseen. The gaseous detectors must be able to withstand particle fluences of up to 10^{15}–$10^{16}\,\mathrm{cm}^{-2}$ and charge deposits on chamber wires of $\approx 1\,\mathrm{C/cm}$ per year.

In standard multiwire proportional chambers this margin can be reached with certain gas mixtures and carefully designed chambers

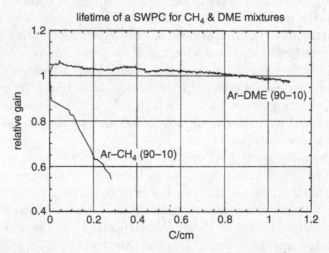

Fig. 12.1. Comparison of the gain variation of a clean single-wire proportional chamber filled with argon–methane or argon–dimethyl ether under irradiation [15].

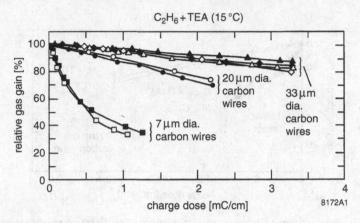

Fig. 12.2. Wire-chamber ageing in multiwire proportional chambers filled with C_2H_6 + TEA for different anode-wire diameters as a function of the deposited charge on the anode [16].

constructed of selected materials. Figure 12.1 shows the gain variation of a clean single-wire proportional chamber under irradiation for two different gas mixtures [15]. While the gain loss in standard argon–methane mixtures (90:10) is substantial already after $0.2\,C/cm$, an argon–dimethyl-ether $((CH_3)_2O)$ filling (90:10) loses less than 10% in gain after an accumulated charge of $1\,C/cm$.

Figure 12.2 shows the ageing properties of a multiwire proportional chamber filled with a mixture of ethane (C_2H_6) and the photosensitive gas TEA (triethylamine, $(C_2H_5)_3N$) often used for the detection of Cherenkov photons in gaseous ring-imaging Cherenkov counters (RICH) [16]. The gain change correlates with the diameter of the anode wire, which is not a surprise because depositions have the largest effect for low-diameter wires. Already for charge doses as low as $1\,mC/cm$ anode wire significant gain losses are experienced.

Figure 12.3 shows a comparison between TEA and TMAE (Tetrakis-dimethylamino ethylene, $[(CH_3)_2N]_2C = C[N(CH_3)_2]_2$) for a wire diameter of $20\,\mu m$ [16]. The ageing rate with TMAE is considerably faster than that with TEA. Electrode coatings observed in TMAE can also induce the Malter effect, as indicated in the figure. Detectors of this type can only be used in gaseous Cherenkov counters in relatively low radiation environments.

Apart from multiwire proportional and drift chambers, micropattern detectors or gas electron multipliers can also be used. Here it is not only important to reduce ageing effects such as gain losses, but also to maintain best spatial resolution and best timing properties.

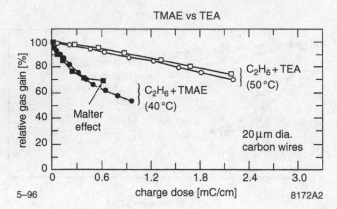

Fig. 12.3. Comparison of wire-chamber ageing in multiwire proportional chambers filled with C_2H_6+TEA versus C_2H_6+TMAE for a fixed anode-wire diameter in its dependence on the deposited charge on the anode wire [16].

Naturally, prototype detectors are tested for ageing under laboratory conditions with X rays, γ rays or electrons. In addition, there is also experience of ageing from low-rate colliders, such as from the Large Electron–Positron collider LEP at CERN. The detectors operated under these conditions may not function in high-intensity beams of hadrons where the ionisation densities can exceed those obtained with electrons by a large margin (up to 100 times of minimum-ionising particles). Especially α particles and nuclear recoils with high Z and low velocities lead to huge charge densities which might initiate streamer or even spark formation. High intensities will also create *space-charge effects*, which are sure to lead to gain losses.

Sparks are particularly dangerous because they can cause local damage to the electrodes which might introduce low-resistivity channels thereby leading the way for further discharges. This is because local enhancements of the field are formed at the edges of the imperfections on the electrodes created by the spark.

Gases of interest for high-rate applications are, e.g., $Ar(Xe)/CO_2$ or $Ar(Xe)/CO_2/O_2$. CF_4 mixtures also show little ageing, but CF_4 is quite aggressive and limits the choice of construction materials for detectors and gas systems. Under high irradiation the CF_4 molecule might be decomposed thereby creating fluorine radicals and hydrofluoric acid (HF) which might attack the chamber body (Al, Cu, glass, G-10, etc.). There is no final proof that CF_4 is a reliable chamber gas for harsh radiation environments. To be on the safer side, hydrocarbons should be avoided.

For the expensive xenon-based gas mixtures in large detector systems one is forced to use recirculation systems where special purification elements are required to remove long-lived radicals. Also cleaning runs with Ar/CO_2 might be helpful.

Apart from undesired contaminants, additions of atomic or molecular oxygen and/or water may take a positive influence on ageing phenomena.

Special care has to be taken for possible *silicon contaminants*. Silicon – as one of the most frequently occurring elements on Earth – is contained in many materials which are used for chamber construction (like G-10 (glass fibre–reinforced epoxy resin), various oils, lubricants, rubber and adhesives, grease, O-rings, molecular sieves) and in dust. Silicon is frequently contained in gas bottles in the form of silane (SiH_4) or tetrafluorsilane (SiF_4). Silicon can, together with hydrocarbon contaminants, form silicon carbide; this, together with oxygen silicates, which have – because of their high mass – a low volatility and almost are impossible to remove from the chamber volume, will be preferentially deposited on the electrodes.

Apart from avoiding unfavourable contaminants in the chamber gas, and by carefully selecting components for chamber construction and the gas system, some constructional features can also be recommended to suppress ageing affects.

Larger cathode surfaces normally have smaller electric fields at their surface compared to layers of cathode wires. Therefore, continuous cathodes have a reduced tendency for deposition compared to cathode wires. The effect of deposits on thin anode wires is quite obviously enhanced compared to thick anode wires. Also careful selection of the electrode material can be of major influence on the lifetime of the chamber. Gold-plated tungsten wires are quite resistant against contaminants, while wires of high-resistance material (Ni/Cr/Al/Cu alloys) tend to react with contaminants or their derivatives, and this may lead to drastic ageing effects.

Certain contaminants and deposits can be dissolved at least partially by additions of, e.g., water vapour or acetone. Macroscopic deposits on wires can be 'burnt off' by deliberately causing sparks. On the other hand, sparking may also lead to the formation of carbon fibres (*whiskers*) which significantly reduce the lifetime of chambers, and can even induce wire breaking.

Figure 12.4 shows some examples of deposits on anode wires [3]. On the one hand, one can see more or less continuous anode coatings which may alter the surface resistance of the anode. On the other hand, also hair-like polymerisation structures are visible which will decisively deteriorate the field quality in the vicinity of the anode wire and also may lead to sparking.

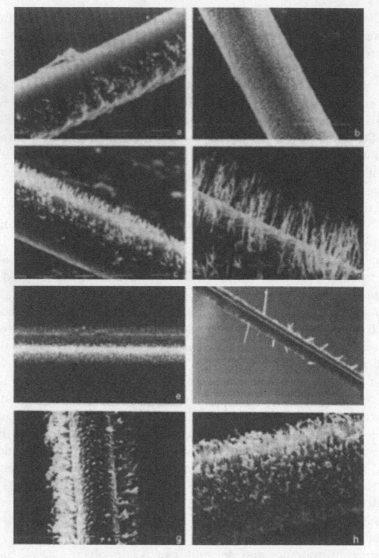

Fig. 12.4.   Examples of depositions on anode wires [3].

12.2   Radiation hardness of scintillators

Scintillators, Cherenkov media, wavelength shifters and readout fibres
are susceptible to degradation due to both natural ageing and radiation effects. High radiation fields will reduce the light output and
transmission of the transparent media. If scintillators are used in calorimeters as sampling element, non-uniformities of response might be
created as a consequence of non-uniform irradiation. Light losses and

reduction in *transparency* will increase the constant term in the relative energy resolution, in particular, for hadron calorimeters in harsh radiation environments.

A possible radiation damage is also sensitive to details of the detector construction (choice of the scintillator material and the thickness of the scintillator plates) and its operation characteristics (e.g. selection of the wavelength to be used). It also makes a difference whether the scintillator sheets are sealed in vacuum or exposed to air (oxygen). Tests of radiation hardness with X rays, γ rays or electrons may not give conclusive information of the behaviour in high-radiation environments of hadrons or heavily ionising particles. It is also problematic to extrapolate from a short-term irradiation test with high doses to long-term operation with comparable doses but moderate dose rates.

At high absorbed doses plastic scintillators suffer a deterioration of both light output and transparency. This effect is poorly understood but usually ascribed to the creation of colour centres caused by the radiation. Since the transparency decreases, the total light output reduction of the scintillator depends strongly on the counter size and shape. Samples of the popular scintillators, BC-408, BC-404 and EJ-200 of 6 cm length, were studied in [17]. A light output decrease of 10%–14% was found after 600 Gy (60 krad) of absorbed dose. For a large number of new and known scintillators the damage under γ irradiation was studied in detail in [18]. The well-known NE-110 polyvinyltoluene scintillator retained about 60% of the initial light output at an absorbed dose of 34 kGy (3.4 Mrad). The deterioration effect strongly depends on the chemical composition of the material. The best scintillators based on polystyrene kept 70%–80% of light output after absorption of 100 kGy (10 Mrad). However, a noticeable transparency deterioration was observed already at 2–3 kGy (200–300 krad).

It should be noted that the radiation damage does not only depend on the total absorbed dose but from the dose rate as well. A certain recovery of the light output was observed after several weeks for some scintillators.

The radiation tolerance of inorganic scintillation crystals used in electromagnetic calorimeters (see Chap. 8) varies within a wide range [19]. The main radiation-induced effect is a transparency reduction exhibiting a strong dependence of the output signal on the shape and size of the counter. It should be noted that a partial recovery some time after irradiation is observed for some of the materials.

Widely used alkali-halide crystals, e.g. CsI and NaI, have a moderate radiation resistance [20, 21]. A typical dose dependence of the light output for several CsI(Tl) crystals of 30 cm length is presented in Fig. 12.5 [21]. These crystals can be used up to several tens of Gray (a few krad) of absorbed dose which is usually sufficient for low-energy experiments.

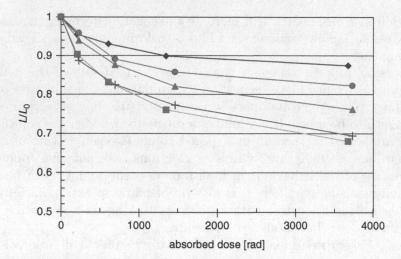

Fig. 12.5.   Relative light output, L/L_0, in its dependence on the absorbed dose for several crystals of 30 cm length [21].

Some of the oxide scintillation crystals show a much improved radiation resistance. It was reported [22, 23] that BGO crystals grown according to a special technology can be used under γ radiation up to 0.8–1 MGy (80–100 Mrad) of absorbed dose. Lead-tungstate crystals will be used in the CMS electromagnetic calorimeter [24] at an expected absorbed dose of up to 10 kGy (1 Mrad) per year.

In contrast to solid detector materials, liquid scintillators have shown excellent levels of radiation resistance. This can be due to the fact that *dislocations* caused by impacts of heavily ionising particles are more easily repaired in liquids.

12.3   Radiation hardness of Cherenkov counters

For Cherenkov media in general – apart from the global reduction of transparency under irradiation – also the change of the frequency-dependent transparency resulting in a modification of the average effective index of refraction, and the introduction of non-uniformities can degrade the performance of this type of detector significantly.

As an example, lead glass, frequently used for calorimeters, loses transparency up to a considerable extent at an absorbed dose of several tens of Gy (several krad). However, cerium admixtures improve the radiation hardness substantially, and cerium-doped glasses can withstand an irradiation up to 100 Gy (10 krad) [25].

To build a Cherenkov detector of high radiation resistance one can use quartz as radiator. Tests of the prototype of the CMS hadronic forward calorimeter using quartz fibres embedded into an iron absorber showed that an absorbed dose of 1 MGy (100 Mrad) induces an additional light attenuation of about 30% per m at a wavelength of 450 nm [26].

12.4 Radiation hardness of silicon detectors

The performance of silicon detectors depends also on the radiation environment. Heavily ionising particles or neutrons may displace atoms in the silicon lattice producing *interstitials* and thereby affecting their function. Strongly ionising particles will deposit a large amount of charge locally thus producing space-charge effects. This can also happen when energetic hadrons generate nuclear recoils by nuclear reactions within the silicon detector.

The radiation damage in silicon can be subdivided into bulk and surface damage. The displacement effect in the bulk leads to increased *leakage currents*. Charge carriers produced by the signal particles can be trapped in these defects and space charge can build up which might require to change the operating voltage. If sufficient energy is transferred to recoil atoms, they can generate dislocations themselves thus creating dislocation clusters.

Radiation damage at the surface can lead to charge build-up in the surface layers with the consequence of increased surface currents. In silicon pixel detectors also the interpixel isolation is affected.

Bulk damage leads to an increase in the reverse-bias current. Since this is strongly temperature-dependent, even a modest cooling can reduce this effect. Due to the build-up of space charge in the detector, the required operating voltage to collect the generated signal charge drops initially with fluence until the positive and negative space charges balance. At large fluences the negative space charge starts to dominate and the required operating voltage increases. Silicon pixel or strip detectors can stand applied voltages up to $\approx 500\,\text{V}$.

A radiation-induced increase of the reverse current of many types of silicon devices is presented in Fig. 12.6 as a function of the radiation intensity equivalent to 1 MeV neutrons, Φ_{eq} [27, 28]. This dependence can be expressed by the simple equation

$$I\{\text{A}\} = \alpha \cdot \Phi_{\text{eq}}\{\text{cm}^{-2}\}\, V\{\text{cm}^3\}\;, \qquad (12.1)$$

where $\alpha = (3.99 \pm 0.03) \times 10^{-17}\,\text{A cm}$, when the silicon device has undergone a certain annealing after irradiation.

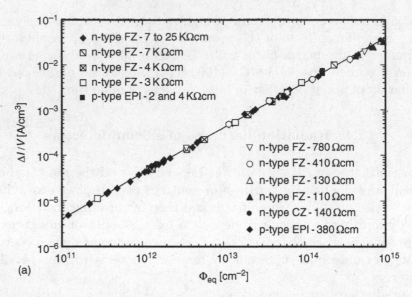

Fig. 12.6. Reverse current of different silicon detectors produced by various technologies induced by the exposition to an equivalent fluence Φ_{eq}. The measurements were performed after a heat treatment for 80 min at 60 °C [28].

The mobility of defects in silicon can be substantially suppressed by strong cooling. On the other hand, additions of oxygen have the effect of 'capturing' vacancies in the silicon lattice and it could also capture interstitials. These *oxygenated silicon detectors* – even at fairly moderate cooling – are significantly radiation harder compared to standard silicon devices without additions of oxygen [29].

It has to be mentioned that the radiation damage of a silicon detector, as seen in an increase in leakage current, effective doping change or the creation of trapped states, decreases with time after the end of the irradiation. This 'improvement' of the properties of the damage in silicon detectors depends critically on the temperature at which the counter is stored. This partial disappearance of the radiation damage has been called 'annealing'. The crystal may even become perfect again, for example, if the vacancies created by irradiation are filled in again by silicon interstitials. Frequently, the defects may also be transformed to more stable defect types which may have less harmful properties [30]. The term 'annealing' already indicates that the defects are in general quite stable up to a certain temperature. They may disappear if a given 'annealing temperature' is exceeded. The characteristic properties of such defects can be reduced at higher temperatures, and this may even happen over a very extended period of up to a year after the irradiation has ended. The details of the annealing process appear to be rather complicated and poorly understood.

On the other hand, silicon detectors can also be trained to tolerate high radiation levels. The idea of this *radiation hardening* is to use techniques that result in silicon material whose properties are not significantly altered when they are exposed to radiation. To achieve this, different approaches can be followed: stable defects can be created by controlled doping or by the design of radiation-tolerant device structures of the silicon material.

When trying to improve the radiation hardness of silicon detectors, one has always to keep in mind that the associated readout electronics, often integrated onto the silicon chip, is exposed in the same way to the radiation as the detector. Therefore, in designing a radiation-hard silicon detector, both aspects, the detector and the readout, have to be considered together.

For silicon detectors and readout components used in a harsh radiation environment in locations where they are not easily accessible, like in the running of LHC experiments, sufficient safety factors have to be foreseen to guarantee the proper functioning of the devices [31, 32].

12.5 Problems

12.1 Defects produced by irradiation decrease in the quiet phase after irradiation according to

$$N_\mathrm{d}(t) = N_\mathrm{d}(0)\,\mathrm{e}^{-t/\tau}\ ,$$

where the decay time τ depends on the activation energy E_a and on the annealing temperature T like

$$\tau(T) = \tau_0\,\mathrm{e}^{E_\mathrm{a}/kT}\ .$$

In room-temperature annealing the decay time can easily be one year ($E_\mathrm{a} = 0.4\,\mathrm{eV}$, $kT = 1/40\,\mathrm{eV}$). If the annealing time should be reduced to one month, by how much should the ambient temperature be increased?

12.2 The electron signal in a proportional tube is given by

$$\Delta U^- = -\frac{N\,e}{C\ln(r_\mathrm{a}/r_\mathrm{i})}\ln(r_0/r_\mathrm{i})\ ,$$

where r_0 is the position, where the charge has been created, and $r_\mathrm{a}, r_\mathrm{i}$ are the outer radius of the counter and the anode-wire radius, respectively ($r_\mathrm{a}/r_\mathrm{i} = 100$, $r_0/r_\mathrm{i} = 2$). Work out the gain loss if the anode-wire diameter is increased by 10% due to a conductive deposition.

References

[1] J.A. Kadyk, J. Va'vra & J. Wise, Use of Straw Tubes in High Radiation Environments, *Nucl. Instr. Meth.* **A300** (1991) 511–17

[2] J. Va'vra, Review of Wire Chamber Ageing, *Nucl. Instr. Meth.* **A252** (1986) 547–63, and references therein

[3] J.A. Kadyk, Wire Chamber Aging, *Nucl. Instr. Meth.* **300** (1991) 436–79

[4] R. Bouclier *et al.*, Ageing of Microstrip Gas Chambers: Problems and Solutions, *Nucl. Instr. Meth.* **A381** (1996) 289–319

[5] *Proceedings of the Workshop on Radiation Damage to Wire Chambers*, LBL-21170, Lawrence-Berkeley Laboratory (1986) 1–344

[6] Proceedings of the International Workshop on Ageing Phenomena in Gaseous Detectors, *Nucl. Instr. Meth.* **A515** (2003) 1–385

[7] R. Kotthaus, A Laboratory Study of Radiation Damage to Drift Chambers, *Nucl. Instr. Meth.* **A252** (1986) 531–44

[8] A. Algeri *et al.*, *Anode Wire Ageing in Proportional Chambers: The Problem of Analog Response*, CERN-PPE-93-76 (1993); *Nucl. Instr. Meth.* **A338** (1994) 348–67

[9] J. Wise, *Chemistry of Radiation Damage to Wire Chambers*, Thesis, LBL-32500 (92/08), Lawrence-Berkeley Laboratory (1992)

[10] M.M. Fraga *et al.*, Fragments and Radicals in Gaseous Detectors, *Nucl. Instr. Meth.* **A323** (1992) 284–8

[11] M. Capéans *et al.*, *Ageing Properties of Straw Proportional Tubes with a Xe-CO_2-CF_4 Gas Mixture*, CERN-PPE-93-136 (1993); *Nucl. Instr. Meth.* **A337** (1994) 122–6

[12] M. Titov *et al.*, *Summary and Outlook of the International Workshop on Aging Phenomena in Gaseous Detectors*, DESY, Hamburg, Germany, 2–5 October 2001, hep-ex/0204005, *ICFA Instrum. Bull.* **24** (2002) 22–53

[13] M. Titov, *Radiation Damage and Long-Term Aging in Gas Detectors*, physics/0403055 (2004)

[14] L. Malter, Thin Film Field Emission, *Phys. Rev.* **50** (1936) 48–58

[15] M. Capeans, Aging and Materials: Lessons for Detectors and Gas Systems, International Workshop on Aging Phenomena in Gaseous Detectors, DESY, Hamburg, Germany, 2–5 October 2001, *Nucl. Instr. Meth.* **A515** (2003) 73–88

[16] J. Va'vra, *Wire Ageing with the TEA Photocathode*, SLAC-Pub-7168 (1996); *Nucl. Instr. Meth.* **A387** (1997) 183–5

[17] Zhao Li *et al.*, Properties of Plastic Scintillators after Irradiation, *Nucl. Instr. Meth.* **A552** (2005) 449–55

[18] V.G. Vasil'chenko *et al.*, New Results on Radiation Damage Studies of Plastic Scintillators, *Nucl. Instr. Meth.* **A369** (1996) 55–61

[19] R.Y. Zhu, Radiation Damage in Scintillating Crystals, *Nucl. Instr. Meth.* **A413** (1998) 297–311

[20] T. Hryn'ova *et al.*, A Study of the Impact of Radiation Exposure on Uniformity of Large CsI(Tl) Crystals for the BABAR Detector, *Nucl. Instr. Meth.* **A535** (2004) 452–6

[21] D.M. Beylin *et al.*, Study of the Radiation Hardness of CsI(Tl) Scintillation Crystals, *Nucl. Instr. Meth.* **A541** (2005) 501–15

[22] Ya.V. Vasiliev *et al.*, BGO Crystals Grown by a Low Thermal Gradient Czochralski Technique, *Nucl. Instr. Meth.* **A379** (1996) 533–5

[23] K.C. Peng *et al.*, Performance of Undoped BGO Crystals under Extremely High Dose Conditions, *Nucl. Instr. Meth.* **A427** (1999) 524–7

[24] *The Compact Muon Solenoid Technical Proposal*, CERN/LHCC 94-38 (1994)

[25] M. Kobayashi *et al.*, Radiation Hardness of Lead Glasses TF1 and TF101, *Nucl. Instr. Meth.* **A345** (1994) 210–2

[26] I. Dumanoglu *et al.*, *Radiation Hardness Studies of High OH$^-$ Quartz Fibers for a Hadronic Forward Calorimeter of the Compact Muon Solenoid Experiment at the Large Hadron Collider*, pp. 521–5, in Proc. 10th Int. Conf. on Calorimetry in Particle Physics, Pasadena, California, USA, 2002, World Scientific, Singapore (2002)

[27] M. Turala, Silicon Tracking Detectors – Historical Overview, *Nucl. Instr. Meth.* **A541** (2005) 1–14

[28] G. Lindström *et al.*, Radiation Hard Silicon Detectors Developments by the RD48 (ROSE) Collaboration, *Nucl. Instr. Meth.* **A466** (2001) 308–26

[29] H. Spieler, pp. 262–3, in S. Eidelman *et al.*, *Phys. Lett.* **B592 Vol. 1–4** (2004) 1–1109

[30] G. Lutz, *Semiconductor Radiation Detectors*, Springer, Berlin (1999)

[31] *Radiation Hardness Assurance* http://lhcb-elec.web.cern.ch/lhcb-elec/html/radiation_hardness.htm; C. Civinini & E. Focardi (eds.), Procs. 7th Int. Conf. on Large Scale Applications and Radiation Hardness of Semiconductor Detectors, *Nucl. Instr. Meth.* **A570** (2007) 225–350

[32] C. Leroy & P.-G. Rancoita, Particle Interaction and Displacement Damage in Silicon Devices Operated in Radiation Environments, *Rep. Proc. Phys.* **70** (2007) 493–625

13

Example of a general-purpose detector: Belle

Our job in physics is to see things simply, to understand a great many complicated phenomena, in terms of a few simple principles.

Steven Weinberg

A present-day experiment in high energy physics usually requires a multipurpose experimental setup consisting of at least several (or many) subsystems. This setup (called commonly 'detector') contains a multitude of sensitive channels which are necessary to measure the characteristics of particles produced in collisions or decays of the initial particles. A typical set of detector properties includes abilities of tracking, i.e. measurement of vertex coordinates and charged-particle angles, measurements of charged-particle momenta, particle energy determination and particle identification. A very important system is the trigger which detects the occurrence of an event of interest and produces a signal to start the readout of the information from the relevant channels. Since high energy physics experiments are running for months or years, the important task is to monitor and control the parameters of the detector and to keep them as stable as possible. To fulfil this task the detector is usually equipped with a so-called *slow control system*, which continuously records hundreds of experimental parameters and warns experimentalists if some of them are beyond certain boundaries.

To control the process of accumulating statistics and calculating the cross sections and decay rates, a luminosity measurement system is mandatory (for the term definition, see Chap. 4).

One of the general-purpose detectors, Belle, is discussed in this chapter. The Belle detector for experiments at the KEKB, an energy-asymmetric B factory with high luminosity, has been constructed at KEK in Tsukuba, Japan, to study CP violation in B-meson decays. The KEKB e^+e^- collider [1] is based on two separate rings for electrons (8 GeV) and positrons (3.5 GeV) installed in a tunnel with a circumference of about 3 km.

The present luminosity is about $1.6 \cdot 10^{34}\,\mathrm{cm}^{-2}\,\mathrm{s}^{-1}$ which is achieved at a 2 A positron current and a 1.5 A electron current.

The Belle detector was constructed in 1994–8. Since 1999, the detector has been running and by now (beginning of 2007) this experiment has collected about $700\,\mathrm{fb}^{-1}$ of integrated luminosity.

13.1 Detector components

The detector layout is shown in Fig. 13.1 and its detailed description can be found in [2]. Most results reported in this chapter are taken from [2].

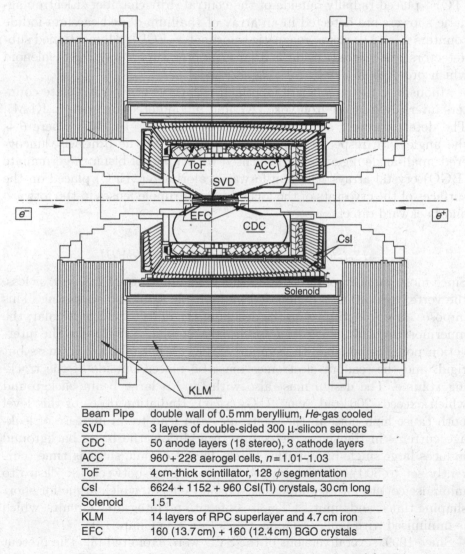

Beam Pipe	double wall of 0.5 mm beryllium, *He*-gas cooled
SVD	3 layers of double-sided 300 μ-silicon sensors
CDC	50 anode layers (18 stereo), 3 cathode layers
ACC	960 + 228 aerogel cells, $n = 1.01$–1.03
ToF	4 cm-thick scintillator, 128 ϕ segmentation
CsI	6624 + 1152 + 960 CsI(Tl) crystals, 30 cm long
Solenoid	1.5 T
KLM	14 layers of RPC superlayer and 4.7 cm iron
EFC	160 (13.7 cm) + 160 (12.4 cm) BGO crystals

Fig. 13.1. Schematic view of the Belle detector [2].

The electron and positron beams cross at an angle of ±11 mrad inside the beam pipe. The central part ($-4.6\,$cm $\leq z \leq 10.1\,$cm) of the beam pipe is a double-wall beryllium cylinder with an inner diameter of 30 mm. A 2.5 mm gap between the inner and outer walls of the cylinder provides a helium gas channel for cooling. Each wall has a thickness of 0.5 mm.

B-meson decay vertices are measured by a silicon vertex detector (SVD) situated just outside the beam pipe. Charged-particle tracking is provided by a 50-layer wire drift chamber (central drift chamber, CDC). Particle identification is based on $\mathrm{d}E/\mathrm{d}x$ measurements in the CDC, in aerogel Cherenkov counters (ACC) and time-of-flight counters (TOF) placed radially outside of the central drift chamber. Electromagnetic showers are detected in an array of thallium-doped caesium-iodide counters (CsI(Tl); electromagnetic calorimeter, ECL). All mentioned sub-detectors are located inside a 3.4 m-diameter superconducting solenoid which provides an axial 1.5 T magnetic field.

Muons and K_L mesons are identified by arrays of resistive-plate counters interleaved in the iron yoke (K_L and muon detection system, KLM). The detector covers a θ region extending from 17° to 150°, where θ is the angle with respect to the beam axis. Part of the otherwise uncovered small-angle region is instrumented with a pair of bismuth-germanate (BGO) crystal arrays (electron forward calorimeter, EFC) placed on the surfaces of the cryostats of the focussing quadrupole lenses in the forward and backward directions.

13.1.1   *The silicon vertex detector (SVD)*

Since most particles of interest in Belle have momenta of $1\,$GeV$/c$ or less, the vertex resolution is dominated by multiple Coulomb scattering. This imposes strict constraints on the design of the detector. In particular, the innermost layer of the *vertex detector* must be placed as close to the inter-action point as possible, the support structure must be low in mass but rigid, and the readout electronics must be placed outside of the tracking volume. The design must also withstand a large beam background which exceeds 200 krad/year (2 kGy/year). Radiation doses of this level both cause high noise in the electronics and lead to an increase of leakage currents in the silicon detectors. In addition, the beam background induces large single-hit counting rates. The electronic shaping time (currently set to 500 ns) is determined by a trade-off between the desire to minimise counting-rate and leakage-current effects, which argue for short shaping times, and input-FET noise of front-end integrated circuits, which is minimised with longer shaping times (see also Chap. 14).

Since 1999, several versions of the SVD were exploited [3]. The present detector, SVD-2.0 [4], is shown in Fig. 13.2. It consists of four layers in a

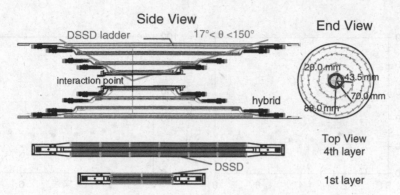

Fig. 13.2. Silicon vertex detector SVD-2.0.

barrel-only design and covers a polar angle of $17° < \theta < 150°$. Each layer is constructed of independent ladders. The ladders comprise *double-sided silicon strip detectors* (DSSDs) reinforced by boron-nitride support ribs.

The double-sided silicon strip detector used in the present SVD is an S4387 microstrip detector manufactured by Hamamatsu Photonics especially for Belle. The thickness of the depleted area is $300\,\mu$m, the bias potentials applied to p and n sides are $-40\,$V and $+40\,$V, respectively. Strip pitches are $75\,\mu$m for the p side, which is used for the z-coordinate measurement, and $50\,\mu$m for the n side that measures φ. The total number of readout channels is 110 592.

The readout chain for this detector is based on the VA1TA integrated circuit which is placed to the ceramic hybrid and connected to the DSSD [5]. The VA1TA chip is manufactured using $0.35\,\mu$m basic-size elements. It comprises 128 readout channels. The small size of the elements provides good radiation tolerance. Each channel contains a charge-sensitive preamplifier followed by a CR–RC shaping amplifier. The outputs of the shapers are fed to track and hold circuits, which consist of capacitors and CMOS switches. The trigger signal initiates the analogue information to be transferred to storage capacitors which then can be sequentially read out. The important feature of this chip is its excellent radiation tolerance of more than $20\,$Mrad ($200\,$kGy).

The back-end electronics is a system of flash analogue-to-digital converters (FADCs), digital signal processors (DSPs) and field-programmable gate arrays (FPGAs), mounted on standard 6U VME boards. The digital signal processors perform on-line common-mode noise subtraction, data sparsification and data formatting.

The impact-parameter resolution depending on the particle momentum is presented in Fig. 13.3. The resolution of the present silicon vertex detector is approximated by the formulae

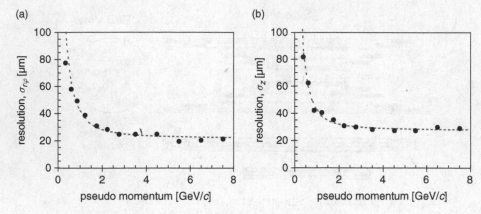

Fig. 13.3. Impact-parameter resolution depending on the particle momentum for the '$r\varphi$' projection (a) and in the 'z' coordinate (b). The measurement was performed with cosmic-ray muons. The dashed lines are approximations according to the quoted formulae. The 'pseudo momentum' is $p\beta \sin^{3/2}\theta$ for the $r\varphi$ projection and $p\beta \sin^{5/2}\theta$ for the z projection.

$$\sigma_{r\varphi}\;[\mu\mathrm{m}] = \sqrt{(22)^2 + \left(\frac{36}{p\;[\mathrm{GeV}/c]\,\beta\,\sin^{3/2}\theta}\right)^2}\;, \qquad (13.1)$$

$$\sigma_{z}\;[\mu\mathrm{m}] = \sqrt{(28)^2 + \left(\frac{32}{p\;[\mathrm{GeV}/c]\,\beta\,\sin^{5/2}\theta}\right)^2}\;. \qquad (13.2)$$

13.1.2   The central drift chamber (CDC)

The *CDC* geometry can be seen in Fig. 13.1 [6]. It is asymmetric in the z direction in order to provide an angular coverage of $17° < \theta < 150°$. The longest wires are 2400 mm long. The inner and outer CDC radii are 102 mm and 874 mm, respectively. The forward and backward small-r regions have conical shapes in order to get clear of the accelerator components while maximising the acceptance.

The chamber has 50 cylindrical layers of anode wires and 8400 drift cells that are organised into 6 axial and 5 small-angle stereo superlayers. The stereo angles in each stereo superlayer are determined by maximising the z measurement capability while keeping the gain variations along the wire below 10%. Thus, the stereo angles vary from -57 mrad to $+74$ mrad.

The individual drift cells are nearly quadratic and, except for the inner two layers, have a maximum drift distance between 8 mm and 10 mm and a

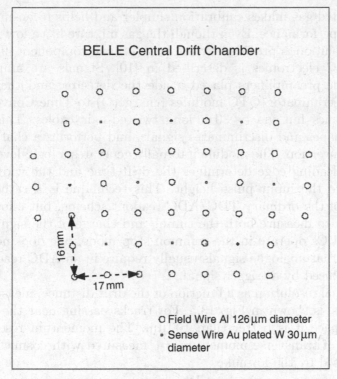

Fig. 13.4. The Belle CDC cell structure.

radial width that ranges from 15.5 mm to 17 mm (Fig. 13.4). The drift cells in the inner layers are smaller than the others having sizes about 5 mm × 5 mm. The sense wires are gold-plated tungsten wires of 30 μm diameter to maximise the electric drift field. To reduce the material, the field wires are made of unplated aluminium. The 126 μm diameter was chosen to keep the electric field on the surface of the wires below 20 kV/cm, the limit for avoiding radiation damage (see Chap. 12). The total wire tension of 3.5 tons is supported by aluminium end plates and carbon-fibre-reinforced-plate (CFRP) cylinder structures that extend between the end plates.

The use of a low-Z gas is important for minimising multiple-Coulomb-scattering contributions to the momentum resolution. Since low-Z gases have a smaller photoelectric cross section than argon-based gases, they have the additional advantage of reduced background from synchrotron radiation. A gas mixture of 50% helium and 50% ethane was selected for the CDC filling. This mixture has a long radiation length (640 m), and a drift velocity that saturates at 4 cm/μs at a relatively low electric field [7, 8]. This is important for operating square-cell drift chambers because of large field non-uniformities inherent to their geometry. The use of a saturated gas makes calibration simpler and helps to ensure reliable and stable performance. Even though the gas mixture has a low Z, a good dE/dx resolution is provided by the large ethane component [9].

The CDC electronics is described in [10]. Signals are amplified by Radeka-type preamplifiers placed inside the detector, and then sent to Shaper–Discriminator–QTC modules (charge(Q)-to-Time Conversion) in the electronics hut via $\approx 30\,\mathrm{m}$ long twisted-pair cables. This module receives, shapes and discriminates signals, and performs a charge(Q)-to-time(T) conversion. The module internally generates a logic-level output, where the leading edge determines the drift time and the width is proportional to the input pulse height. This technique is a rather simple extension of the ordinary TDC/ADC readout scheme, but allows to use only TDCs to measure both the timing and charge of the signals. Since multihit TDCs operate in the common stop mode, one does not need a long delay that analogue signals usually require in an ADC readout with a gate produced by a trigger signal.

The spatial resolution as a function of the drift distance, measured with cosmic rays, is shown in Fig. 13.5. For tracks passing near the centre of the drift space it is better than $100\,\mu\mathrm{m}$. The momentum resolution as a function of transverse momentum p_T measured with cosmic muons is approximated by the formula

$$\frac{\sigma_{p_\mathrm{T}}}{p_\mathrm{T}}[\%] = \sqrt{((0.201 \pm 0.003)p_\mathrm{T}\,[\mathrm{GeV}/c])^2 + ((0.290 \pm 0.006)/\beta)^2} \;.$$

$$(13.3)$$

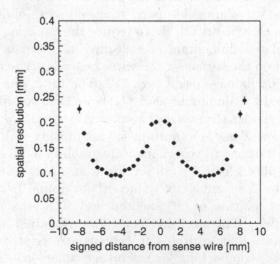

Fig. 13.5.   Spatial resolution as a function of the drift distance.

No apparent systematic effects due to different particle charge were observed. The momentum resolution measured in the experiment for muons produced in the process $e^+e^- \to \mu^+\mu^-$ is $\sigma_{p_T}/p_T = (1.64\pm0.04)\%$ for the momentum range 4–5.2 GeV/c. This is somewhat worse compared to the Monte Carlo expectations.

The dE/dx measurements in the central drift chamber are used for particle identification. The truncated-mean method was employed to estimate the most probable energy loss. The largest 20% of the measured dE/dx values for each track were discarded and the remaining data were averaged. Such a procedure minimises an influence of the tail of the dE/dx Landau distribution. The $\langle$dE/d$x\rangle$ resolution was measured to be 7.8% in the momentum range from 0.4 GeV/c to 0.6 GeV/c, while the resolution for Bhabha and μ-pair events is about 6%.

13.1.3 The aerogel Cherenkov-counter system (ACC)

Particle identification, specifically the ability to distinguish $\pi^\pm$ from $K^\pm$, plays a key rôle in the understanding of CP violation in the B system. An array of silica-aerogel threshold Cherenkov counters has been selected as part of the Belle particle-identification system to extend the momentum coverage beyond the reach of dE/dx measurements by the CDC and by time-of-flight measurements by the time-of-flight system (TOF).

The configuration of the *silica-aerogel Cherenkov-counter system* (ACC), in the central part of the Belle detector, is shown in Fig. 13.6 [11, 12]. The ACC consists of 960 counter modules segmented into 60 cells in the φ direction for the barrel part and 228 modules arranged in 5

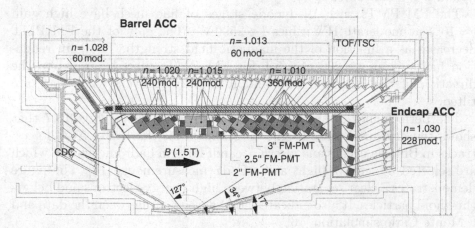

Fig. 13.6.   The arrangement of the ACC system.

concentric layers for the forward endcap part of the detector. All counters are arranged in a projective semi-tower geometry, pointing to the interaction point. In order to obtain good pion/kaon separation for the whole kinematical range, the refractive indices of the aerogels are selected to be between 1.01 and 1.03, depending on their polar-angle region.

The counter contains five aerogel tiles stacked in a thin (0.2 mm thick) aluminium box of approximate dimensions $12 \times 12 \times 12 \, cm^3$. Since the ACC is operated in a high magnetic field of 1.5 T in the Belle detector, *fine-mesh photomultiplier tubes* (FM PMTs), attached directly to the aerogels at the sides of the box, are used for the detection of Cherenkov light, taking advantage of their large effective area and high gain [13].

Silica aerogel has been used in several experiments, but its transparency became worse within a few years of use. This phenomenon may be attributed to the hydrophilic property of silica aerogel. In order to prevent such effects, a special type of this material was developed and produced, highly hydrophobic by changing the surface hydroxyl groups into trimethylsilyl groups [14]. As a result of this treatment, the silica aerogel used for the Belle's ACC remains transparent even four years after it has been manufactured.

All aerogel tiles thus produced have been checked for optical transparency, transmittance of unscattered light, refractive index, dimensions, etc.

The FM PMTs were produced by Hamamatsu Photonics. Each FM PMT has a borosilicate glass window, a bialkali photocathode, 19 fine-mesh dynodes and an anode. Three types of FM PMTs of 2, 2.5 and 3 inches in diameter are used in the ACC. The average quantum efficiency of the photocathode is 25% at 400 nm wavelength. The optical acceptance, that is, the ratio of the total area of the holes to the total area of mesh dynodes is about 50%.

The FM PMTs with 19 dynode stages of fine mesh have high gain ($\approx 10^8$) at moderate HV values ($< 2500 \, V$). The gain of the FM PMT decreases as a function of the magnetic field strength. The gain reduction factor is about 200 at 1.5 T for FM PMTs placed parallel to the direction of the magnetic field and it slightly recovers when they are tilted.

The performance of the ACC system is illustrated by Fig. 13.7 that shows the measured pulse-height distribution for the barrel ACC for $e^{\pm}$ tracks in Bhabha events and also $K^{\pm}$ candidates in hadronic events, which are selected by time-of-flight and dE/dx measurements [15]. The figure demonstrates a clear separation between high-energy electrons and below-threshold particles. It also indicates good agreement between the data and a Monte Carlo simulation [16].

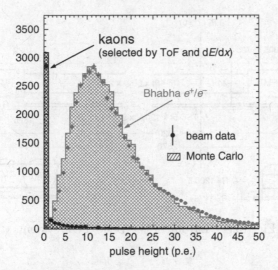

Fig. 13.7. Pulse-height spectra in units of photoelectrons observed by the barrel ACC for electrons and kaons. Kaon candidates were obtained by dE/dx and TOF measurements. The Monte Carlo expectations are superimposed.

13.1.4 Time-of-flight counters (TOF)

For a 1.2 m path, the *TOF system* with 100 ps time resolution is effective for particle momenta below about 1.2 GeV/c, which encompasses 90% of the particles produced in Υ(4S) decays. In addition to particle identification, the TOF counters provide fast signals for the trigger system to generate gate signals for ADCs and stop signals for TDCs.

The TOF system consists of 128 TOF counters and 64 thin trigger scintillation counters (TSCs). Two trapezoidally shaped TOF counters and one TSC counter, with a 1.5 cm intervening radial gap, form one module. In total 64 TOF/TSC modules located at a radius of 1.2 m from the interaction point cover a polar-angle range from 34° to 120°. The minimum transverse momentum to reach the TOF counters is about 0.28 GeV/c. The module dimensions are given in Fig. 13.8. These modules are individually mounted on the inner wall of the barrel ECL container. The 1.5 cm gap between the TOF counters and TSC counters was introduced to isolate the time-of-flight system from photon-conversion backgrounds by requiring a coincidence between the TOF and TSC counters. Electrons and positrons created in the TSC layer are prevented from reaching the TOF counters due to spiralling in this gap in the 1.5 T field.

Hamamatsu (HPK) type R6680 fine-mesh photomultipliers (FM PMTs), with 2 inch diameter and 24 dynode stages, have been selected for the TOF counter. These FM PMTs provide a gain of 3×10^6 at a high voltage below 2800 V in a magnetic field of 1.5 T. The bialkali photocathode

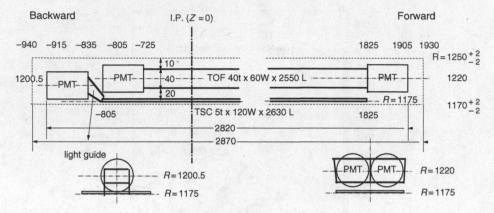

Fig. 13.8.   Layout of the Belle TOF system.

with an effective diameter of 39 mm covers 50% of the end area of each TOF counter. The transit-time spread is 320 ps (rms), the rise and fall times are 3.5 ns and 4.5 ns, respectively, and the pulse width is about 6 ns at FWHM. FM PMTs were attached to the TOF counter ends with an air gap of ≈ 0.1 mm. In the case of the TSC counters the tubes were glued to the light guides at the backward ends. The air gap between the scintillator and PMT in the TOF counter helps to select earlier-arrival photons and reduces the gain-saturation effect of the FM PMT that might arise due to large pulses at a very high rate. Photons arriving at large angles to the counter axis cannot leave the scintillator due to internal reflection and reach the PMT. In this way only photons with shorter travel times and smaller time spread are selected.

The TOF and TSC scintillators (BC408, Bicron) were wrapped with one layer of 45 μm thick polyvinyl film (Tedlar) for light tightness and surface protection. This thin wrapping minimises the dead space between adjacent TOF counters. The effective light attenuation length is about 3.9 m while the effective light propagation velocity is 14.4 cm/ns. The number of photoelectrons received from the TSC counter per one minimum-ionising particle (MIP) crossing the counter depends considerably on the crossing point of the particle, but it exceeds 25 photoelectrons over the whole counter. This ensures a high efficiency of 98% for the TOF trigger even at a nominal discrimination level of 0.5 MIPs.

A block diagram of a single channel of the TOF front-end electronics is shown in Fig. 13.9. Each photomultiplier signal is split into two. One is sent to the charge-to-time converter and then to a multihit TDC for charge measurement. The other generates signals corresponding to two different threshold levels: a high level (HL) and a low level (LL). Two LeCroy MVL107s (Monolithic Voltage Comparators) are used for discriminators,

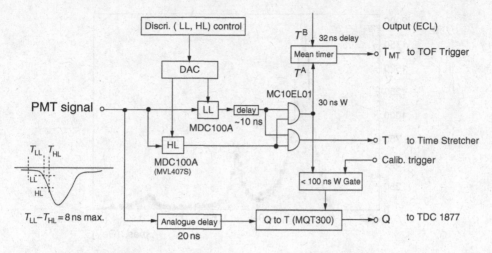

Fig. 13.9. Block diagram of a single channel of the TOF front-end electronics.

with threshold levels set between 0.3–0.5 MIPs for HL and 0.05–0.1 MIPs for LL. The LL output provides the TOF timing and the HL output provides a trigger signal. The HL is used to generate a self gate for the LeCroy MQT300A Q-to-T conversion and also to gate the LL output. A common trigger is prepared for pedestal calibration of the MQT300A. The signal T is further processed in a time stretcher for readout by a TDC 1877S. The MQT (monolithic charge-to-time converter) output Q is a timing signal corresponding to the charge, which is directly recorded with the TDC 1877S.

To achieve high time resolution the following time-walk correction formulae are applied in the off-line data processing:

$$T_{\text{obs}}^{\text{twc}} = T_{\text{raw}} - \left(\frac{z}{V_{\text{eff}}} + \frac{S}{\sqrt{Q}} + F(z) \right) , \qquad (13.4)$$

where T_{raw} is the PMT signal time, z is the particle's hit position on a TOF counter, V_{eff} is the effective velocity of light in the scintillator, Q is the charge of the signal, S is the coefficient of time walk, $T_{\text{obs}}^{\text{twc}}$ the time-walk-corrected observed time, and

$$F(z) = \sum_{n=0}^{n=5} A_n z^n . \qquad (13.5)$$

The coefficients, $1/V_{\text{eff}}$, S and A_n for $n = 0$ to 5, were determined from experimental data.

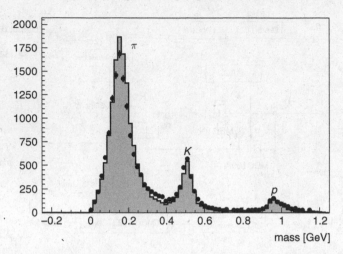

Fig. 13.10. Mass distribution from TOF measurements for particle momenta below 1.2 GeV/c.

The counter time resolution measured for muons from the process $e^+e^- \to \mu^+\mu^-$ is about 100 ps with a small z dependence. This satisfies the design goal. This value is reached even though the transit-time spread is 320 ps, because the number of photoelectrons is about 200 per minimum-ionising particle. Therefore – due to large photoelectron statistics – the contribution of the transit-time spread to the time resolution is considerably reduced. Figure 13.10 shows the mass distribution for tracks with momenta below 1.2 GeV/c in hadronic events. The mass m is calculated using the equation

$$m^2 = \left(\frac{1}{\beta^2} - 1 \right) P^2 = \left[\left(\frac{c T_{\mathrm{obs}}^{\mathrm{twc}}}{L_{\mathrm{path}}} \right)^2 - 1 \right] P^2 , \qquad (13.6)$$

where P and L_{path} are the momentum and path length of the particle determined from the CDC track fit. Clear peaks corresponding to $\pi^\pm, K^\pm$ and protons are seen. The data points are in good agreement with a Monte Carlo prediction (histogram) of $\sigma_{\mathrm{TOF}} = 100$ ps.

13.1.5 Electromagnetic calorimetry (ECL)

Since one third of B decay products are π^0s and other neutral particles providing photons in a wide range from 20 MeV to 4 GeV, a *high-resolution calorimeter* is a very important part of the detector. CsI(Tl) scintillation crystals were chosen as a material for the calorimeter due to the high CsI(Tl) light output, its short radiation length, good mechanical properties and moderate price. The main tasks of the calorimeter are:

- detection of γ quanta with high efficiency,
- precise determination of the photon energy and coordinates,
- electron/hadron separation,
- generation of a proper signal for the trigger,
- on-line and off-line luminosity measurement.

The electromagnetic calorimeter (ECL) consists of a barrel section of 3.0 m in length with an inner radius of 1.25 m and annular endcaps at $z = 2.0$ m (forward part) and $z = -1.0$ m (backward part) from the interaction point. The calorimeter covers the polar-angle region of $12.4° < \theta < 155.1°$ except for two gaps $\approx 1°$ wide between the barrel and endcaps.

The barrel part has a tower structure projected to the vicinity of the interaction point. It contains 6624 CsI(Tl) modules of 29 different types. Each crystal is a truncated pyramid of average size of about $6 \times 6 \, \text{cm}^2$ in cross section and 30 cm ($16.2 \, X_0$) in length. The endcaps contain altogether 2112 CsI crystals of 69 types. The total number of the crystals is 8736 with a total mass of about 43 tons.

Each crystal is wrapped with a layer of 200 µm thick Gore-Tex porous Teflon and covered by a 50 µm thick aluminised polyethylene. For light readout two $10 \times 20 \, \text{mm}^2$ Hamamatsu S2744-08 photodiodes are glued to the rear surface of the crystal via an intervening 1 mm thick acrylic plate. The acrylic plate is used because direct glue joints between the photodiode and the CsI were found to fail after temperature cycling, probably due to the different thermal expansion coefficients of silicon and CsI. The LED attached to the plate can inject light pulses to the crystal volume to monitor the stability of the optical condition. Two preamplifiers are attached to the photodiodes. For electronic channel monitoring and control, test pulses are fed to the inputs of the preamplifier. An aluminium-shielded preamplifier box is attached to the aluminium base plate with screws. The mechanical assembly of a single CsI(Tl) counter is shown in Fig. 13.11. The signal yield of this counter was measured to be about 5000 photoelectrons per 1 MeV of energy deposited in the crystal. The noise level is equal to about 200 keV in absence of a beam.

The barrel crystals were installed in a honeycomb-like structure formed by 0.5 mm-thick aluminium septum walls stretched between the inner and outer cylinders. The outer cylinder, the two end rings and the reinforcing bars are made of stainless steel and form a rigid structure that supports the weight of the crystals. The inner cylinder is made of 1.6 mm-thick aluminium to minimise the inactive material in front of the calorimeter. The overall support structure is made gas tight and filled with dry air to provide a low-humidity (5%) environment for the CsI(Tl)

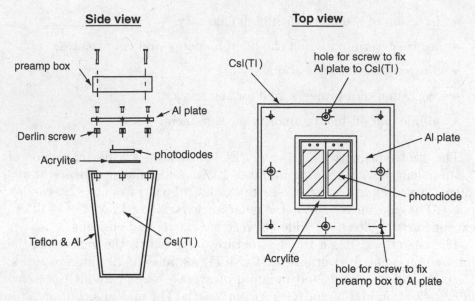

Fig. 13.11.   Mechanical assembly of the ECL counter.

crystals. The preamplifier heat, a total of $3\,\mathrm{kW}$, is removed by a liquid-based cooling system. An operating temperature of lower than $30\,^\circ\mathrm{C}$ with $\pm 1\,^\circ\mathrm{C}$ stability is required for the stable operation of the electronics. The endcap support structure is similar to that of the barrel.

A block diagram of the readout electronics is shown in Fig. 13.12. The preamplifier output is transmitted by $10\,\mathrm{m}$ long, $50\,\Omega$ twisted-pair cables to a shaping circuit where the two signals from the same crystal are summed. The summed signal is then split into two streams: one for the main data acquisition for energy measurements and the other for the trigger electronics. The main signals for energy measurements are shaped with a $\tau = 1\,\mu\mathrm{s}$ time constant and fed into a charge-to-time (Q-to-T) converter, LeCroy MQT300A, installed on the same card. The output of the Q-to-T converter is transmitted via a twisted pair to a multi-hit TDC module (LeCroy 1877S) in the electronics hut for digitisation. The trigger signal is shaped with a shorter time constant and ≈ 16 lines are combined to form an analogue sum for the level-1 trigger.

The absolute energy calibration has been carried out by using Bhabha ($e^+e^- \to e^+e^-$) and annihilation ($e^+e^- \to \gamma\gamma$) events. With a sample containing $N_{e^+e^-}$ Bhabha events, the calibration constant of the jth counter g_j is obtained by minimising χ^2 defined as

$$\chi^2 = \sum_{k}^{2N_{e^+e^-}} \left(\frac{E_k(\theta)f(\theta) - \sum_j g_j E_j}{\sigma} \right)^2 , \qquad (13.7)$$

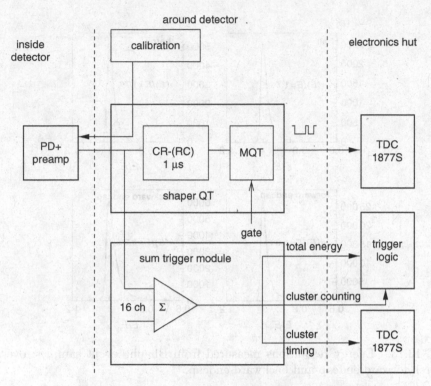

Fig. 13.12. Block diagram of the ECL readout electronics.

where E_k is the expected energy of the scattered electrons. All electrons and positrons are included in the sum. This value is a function of θ in the asymmetric collider. The function $f(\theta)$ is the correction factor due to shower leakage and the effect of the front material, which was determined by a Monte Carlo simulation. The χ^2 minimisation is carried out by taking a $\approx 8000 \times 8000$ sparse-matrix inversion into account. Approximately 100 events per counter are used for this calibration.

An energy resolution for electrons of about 1.7% for the energy range from 4 GeV to 7 GeV was achieved from Bhabha scattering ($e^+e^- \rightarrow e^+e^-$) averaged over the whole calorimeter as shown in Fig. 13.13. The resolution does not change too much in this energy region since the leakage of the calorimeter increases with energy while the number of particles in the shower is also increasing and these effects compensate each other up to certain extent.

Two-photon invariant-mass distributions in hadronic events are shown in Figs. 13.14 (a) and (b). The clear peaks of π^0 and η mesons are seen at each nominal mass, and a mass resolution has been achieved to be 4.8 MeV for π^0 and about 12 MeV for η.

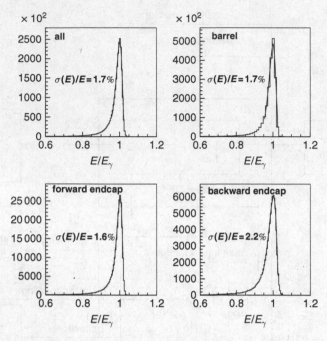

Fig. 13.13.   Energy resolutions measured from Bhabha event samples: overall, barrel, forward endcap and backward endcap.

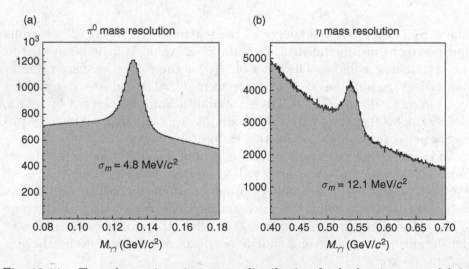

Fig. 13.14.   Two-photon invariant-mass distribution for hadronic events (a) in the $\pi^0 \to \gamma\gamma$ and (b) the $\eta \to \gamma\gamma$ range, where each photon energy was required to be greater than 30 MeV in the barrel region.

13.1.6 The K_L and muon detection system (KLM)

The *KLM detection system* was designed to identify K_Ls and muons with high efficiency over a broad momentum range greater than 600 MeV/c. The barrel-shaped region around the interaction point covers an angular range from 45° to 125° in polar angle, and the endcaps in the forward and backward directions extend this range to 20° and 155°.

The KLM detection system consists of alternating layers of charged-particle detectors and 4.7 cm-thick iron plates. There are 15 detector layers and 14 iron layers in the octagonal barrel region and 14 detector layers in each of the forward and backward endcaps. The iron plates provide a total of 3.9 interaction lengths of material for a particle travelling normal to the detector planes. In addition, the electromagnetic calorimeter, ECL, provides another 0.8 interaction length of material to convert K_Ls. K_Ls which interact in the iron or ECL produce a shower of ionising particles. The location of this shower determines the direction of the K_L, but the size of the shower does not permit a useful measurement of the K_L energy. The multiple layers of charged-particle detectors and iron allow a discrimination between muons and charged hadrons ($\pi^\pm$ or $K^\pm$) based upon their range and transverse scattering. Muons pass a much longer distance with smaller deflections on average than strongly interacting hadrons.

Charged particles are detected in the KLM system by glass–electrode resistive-plate counters (RPCs) [17, 18]. Resistive-plate counters have two parallel plate electrodes with high bulk resistivity ($\geq 10^{10}\,\Omega \cdot \mathrm{cm}$) separated by a gas-filled gap. In the streamer mode, an ionising particle traversing the gap initiates a streamer in the gas that results in a local discharge of the plates. The discharge induces a signal on external pickup strips, which is used to record the location and the time of the ionisation.

Figure 13.15 shows the cross section of a superlayer, in which two RPCs are sandwiched between the orthogonal θ and φ pickup strips with ground planes for signal reference and proper impedance. This unit structure of two RPCs and two readout planes is enclosed in an aluminium box and is less than 3.7 cm thick. Signals from both RPCs are picked up by copper strips above and below the pair of RPCs, providing a three-dimensional space point for particle tracking. Multiple scattering of particles as they travel through the iron is typically a few centimetres. This sets the scale for the projected spatial resolution of the KLM. The pickup strips in the barrel vary in width from layer to layer but are approximately 50 mm wide with lengths from 1.5 m to 2.7 m.

The double-gap design provides redundancy and results in high superlayer efficiency of $\geq 98\%$, despite the relatively low single-layer RPC efficiency of 90% to 95%.

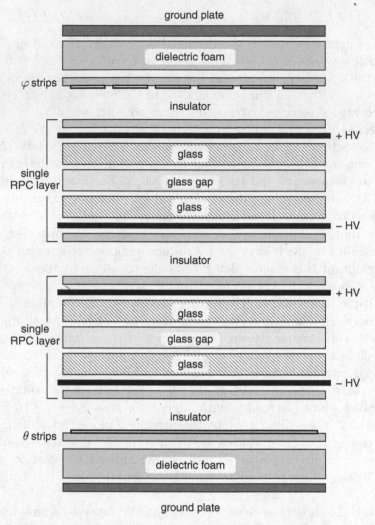

Fig. 13.15.   Cross section of a KLM superlayer.

Cosmic rays were used to measure the detection efficiency and resolution of the superlayers. The momenta of cosmic muons were measured by the central drift chamber using the solenoidal field of 1.5 T. Below 500 MeV/c, the muons do not reach the KLM system. A comparison of the measured range of a particle with the predicted range for a muon allows us to assign a likelihood of being a muon. In Fig. 13.16a the muon detection efficiency versus momentum is shown for a likelihood cut of 0.7. Some fraction of charged pions and kaons will be misidentified as muons. A sample of $K_S \rightarrow \pi^+\pi^-$ events in the e^+e^- collision data was used to determine this fake rate. The fraction of pions which is misidentified as

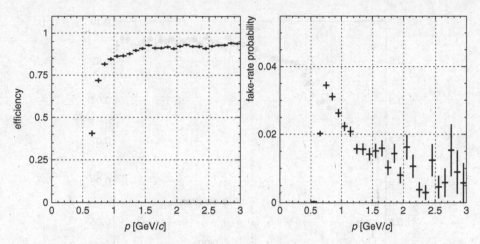

Fig. 13.16. Muon detection efficiency (a) and fake-rate probability (b) versus momentum in the K_L and muon detection system.

muons is shown in Fig. 13.16b again with the same muon likelihood cut. Above $1.5\,\mathrm{GeV}/c$ one finds a muon-identification efficiency of better than 90% with a fake-rate probability of less than 2%.

13.2 Particle identification

For *particle identification* at the Belle detector the information from all subsystems is used.

Electrons are identified by using the following discriminants:

- the ratio of energy deposited in the electromagnetic calorimeter (ECL) and charged-track momentum measured by the central drift chamber (CDC),
- the transverse shower shape in the ECL,
- the matching between a cluster in the ECL and the charged-track position extrapolated to the ECL,
- $\mathrm{d}E/\mathrm{d}x$ measured by the CDC,
- the light yield in the aerogel Cherenkov-counter system (ACC), and
- time of flight measured by the time-of-flight system (TOF).

The probability density functions (PDFs) for the discriminants were formed beforehand. Based on each PDF, likelihood probabilities are calculated on a track-by-track basis and combined into a final likelihood output. This likelihood calculation is carried out by taking into account the momentum and angular dependence. The efficiency and fake-rate probability are displayed in Fig. 13.17 using electrons in real

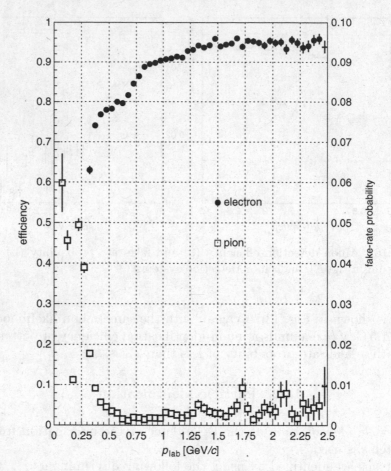

Fig. 13.17.   Electron identification efficiency (circles) and fake-rate probability for charged pions (squares). Note the different scales for the efficiency and fake-rate probability.

$e^+e^- \to e^+e^-e^+e^-$ events for the efficiency measurement and $K_S \to \pi^+\pi^-$ decays in real data for the fake-rate evaluation. For momenta greater than $1\,\mathrm{GeV}/c$, the electron identification efficiency is maintained to be above 90% while the fake-rate probability is kept at around 0.2% to 0.3%.

The K/π identification is carried out by combining information from three nearly independent measurements:

- the $\mathrm{d}E/\mathrm{d}x$ measurement by the CDC,
- the TOF measurement, and
- the measurement of the number of photoelectrons (N_{pe}) in the ACC.

As in the case of electron identification (EID), the likelihood function for each measurement was calculated, and the product of the three likelihood functions yields the overall likelihood probability for being a kaon or a pion, P_K or P_π. A particle is then identified as a kaon or a pion by cutting on the likelihood ratio (PID):

$$\text{PID}(K) = \frac{P_K}{P_K + P_\pi} \ , \quad \text{PID}(\pi) = 1 - \text{PID}(K) \ . \tag{13.8}$$

The validity of the K/π identification has been demonstrated using the charm decay, $D^{*+} \to D^0\pi^+$, followed by $D^0 \to K^-\pi^+$. The characteristically slow π^+ from the D^{*+} decay allows these decays to be selected with a good signal/background ratio (better than 30), without relying on particle identification. Therefore, the detector performance can be directly probed with the daughter K and π mesons from the D decay, which can be tagged by their relative charge with respect to the slow pion. The measured K efficiency and π fake-rate probability in the barrel region are plotted as functions of the track momentum from $0.5\,\text{GeV}/c$ to $4.0\,\text{GeV}/c$ in Fig. 13.18. A likelihood-ratio cut, $\text{PID}(K) \geq 0.6$, is applied

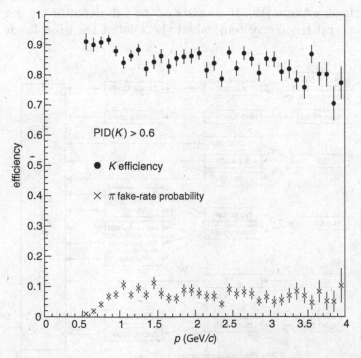

Fig. 13.18. K efficiency and π fake-rate probability, measured with $D^{*+} \to D^0(K\pi) + \pi^+$ decays, for the barrel region. A likelihood-ratio cut $\text{PID}(K) \geq 0.6$ is applied.

in this figure. For most of the region, the measured K efficiency exceeds 80%, while the π fake-rate probability is kept below 10%.

13.3 Data-acquisition electronics and trigger system

The total rate of the physical processes of interest at a luminosity of $10^{34}\,\mathrm{cm}^{-2}\,\mathrm{s}^{-1}$ is about 100 Hz. Samples of Bhabha and $e^+e^- \rightarrow \gamma\gamma$ events are accumulated as well to measure the luminosity and to calibrate the detector responses but, since their rates are very large, these trigger rates must be prescaled by a factor of ≈ 100. Because of the high beam current, the studied events are accompanied by a high beam-related background, which is dominated by lost electrons and positrons. Thus the trigger conditions should be such that background rates are kept within the tolerance of the data-acquisition system (max. 500 Hz), while the efficiency for physics events of interest is kept high. It is important to have redundant triggers to keep the efficiency high even for varying conditions. The Belle trigger system has been designed and developed to satisfy these requirements.

The Belle *trigger system* consists of the Level-1 hardware trigger and a software trigger. Figure 13.19 shows the schematic view of the Belle Level-1 trigger system [19]. It consists of the sub-detector trigger systems and the central trigger system called the Global Decision Logic (GDL).

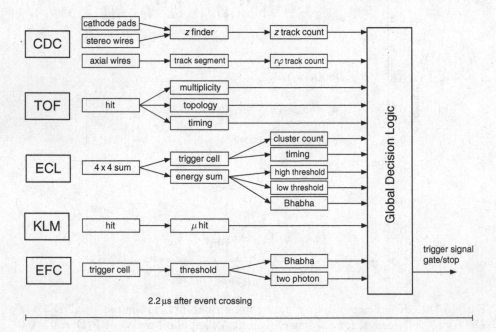

Fig. 13.19. The Level-1 trigger system for the Belle detector.

The sub-detector trigger systems are based on two categories: track triggers and energy triggers. The central drift chamber and the time-of-flight system are used to yield trigger signals for charged particles. The CDC provides $r\varphi$- and rz-track trigger signals. The ECL trigger system provides triggers based on total energy deposit and cluster counting of crystal hits. These two categories allow sufficient redundancy. The KLM trigger gives additional information on muons. The sub-detectors process event signals in parallel and provide trigger information to the GDL, where all information is combined to characterise an event type.

The trigger system provides a trigger signal within a fixed time of $2.2\,\mu s$ after the event occurrence. The trigger signal is used for the gate signal of the ECL readout and the stop signal of the TDCs for the CDC, providing T_0. Therefore, it is important to have good timing accuracy. The timing of the trigger is primarily determined by the TOF trigger which has a time jitter less than $10\,ns$. ECL trigger signals are also used as timing signals for events in which the TOF trigger is not available. In order to maintain the $2.2\,\mu s$ latency, each sub-detector trigger signal is required to be available at the GDL input with a maximum delay of $1.85\,\mu s$. Timing adjustments are done at the input of the global decision logic. As a result, the GDL is left with a fixed $350\,ns$ processing time to form the final trigger signal. The Belle trigger system, including most of the sub-detector trigger systems, is operated in a pipelined manner with clocks synchronised to the KEKB accelerator RF signal. The base system clock is $16\,MHz$ which is obtained by subdividing $509\,MHz$ RF by 32. The higher-frequency clocks, $32\,MHz$ and $64\,MHz$, are also available for systems requiring faster processing.

The Belle trigger system extensively utilises programmable logic chips, Xilinx Field Programmable Gate Array (FPGA) and Complex Programmable Logic Device (CPLD) chips, which provide the majority of the trigger logic and reduce the number of types of hardware modules.

In order to satisfy the data-acquisition requirements so that it works at $500\,Hz$ with a dead-time fraction of less than 10%, a distributed parallel system has been developed. The global scheme of the system is shown in Fig. 13.20. The entire system is segmented into seven subsystems running in parallel, each handling the data from a sub-detector. Data from each subsystem are combined into a single event record by an event builder, which converts 'detector-by-detector' parallel data streams to an 'event-by-event' data flow. The event-builder output is transferred to an on-line computer farm, where another level of event filtering is done after fast event reconstruction. The data are then sent to a mass-storage system located at the computer centre via optical fibres. A typical data size of a hadronic event by $B\bar{B}$ or $q\bar{q}$ production is measured to be about $30\,kB$, which corresponds to a maximum data transfer rate of $15\,MB/s$.

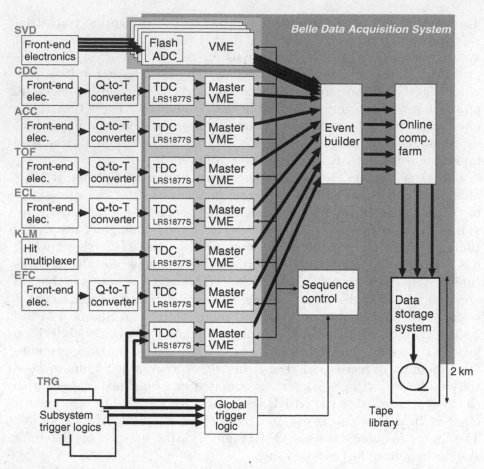

Fig. 13.20.   Belle DAQ system overview.

A charge-to-time (Q-to-T) technique was adopted to read out signals from most of the detectors. Instead of using ADCs to digitise the amplitudes of signals, the charge is once stored in a capacitor and discharged at a constant rate. Two pulses, the separation of which is proportional to the signal amplitude, are generated at the start and stop moments of the discharge. By digitising the time interval of the two timing pulses with respect to a common stop timing, both the timing and the amplitude of the input signal are determined. For time digitisation a multi-hit FASTBUS TDC module, LeCroy LRS1877S, is used. Up to 16 timing pulses are recorded for 96 channels in a single-width module with a sparsification capability. The least significant bit is 500 ps. A programmable time window has a 16-bit range, which corresponds to a full scale of 32 µs.

Most of the detectors, CDC, ACC, TOF, ECL and EFC, are read out by using the Q-to-T and TDC technique. The use of the Q-to-T technique reduces the number of cables for the CDC wires by one half. In the case of the TOF, the time resolution of 100 ps is achieved by using a time stretcher which expands the pulse width by a factor of 20. In the case of the ECL, a 16-bit dynamic range is achieved by using 3 ranges. The signal is split and fed into three preamplifiers of different gain. Then each signal feeds a Q-to-T circuit. When the signal is small, one gets four output signals: the trigger signal and three signals from the three channels. The time between each signal and the trigger is proportional to the pulse height. When the amplitude of the signal exceeds a certain value, the corresponding time exceeds a preselected gate width (overflow) and no output pulse is generated. So, for large-amplitude signals one can only see the trigger pulse and a time pulse from the low-gain channel. After digitisation one can therefore identify the analogue information by the number of time signals for the ECL pulse.

The KLM strip information is also read out by using the same type of TDC. Strip signals are multiplexed into serial lines and recorded by the TDC as time pulses. These pulses are decoded to reconstruct hit strips. Similarly, trigger signals from each subdetector including those for the intermediate stages are recorded using a TDC. A full set of trigger signals gives us complete information for the trigger studies.

A unified FASTBUS TDC readout subsystem developed for Belle is applicable to all the detectors except the Silicon Vertex Detector (SVD). A FASTBUS processor interface (FPI) controls these TDC modules, and the FPI is controlled by a readout-system controller in a master VME crate. Readout software runs on the VxWorks real-time operating system on a Motorola 68040 CPU module, MVME162. Data are passed to an event-builder transmitter in the same VME crate. The overall transfer rate of the subsystem is about 3.5 MB/s.

13.4   Luminosity measurement and the detector performance

The Belle detector started operation in 1999 and an integrated luminosity of about 700 fb^{-1} has been collected by the beginning of 2007.

The *luminosity* of the collider is an important parameter of the experiment. It should be measured continuously during the experiment running for collider-operation monitoring and tuning.

For the Belle detector the on-line luminosity is measured by counting $e^+e^- \to e^+e^-$ events when two final particles constitute a back-to-back configuration at the endcap parts of the calorimeter with an energy deposition exceeding a high threshold. The rate of these events is about $300\,\mathrm{Hz}$ at a luminosity of $\approx 10^{34}\,\mathrm{cm}^{-2}\,\mathrm{s}^{-1}$ which provides a reasonable statistical accuracy at a $10\,\mathrm{s}$ integration time.

The total luminosity integrated over a certain period is determined in off-line data analysis using the same process but detected in the barrel calorimeter.

Some examples of the event reconstruction are given in the event displays of Figs. 13.21 and 13.22.

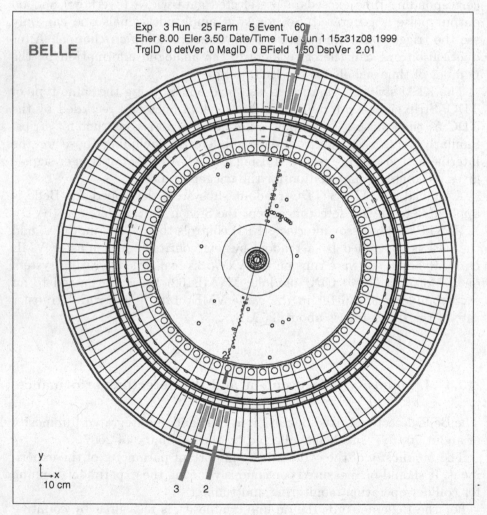

Fig. 13.21.   An example of an e^+e^- elastic scattering event ($r\varphi$ projection).

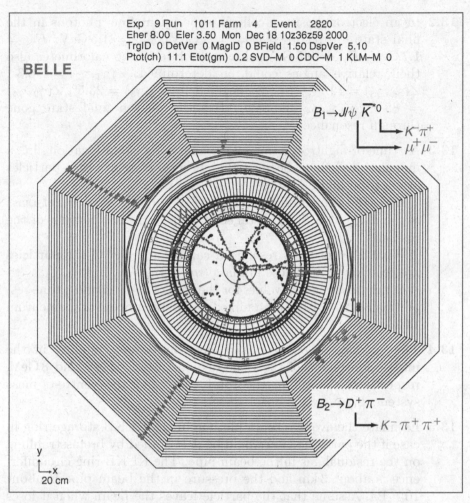

Fig. 13.22. An example of a fully reconstructed $e^+e^- \rightarrow B\bar{B}$ event.

The Belle experiment produced a large amount of physics data. The most important results are the observation of CP violation in B-meson decays [20] and the discovery of various new particles [21].

13.5 Problems

13.1 A pion beam passes through a 1.5 m Cherenkov counter that contains CO_2 at 3 atm pressure. At what momentum should the efficiency of the counter reach 50%?

Assume: quantum efficiency of the photomultiplier: 20%, geometrical light collection efficiency: 10%, transfer probability cathode → dynode: 80%.

13.2 In an electron–positron-collider experiment four photons in the final state are recorded ($E_{\gamma_1} = 1\,\text{GeV}$, $E_{\gamma_2} = 1.5\,\text{GeV}$, $E_{\gamma_3} = 1.7\,\text{GeV}$, $E_{\gamma_4} = 0.5\,\text{GeV}$). From the hits in the calorimeter also their relative angles could be determined: $\angle(\gamma_1, \gamma_2) = 6.3°$, $\angle(\gamma_1, \gamma_3) = 12.7°$, $\angle(\gamma_1, \gamma_4) = 160.0°$, $\angle(\gamma_2, \gamma_3) = 23.5°$, $\angle(\gamma_2, \gamma_4) = 85.0°$, $\angle(\gamma_3, \gamma_4) = 34.6°$. Has the photon final state gone through resonances?

13.3 A time-of-flight system allows particles identification in a momentum-defined beam if the energies of relativistic particles are not too high: $\Delta t = \frac{L \cdot c}{2 \cdot p^2} \cdot (m_2^2 - m_1^2)$.

Δt is the time-of-flight difference of the two particles of mass m_1 and m_2 at a flight distance of L. p is the momentum of the particles.

What kind of mass resolution can be achieved if the particles are relativistic and close in mass ($m_1 \approx m_2$)?

Estimate numerical values for the case of $L = 1\,\text{m}$ and a time resolution of $\Delta t = 10\,\text{ps}$ for a mixed muon/pion beam of momentum $1\,\text{GeV}/c$.

13.4 The energies of positron and electron beams colliding in the interaction region of the Belle detector are $3.5\,\text{GeV}$ and $8\,\text{GeV}$, respectively. Calculate the total energy in the centre-of-mass system, E_{CM}.

13.5 Estimate the average beam lifetime in the KEKB storage ring in case if the particle losses would be determined by bremsstrahlung on the residual gas in the beam pipe. The KEKB ring circumference is about $3\,\text{km}$ and the pressure in the beam pipe is about $10^{-7}\,\text{Pa}$. Assume that the particle leaves the beam when it loses more than 1% of its energy.

13.6 Estimate the rate of the muon-pair production, $e^+e^- \to \mu^+\mu^-$, at the Belle detector for a luminosity of $L = 10^{34}\,\text{cm}^{-2}\,\text{s}^{-1}$ if the muons are detected in a fiducial volume limited by the polar angles ranging from $\theta_0 = 30°$ to $\pi - \theta_0 = 150°$ in the centre-of-mass frame.

References

[1] S. Kurokawa & E. Kikutani, Overview of the KEKB Accelerators, *Nucl. Instr. Meth.* **A499** (2003) 1–7

[2] A. Abashian *et al.* (Belle collaboration), The Belle Detector, *Nucl. Instr. Meth.* **A479** (2002) 117–232

[3] T. Kawasaki, The Belle Silicon Vertex Detector, *Nucl. Instr. Meth.* **A494** (2002) 94–101

[4] R. Abe *et al.*, Belle/SVD2 Status and Performance, *Nucl. Instr. Meth.* **A535** (2004) 379–83

[5] M. Friedl *et al.*, Readout, First- and Second-Level Triggers of the New Belle Silicon Vertex Detector, *Nucl. Instr. Meth.* **A535** (2004) 491–6

[6] H. Hirano *et al.*, A High-Resolution Cylindrical Drift Chamber for the KEK B-factory, *Nucl. Instr. Meth.* **A455** (2000) 294–304

[7] S. Uno *et al.*, Study of a Drift Chamber Filled with a Helium-Ethane Mixture, *Nucl. Instr. Meth.* **A330** (1993) 55–63

[8] O. Nitoh *et al.*, Drift Velocity of Electrons in Helium-based Gas Mixtures Measured with a UV Laser, *Jpn. J. Appl. Phys* **33** (1994) 5929–32

[9] K. Emi *et al.*, Study of a dE/dx Measurement and the Gas-Gain Saturation by a Prototype Drift Chamber for the BELLE-CDC, *Nucl. Instr. Meth.* **A379** (1996) 225–31

[10] Y. Fujita *et al.*, Test of Charge-to-Time Conversion and Multi-hit TDC Technique for the BELLE CDC Readout, *Nucl. Instr. Meth.* **A405** (1998) 105–10

[11] T. Iijima *et al.*, Aerogel Cherenkov Counter for the BELLE Experiment, *Nucl. Instr. Meth.* **A379** (1996) 457–9

[12] T. Sumiyoshi *et al.*, Silica Aerogel Cherenkov Counter for the KEK B-factory Experiment, *Nucl. Instr. Meth.* **A433** (1999) 385–91

[13] T. Iijima *et al.*, Study on fine-mesh PMTs for Detection of Aerogel Cherenkov Light, *Nucl. Instr. Meth.* **A387** (1997) 64–8

[14] H. Yokoyama & M. Yokogawa, Hydrophobic Silica Aerogels, *J. Non-Cryst. Solids* **186** (1995) 23–9

[15] T. Iijima *et al.*, Aerogel Cherenkov Counter for the BELLE Detector, *Nucl. Instr. Meth.* **A453** (2000) 321–5

[16] R. Suda *et al.*, Monte-Carlo Simulation for an Aerogel Cherenkov Counter, *Nucl. Instr. Meth.* **A406** (1998) 213–26

[17] R. Cardarelli *et al.*, Progress in Resistive Plate Counters, *Nucl. Instr. Meth.* **A263** (1988) 20–5

[18] L. Antoniazzi *et al.*, Resistive Plate Counters Readout System, *Nucl. Instr. Meth.* **A307** (1991) 312–15

[19] Y. Ushiroda *et al.*, Development of the Central Trigger System for the BELLE Detector at the KEK B-factory, *Nucl. Instr. Meth.* **A438** (1999) 460–71

[20] K. Abe *et al.* (Belle Collaboration), Observation of Large CP Violation in the Neutral B Meson System, *Phys. Rev. Lett.* **87** (2001) 091802, 1–7; Evidence for CP-Violating Asymmetries in $B^0 \to \pi^+\pi^-$ Decays and Constraints on the CKM Angle φ_2, *Phys. Rev.* **D68** (2003) 012001, 1–15

[21] S.-K. Choi, S.L. Olsen *et al.* (Belle Collaboration), Observation of a New Narrow Charmonium State in Exclusive $B^+ \to K^+\pi^+\pi^- J/\psi$ Decays, *Phys. Rev. Lett.* **91** (2003) 262001, 1–6, hep-ex/0309032 (2003); Observation of the $\eta_c(2S)$ in Exclusive $B \to KK_sK^-\pi^+$ Decays, *Phys. Rev. Lett.* **89** (2002) 102001

14

Electronics*

Everything should be made as simple as possible, but not simpler.

Albert Einstein

14.1 Introduction

Electronics are a key component of all modern detector systems. Although experiments and their associated electronics can take very different forms, the same basic principles of the electronic readout and optimisation of signal-to-noise ratio apply to all. This chapter gives an introduction to electronic noise, signal processing and digital electronics. Because of space limitations, this can only be a brief overview. A more detailed discussion of electronics with emphasis on semiconductor detectors is given elsewhere [1]. Tutorials on detectors, signal processing and electronics are also available on the worldwide web [2].

The purpose of front-end electronics and signal processing systems is to

(i) acquire an electrical signal from the sensor. Typically, this is a short current pulse.

(ii) tailor the time response of the system to optimise

   (a) the minimum detectable signal (detect hit/no hit),

   (b) energy measurement,

   (c) event rate,

   (d) time of arrival (timing measurement),

---

* This chapter was contributed by Helmuth Spieler, Lawrence Berkeley National Laboratory, Berkeley, California, USA.

(e) insensitivity to sensor pulse shape,

(f) or some combination of the above.

(iii) digitise the signal and store for subsequent analysis.

Generally, these properties cannot be optimised simultaneously, so compromises are necessary. In addition to these primary functions of an electronic readout system, other considerations can be equally or even more important, for example radiation resistance, low power (portable systems, large detector arrays, satellite systems), robustness, and – last, but not least – cost.

14.2   Example systems

Figure 14.1 illustrates the components and functions of a radiation detector system. The sensor converts the energy deposited by a charged particle (or photon) to an electrical signal. This can be achieved in a variety of ways. In direct detection – semiconductor detectors, wire chambers, or other types of ionisation chambers – energy is deposited in an absorber and converted into charge pairs, whose number is proportional to the absorbed energy. The signal charge can be quite small, in semiconductor sensors about $50\,\text{aC}$ ($5 \cdot 10^{-17}\,\text{C}$) for $1\,\text{keV}$ X rays and $4\,\text{fC}$ ($4 \cdot 10^{-15}\,\text{C}$) in a typical high-energy tracking detector, so the sensor signal must be amplified. The magnitude of the sensor signal is subject to statistical fluctuations, and electronic noise further 'smears' the signal. These fluctuations will be discussed below, but at this point we note that the sensor and preamplifier must be designed carefully to minimise electronic noise. A critical parameter is the total capacitance in parallel with the input, i.e. the sensor capacitance and input capacitance of the amplifier. The *signal-to-noise ratio* increases with decreasing capacitance. The contribution of electronic noise also relies critically on the next stage, the pulse shaper, which determines the bandwidth of the system and hence the

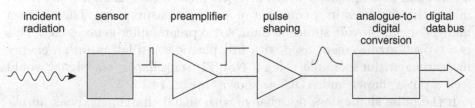

incident   sensor   preamplifier   pulse   analogue-to-   digital
radiation   shaping   digital   data bus
                                                        conversion

Fig. 14.1.   Basic detector functions: radiation is absorbed in the sensor and converted into an electrical signal. This low-level signal is integrated in a preamplifier, fed to a pulse shaper, and then digitised for subsequent storage and analysis.

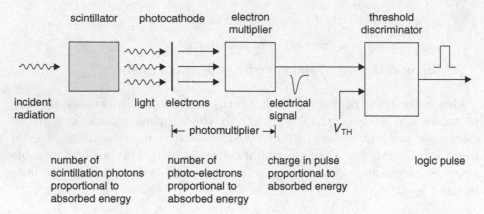

Fig. 14.2. In a scintillation detector absorbed energy is converted into visible light. The scintillation photons are commonly detected by a photomultiplier, which can provide sufficient gain to directly drive a threshold discriminator.

overall electronic noise contribution. The shaper also limits the duration of the pulse, which sets the maximum signal rate that can be accommodated. The shaper feeds an *analogue-to-digital converter* (ADC), which converts the magnitude of the analogue signal into a bit pattern suitable for subsequent digital storage and processing.

A scintillation detector (Fig. 14.2) utilises indirect detection, where the absorbed energy is first converted into visible light. The number of scintillation photons is proportional to the absorbed energy. The scintillation light is detected by a photomultiplier (PMT), consisting of a photocathode and an electron multiplier. Photons absorbed in the photocathode release electrons, whose number is proportional to the number of incident scintillation photons. At this point energy absorbed in the scintillator has been converted into an electrical signal whose charge is proportional to the energy. Increased in magnitude by the electron multiplier, the signal at the PMT output is a current pulse. Integrated over time this pulse contains the signal charge, which is proportional to the absorbed energy. Figure 14.2 shows the PMT output pulse fed directly to a threshold discriminator, which fires when the signal exceeds a predetermined threshold, as in a counting or timing measurement. The electron multiplier can provide sufficient gain, so no preamplifier is necessary. This is a typical arrangement used with fast plastic scintillators. In an energy measurement, for example using a NaI(Tl) scintillator, the signal would feed a pulse shaper and ADC, as shown in Fig. 14.1.

If the pulse shape does not change with signal charge, the peak amplitude – the pulse height – is a measure of the signal charge, so this measurement is called *pulse-height analysis*. The pulse shaper can serve multiple functions, which are discussed below. One is to tailor the pulse shape to the ADC. Since the ADC requires a finite time to acquire the signal, the input pulse may not be too short and it should have a gradually rounded peak. In scintillation detector systems the shaper is frequently an integrator and implemented as the first stage of the ADC, so it is invisible to the casual observer. Then the system appears very simple, as the PMT output is plugged directly into a charge-sensing ADC.

A detector array combines the sensor and the analogue signal-processing circuitry together with a readout system. Figure 14.3 shows the circuit blocks in a representative *readout integrated circuit* (IC). Individual sensor electrodes connect to parallel channels of analogue signal-processing circuitry. Data are stored in an analogue pipeline[‡] pending a readout command. Variable write and read (R/W) pointers are used to allow simultaneous read and write. The signal in the time slot of interest is digitised, compared with a digital threshold, and read out. Circuitry is included to generate test pulses that are injected into the input to simulate a detector signal. This is a very useful feature in setting up the system and is also a key function in chip testing prior to assembly. Analogue control levels are set by *digital-to-analogue converters* (DACs). Multiple ICs

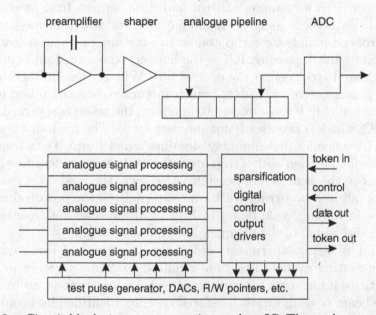

Fig. 14.3. Circuit blocks in a representative readout IC. The analogue processing chain is shown at the top. Control is passed from chip to chip by token passing.[†]

[‡] A form of parallel processing enabling streams of instructions or data to be executed or handled concurrently.

[†] In the technique of token passing only that part of an electronic circuitry is allowed to communicate that has a 'token'. When the information of that part is read out, the token is passed on to the next element which can then communicate.

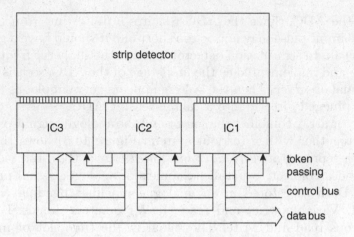

Fig. 14.4.   Multiple ICs are ganged to read out a strip detector. The rightmost chip IC1 is the master. A command on the control bus initiates the readout. When IC1 has written all of its data, it passes the token to IC2. When IC2 has finished, it passes the token to IC3, which in turn returns the token to the master IC1.

are connected to a common control and data output bus, as shown in Fig. 14.4. Each IC is assigned a unique address, which is used in issuing control commands for setup and *in situ* testing. Sequential readout is controlled by token passing. IC1 is the master, whose readout is initiated by a command (trigger) on the control bus. When it has finished writing data, it passes the token to IC2, which in turn passes the token to IC3. When the last chip has completed its readout, the token is returned to the master IC, which is then ready for the next cycle. The readout bit stream begins with a header that uniquely identifies a new frame. Data from individual ICs are labelled with a chip identifier and channel identifiers. Many variations on this scheme are possible. As shown, the readout is event oriented, i.e. all hits occurring within an externally set exposure time (e.g. time slice in the analogue buffer in Fig. 14.3) are read out together. For a concise discussion of data acquisition systems see [3].

In colliding beam experiments only a small fraction of beam crossings yields interesting events. The time required to assess whether an event is potentially interesting is typically of order microseconds, so hits from multiple beam crossings must be stored on-chip, identified by beam crossing or time stamp. Upon receipt of a trigger only the interesting data are digitised and read out ('sparsification').[§] This allows use of a digitiser that is slower than the collision rate. It is also possible to read out analogue signals and digitise them externally. Then the output stream is a

---

[§] Sparsification is a tool where only those elements of an electronic system that have relevant information are read out, thus providing a technique for speeding up the readout.

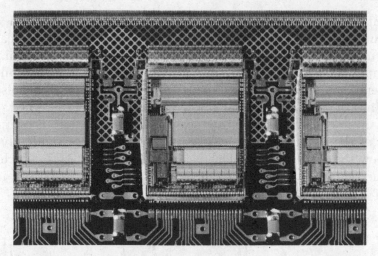

Fig. 14.5. Closeup of ICs mounted on a hybrid utilising a flexible polyimide substrate [4]. The three large rectangular objects are the readout chips, each with 128 channels. Within the chips the structure of the different circuit blocks is clearly visible, e.g. the 128 parallel analogue processing chains at the upper end. The 128 inputs at the upper edge are wire bonded to a pitch adapter to make the transition from the approximately 50 μm pitch of the readout to the 80 μm pitch of the silicon strip detector. The power, data and control lines are wire bonded at the lower edge. Bypass capacitors (the small rectangular objects with shiny contacts at the top and bottom) positioned between the readout chips connect to bond pads on the chip edges to reduce the series inductance to the on-chip circuitry. The ground plane is patterned as a diamond grid to reduce material. (Photograph courtesy of A. Ciocio.)

sequence of digital headers and analogue pulses. An alternative scheme only records the presence of a hit. The output of a threshold comparator signifies the presence of a signal and is recorded in a digital pipeline that retains the crossing number.

Figure 14.5 shows a closeup of ICs mounted on a hybrid circuit using a flexible polyimide substrate [4]. The wire bonds connecting the IC to the hybrid are clearly visible. Channels on the IC are laid out on a pitch of about 50 μm and pitch adapters fan out to match the 80 μm pitch of the silicon strip detector. The space between chips accommodates bypass capacitors and connections for control busses carrying signals from chip to chip.

14.3 Detection limits

The *minimum detectable signal* and the precision of the amplitude measurement are limited by fluctuations. The signal formed in the sensor fluctuates, even for a fixed energy absorption. In addition, electronic noise introduces baseline fluctuations, which are superimposed on the signal and alter the peak amplitude. Figure 14.6a shows a typical noise waveform. Both the amplitude and time distributions are random. When superimposed on a signal, the noise alters both the amplitude and time dependence, as shown in Fig. 14.6b. As can be seen, the noise level determines the minimum signal whose presence can be discerned.

In an optimised system, the time scale of the fluctuations is comparable to that of the signal, so the peak amplitude fluctuates randomly above and below the average value. This is illustrated in Fig. 14.7, which shows the same signal viewed at four different times. The fluctuations in peak

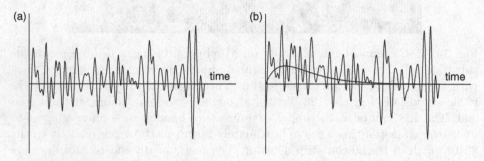

Fig. 14.6. Waveforms of random noise (a) and signal + noise (b), where the peak signal is equal to the rms noise level ($S/N = 1$). The noiseless signal is shown for comparison.

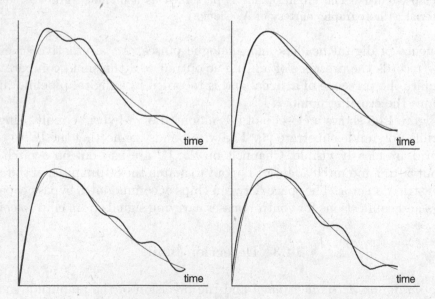

Fig. 14.7. Signal plus noise for four different pulse examples, shown for a signal-to-noise ratio of about 20. The noiseless signal is superimposed for comparison.

amplitude are obvious, but the effect of noise on timing measurements can also be seen. If the timing signal is derived from a threshold discriminator, where the output fires when the signal crosses a fixed threshold, amplitude fluctuations in the leading edge translate into time shifts. If one derives the time of arrival from a centroid analysis, the timing signal also shifts (compare the top and bottom right figures). From this one sees that signal-to-noise ratio is important for all measurements – sensing the presence of a signal or the measurement of energy, timing or position.

14.4 Acquiring the sensor signal

The sensor signal is usually a short current pulse $i_s(t)$. Typical durations vary widely, from 100 ps for thin Si sensors to several µs for inorganic scintillators. However, the physical quantity of interest is often the deposited energy, so one has to integrate over the current pulse,

$$E \propto Q_s = \int i_s(t)\,dt \ . \tag{14.1}$$

This integration can be performed at any stage of a linear system, so one can

  (i) integrate on the sensor capacitance,

  (ii) use an integrating preamplifier ('charge-sensitive' amplifier),

  (iii) amplify the current pulse and use an integrating ADC ('charge-sensing' ADC),

  (iv) rapidly sample and digitise the current pulse and integrate numerically.

In large systems the first three options tend to be most efficient.

14.4.1 *Signal integration*

Figure 14.8 illustrates signal formation in an ionisation chamber connected to an amplifier with a very high input resistance. The ionisation chamber volume could be filled with gas, liquid or a solid, as in a silicon sensor. As mobile charge carriers move towards their respective electrodes, they change the induced charge on the sensor electrodes, which form a capacitor C_d. If the amplifier has a very small input resistance R_i, the time constant $\tau = R_i(C_d + C_i)$ for discharging the sensor is small, and the amplifier will sense the signal current (C_i is the dynamic input capacitance). However, if the input time constant is large compared to the duration of the current

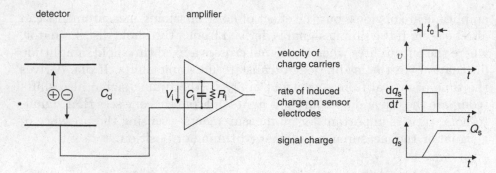

Fig. 14.8.   Charge collection and signal integration in an ionisation chamber.

pulse, the current pulse will be integrated on the capacitance and the resulting voltage at the amplifier input is

$$V_i = \frac{Q_s}{C_d + C_i} \; . \tag{14.2}$$

The magnitude of the signal is dependent on the sensor capacitance. In a system with varying sensor capacitances, a Si tracker with varying strip lengths, for example, or a partially depleted semiconductor sensor, where the capacitance varies with the applied bias voltage, one would have to deal with additional calibrations. However, the charge-sensitive amplifiers, widely used for detector readout, would overcome this problem by providing an output signal height practically independent of the input capacity.

Figure 14.9 shows the principle of a *feedback amplifier* that performs integration. It consists of an inverting amplifier with voltage gain $-A$ and a feedback capacitor C_f connected from the output to the input. To simplify the calculation, let the amplifier have an infinite input impedance, so no current flows into the amplifier input. If an input signal produces a voltage v_i at the amplifier input, the voltage at the amplifier output is $-Av_i$. Thus, the voltage difference across the feedback capacitor is $v_f = (A+1)v_i$ and the charge deposited on C_f is $Q_f = C_f v_f = C_f(A+1)v_i$. Since no current can flow into the amplifier, all of the signal current must charge

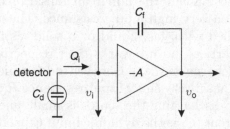

Fig. 14.9.   Principle of a *charge-sensitive amplifier*.

up the feedback capacitance, so $Q_f = Q_i$. The amplifier input appears as a 'dynamic' input capacitance

$$C_i = \frac{Q_i}{v_i} = C_f(A + 1) \ . \qquad (14.3)$$

The voltage output per unit input charge is

$$A_Q = \frac{dv_o}{dQ_i} = \frac{Av_i}{C_i v_i} = \frac{A}{C_i} = \frac{A}{A + 1} \cdot \frac{1}{C_f} \approx \frac{1}{C_f} \quad (A \gg 1) \ , \qquad (14.4)$$

so the charge gain is determined by a well-controlled component, the feedback capacitor.

The signal charge Q_s will be distributed between the sensor capacitance C_d and the dynamic input capacitance C_i. The ratio of measured charge to signal charge is

$$\frac{Q_i}{Q_s} = \frac{Q_i}{Q_d + Q_i} = \frac{C_i}{C_d + C_i} = \frac{1}{1 + C_d/C_i} \ , \qquad (14.5)$$

so the dynamic input capacitance must be large compared to the sensor capacitance.

Another very useful byproduct of the integrating amplifier is the ease of charge calibration. By adding a test capacitor as shown in Fig. 14.10, a voltage step injects a well-defined charge into the input node. If the dynamic input capacitance C_i is much larger than the test capacitance C_T, the voltage step at the test input will be applied nearly completely across the test capacitance C_T, thus injecting a charge $C_T \Delta V$ into the input.

The preceding discussion assumed that the amplifiers are infinitely fast, so they respond instantaneously to the applied signal. In reality amplifiers

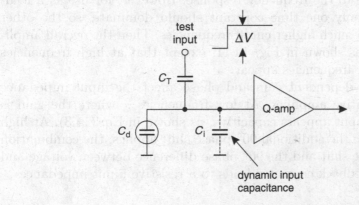

Fig. 14.10. Adding a test input to a charge-sensitive amplifier provides a simple means of absolute charge calibration.

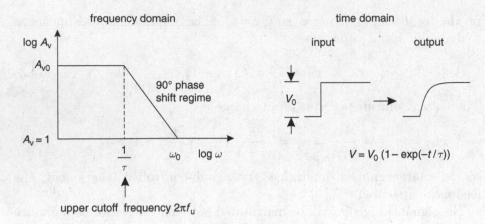

Fig. 14.11.   The time constant of an amplifier τ affects both the frequency and the time response. The amplifier's cutoff frequency is $\omega_u = 1/\tau = 2\pi f_u$. Both the time and frequency response are fully equivalent representations.

have a limited bandwidth, which translates into a time response. If a voltage step is applied to the input of the amplifier, the output does not respond instantaneously, as internal capacitances must first charge up. This is shown in Fig. 14.11. In a simple amplifier the time response is determined by a single time constant τ, corresponding to a cutoff (corner) frequency $\omega_u = 1/\tau = 2\pi f_u$. In the frequency domain the gain of a simple single-stage amplifier is constant up to the cutoff frequency f_u and then decreases inversely proportional to frequency with an additional phase shift of $90°$. In this regime the product of gain and bandwidth is constant, so extrapolation to unity gain yields the gain–bandwidth product $\omega_0 = A_{v0} \cdot \omega_u$. In practice, amplifiers utilise multiple stages, all of which contribute to the frequency response. However, for use as a feedback amplifier, only one time constant should dominate, so the other stages must have much higher cutoff frequencies. Then the overall amplifier response is as shown in Fig. 14.11, except that at high frequencies additional corner frequencies appear.

The frequency-dependent gain and phase affect the input impedance of a charge-sensitive amplifier. At low frequencies – where the gain is constant – the input appears capacitive, as shown in Eq. (14.3). At high frequencies where the additional $90°$ phase shift applies, the combination of amplifier phase shift and the $90°$ phase difference between voltage and current in the feedback capacitor leads to a resistive input impedance

$$Z_i = \frac{1}{\omega_0 C_f} \equiv R_i \; . \tag{14.6}$$

Thus, at low frequencies $f \ll f_u$ the input of a charge-sensitive amplifier appears capacitive, whereas at high frequencies $f \gg f_u$ it appears resistive.

Suitable amplifiers invariably have corner frequencies well below the frequencies of interest for radiation detectors, so the input impedance is resistive. This allows a simple calculation of the time response. The sensor capacitance is discharged by the resistive input impedance of the feedback amplifier with the time constant

$$\tau_i = R_i C_d = \frac{1}{\omega_0 C_f} \cdot C_d \; . \tag{14.7}$$

From this we see that the rise time of the charge-sensitive amplifier increases with sensor capacitance. The amplifier response can be slower than the duration of the current pulse from the sensor, as charge is initially stored on the detector capacitance, but the amplifier should respond faster than the peaking time of the subsequent pulse shaper. The feedback capacitance should be much smaller than the sensor capacitance. If $C_f = C_d/100$, the amplifier's gain–bandwidth product must be $100/\tau_i$, so for a rise time constant of 10 ns the gain–bandwidth product must be $\omega = 10^{10}\,\mathrm{s}^{-1} \cong 1.6\,\mathrm{GHz}$. The same result can be obtained using conventional operational amplifier feedback theory.

The mechanism of reducing the input impedance through shunt feedback leads to the concept of the *virtual ground*. If the gain is infinite, the input impedance is zero. Although very high gains (of order 10^5–10^6) are achievable in the kHz range, at the frequencies relevant for detector signals the gain is much smaller. The input impedance of typical charge-sensitive amplifiers in strip-detector systems is of order kΩ. Fast amplifiers designed to optimise power dissipation achieve input impedances of 100–500 Ω [5]. None of these qualifies as a 'virtual ground', so this concept should be applied with caution.

Apart from determining the signal rise time, the input impedance is critical in position-sensitive detectors. Figure 14.12 illustrates a silicon-strip sensor read out by a bank of amplifiers. Each strip electrode has a capacitance C_b to the backplane and a fringing capacitance C_{ss} to the neighbouring strips. If the amplifier has an infinite input impedance, charge induced on one strip will capacitively couple to the neighbours and the signal will be distributed over many strips (determined by C_{ss}/C_b). If, on the other hand, the input impedance of the amplifier is low compared to the interstrip impedance, practically all of the charge will flow into the amplifier, as current seeks the path of least impedance, and the neighbours will show only a small signal.

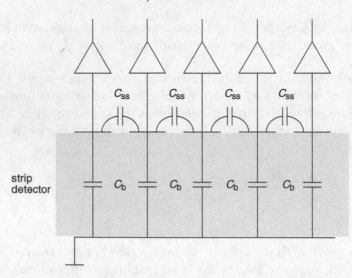

Fig. 14.12.   To preserve the position resolution of strip detectors the readout amplifiers must have a low input impedance to prevent spreading of signal charge to the neighbouring electrodes.

14.5   Signal processing

As noted in the introduction, one of the purposes of signal processing is to improve the signal-to-noise ratio by tailoring the spectral distributions of the signal and the electronic noise. However, for many detectors electronic noise does not determine the resolution. This is especially true for counters using photomultipliers for signal readout and amplification. For example, in a NaI(Tl) scintillation detector measuring 511 keV gamma rays, say in a positron-emission tomography system, 25 000 scintillation photons are produced. Because of reflective losses, about 15 000 reach the photocathode. This translates to about 3000 electrons reaching the first dynode. The gain of the electron multiplier will yield about $3 \cdot 10^9$ electrons at the anode. The statistical spread of the signal is determined by the smallest number of electrons in the chain, i.e. the 3000 electrons reaching the first dynode, so the resolution is $\sigma_E/E = 1/\sqrt{3000} \approx 2\%$, which at the anode corresponds to about $3 \cdot 10^9 \cdot 2\% \approx 6 \cdot 10^7$ electrons. This is much larger than the electronic noise in any reasonably designed system. This situation is illustrated in Fig. 14.13a. In this case, signal acquisition and count-rate capability may be the prime objectives of the pulse processing system. Figure 14.13b shows the situation for high-resolution sensors with small signals. Examples are semiconductor detectors, photodiodes or ionisation chambers. In this case, low noise is critical. Baseline fluctuations

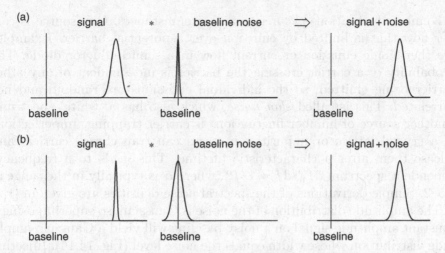

Fig. 14.13. Signal and baseline fluctuations add in quadrature. For large signal variance (a) as in scintillation detectors or proportional chambers the baseline noise is usually negligible, whereas for small signal variance as in semiconductor detectors or liquid-Ar ionisation chambers, baseline noise is critical.

can have many origins, external interference, artifacts due to imperfect electronics, etc., but the fundamental limit is electronic noise.

14.6 Electronic noise

Consider a current flowing through a sample bounded by two electrodes, i.e. n electrons moving with velocity v. The induced current depends on the spacing l between the electrodes (following 'Ramo's theorem' [1, 6]), so

$$i = \frac{nev}{l} \ . \tag{14.8}$$

The fluctuation of this current is given by the total differential

$$\langle \mathrm{d}i \rangle^2 = \left(\frac{ne}{l} \langle \mathrm{d}v \rangle \right)^2 + \left(\frac{ev}{l} \langle \mathrm{d}n \rangle \right)^2 , \tag{14.9}$$

where the two terms add in quadrature, as they are statistically uncorrelated. From this one sees that two mechanisms contribute to the total noise, velocity and number fluctuations.

Velocity fluctuations originate from thermal motion. Superimposed on the average drift velocity are random velocity fluctuations due to thermal excitations. This *thermal noise* is described by the long-wavelength limit of Planck's blackbody spectrum, where the spectral density, i.e. the power per unit bandwidth, is constant (*white noise*).

Number fluctuations occur in many circumstances. One source is carrier flow that is limited by emission over a potential barrier. Examples are thermionic emission or current flow in a semiconductor diode. The probability of a carrier crossing the barrier is independent of any other carrier being emitted, so the individual emissions are random and not correlated. This is called *shot noise*, which also has a 'white' spectrum. Another source of number fluctuations is carrier trapping. Imperfections in a crystal lattice or impurities in gases can trap charge carriers and release them after a characteristic lifetime. This leads to a frequency-dependent spectrum $dP_n/df = 1/f^\alpha$, where α is typically in the range of 0.5–2. Simple derivations of the spectral noise densities are given in [1].

The amplitude distribution of the noise is Gaussian, so superimposing a constant amplitude signal on a noisy baseline will yield a Gaussian amplitude distribution whose width equals the noise level (Fig. 14.14). Injecting a pulser signal and measuring the width of the amplitude distribution yields the noise level.

14.6.1 Thermal (Johnson) noise

The most common example of noise due to velocity fluctuations is the noise of resistors. The spectral noise power density versus frequency is

$$\frac{dP_n}{df} = 4kT \; , \tag{14.10}$$

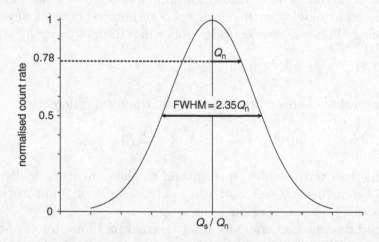

Fig. 14.14. Repetitive measurements of the signal charge yield a Gaussian distribution whose standard deviation equals the rms noise level Q_n. Often the width is expressed as the full width at half maximum (FWHM), which is 2.35 times the standard deviation.

where k is the Boltzmann constant and T the absolute temperature. Since the power in a resistance R can be expressed through either voltage or current,

$$P = \frac{V^2}{R} = I^2 R , \qquad (14.11)$$

the spectral voltage and current noise densities are

$$\frac{\mathrm{d}V_n^2}{\mathrm{d}f} \equiv e_n^2 = 4kTR \quad \text{and} \quad \frac{\mathrm{d}I_n^2}{\mathrm{d}f} \equiv i_n^2 = \frac{4kT}{R} . \qquad (14.12)$$

The total noise is obtained by integrating over the relevant frequency range of the system, the bandwidth, so the total noise voltage at the output of an amplifier with a frequency-dependent gain $A(f)$ is

$$v_{on}^2 = \int_0^\infty e_n^2 A^2(f) \, \mathrm{d}f . \qquad (14.13)$$

Since the spectral noise components are non-correlated, one must integrate over the noise power, i.e. the voltage squared. The total noise increases with bandwidth. Since small bandwidth corresponds to large rise times, increasing the speed of a pulse measurement system will increase the noise.

14.6.2   Shot noise

The spectral density of shot noise is proportional to the average current I,

$$i_n^2 = 2eI , \qquad (14.14)$$

where e is the electron charge. Note that the criterion for shot noise is that carriers are injected independently of one another, as in thermionic or semiconductor diodes. Current flowing through an ohmic conductor does not carry shot noise, since the fields set up by any local fluctuation in charge density can easily draw in additional carriers to equalise the disturbance.

14.7   Signal-to-noise ratio versus sensor capacitance

The basic noise sources manifest themselves as either voltage or current fluctuations. However, the desired signal is a charge, so to allow a comparison we must express the signal as a voltage or current. This was illustrated for an ionisation chamber in Fig. 14.8. As was noted, when the input time constant $R_i(C_d + C_i)$ is large compared to the duration of the sensor current pulse, the signal charge is integrated on the input capacitance, yielding the signal voltage $V_s = Q_s/(C_d + C_i)$. Assume that the amplifier has an input noise voltage V_n. Then the signal-to-noise ratio is

$$\frac{V_s}{V_n} = \frac{Q_s}{V_n(C_d + C_i)} \ . \tag{14.15}$$

This is a very important result – the signal-to-noise ratio for a given signal charge is inversely proportional to the total capacitance at the input node. Note that zero input capacitance does not yield an infinite signal-to-noise ratio. As shown in [1], this relationship only holds when the input time constant is about ten times greater than the sensor current pulse width. The dependence of signal-to-noise ratio on capacitance is a general feature that is independent of amplifier type. Since feedback cannot improve signal-to-noise ratio, Eq. (14.15) holds for charge-sensitive amplifiers, although in that configuration the charge signal is constant, but the noise increases with total input capacitance (see [1]). In the noise analysis the feedback capacitance adds to the total input capacitance (the passive capacitance, not the dynamic input capacitance), so C_f should be kept small.

14.8   Pulse shaping

Pulse shaping has two conflicting objectives. The first is to limit the bandwidth to match the measurement time. Too large a bandwidth will increase the noise without increasing the signal. Typically, the pulse shaper transforms a narrow sensor pulse into a broader pulse with a gradually rounded maximum at the peaking time. This is illustrated in Fig. 14.15. The signal amplitude is measured at the peaking time T_P.

The second objective is to constrain the pulse width so that successive signal pulses can be measured without overlap (pile-up), as illustrated in Fig. 14.16. Reducing the pulse duration increases the allowable signal rate, but at the expense of electronic noise.

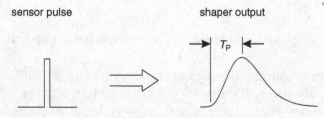

Fig. 14.15.   In energy measurements a pulse processor typically transforms a short sensor current pulse to a broader pulse with a peaking time T_P.

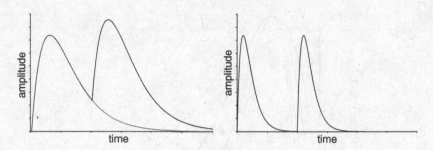

Fig. 14.16. Amplitude pile-up occurs when two pulses overlap (left). Reducing the shaping time allows the first pulse to return to the baseline before the second pulse arrives.

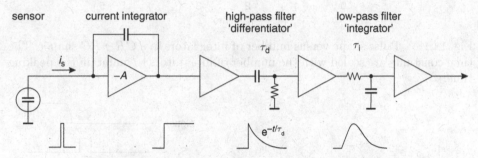

Fig. 14.17. Components of a pulse shaping system. The signal current from the sensor is integrated to form a step impulse with a long decay. A subsequent high-pass filter ('differentiator') limits the pulse width and the low-pass filter ('integrator') increases the rise time to form a pulse with a smooth cusp.

In designing the shaper it is necessary to balance these conflicting goals. Usually, many different considerations lead to a 'non-textbook' compromise; 'optimum shaping' depends on the application.

A simple shaper is shown in Fig. 14.17. A high-pass filter sets the duration of the pulse by introducing a decay time constant τ_d. Next a low-pass filter with a time constant τ_i increases the rise time to limit the noise bandwidth. The high-pass filter is often referred to as a 'differentiator', since for short pulses it forms the derivative. Correspondingly, the low-pass filter is called an 'integrator'. Since the high-pass filter is implemented with a CR section and the low-pass with an RC, this shaper is referred to as a CR–RC shaper. Although pulse shapers are often more sophisticated and complicated, the CR–RC shaper contains the essential features of all pulse shapers, a lower frequency bound and an upper frequency bound.

After peaking the output of a simple CR–RC shaper returns to baseline rather slowly. The pulse can be made more symmetrical, allowing higher signal rates for the same peaking time. Very sophisticated circuits have

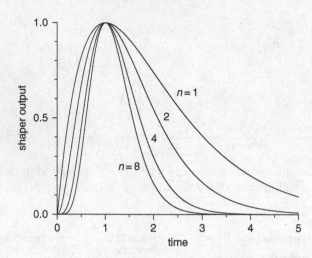

Fig. 14.18.   Pulse shape versus number of integrators in a CR–nRC shaper. The time constants are scaled with the number of integrators to maintain the peaking time.

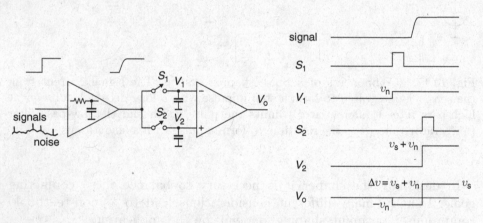

Fig. 14.19.   Principle of a shaper using correlated double sampling. v_s and v_n are the signal and noise voltages.

been developed towards this goal, but a conceptually simple way is to use multiple integrators, as illustrated in Fig. 14.18. The integration and differentiation time constants are scaled to maintain the peaking time. Note that the peaking time is a key design parameter, as it dominates the noise bandwidth and must also accommodate the sensor response time.

Another type of shaper is the correlated double sampler, illustrated in Fig. 14.19. This type of shaper is widely used in monolithically integrated circuits, as many CMOS processes (see Sect. 14.11.1) provide only capacitors and switches, but no resistors. This is an example of a time-variant filter. The CR–nRC filter described in Fig. 14.18 acts continuously on the signal, whereas the correlated double sample changes filter parameters versus time. Input signals are superimposed on a slowly fluctuating baseline. To remove the baseline fluctuations the baseline is sampled prior to the signal. Next, the signal plus baseline is sampled and the previous baseline sample subtracted to obtain the signal. The prefilter is critical to limit the noise bandwidth of the system. Filtering after the sampler is useless, as noise fluctuations on time scales shorter than the sample time will not be removed. Here the sequence of filtering is critical, unlike a time-invariant linear filter, where the sequence of filter functions can be interchanged.

14.9   Noise analysis of a detector and front-end amplifier

To determine how the pulse shaper affects the signal-to-noise ratio, consider the detector front end in Fig. 14.20. The detector is represented by the capacitance C_d, a relevant model for many radiation sensors. Sensor bias voltage is applied through the resistor R_b. The bypass capacitor C_b shunts any external interference coming through the bias supply line to ground. For high-frequency signals this capacitor appears as a low impedance, so for sensor signals the 'far end' of the bias resistor is connected to ground. The coupling capacitor C_c blocks the sensor bias voltage from the amplifier input, which is why a capacitor serving this rôle is also called a 'blocking capacitor'. The series resistor R_s represents any resistance present in the connection from the sensor to the amplifier input. This includes the resistance of the sensor electrodes, the resistance of the connecting wires or traces, any resistance used to protect the amplifier against large voltage transients ('input protection'), and parasitic resistances in the input transistor.

The following implicitly includes a constraint on the bias resistance, whose rôle is often misunderstood. It is often thought that the signal current generated in the sensor flows through R_b and the resulting voltage drop is measured. If the time constant $R_b C_d$ is small compared to the

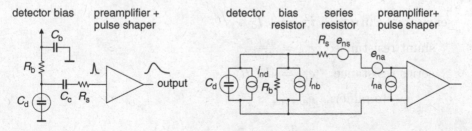

Fig. 14.20.   A detector front-end circuit and its equivalent circuit for noise calculations.

peaking time of the shaper T_P, the sensor will have discharged through R_b and much of the signal will be lost. Thus, we have the condition $R_b C_d \gg T_P$, or $R_b \gg T_P/C_d$. The bias resistor must be sufficiently large to block the flow of signal charge, so that all of the signal is available for the amplifier.

To analyse this circuit a voltage amplifier will be assumed, so all noise contributions will be calculated as a noise voltage appearing at the amplifier input. Steps in the analysis include: (1) determine the frequency distribution of all noise voltages presented to the amplifier input from all individual noise sources, (2) integrate over the frequency response of the shaper (for simplicity a CR–RC shaper) and determine the total noise voltage at the shaper output, and (3) determine the output signal for a known input signal charge. The equivalent noise charge (ENC) is the signal charge for whioh $S/N = 1$.

The *equivalent circuit* for the noise analysis (second panel of Fig. 14.20) includes both current and voltage noise sources. The 'shot noise' i_{nd} of the sensor leakage current is represented by a current noise generator in parallel with the sensor capacitance. As noted above, resistors can be modelled either as a voltage or current generator. Generally, resistors shunting the input act as noise current sources and resistors in series with the input act as noise voltage sources (which is why some in the detector community refer to current and voltage noise as *parallel* and *series noise*). Since the bias resistor effectively shunts the input, as the capacitor C_b passes current fluctuations to ground, it acts as a current generator i_{nb} and its noise current has the same effect as the shot-noise current from the detector. The shunt resistor can also be modelled as a noise voltage source, yielding the result that it acts as a current source. Choosing the appropriate model merely simplifies the calculation. Any other shunt resistances can be incorporated in the same way. Conversely, the series resistor R_s acts as a voltage generator. The electronic noise of the amplifier is described fully by a combination of voltage and current sources at its input, shown as e_{na} and i_{na}.

Thus, the noise sources are

$$\text{sensor bias current:} \quad i_{nd}^2 = 2eI_d \ ,$$
$$\text{shunt resistance:} \quad i_{nb}^2 = \frac{4kT}{R_b} \ ,$$
$$\text{series resistance:} \quad e_{ns}^2 = 4kTR_s \ ,$$
$$\text{amplifier:} \quad e_{na}, i_{na} \ ,$$

where e is the electron charge, I_d the sensor bias current, k the Boltzmann constant, and T the temperature. Typical amplifier noise parameters e_{na}

and i_{na} are of order $nV/\sqrt{Hz}$ and $fA/\sqrt{Hz}$ (FETs)[¶] to $pA/\sqrt{Hz}$ (bipolar transistors). Amplifiers tend to exhibit a 'white' noise spectrum at high frequencies (greater than order kHz), but at low frequencies show excess noise components with the spectral density

$$e_{nf}^2 = \frac{A_f}{f} \; , \qquad (14.16)$$

where the noise coefficient A_f is device specific and of order 10^{-10}–$10^{-12} \, V^2$.

The noise voltage generators are in series and simply add in quadrature. White noise distributions remain white. However, a portion of the noise currents flows through the detector capacitance, resulting in a frequency-dependent noise voltage $i_n/(\omega C_d)$, so the originally white spectrum of the sensor shot noise and the bias resistor now acquires a $1/f$ dependence. The frequency distribution of all noise sources is further altered by the combined frequency response of the amplifier chain $A(f)$. Integrating over the cumulative noise spectrum at the amplifier output and comparing to the output voltage for a known input signal yields the signal-to-noise ratio. In this example the shaper is a simple CR–RC shaper, where for a given differentiation time constant the signal-to-noise ratio is maximised when the integration time constant equals the differentiation time constant, $\tau_i = \tau_d \equiv \tau$. Then the output pulse assumes its maximum amplitude at the time $T_P = \tau$.

Although the basic noise sources are currents or voltages, since radiation detectors are typically used to measure charge, the system's noise level is conveniently expressed as an equivalent noise charge Q_n. As noted previously, this is equal to the detector signal that yields a signal-to-noise ratio of 1. The equivalent noise charge is commonly expressed in Coulombs, the corresponding number of electrons, or the equivalent deposited energy (eV). For the above circuit the equivalent noise charge is

$$Q_n^2 = \left(\frac{e^2}{8}\right) \left[\left(2eI_d + \frac{4kT}{R_b} + i_{na}^2 \right) \cdot \tau + \left(4kTR_s + e_{na}^2 \right) \cdot \frac{C_d^2}{\tau} + 4A_f C_d^2 \right] \; .$$
$$(14.17)$$

The prefactor $e^2/8 = \exp(2)/8 = 0.924$ normalises the noise to the signal gain. The first term combines all noise current sources and increases with shaping time. The second term combines all noise voltage sources and decreases with shaping time, but increases with sensor capacitance. The third term is the contribution of amplifier $1/f$ *noise* and, as a voltage source, also increases with sensor capacitance. The $1/f$ term is

---

[¶] FET – field effect transistor.

independent of shaping time, since for a $1/f$ spectrum the total noise depends on the ratio of upper to lower cutoff frequency, which depends only on shaper topology, but not on the shaping time.

The equivalent noise charge can be expressed in a more general form that applies to all types of pulse shapers:

$$Q_n^2 = i_n^2 F_i T_S + e_n^2 F_v \frac{C^2}{T_S} + F_{vf} A_f C^2 \, , \qquad (14.18)$$

where F_i, F_v and F_{vf} depend on the shape of the pulse determined by the shaper and T_S is a characteristic time, for example, the peaking time of a CR–nRC shaped pulse or the prefilter time constant in a correlated double sampler. C is the total parallel capacitance at the input, including the amplifier input capacitance. The shape factors F_i and F_v are easily calculated,

$$F_i = \frac{1}{2T_S} \int_{-\infty}^{\infty} [W(t)]^2 \, \mathrm{d}t \, , \quad F_v = \frac{T_S}{2} \int_{-\infty}^{\infty} \left[\frac{\mathrm{d}W(t)}{\mathrm{d}t} \right]^2 \, \mathrm{d}t \, . \qquad (14.19)$$

For time-invariant pulse shaping $W(t)$ is simply the system's impulse response (the output signal seen on an oscilloscope) with the peak output signal normalised to unity. For a time-variant shaper the same equations apply, but $W(t)$ is determined differently. See [7–10] for more details.

A CR–RC shaper with equal time constants $\tau_i = \tau_d$ has $F_i = F_v = 0.9$ and $F_{vf} = 4$, independent of the shaping time constant, so for the circuit in Fig. 14.17 Eq. (14.18) becomes

$$Q_n^2 = \left(2q_e I_d + \frac{4kT}{R_b} + i_{na}^2 \right) F_i T_S + \left(4kT R_s + e_{na}^2 \right) F_v \frac{C^2}{T_S} + F_{vf} A_f C^2 \, .$$

$$(14.20)$$

Pulse shapers can be designed to reduce the effect of current noise, to mitigate radiation damage, for example. Increasing pulse symmetry tends to decrease F_i and increase F_v, e.g. to $F_i = 0.45$ and $F_v = 1.0$ for a shaper with one CR differentiator and four cascaded RC integrators.

Figure 14.21 shows how equivalent noise charge is affected by shaping time. At short shaping times the voltage noise dominates, whereas at long shaping times the current noise takes over. Minimum noise is obtained where the current and voltage contributions are equal. The noise minimum is flattened by the presence of $1/f$ noise. Also shown is that increasing the detector capacitance will increase the voltage noise contribution and shift the noise minimum to longer shaping times, albeit with an increase in minimum noise.

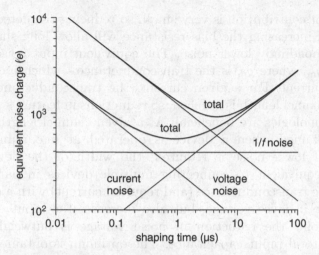

Fig. 14.21. Equivalent noise charge versus shaping time. At small shaping times (large bandwidth) the equivalent noise charge is dominated by voltage noise, whereas at long shaping times (large integration times) the current noise contributions dominate. The total noise assumes a minimum where the current and voltage contributions are equal. The '$1/f$' noise contribution is independent of shaping time and flattens the noise minimum. Increasing the voltage or current noise contribution shifts the noise minimum. Increased voltage noise is shown as an example.

For quick estimates one can use the following equation, which assumes a field effect transistor (FET) amplifier (negligible i_{na}) and a simple CR–RC shaper with peaking time τ. The noise is expressed in units of the electron charge e and C is the total parallel capacitance at the input, including C_d, all stray capacitances and the amplifier's input capacitance

$$Q_n^2 = 12 \left[\frac{e^2}{\text{nA ns}} \right] I_d \tau + 6 \cdot 10^5 \left[\frac{e^2 \, \text{k}\Omega}{\text{ns}} \right] \frac{\tau}{R_b}$$

$$+ \, 3.6 \cdot 10^4 \left[\frac{e^2 \, \text{ns}}{(\text{pF})^2 (\text{nV})^2/\text{Hz}} \right] e_n^2 \frac{C^2}{\tau} \, . \quad (14.21)$$

The noise charge is improved by reducing the detector capacitance and leakage current, judiciously selecting all resistances in the input circuit, and choosing the optimum shaping time constant. The noise parameters of a well-designed amplifier depend primarily on the input device. Fast, high-gain transistors are generally best.

In field effect transistors, both junction field effect transistors (JFETs) and metal oxide semiconductor field effect transistors (MOSFETs), the noise current contribution is very small, so reducing the detector leakage current and increasing the bias resistance will allow long shaping times with correspondingly lower noise. The equivalent input noise voltage is $e_n^2 \approx 4kT/g_m$, where g_m is the transconductance,[||] which increases with operating current. For a given current, the transconductance increases when the channel length is reduced, so reductions in feature size with new process technologies are beneficial. At a given channel length, minimum noise is obtained when a device is operated at maximum transconductance. If lower noise is required, the width of the device can be increased (equivalent to connecting multiple devices in parallel). This increases the transconductance (and required current) with a corresponding decrease in noise voltage, but also increases the input capacitance. At some point the reduction in noise voltage is outweighed by the increase in total input capacitance. The optimum is obtained when the FET's input capacitance equals the external capacitance (sensor + stray capacitance). Note that this capacitive matching criterion only applies when the input-current noise contribution of the amplifying device is negligible.

Capacitive matching comes at the expense of power dissipation. Since the minimum is shallow, one can operate at significantly lower currents with just a minor increase in noise. In large detector arrays power dissipation is critical, so FETs are hardly ever operated at their minimum noise. Instead, one seeks an acceptable compromise between noise and power dissipation (see [1] for a detailed discussion). Similarly, the choice of input devices is frequently driven by available fabrication processes. High-density integrated circuits tend to include only MOSFETs, so this determines the input device, even where a bipolar transistor would provide better performance.

In bipolar transistors the shot noise associated with the base current I_B is significant, $i_{nB}^2 = 2eI_B$. Since $I_B = I_C/\beta_{DC}$, where I_C is the collector current and β_{DC} the direct current gain, this contribution increases with device current. On the other hand, the equivalent input noise voltage

$$e_n^2 = \frac{2(kT)^2}{eI_C} \tag{14.22}$$

decreases with collector current, so the noise assumes a minimum at a specific collector current,

$$Q_{n,min}^2 = 4kT\frac{C}{\sqrt{\beta_{DC}}}\sqrt{F_i F_v} \quad \text{at} \quad I_C = \frac{kT}{e}C\sqrt{\beta_{DC}}\sqrt{\frac{F_v}{F_i}}\frac{1}{T_S} \; . \tag{14.23}$$

[||] Transconductance is the ratio of the current change at the output and the corresponding voltage change at the input, $g_m = \Delta I_{out}/\Delta V_{in}$.

For a *CR–RC* shaper and $\beta_{DC} = 100$,

$$Q_{n,min} \approx 250 \left[\frac{e}{\sqrt{pF}} \right] \cdot \sqrt{C} \quad \text{at} \quad I_C = 260 \left[\frac{\mu A \, ns}{pF} \right] \cdot \frac{C}{T_S} \, . \tag{14.24}$$

The minimum obtainable noise is independent of shaping time (unlike FETs), but only at the optimum collector current I_C, which does depend on shaping time.

In bipolar transistors the input capacitance is usually much smaller than the sensor capacitance (of order $1 \, pF$ for $e_n \approx 1 \, nV/\sqrt{Hz}$) and substantially smaller than in FETs with comparable noise. Since the transistor input capacitance enters into the total input capacitance, this is an advantage. Note that capacitive matching does not apply to bipolar transistors, because their noise current contribution is significant. Due to the base current noise bipolar transistors are best at short shaping times, where they also require lower power than FETs for a given noise level.

When the input noise current is negligible, the noise increases linearly with sensor capacitance. The noise slope

$$\frac{dQ_n}{dC_d} \approx 2e_n \cdot \sqrt{\frac{F_v}{T}} \tag{14.25}$$

depends both on the preamplifier (e_n) and the shaper (F_v, T). The zero intercept can be used to determine the amplifier input capacitance plus any additional capacitance at the input node.

Practical noise levels range from $< 1 \, e$ for charge-coupled devices (CCDs) at long shaping times to $\approx 10^4 \, e$ in high-capacitance liquid-Ar calorimeters. Silicon strip detectors typically operate at $\approx 10^3$ electrons, whereas pixel detectors with fast readout provide noise of 100–200 electrons. Transistor noise is discussed in more detail in [1].

14.10 Timing measurements

Pulse-height measurements discussed up to now emphasise measurement of signal charge. *Timing measurements* seek to optimise the determination of the time of occurrence. Although, as in amplitude measurements, signal-to-noise ratio is important, the determining parameter is not signal-to-noise, but slope-to-noise ratio. This is illustrated in Fig. 14.22, which shows the leading edge of a pulse fed into a threshold discriminator (comparator), a *leading-edge trigger*. The instantaneous signal level is modulated by noise, where the variations are indicated by the

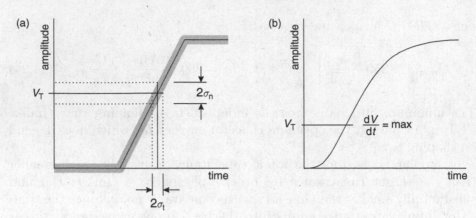

Fig. 14.22.   Fluctuations in signal amplitude crossing a threshold translate into timing fluctuations (a). With realistic pulses the slope changes with amplitude, so minimum timing jitter occurs with the trigger level at the maximum slope.

shaded band. Because of these fluctuations, the time of threshold crossing fluctuates. By simple geometrical projection, the timing variance or *jitter* is

$$\sigma_t = \frac{\sigma_n}{(\mathrm{d}S/\mathrm{d}t)_{S_T}} \approx t_r \frac{\sigma_n}{S} \; , \tag{14.26}$$

where σ_n is the rms noise and the derivative of the signal $\mathrm{d}S/\mathrm{d}t$ is evaluated at the trigger level S_T. To increase $\mathrm{d}S/\mathrm{d}t$ without incurring excessive noise, the amplifier bandwidth should match the rise time of the detector signal. The 10%–90% rise time of an amplifier with bandwidth f_u (see Fig. 14.11) is

$$t_r = 2.2\,\tau = \frac{2.2}{2\pi f_u} = \frac{0.35}{f_u} \; . \tag{14.27}$$

For example, an oscilloscope with 350 MHz bandwidth has a 1 ns rise time. When amplifiers are cascaded, which is invariably necessary, the individual rise times add in quadrature:

$$t_r \approx \sqrt{t_{r1}^2 + t_{r2}^2 + \ldots + t_{rn}^2} \; . \tag{14.28}$$

Increasing signal-to-noise ratio improves time resolution, so minimising the total capacitance at the input is also important. At high signal-to-noise ratios the time jitter can be much smaller than the rise time.

The second contribution to time resolution is time walk, where the timing signal shifts with amplitude as shown in Fig. 14.23. This can be corrected by various means, either in hardware or software. For more detailed tutorials on timing measurements, see [1, 11].

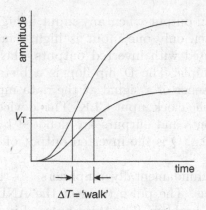

Fig. 14.23. The time at which a signal crosses a fixed threshold depends on the signal amplitude, leading to 'time walk'.

14.11 Digital electronics

Analogue signals utilise continuously variable properties of the pulse to impart information, such as the pulse amplitude or pulse shape. Digital signals have constant amplitude, but the presence of the signal at specific times is evaluated, i.e. whether the signal is in one of two states, 'low' or 'high'. However, this still involves an analogue process, as the presence of a signal is determined by the signal level exceeding a threshold at the proper time.

14.11.1 Logic elements

Figure 14.24 illustrates several functions utilised in digital circuits ('logic' functions). An *AND gate* provides an output only when all inputs are

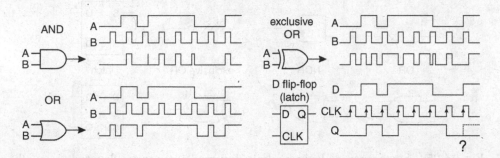

Fig. 14.24. Basic logic functions include gates (AND, OR, Exclusive OR) and flip-flops. The outputs of the AND and D flip-flop show how small shifts in relative timing between inputs can determine the output state.

high. An *OR* gives an output when any input is high. An *eXclusive OR (XOR)* responds when only one input is high. The same elements are commonly implemented with inverted outputs, then called NAND and NOR gates, for example. The D flip-flop is a bistable memory circuit that records the presence of a signal at the data input D when a signal transition occurs at the clock input CLK. This device is commonly called a *latch*. Inverted inputs and outputs are denoted by small circles or by superimposed bars, e.g. $\overline{Q}$ is the inverted output of a flip-flop, as shown in Fig. 14.25.

Logic circuits are fundamentally amplifiers, so they also suffer from bandwidth limitations. The pulse train of the AND gate in Fig. 14.24 illustrates a common problem. The third pulse of input B is going low at the same time that input A is going high. Depending on the time overlap, this can yield a narrow output that may or may not be recognised by the following circuit. In an XOR this can occur when two pulses arrive nearly at the same time. The D flip-flop requires a minimum setup time for a level change at the D input to be recognised, so changes in the data level may not be recognised at the correct time. These marginal events may be extremely rare and perhaps go unnoticed. However, in complex systems the combination of 'glitches' can make the system 'hang up', necessitating a system reset. Data transmission protocols have been developed to detect such errors (parity checks, Hamming codes, etc.), so corrupted data can be rejected.

Some key aspects of logic systems can be understood by inspecting the circuit elements that are used to form logic functions. In an *n*-channel metal oxide semiconductor (NMOS) transistor a conductive channel is formed when the input electrode is biased positive with respect to the

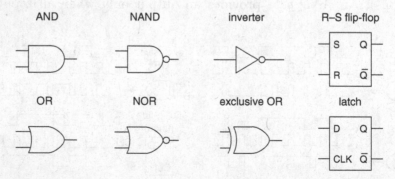

Fig. 14.25.   Some common logic symbols. Inverted outputs are denoted by small circles or by a superimposed bar, as for the latch output $\overline{Q}$. Additional inputs can be added to gates as needed. An R–S flip-flop (R–S: Reset–Set) sets the Q output high in response to an S input. An R input resets the Q output to low.

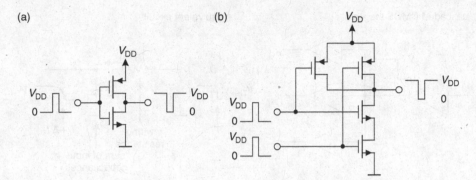

Fig. 14.26. A CMOS inverter (a) and NAND gate (b).

channel. The input, called the 'gate', is capacitively coupled to the output channel connected between the 'drain' and 'source' electrodes. A p-channel (PMOS) transistor is the complementary device, where a conductive channel is formed when the gate is biased negative with respect to the source.

Complementary MOS (CMOS) logic utilises both NMOS and PMOS transistors as shown in Fig. 14.26. In the inverter the lower (NMOS) transistor is turned off when the input is low, but the upper (PMOS) transistor is turned on, so the output is connected to V_{DD}, taking the output high. Since the current path from V_{DD} to ground is blocked by either the NMOS or PMOS device being off, the power dissipation is zero in both the high and low states. Current only flows during the level transition when both devices are on as the input level is at approximately $V_{DD}/2$. As a result, the power dissipation of CMOS logic is significantly lower than in NMOS or PMOS circuits, which draw current in either one or the other logic state. However, this reduction in power is obtained only in logic circuitry. CMOS analogue amplifiers are not fundamentally more power efficient than NMOS or PMOS circuits, although CMOS allows more efficient circuit topologies.

14.11.2 Propagation delays and power dissipation

Logic elements always operate in conjunction with other circuits, as illustrated in Fig. 14.27. The wiring resistance together with the total load capacitance increases the rise time of the logic pulse and as a result delays the time when the transition crosses the logic threshold. The energy dissipated in the wiring resistance R is

$$E = \int i^2(t) R \, dt \, . \tag{14.29}$$

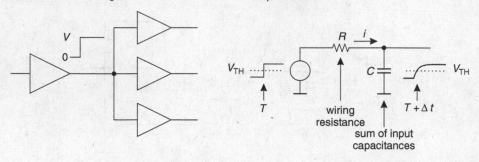

Fig. 14.27.   The wiring resistance together with the distributed load capacitance delays the signal.

The current flow during one transition is

$$i(t) = \frac{V}{R} \exp\left(-\frac{t}{RC}\right) , \qquad (14.30)$$

so the dissipated energy per transition (either positive or negative)

$$E = \frac{V^2}{R} \int\limits_0^\infty \exp\left(-\frac{2t}{RC}\right) \mathrm{d}t = \frac{1}{2}CV^2 . \qquad (14.31)$$

When pulses occur at a frequency f, the power dissipated in both the positive and negative transitions is

$$P = fCV^2 . \qquad (14.32)$$

Thus, the power dissipation increases with clock frequency and the square of the logic swing.

   Fast logic is time-critical. It relies on logic operations from multiple paths coming together at the right time. Valid results depend on maintaining minimum allowable overlaps and set-up times as illustrated in Fig. 14.24. Each logic circuit has a finite propagation delay, which depends on circuit loading, i.e. how many loads the circuit has to drive. In addition, as illustrated in Fig. 14.27 the wiring resistance and capacitive loads introduce delay. This depends on the number of circuits connected to a wire or trace, the length of the trace and the dielectric constant of the substrate material. Relying on control of circuit and wiring delays to maintain timing requires great care, as it depends on circuit variations and temperature. In principle all of this can be simulated, but in complex systems there are too many combinations to test every one. A more robust solution is to use synchronous systems, where the timing of all transitions

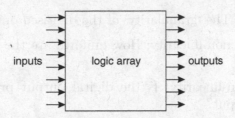

Fig. 14.28. Complex logic circuits are commonly implemented using logic arrays that as an integrated block provide the desired outputs in response to specific input combinations.

is determined by a master clock. Generally, this does not provide the utmost speed and requires some additional circuitry, but increases reliability. Nevertheless, clever designers frequently utilise asynchronous logic. Sometimes it succeeds ... and sometimes it does not.

14.11.3 Logic arrays

Commodity integrated circuits with basic logic blocks are readily available, e.g. with four NAND gates or two flip-flops in one package. These can be combined to form simple digital systems. However, complex logic systems are no longer designed using individual gates. Instead, logic functions are described in a high-level language (e.g. VHDL**), synthesised using design libraries, and implemented as custom ICs – *application-specific IC*s (ASICs) – or programmable logic arrays. In these implementations the digital circuitry no longer appears as an ensemble of inverters, gates and flip-flops, but as an integrated logic block that provides specific outputs in response to various input combinations. This is illustrated in Fig. 14.28. Field Programmable Gate or logic Arrays (FPGAs) are a common example. A representative FPGA has 512 pads usable for inputs and outputs, $\approx 10^6$ gates, and $\approx 100\,\mathrm{K}$ of memory. Modern design tools also account for propagation delays, wiring lengths, loads and temperature dependence. The design software also generates 'test vectors' that can be used to test finished parts. Properly implemented, complex digital designs can succeed on the first pass, whether as ASICs or as logic or gate arrays.

14.12 Analogue-to-digital conversion

For data storage and subsequent analysis the analogue signal at the shaper output must be digitised. Important parameters for *analogue-to-digital converters* (ADCs or A/Ds) used in detector systems are as follows:

** VHDL – VHSIC Hardware Description Language; VHSIC – Very High Speed Integrated Circuit.

(i) Resolution: The 'granularity' of the digitised output.

(ii) Differential non-linearity: How uniform are the digitisation increments?

(iii) Integral non-linearity: Is the digital output proportional to the analogue input?

(iv) Conversion time: How much time is required to convert an analogue signal to a digital output?

(v) Count-rate performance: How quickly can a new conversion commence after completion of a prior one without introducing deleterious artifacts?

(vi) Stability: Do the conversion parameters change with time?

Instrumentation ADCs used in industrial data acquisition and control systems share most of these requirements. However, detector systems place greater emphasis on differential non-linearity and count-rate performance. The latter is important, as detector signals often occur randomly, in contrast to systems where signals are sampled at regular intervals. As in amplifiers, if the DC gain is not precisely equal to the high-frequency gain, the baseline will shift. Furthermore, following each pulse it takes some time for the baseline to return to its quiescent level. For periodic signals of roughly equal amplitude these baseline deviations will be the same for each pulse, but for a random sequence of pulses with varying amplitudes, the instantaneous baseline level will be different for each pulse and broaden the measured signal.

Conceptually, the simplest technique is *flash conversion*, illustrated in Fig. 14.29. The signal is fed in parallel to a bank of threshold comparators. The individual threshold levels are set by a resistive divider. The comparator outputs are encoded such that the output of the highest-level comparator that fires yields the correct bit pattern. The threshold levels can be set to provide a linear conversion characteristic where each bit corresponds to the same analogue increment, or a non-linear characteristic to provide increments proportional to the absolute level, which provides constant relative resolution over the range, for example.

The big advantage of this scheme is speed; conversion proceeds in one step and conversion times $< 10\,\mathrm{ns}$ are readily achievable. The drawbacks are component count and power consumption, as one comparator is required per conversion bin. For example, an 8-bit converter requires 256 comparators. The conversion is always monotonic and differential non-linearity is determined by the matching of the resistors in the threshold divider. Only relative matching is required, so this topology is a good match for monolithic integrated circuits. *Flash ADCs* are available with

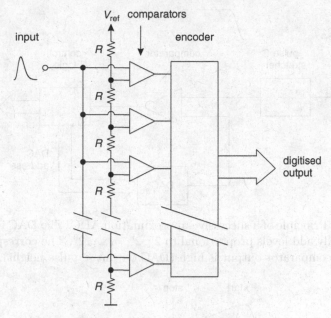

Fig. 14.29.   Block diagram of a flash ADC.

conversion rates $> 500\,\text{MS/s}$ (megasamples per second) at 8-bit resolution and a power dissipation of about 5 W.

The most commonly used technique is the successive-approximation ADC, shown in Fig. 14.30. The input pulse is sent to a pulse stretcher, which follows the signal until it reaches its cusp and then holds the peak value. The stretcher output feeds a comparator, whose reference is provided by a digital-to-analogue converter (DAC). The DAC is cycled beginning with the most significant bits. The corresponding bit is set when the comparator fires, i.e. the DAC output becomes less than the pulse height. Then the DAC cycles through the less significant bits, always setting the corresponding bit when the comparator fires. Thus, n-bit resolution requires n steps and yields 2^n bins. This technique makes efficient use of circuitry and is fairly fast. High-resolution devices (16–20 bits) with conversion times of order μs are readily available. Currently a 16-bit ADC with a conversion time of $1\,\mu\text{s}$ (1 MS/s) requires about 100 mW.

A common limitation is differential non-linearity (DNL), since the resistors that set the DAC levels must be extremely accurate. For DNL $< 1\%$ the resistor determining the 2^{12}-level in a 13-bit ADC must be accurate to $< 2.4 \cdot 10^{-6}$. As a consequence, differential non-linearity in high-resolution successive-approximation converters is typically 10%–20% and often exceeds the 0.5 LSB (least significant bit) required to ensure monotonic response.

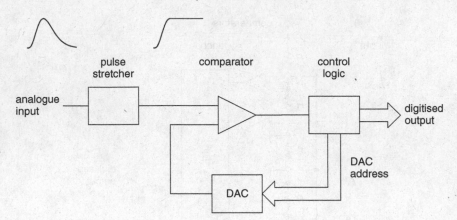

Fig. 14.30. Principle of a successive-approximation ADC. The DAC is controlled to sequentially add levels proportional to $2^n, 2^{n-1}, \ldots, 2^0$. The corresponding bit is set if the comparator output is high (DAC output < pulse height).

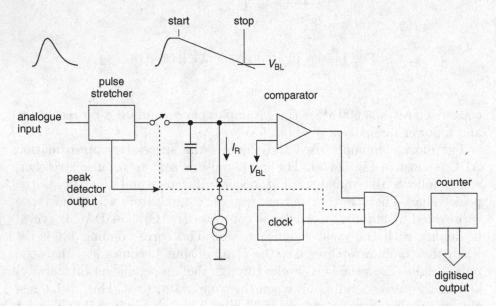

Fig. 14.31. Principle of a Wilkinson ADC. After the peak amplitude has been acquired, the output of the peak detector initiates the conversion process. The memory capacitor is discharged by a constant current while counting the clock pulses. When the capacitor is discharged to the baseline level V_{BL}, the comparator output goes low and the conversion is complete.

The *Wilkinson ADC* [12] has traditionally been the mainstay of precision pulse digitisation. The principle is shown in Fig. 14.31. The peak signal amplitude is acquired by a combined peak detector/pulse stretcher and transferred to a memory capacitor. The output of the peak detector initiates the conversion process:

(i) The memory capacitor is disconnected from the stretcher,

(ii) a current source is switched on to linearly discharge the capacitor with current I_R, and simultaneously

(iii) a counter is enabled to determine the number of clock pulses until the voltage on the capacitor reaches the baseline level V_{BL}.

The time required to discharge the capacitor is a linear function of pulse height, so the counter content provides the digitised pulse height. The clock pulses are provided by a crystal oscillator, so the time between pulses is extremely uniform and this circuit inherently provides excellent differential linearity. The drawback is the relatively long conversion time T_C, which is proportional to the pulse height, $T_C = n \cdot T_{clk}$, where the channel number n corresponds to the pulse height. For example, a clock frequency of 100 MHz provides a clock period $T_{clk} = 10$ ns and a maximum conversion time $T_C = 82$ μs for 13 bits ($n = 8192$). Clock frequencies of 100 MHz are typical, but > 400 MHz have been implemented with excellent performance (DNL $< 10^{-3}$). This scheme makes efficient use of circuitry and allows low power dissipation. Wilkinson ADCs have been implemented in 128-channel readout ICs for silicon strip detectors [13]. Each ADC added only 100 μm to the length of a channel and a power of 300 μW per readout channel.

14.13 Time-to-digital converters (TDCs)

The combination of a clock generator with a counter is the simplest technique for *time-to-digital conversion*, as shown in Fig. 14.32. The clock pulses are counted between the start and stop signals, which yields a direct readout in real time. The limitation is the speed of the counter, which in current technology is limited to about 1 GHz, yielding a time resolution of 1 ns. Using the stop pulse to strobe the instantaneous counter status into a register provides multi-hit capability.

Analogue techniques are commonly used in high-resolution digitisers to provide resolution in the range of ps to ns. The principle is to convert a time interval into a voltage by charging a capacitor through a switchable current source. The start pulse turns on the current source and the stop pulse turns it off. The resulting voltage on the capacitor C is $V = Q/C = I_T(T_{stop} - T_{start})/C$, which is digitised by an ADC. A convenient implementation switches the current source to a smaller discharge current I_R and uses a Wilkinson ADC for digitisation, as illustrated in Fig. 14.33. This technique provides high resolution, but at the expense of dead time and multi-hit capability.

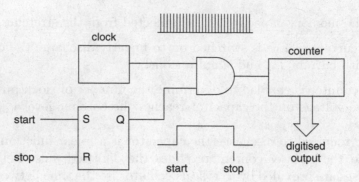

Fig. 14.32.   The simplest form of time digitiser counts the number of clock pulses between the start and stop signals.

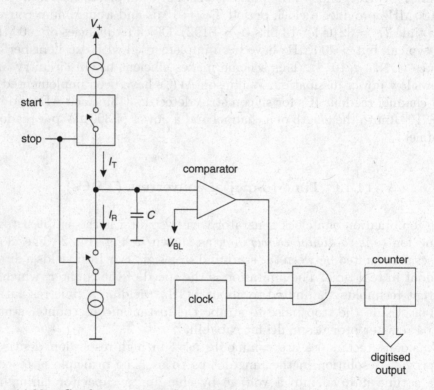

Fig. 14.33.   Combining a time-to-amplitude converter with an ADC forms a time digitiser capable of ps resolution. The memory capacitor C is charged by the current I_T for the duration $T_{start} - T_{stop}$ and subsequently discharged by a Wilkinson ADC.

14.14   Signal transmission

Signals are transmitted from one unit to another through *transmission lines*, often coaxial cables or ribbon cables. When transmission lines are not terminated with their characteristic impedance, the signals are reflected. As a signal propagates along the cable, the ratio of instantaneous voltage to current equals the cable's characteristic impedance $Z_0 = \sqrt{L/C}$, where L and C are the inductance and capacitance per unit length. Typical impedances are $50\,\Omega$ or $75\,\Omega$ for coaxial cables and $\approx 100\,\Omega$ for ribbon cables. If at the receiving end the cable is connected to a resistance different from the cable impedance, a different ratio of voltage to current must be established. This occurs through a reflected signal. If the termination is less than the line impedance, the voltage must be smaller and the reflected voltage wave has the opposite sign. If the termination is greater than the line impedance, the voltage wave is reflected with the same polarity. Conversely, the current in the reflected wave is of like sign when the termination is less than the line impedance and of opposite sign when the termination is greater. Voltage reflections are illustrated in Fig. 14.34. At the sending end the reflected pulse appears after twice the propagation delay of the cable. Since in the presence of a dielectric the velocity of propagation is $v = c/\sqrt{\varepsilon}$, in typical coaxial and ribbon cables the delay is $5\,\mathrm{ns/m}$.

Cable drivers often have a low output impedance, so the reflected pulse is reflected again towards the receiver, to be reflected again, etc. This is

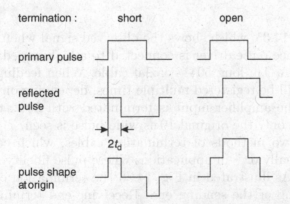

Fig. 14.34.   Voltage pulse reflections on a transmission line terminated either with a short (left) or open circuit (right). Measured at the sending end, the reflection from a short at the receiving end appears as a pulse of opposite sign delayed by the round-trip delay of the cable. If the total delay is less than the pulse width, the signal appears as a bipolar pulse. Conversely, an open circuit at the receiving end causes a reflection of like polarity.

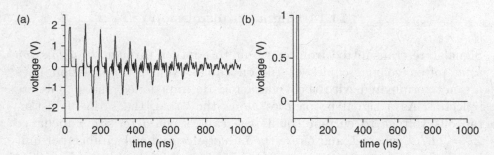

Fig. 14.35. (a) Signal observed in an amplifier when a low-impedance driver is connected to an amplifier through a 4 m long coaxial cable. The cable impedance is 50 Ω and the amplifier input appears as 1 kΩ in parallel with 30 pF. When the receiving end is properly terminated with 50 Ω, the reflections disappear (b).

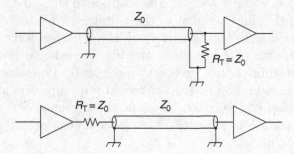

Fig. 14.36. Cables may be terminated at the receiving end (top, shunt termination) or sending end (bottom, series termination).

shown in Fig. 14.35, which shows the observed signal when the output of a low-impedance pulse driver is connected to a high-impedance amplifier input through a 4 m long 50 Ω coaxial cable. When feeding a counter, a single pulse will be registered multiple times, depending on the threshold level. When the amplifier input is terminated with 50 Ω, the reflections disappear and only the original 10 ns wide pulse is seen.

There are two methods of terminating cables, which can be applied either individually or – in applications where pulse fidelity is critical – in combination. As illustrated in Fig. 14.36 the termination can be applied at the receiving or the sending end. Receiving-end termination absorbs the signal pulse when it arrives at the receiver. With sending-end termination the pulse is reflected at the receiver, but since the reflected pulse is absorbed at the sender, no additional pulses are visible at the receiver. At the sending end the original pulse is attenuated two-fold by the voltage divider formed by the series resistor and the cable impedance. However, at the receiver the pulse is reflected with the same polarity, so the superposition of the original and the reflected pulses provides the original amplitude.

This example uses voltage amplifiers, which have low output and high input impedances. It is also possible to use current amplifiers, although this is less common. Then, the amplifier has a high output impedance and low input impedance, so shunt termination is applied at the sending end and series termination at the receiving end.

Terminations are never perfect, especially at high frequencies where stray capacitance becomes significant. For example, the reactance of 10 pF at 100 MHz is 160 Ω. Thus, critical applications often use both series and parallel termination, although this does incur a 50% reduction in pulse amplitude. In the μs regime, amplifier inputs are usually designed as high impedance, whereas timing amplifiers tend to be internally terminated, but one should always check if this is the case. As a rule of thumb, whenever the propagation delay of cables (or connections in general) exceeds a few per cent of the signal rise time, proper terminations are required.

14.15   Interference and pickup

The previous discussion analysed random noise sources inherent to the sensor and front-end electronics. In practical systems external noise often limits the obtainable detection threshold or energy resolution. As with random noise, external *pickup* introduces baseline fluctuations. There are many possible sources, radio and television stations, local radio frequency (RF) generators, system clocks, transients associated with trigger signals and data readout, etc. Furthermore, there are many ways through which these undesired signals can enter the system. Again, a comprehensive review exceeds the allotted space, so only a few key examples of pickup mechanisms will be shown. A more detailed discussion is given in [1, 2]. Ott [14] gives a more general treatment and texts by Johnson and Graham [15, 16] give useful details on signal transmission and design practices.

14.15.1   Pickup mechanisms

The most sensitive node in a detector system is the input. Figure 14.37 shows how very small spurious signals coupled to the sensor backplane can inject substantial charge. Any change in the bias voltage ΔV directly at the sensor backplane will inject a charge $\Delta Q = C_d\,\Delta V$. Assume a silicon strip sensor with 10 cm strip length. Then the capacitance C_d from the backplane to a single strip is about 1 pF. If the noise level is 1000 electrons ($1.6 \cdot 10^{-16}$ C), ΔV must be much smaller than $Q_n/C_d = 160\,\mu\text{V}$. This can be introduced as noise from the bias supply (some voltage supplies are quite noisy; switching power supplies can be clean, but most are

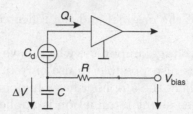

Fig. 14.37. Noise on the detector bias line is coupled through the detector capacitance to the amplifier input.

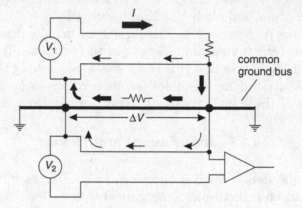

Fig. 14.38. Shared current paths introduce common voltage drops to different circuits.

not) or noise on the ground plane can couple through the capacitor C. Naively, one might assume the ground plane to be 'clean', but it can carry significant interference for the following reason.

One of the most common mechanisms for cross-coupling is shared current paths, often referred to as 'ground loops'. However, this phenomenon is not limited to grounding. Consider two systems: The first is transmitting large currents from a source to a receiver. The second is similar, but is attempting a low-level measurement. Following the prevailing lore, both systems are connected to a massive ground bus, as shown in Fig. 14.38. Current seeks the path of least resistance, so the large current from source V_1 will also flow through the ground bus. Although the ground bus is massive, it does not have zero resistance, so the large current flowing through the ground system causes a voltage drop ΔV.

In system 2 (source V_2) both signal source and receiver are also connected to the ground system. Now the voltage drop ΔV from system 1 is in series with the signal path, so the receiver measures $V_2 + \Delta V$. The cross-coupling has nothing to do with grounding per se, but is due to the common return path. However, the common ground caused the problem by establishing the shared path. This mechanism is not limited to large systems with external ground busses, but also occurs on the scale of printed circuit boards and micron-scale integrated circuits. At high frequencies the impedance is increased due to skin effect and inductance. Note that for high-frequency signals the connections can be made capacitively, so even if there is no DC path, the parasitic capacitance due to mounting structures or adjacent conductor planes can be sufficient to close the loop.

The traditional way of dealing with this problem is to reduce the impedance of the shared path, which leads to the *copper braid syndrome*. However, changes in the system will often change the current paths, so this 'fix' is not very reliable. Furthermore, in many detector systems – tracking detectors, for example – the additional material would be prohibitive. Instead, it is best to avoid the root cause.

14.15.2 Remedial techniques

Figure 14.39 shows a sensor connected to a multistage amplifier. Signals are transferred from stage to stage through definite current paths. It is critical to maintain the integrity of the signal paths, but this does not depend on grounding – indeed Fig. 14.39 does not show any ground connection at all. The most critical parts of this chain are the input, which is the most sensitive node, and the output driver, which tends to circulate the largest current. Circuit diagrams usually are not drawn like Fig. 14.39; the bottom common line is typically shown as ground. For example, in Fig. 14.37 the sensor signal current flows through capacitor C and reaches

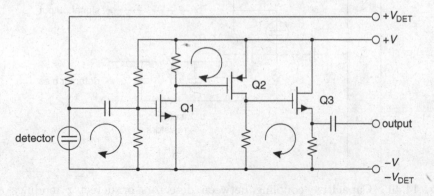

Fig. 14.39.   The signal is transferred from the sensor to the input stage and from stage to stage via local current loops.

the return node of the amplifier through 'ground'. Clearly, it is critical to control this path and keep deleterious currents from this area.

However superfluous *grounding* may be, one cannot let circuit elements simply float with respect to their environment. Capacitive coupling is always present and any capacitive coupling between two points of different potential will induce a signal. This is illustrated in Fig. 14.40, which represents individual detector modules mounted on a support/cooling structure. Interference can couple through the parasitic capacitance of the mount, so it is crucial to reduce this capacitance and control the potential

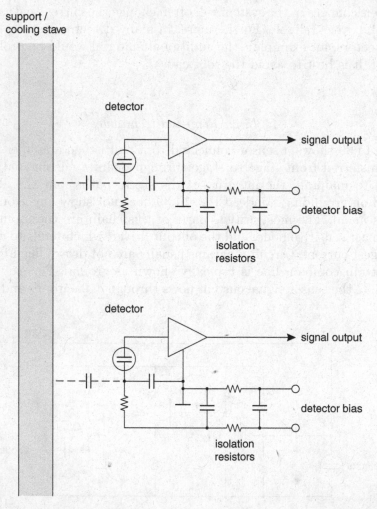

Fig. 14.40.   Capacitive coupling between detectors or detector modules and their environment introduces interference when relative potentials and stray capacitance are not controlled.

of the support structure relative to the detector module. Attaining this goal in reality is a challenge, which is not always met successfully. Nevertheless, paying attention to signal paths and potential references early on is much easier than attempting to correct a poor design after it is done. *Troubleshooting* is exacerbated by the fact that current paths interact, so doing the 'wrong' thing sometimes brings improvement. Furthermore, only one mistake can ruin system performance, so if this has been designed into the system from the outset, one is left with compromises. Nevertheless, although this area is rife with myths, basic physics still applies.

14.16   Conclusion

Signal processing is a key part of modern detector systems. Proper design is especially important when signals are small and electronic noise determines detection thresholds or resolution. Optimisation of noise is well understood and predicted noise levels can be achieved in practical experiments within a few per cent of predicted values. However, systems must be designed very carefully to avoid extraneous pickup.

14.17   Problems

14.1 In a time-of-flight system the start detector has a time resolution of 100 ps and the stop detector has 50 ps resolution. What is the overall time resolution?

14.2 Consider a spectroscopy system whose resolution is determined by electronic noise.

   (a) The current noise contribution is 120 eV and the voltage noise contribution is 160 eV. What is the total noise?

   (b) After cooling the detector, the current noise is 10 eV and the voltage noise remains unchanged at 160 eV. What is the total noise?

14.3 An X-ray spectroscopy system is to resolve the Tl $K_{\alpha 1}$ and $K_{\alpha 2}$ emissions from a ^{203}Hg source. The $K_{\alpha 1}$ and $K_{\alpha 2}$ energies are 72.87 keV and 70.83 keV, at about equal intensities.

   (a) Determine the energy resolution required to separate the two X-ray peaks.

(b) The intrinsic energy resolution of the detector is $\sigma_{\mathrm{det}} = 160\,\mathrm{eV}$. What is the allowable electronic noise contribution?

14.4 A spectroscopy system has the front-end components shown in Fig. 14.20. The Si detector draws a reverse-bias current of 100 nA and has a capacitance of 100 pF. The bias resistor has $R_{\mathrm{b}} = 10\,\mathrm{M\Omega}$ and the total resistance of the connections between the detector and the preamplifier input is $10\,\Omega$. The preamplifier has an equivalent input noise voltage of $1\,\mathrm{nV}/\sqrt{\mathrm{Hz}}$ with negligible $1/f$ and current noise.

(a) The system utilises a simple CR–RC pulse shaper with integration and differentiation time constants of $1\,\mu\mathrm{s}$. What is the electronic noise expressed in electrons and in eV? How large are the contributions of the individual noise sources?

(b) Assume a CR–RC shaper with adjustable peaking time, where both the integration and differentiation time constants are adjusted simultaneously to be equal. What are the noise current and noise voltage contributions at $1\,\mu\mathrm{s}$ shaping time? Determine the time constant that yields minimum noise.

(c) Using the CR–RC shaper at the optimum shaping time determined in (b), what is the minimum value of bias resistor that will degrade the overall noise by less than 1%?

14.5 The signal at the input of a voltage-sensitive amplifier is a 10 mV pulse with a rise time of 10 ns (10%–90%). The equivalent input noise of the amplifier is $10\,\mu\mathrm{V}$ rms. The amplifier feeds a simple threshold comparator.

(a) Assume a comparator threshold of 5 mV. What is the timing jitter?

(b) Still keeping the threshold at 5 mV, how much does the output of the comparator shift when the signal changes from 10 mV to 50 mV? As an approximation assume a perfectly linear transition:

$$t(V_{\mathrm{T}}) = \frac{V_{\mathrm{T}}}{V_{\mathrm{s}}}t_{\mathrm{r}} + t_0 \;,$$

where t_0 is the time at which the pulse attains 10% of its peak, i.e. 1 ns.

References

[1] H. Spieler, *Semiconductor Detector Systems*, Oxford University Press, Oxford (2005) ISBN 0-19-852784-5

[2] www-physics.lbl.gov/~spieler

[3] J. Butler, Triggering and Data Acquisition General Considerations, *Instrumentation in Elementary Particle Physics, AIP Conf. Proc.* **674** (2003) 101–29

[4] T. Kondo *et al.*, Construction and Performance of the ATLAS Silicon Microstrip Barrel Modules, *Nucl. Instr. Meth.* **A485** (2002) 27–42

[5] I. Kipnis, H. Spieler & T. Collins, A Bipolar Analog Front-End Integrated Circuit for the SDC Silicon Tracker, *IEEE Trans. Nucl. Sci.* **NS-41(4)** (1994) 1095–103

[6] S. Ramo, Currents Induced by Electron Motion, *Proc. IRE* **27** (1939) 584–5

[7] F.S. Goulding, Pulse Shaping in Low-Noise Nuclear Amplifiers: A Physical Approach to Noise Analysis, *Nucl. Instr. Meth.* **100** (1972) 493–504

[8] F.S. Goulding & D.A. Landis, Signal Processing for Semiconductor Detectors, *IEEE Trans. Nucl. Sci.* **NS-29(3)** (1982) 1125–41

[9] V. Radeka, Trapezoidal Filtering of Signals from Large Germanium Detectors at High Rates, *Nucl. Instr. Meth.* **99** (1972) 525–39

[10] V. Radeka, Signal, Noise and Resolution in Position-Sensitive Detectors, *IEEE Trans. Nucl. Sci.* **NS-21** (1974) 51–64

[11] H. Spieler, Fast Timing Methods for Semiconductor Detectors, *IEEE Trans. Nucl. Sci.* **NS-29(3)** (1982) 1142–58

[12] D.H. Wilkinson, A Stable Ninety-Nine Channel Pulse Amplitude Analyser for Slow Counting, *Proc. Cambridge Phil. Soc.* **46(3)** (1950) 508–18

[13] M. Garcia-Sciveres *et al.*, The SVX3D Integrated Circuit for Dead-Timeless Silicon Strip Readout, *Nucl. Instr. Meth.* **A435** (1999) 58–64

[14] H.W. Ott, *Noise Reduction Techniques in Electronic Systems*, 2nd edition, Wiley, New York (1988) ISBN 0-471-85068-3, TK7867.5.087

[15] H. Johnson & M. Graham, *High-Speed Digital Design*, Prentice-Hall PTR, Upper Saddle River (1993) ISBN 0-13-395724-1, TK7868.D5J635

[16] H. Johnson & M. Graham, *High-Speed Signal Propagation*, Prentice-Hall PTR, Upper Saddle River (2002) ISBN 0-13-084408-X, TK5103.15.J64

15

Data analysis*

Without the hard little bits of marble which are called 'facts' or 'data' one cannot compose a mosaic; what matters, however, are not so much the individual bits, but the successive patterns into which you arrange them, then break them up and rearrange them.

Arthur Koestler

15.1 Introduction

The analysis of data and extraction of relevant results are goals of particle physics and astroparticle physics experiments. This involves the processing of *raw detector data* to yield a variety of final-state physics objects followed by the application of selection criteria designed to extract and study a signal process of interest while rejecting (or reducing to a known and manageable level) background processes which may mimic it. Collectively, this is referred to as an *analysis*. Physics analyses are performed either to measure a known physical quantity (e.g. the lifetime of an unstable particle), or to determine if the data are compatible with a physics hypothesis (e.g. the existence of a Higgs boson). At each stage, however, a variety of 'higher-order' issues separate particle physics and astroparticle physics analyses from a brute-force application of signal-processing techniques.

15.2 Reconstruction of raw detector data

All physics analysis starts from the information supplied from the data-acquisition system, the raw detector data. In contemporary collider

---

* Steve Armstrong, CERN, now at Société Générale de Surveillance (SGS), contributed to this chapter. It is an updated version of the original data analysis chapter in the first edition of this book originally written by Armin Böhrer, Siegen.

436

experiments or cosmic-ray experiments, these raw detector data consist of the digitised output of detector electronic signals. Signals are induced in the detector electronics by the passage of particles, which leave 'hits' in active detector elements. Modern detectors are frequently highly granular, resulting in dozens or even hundreds of hits per particle per detector system.

The process referred to as *event reconstruction* aims to produce meaningful physics objects from the binary data associated with these hits while coping with electronic noise from the detectors themselves and the inherent physical processes associated with the passage of the final-state particles through the detector material. Raw detector data must also be merged with other predefined sets of data in the event-reconstruction process. A *detector description* contains detailed information on the geometry, position and orientation of active detector elements. Calibration and alignment data contain intrinsic quantities related to detector components which influence its performance (e.g. gas purity, high voltage values, temperatures, etc.); frequently, these data are assumed to be constant over a specific duration of data-taking time, referred to as a *run*.

In the previous chapters, a wide variety of detector technologies have been reviewed. A few examples of information which can be extracted from each kind of detector are presented for illustration:

- **Silicon microstrip detectors (SMDs)** As described in Sect. 7.5, particles traversing an SMD ionise the bulk silicon material liberating electron–hole pairs which drift to implanted strips generating measurable signals. Raw data include pulses recorded on strips around the signal region to permit the use of interpolation techniques. When combined with knowledge of the local strip positions as well as the global position of the silicon wafer, two- or three-dimensional coordinates may be extracted.

- **Multiwire proportional chambers (MWPCs)** As described in Sect. 7.1, particles traversing a MWPC ionise the gas. The electrons initiate avalanches that create a signal when the charge is collected at the electrodes. Raw data include the drift time, the wire and pad positions as well as the arrival time of the pulse and the charge at both ends of the wire. Combining this input with calibration constants, such as drift velocity and the moment of the intersection t_0, the location of the electron initiating the avalanche both in the plane perpendicular to the wire and along the wire may be obtained. When the global position of the wire is included, three-dimensional coordinates may be extracted.

- **Time-projection chambers (TPCs)**   As described in Sect. 7.3.3, charged particles traversing a TPC volume leave ionisation trails in the gas. Ionisation electrons drift towards an end plate containing an MWPC. Raw data contain both the drift time and the profile of signals on wires and cathode pads to permit determination of the z and ϕ coordinates. When combined with knowledge of the wire and pad position, the r position may also be extracted. The pulse heights of hits may also be included, and, when considered along a putative track trajectory, may yield dE/dx information providing valuable particle-identification discrimination.

- **Electromagnetic and hadron calorimeters** (see Chap. 8)   In many high energy physics experiments, energy measurements are made using a combination of electromagnetic and hadron calorimeters. Usually, sampling calorimeters are used, where absorber material is interspersed with chambers or scintillator material providing analogue information proportional to the energy deposited. The accuracy of the location where a particle has passed through the detector is limited by the number of calorimeter cells, or granularity, of the calorimeter. The granularity is typically determined by the intrinsic nature of the calorimeter (e.g. electromagnetic, hadron, compensating hadron) as well as the number of readout channels that can be handled (e.g., the liquid-argon electromagnetic calorimeter for the ATLAS experiment at LHC will have more than $2 \cdot 10^5$ calorimeter cells). Each cell which exceeds a certain threshold is read out and becomes part of the detector raw data. Once position information of the cells is factored in, along with calibration information, cells may be clustered together to determine localised energy depositions.

- **Time-of-flight (TOF) detectors**   Scintillators, resistive-plate chambers (RPCs), planar spark counters or spark chambers frequently have the ability to record with high precision the time of passage of a charged particle. When placed at sufficient distance from each other or from a known interaction point, they can provide valuable timing information, which, when used in conjunction with a measurement of the particles' momentum, may be used for particle-identification discrimination. These types of detectors are also often used as trigger counters.

- **Specialised particle identification detectors**   In addition to the TPC and TOF detectors discussed above, additional particle-identification information can be obtained from Cherenkov counters in which the position and diameter of Cherenkov rings are recorded, or transition radiation detectors in which the yield of X-ray photons from charged particles traversing media of differing dielectric constants is measured.

High *granularity* and *hermeticity* are required for modern collider-detector experiments. Hence, the total number of electronic channels can easily exceed 10^8, and, when processed by front-end and intermediate electronics, can lead to raw event sizes in the range of several Mbytes. The demand for large amounts of data in which to search for rare signatures of known or new physics requires high interaction rates; for example, the LHC interaction rate will be 40 MHz. To reduce this to the range of 100 Hz for offline storage requires sophisticated trigger systems, frequently consisting of multiple *levels* of hardware and software.

Once written to storage, events must undergo further processing. Raw event data are fed into reconstruction algorithms which process the events. The previous generation of experiments (i.e. the LEP experiments) used the FORTRAN programming language. Current and future experiments have migrated their reconstruction software to object-oriented languages such as C++ and Java. In either case, event reconstruction yields basic physics objects such as charged-particle trajectories in a tracking detector or clusterised energy depositions (*energy-flow objects*) in a calorimeter. These basic physics objects are the fundamental building blocks of analysis, and are discussed further below.

15.3 Analysis challenges

Once the reconstructed physics objects are available for analysis, *selection criteria* must be designed and applied to them. Most often, these selection criteria are designed on the basis of a *Monte Carlo simulation* of relevant underlying physics processes as well as a simulation of the response of the detector to final-state physics objects. The choice of selection criteria is often a balance between many complementary challenges.

The first challenge is the optimisation and enhancement of the statistical significance of a signal process which involves achieving a high efficiency for signal as well as a high rejection power for background. Frequently, these selection criteria consist of *cuts* on a variety of kinematic features derived from the four-momenta of the final-state physics objects. Additional criteria may be placed upon other characteristics such as particle-identification information. The higher the signal efficiency and background rejection, the fewer data are required to achieve a physics result. In the era of expensive or limited data-taking opportunities, this endeavour has recruited advanced multivariate techniques to exploit fully the information within the data.

The second challenge is the understanding of systematic uncertainties induced into the physics result by the choice of selection criteria. If the modelling of the kinematic behaviour of the underlying physics processes is flawed within the simulation upon which selection criteria are based, a bias is introduced into the final result. Uncertainties associated with this type are referred to as *theoretical* systematic uncertainties. If the modelling of the detector response is imperfect, a further bias may be introduced into the final result; this is referred to as *experimental* systematic uncertainties. Finally, in the era of complex and large detectors, the size of simulated event samples is frequently limited by available computing resources. Frequently, in simulation, only a handful of important events of a relevant physics process are retained after selection criteria are imposed. This induces a *statistical* systematic uncertainty on the final result.

In contemporary high-statistics experiments special attention should be paid to a careful estimate of possible systematic uncertainties. Let us consider as an example a recent publication of the Belle Collaboration describing the first observation of a rare decay of the τ lepton: $\tau^- \to \phi K^- \nu_\tau$ [1] (see also Chap. 13). The analysis is based on a data sample of $401\,\text{fb}^{-1}$ corresponding to 3.58×10^8 events of the process $e^+ e^- \to \tau^+ \tau^-$ produced at a centre-of-mass energy of $10.58\,\text{GeV}$. For this study events are selected where one τ lepton decays purely leptonically (tag side) while the other one decays into the $K^+ K^- K^\pm \nu_\tau$ final state (signal side). To obtain the number of decays under study, the $K^+ K^-$ invariant-mass spectrum, containing the ϕ-meson peak smeared by the detector resolution and a smooth background, was fit. Then the number of signal events (often referred to as a signal yield), $N_{\text{sig}} = 573 \pm 32$, is derived after subtracting the peaking backgrounds coming from the $\tau^- \to \phi \pi^- \nu_\tau$ decay and the $q\bar{q}$ continuum. As can be seen, the statistical uncertainty of N_{sig} is higher than just $\sqrt{N_{\text{sig}}}$. Figure 15.1 shows the result of this fitting procedure.

In this way the branching ratio is obtained from the formula

$$\mathcal{B} = \frac{N_{\text{sig}}}{2N_{\tau^+\tau^-}\varepsilon} \,, \tag{15.1}$$

where $N_{\tau^+\tau^-}$ is the number of produced $\tau^+\tau^-$ pairs and ε is the detection efficiency obtained from a Monte Carlo simulation. The result is

$$\mathcal{B} = (4.05 \pm 0.25) \cdot 10^{-5} \,, \tag{15.2}$$

where the error is statistical, only determined by the number of selected signal events and that of subtracted background events. Systematic uncertainties are estimated as follows. The systematic error on the signal yield

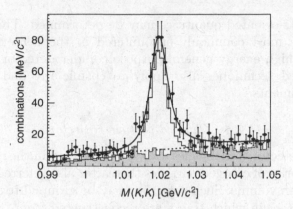

Fig. 15.1. K^+K^- invariant-mass distributions for $\tau^- \to \phi K^- \nu_\tau$. Points with error bars indicate the data. The shaded histogram shows the expectations from $\tau^+\tau^-$ and $q\bar{q}$ background MC simulations. The open histogram is the signal MC with $\mathcal{B}(\tau^- \to \phi K^- \nu_\tau) = 4 \cdot 10^{-5}$ [1].

in the numerator of Eq. (15.1) equals 0.2% and is determined by varying the value of the ϕ-meson width and the shape of background parametrisation. The systematic uncertainty of $N_{\tau+\tau-}$ originates from the uncertainty of the integrated luminosity (1.4%) and our inexact knowledge of the theoretical cross section of the process $e^+e^- \to \tau^+\tau^-(\gamma)$ (1.3%). The dominant uncertainty is due to the detection efficiency, which is affected by various factors: trigger efficiency (1.1%), track-finding efficiency (4%), lepton and kaon identification (3.2% and 3.1%, respectively), the branching fraction of the $\phi \to K^+K^-$ decay (1.2%), and Monte Carlo statistics (0.5%). A total systematic uncertainty of 6.5% is obtained by adding all uncertainties in quadrature, assuming that they are not correlated. The resulting branching fraction is

$$\mathcal{B} = (4.05 \pm 0.25 \pm 0.26) \cdot 10^{-5} \ . \tag{15.3}$$

Note that most of the uncertainties listed above are determined by using various control data samples.

15.4 Analysis building blocks

Different physics analyses can frequently be separated and identified by applying specific selection criteria imposed upon quantities related to various standard final-state physics objects provided by reconstruction of raw detector data. After the identification of these objects and the determination of relevant parameters (e.g. four-momenta, impact parameters with respect to an origin, or decay point of a long-lived particle), relevant results-oriented quantities may be determined. This section discusses objects most commonly encountered in the context of analyses of data from high-energy general-purpose collider detector experiments. Most described techniques also apply to cosmic ray and astroparticle physics experiments.

15.4.1   *Charged-particle trajectories*

A variety of detector technologies exist which can aid in the reconstruction of the trajectories of charged particles (hereafter also referred to as *tracks*) through a given volume. Such detectors may be grouped together to form a *tracking system* in which tracks are reconstructed from their measured spatial coordinates. Tracking systems are usually immersed in powerful magnetic fields of known strength at each point in the fiducial volume so that the electric charge and momentum may be measured.

Events resulting from the high-energy collisions or interactions may contain anywhere from several dozen to several thousand charged particles, which leave hits in the tracking system. To illustrate the extreme complexity, an event in the Inner Detector of the future ATLAS experiment is shown in Fig. 15.2. This detector, typical of contemporary

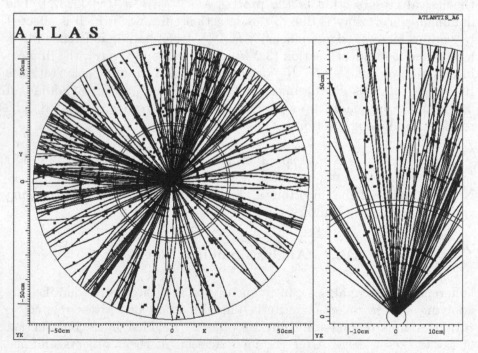

Fig. 15.2.   Reconstruction of charged-particle trajectories in a typical event in the Inner Detector of the ATLAS experiment [2].

tracking chambers, consists of multiple concentric subdetectors utilising different technologies. It combines high-resolution silicon pixel and microstrip detectors at the inner radii (pixel and silicon tracker (SCT)) with a straw-tube tracker at the outer radii (transition radiation tracker (TRT)). Figure 15.2 shows a display in the transverse plane of a simulated event with typical charged-particle multiplicity. The tracking chambers are immersed in a 2 T axial magnetic field parallel to the beam line. Raw measured spatial coordinates are shown with the dots while reconstructed tracks are denoted by the curved lines intersecting relevant spatial coordinates.

Pattern-recognition algorithms attempt to group tracking-detector hits together, first forming two- or three-dimensional coordinates from which track candidates may be found. The challenge then becomes how to group these coordinates together to form tracks. There are two extreme possibilities.

The straightforward method of taking all possible combinations of hits is too time-consuming. The number of combinations for thousands of hits is immense and all possible track candidates must be validated so as not to use hits twice, i.e. by several tracks. The other extreme point of view is the global method where a classification of all tracks is done simultaneously. For points close in space, characteristic values (e.g. coordinates) are entered in an n-dimensional histogram. Hits belonging to the same track should be close in parameter space. A simple example would be the reconstruction of tracks coming from the interaction point without a magnetic field. The ratios $(y_i - y_j)/(x_i - x_j)$ calculated for all i, j pairs of points (Fig. 15.3) and plotted in a histogram would show peaks at the values of the slopes expected for straight tracks.

In practice a method lying between these two approaches is chosen. Its implementation depends heavily on the chamber layout and physics involved.

One method that is commonly used is the *road method*. It is explained most easily for the example of the muon chambers consisting of two double layers of staggered drift tubes (see Fig. 15.3). Reconstructed spatial coordinates for a charged particle outside a magnetic field lie essentially on a straight line. Possible tracks are found from the permutation list of four points lying on a road of a width which corresponds roughly to the spatial resolution (mm or cm).

To all four points on a road with coordinates $x_1, \ldots, x_4$ one has measurements y_i with errors σ_i (here Gaussian errors are assumed). In a straight line fit [3, 4] the expected positions η_i with respect to the measured y_i are linear functions of the x_i:

$$\eta_i = y_i - \epsilon_i = x_i \cdot a_1 + 1 \cdot a_2 \, , \tag{15.4}$$

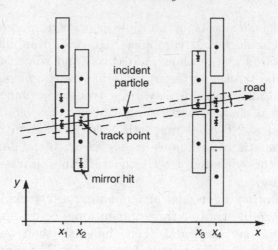

Fig. 15.3. Track finding with the road method and straight line fit. Due to the left–right ambiguity of drift chambers, two coordinates per hit are reconstructed: one being the true track point, the other a mirror hit.

or

$$\vec{\eta} = \vec{y} - \vec{\epsilon} = \mathcal{X} \cdot \vec{a} \ , \tag{15.5}$$

where a_1 is the slope and a_2 is the axis intercept. The matrix $\mathcal{X}$ contains the coordinates x_i in the first and values 1 in the second column. For independent measurements the covariance matrix $\mathcal{C}_y$ is diagonal:

$$\mathcal{C}_y = \begin{pmatrix} \sigma_1^2 & 0 & 0 & 0 \\ 0 & \sigma_2^2 & 0 & 0 \\ 0 & 0 & \sigma_3^2 & 0 \\ 0 & 0 & 0 & \sigma_4^2 \end{pmatrix} =: \mathcal{G}_y^{-1} \ . \tag{15.6}$$

One obtains the values of $\vec{a}$ by the least-squares method, minimising

$$\chi^2 = \vec{\epsilon}^{\mathrm{T}} \mathcal{G}_y \vec{\epsilon} \ , \tag{15.7}$$

which follows a χ^2 distribution with $4 - 2 = 2$ degrees of freedom:

$$\vec{a} = (\mathcal{X}^{\mathrm{T}} \mathcal{G}_y \mathcal{X})^{-1} \mathcal{X}^{\mathrm{T}} \mathcal{G}_y \vec{y} \ , \tag{15.8}$$

and the covariance matrix for $\vec{a}$ is given by

$$\mathcal{C}_a = (\mathcal{X}^{\mathrm{T}} \mathcal{G}_y \mathcal{X})^{-1} =: \mathcal{G}_a^{-1} \ . \tag{15.9}$$

As shown in Fig. 15.3, several track candidates may be fitted to the data points, because of hit ambiguities. To resolve these, the χ^2 can be translated into a confidence limit for the hypothesis of a straight line to be true and one can keep tracks with a confidence level, for example, of more than 99%. The more commonly used choice is to accept the candidate with the smallest χ^2. By this method mirror hits are excluded and ambiguities are resolved.

Points which have been used are marked so that they are not considered for the next track. When all four-point tracks have been found, three-point tracks are searched for to allow for inefficiencies of the drift tube and to account for dead zones between them.

For larger chambers with many tracks, usually in a magnetic field, the following track-finding strategy is adopted. The procedure starts in those places of the drift chamber, where the hit density is lowest, i.e. farthest away from the interaction point. In a first step three consecutive wires with hits are searched for. The expected trajectory of a charged particle in a magnetic field is a helix. As an approximation to a helix, a parabola is fitted to the three hits. This is then extrapolated to the next wire layer or chamber segment. If a hit matching within the errors is found, a new parabola fit is performed. Five to ten consecutive points form a track segment or a chain. In this chain at most two neighbouring wires are allowed not to have a hit. Chain finding is ended when no further points are found or when they do not pass certain quality criteria. When the track-segment finding is complete, the segments are linked by the track-following method. Chains on an arc are joined together and a helix is fitted. Points with large residuals, i.e. points that deviate too much in χ^2, are rejected and the helix fit is redone. The track is extrapolated to the closest approach of the interaction point. In the final fit variations in the magnetic field are included, and a more sophisticated track model is used. In the ALEPH experiment [5], for example, the closest approach to the beam line in $r\varphi$ is denoted by d_0 with the z coordinate at that point z_0 (the z coordinate is measured parallel to the magnetic field along the beam, see Fig. 15.4). The angle φ_0 of the track in the $r\varphi$ plane with respect to the x axis at closest approach, the dip angle λ_0 at that point, and the curvature ω_0 complete the helix parameters: $\vec{H} = (d_0, z_0, \varphi_0, \lambda_0, \omega_0)$. For some applications the set $(d_0, z_0, p_x, p_y, p_z)$ is used, with p_x, p_y, p_z being the components of the track's momentum at closest approach. This procedure also provides the covariance matrix $\mathcal{C}$ for the helix.

The knowledge of the position of the interaction vertex is of particular importance, if one is interested in determining the particles' lifetimes. For colliders the position of the incoming beams is known to be $\approx 200\,\mu\mathrm{m}$ or better, while the length of the colliding bunches may range from a few millimetres to half a metre. The vertex is fitted using all tracks with closest approach to the beam line of less than typically $200\,\mu\mathrm{m}$. This restriction excludes particles not coming from the primary vertex such as

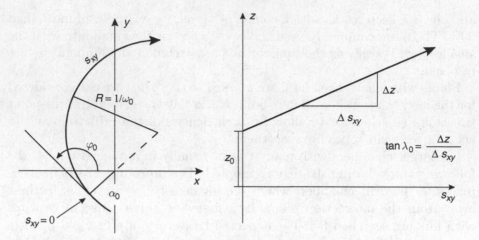

Fig. 15.4.   Definition of helix parameters. On the left, the projection of the helix onto the xy plane orthogonal to the magnetic field and the beam is given. The figure on the right shows the z coordinate versus s_{xy}.

$K_s^0, \Lambda^0, \bar{\Lambda}^0$, called V^0, and photon conversions, which produce a pair of oppositely charged tracks.

15.4.2   Energy reconstruction

Calorimeter detector systems provide not only a measurement of energy deposition, but also position measurements where the energy was deposited. This information may be used in either a local or global sense. When used in a local sense, energy depositions within calorimeters may be grouped together to form clusters of energies to associate to tracks or to neutral particles. Furthermore, the profile of the energy deposition may be used for particle identification (see below).

Aggregate quantities such as total event energy as well as *missing energy* (which might originate from neutrino-like objects) are vital for many physics analyses. This requires the vectorial summation of all visible depositions of energy in calorimeters and the correction for signatures in outer muon systems. This yields energy imbalances with respect to known collision energies or the total event energy balance.

Large deviations from expected imbalances may indicate production of new Weakly Interacting Massive Particles such as supersymmetric neutralinos. On the other hand, detailed examination of expected imbalances helps to identify regions of the event where energetic neutrinos may have escaped detection. To find the energy of a possible neutrino one must detect all other particles in the detector. To each energy deposition in the calorimeter one assigns a vector with length proportional to the measured energy, and its direction given by a line connecting the interaction point with the fired calorimeter cell. A non-zero sum of these vectors in a collider experiment with beams of equal energy and opposite momentum indicates the presence and direction of missing energy. If this is the case, it may be attributed to a neutrino. It must be assumed that no particle escaped, for example, through the beam pipe. Since this cannot be assured, especially for $p\bar{p}$ collider experiments, one usually restricts oneself to the analysis of the momentum transverse to the beam.

In the hard scattering of proton and antiproton only one quark and antiquark collide. The other constituents fragment as jets close to the beam line and partially escape detection. Consequently, the event has a longitudinal imbalance and only the transverse momentum of the neutrino can be used. Certainly, also other corrections have to be taken care of: muons deposit only a small fraction of their energy in the calorimeter. The missing energy must be corrected in this case using the difference between the muon momentum measured in the tracking chamber and its energy seen in the calorimeter.

15.4.3 Quark jets

Quarks produced from or participating in high-energy collisions may manifest themselves as collimated jets of hadrons at sufficiently high energies; this was first observed at centre-of-mass energies near 7 GeV [6]. Quarks may also bremsstrahl gluons thereby creating additional jets in a hadronic event. Primary quarks and any gluons which they may radiate are referred to as *initial partons*. Initial partons carry colour charges and cannot exist in isolation since Nature apparently permits only colour-neutral states to exist freely. Non-perturbative QCD processes convert the coloured initial partons into colour-singlet hadrons. This is referred to as *hadronisation*.

Although the hadronisation process is not well understood, phenomenological models exist. Examples of these models are the string model [7] (implemented in the JETSET Monte Carlo programme [8]) and the cluster model [9] (implemented in the HERWIG Monte Carlo programme [10]). In the string model, for example, the confining nature of the strong interaction dictates that the colour potential of the initial partons becomes proportional to their separation at large distances. As the initial partons fly apart from the interaction point, it becomes energetically favourable for additional quark pairs to be produced from the vacuum. Ultimately, the initial coloured partons are transformed into bound colour-singlet hadronic states.

Jets originating from quarks are qualitatively different than jets from gluons. In Quantum Chromodynamics (QCD), the gluon self-interaction coupling is proportional to the colour factor C_A while the quark–gluon coupling is proportional to the colour factor C_F. The values of the colour factors are determined by the structure of the colour gauge group $SU(3)$. The ratio C_A/C_F is predicted to be $9/4 = 2.25$; this is in good agreement with experimental measurements [11]. Hence gluons are more likely to radiate softer gluons in the hadronisation process, and a gluon jet is consequently broader with a higher particle multiplicity than a light-quark jet of the same energy. These features of gluon jets have been observed in experiment [12].

Although events exhibit jet structure which may have properties qualitatively indicative of the initial parton, jets are intrinsically ill-defined objects. It is impossible to assign all of final-state particles rigorously to a single initial parton. Algorithms exist which cluster charged and neutral particles in an event together to form jets from which an overall four-momentum and other characteristics (e.g. *track multiplicity*, *jet shapes*, etc.) may be determined. These *jet-clustering algorithms* form the basis of most analyses dealing with hadronic events which rely upon the clustered jets to approximate the direction and energies of the initial partons in an event.

Many commonly used jet-clustering schemes are based upon the JADE algorithm [13]. This recursive algorithm begins by considering each instance of energy deposition (e.g. charged particle associated to a calorimeter cluster or candidate neutral particle cluster) in an event to be a pseudo-jet. Pairs of pseudo-jets are then combined according to a metric defined as

$$y_{ij} = \frac{2E_i E_j (1 - \cos\theta_{ij})}{E_{\text{vis}}^2} \, , \qquad (15.10)$$

where i and j are two pseudo-jets, and E_{vis} is the visible energy in the event (i.e. the sum of the energy of all energy-flow objects). The numerator is essentially the invariant mass squared of the two pseudo-jets. The energy and three-momentum of the new pseudo-jets are determined according to a combination scheme from the energy and three-momenta of a previous pseudo-jet and an energy-flow object, yielding a new set of pseudo-jets. In the *E scheme* the simple sum of three-momenta and energy is used. The combination procedure is iterated until all y_{ij} are larger than a specified threshold which is referred to as y_{cut}.

Several variants of the JADE scheme exist and have been extensively studied in the context of QCD-related measurements and predictions [14]. DURHAM, one of the JADE variants, has several advantages (e.g. reduced sensitivity to soft-gluon radiation) [15]. In this scheme, the JADE clustering metric is replaced by

$$y_{ij} = \frac{2 \min(E_i, E_j)^2 (1 - \cos \theta_{ij})}{E_{\text{vis}}^2} . \tag{15.11}$$

The numerator is essentially the square of the lower-energy particle's transverse momentum $k_{\text{T}ij}^2$ with respect to the higher-energy particle.

15.4.4 Stable-particle identification

Another important input for the analysis is the identification of parti- cles. Various methods were described in Chap. 9, such as energy-loss measurements $\mathrm{d}E/\mathrm{d}x$, use of Cherenkov counters and transition-radiation detectors. The different longitudinal and lateral structure of energy depo- sition in calorimeters is used to separate electrons from hadrons. The simplest method is to introduce cuts on the corresponding shape parame- ters. More sophisticated procedures compare the lateral and longitudinal shower shape with a reference using a χ^2 test or *neural networks*. In this case (and in the physics analysis, see below), in contrast to track finding, multilayer feed-forward networks are used. (For pattern recognition feed- back networks are applied.) The input neurons – each neuron represents an energy deposit in a calorimeter cell – are connected with weights to all neurons in a next layer and so forth until one obtains in the last layer one or a few output neurons. The result, which can vary between zero and one, indicates whether the input originated from a pion or an electron. The weights from the neuron connections are adjustable and are obtained by minimising a cost function. This is done by an iterative learning algorithm called *backpropagation* [16–18].

A comparison of these procedures to separate electrons from pions in a calorimeter can be found in [19].

15.4.5 Displaced vertices and unstable-particle reconstruction

The advent of precision tracking detectors with the ability to provide track-impact-parameter resolutions under 50 microns has permitted the use of displaced vertices in a wide spectrum of analysis contexts, most notably in heavy flavour physics. In this method, information from the tracking detectors is extracted not only about the track momentum but also on its precise location. Using an ensemble of tracks, one can fit a hypothesised common origin or *vertex* for these tracks, and compare it to a known collision position or *interaction point*. Vertices with significant displacement from the interaction point result from the decay of beauty and/or charm hadrons. An example of this is shown in Fig. 15.5.

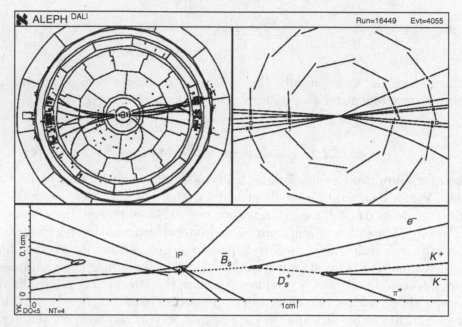

Fig. 15.5. An event display from the ALEPH detector showing the displaced vertex topology of heavy-quark hadron decay [5]. The scale of the upper left event is $\approx 10\,\mathrm{m}$. The upper right-hand display shows tracks in the silicon vertex detector ($\varnothing \approx 20\,\mathrm{cm}$), while the lower event reconstruction shows the decays of B_s and D_s mesons with typical lengths of $\approx 200\,\mathrm{\mu m}$.

A common subclass of displaced vertices are the extremely displaced vertices characterised by only two oppositely charged tracks, referred to as V^0s, which is indicative of, for example, a $\Lambda \to p\pi^-$ decay; in this context, photon conversions to electron–positron pairs may also be thought of as V^0s. A search for V^0s within an event and subsequent calculation of its invariant mass with the charged-track pair is a form of particle identification.

The V^0 decay point, measured in a tracking chamber, is well separated from the primary interaction point. Their decay products are recorded with high precision and allow the reconstruction of the particle's properties. Typical candidates are weakly decaying particles such as B, D, and V^0 (K_s^0, Λ^0) mesons and baryons. Converting photons produce a similar pattern: a photon may convert to an e^+e^- pair in the wall of a tracking chamber, beam pipe, etc. The conversion probability in typical detectors is on the order of a few per cent. Neglecting the masses of the positron and electron and the recoil of the nucleus, the e^+e^- tracks are parallel. This can be seen from the reconstructed photon mass squared: $m_\gamma^2 = 2p_{e^+}p_{e^-}(1 - \cos\theta)$, where θ is the opening angle between electron and positron. Figure 15.6 shows a sketch of a photon conversion in

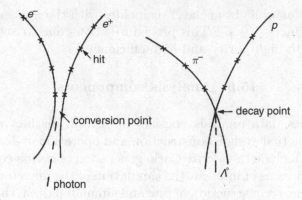

Fig. 15.6.   Sketch of a photon conversion and a Λ^0 decay.

comparison to a Λ^0 decay. The two reconstructed tracks from a photon conversion can either intersect or may not have a common vertex because of measurement errors. The conversion point is found as the point where the two tracks are parallel $(m_\gamma^2 = 0)$ in the plane orthogonal to the magnetic field. The photon momentum is the vector sum of the e^+ and e^- track at or closest to the conversion point.

For massive particles (e.g. Λ^0 with $m = 1.116\,\mathrm{GeV}/c^2$) the opening angle is finite and the trajectories of proton and pion intersect. The closest approach of the two trajectories in space is a good approximation for the decay point. A more precise procedure, however, is to perform a *geometrical fit* using the parameters of the two tracks as obtained from the track fit $(\vec{H}_i)$ and their error matrix $(\mathcal{C}_i)$ and to perform a χ^2 fit. With two tracks, including, for example, ten measurements $\mathcal{H} = (\vec{H}_1, \vec{H}_2)$ (see Fig. 15.4) and nine parameters $\mathcal{Q} = (\vec{D}, \vec{p}_1, \vec{p}_2)$ to be determined (decay point $\vec{D}$, and two momenta $\vec{p}_1$, $\vec{p}_2$), one has a fit with one degree of freedom. The calculation [3, 4, 20] is similar to the straight line fit discussed above. The covariance matrix, however, is non-diagonal as the five track variables are correlated. It is a 10×10 matrix consisting of two submatrices of dimension 5×5. A very important difference is that the expectation values of the nine parameters $\mathcal{Q}$ are not linear functions of the measurements $\mathcal{H}$. Therefore one must obtain the parameters by Taylor series expansion and approximate $\mathcal{X}$, see Eq. (15.5), from the first derivative $\delta\mathcal{H}/\delta\mathcal{Q}$. This matrix is evaluated at an assumed starting value of $\mathcal{Q}_0$, which is derived from an educated guess. Improved parameters $\mathcal{Q}_1$ are obtained using the least-squares method and the procedure is iterated.

With the Λ^0 mass known one can include the mass as a further constraint in the fit. In addition, the origin of the Λ^0 may be known; it is usually the primary vertex. Therefore a kinematical fit can use the fact that the direction of flight of the Λ^0 coincides with the momentum sum of the decay products $\vec{p}_1 + \vec{p}_2$. This procedure allows one to obtain samples of V^0 with both high purity and high efficiency.

15.5   Analysis components

Particle physics data analysis consists of several distinct components, excluding the actual design, construction and operation of detector experiments. These include the Monte Carlo generation of events corresponding to physical processes of interest, the simulation of the detector response to these events, the reconstruction of raw and simulated data, the design and application of selection criteria frequently using multivariate techniques, and the statistical interpretation of results.

15.5.1   Monte Carlo event generators

The generation and study of the four-vectors of final-state particles associated with a physics process of interest is required for developing a particle physics analysis. A wide variety of packages exist which generate a list of particles and their four-vectors associated with well-known, putative or purely hypothetical particle physics processes. These packages build upon decades of theoretical and phenomenological research and constant revision based upon new experimental observations and measurements. At the heart of these packages is the numerical technique developed by Stanislaw Ulam and referred to as *Monte Carlo techniques* [21, 22]. An excellent overview of modern Monte Carlo techniques in particle physics is provided in [23]. A brief summary of some commonly used packages is given in Appendix 4; in each case, the packages have extensive development histories and are continuously subject to revisions and updates.

Beyond the accelerator domain – in the field of astroparticles – particle interactions have also been modelled to describe the propagation of energetic cosmic-ray particles with energies in excess of PeV through the Earth's atmosphere. The measured primary cosmic-ray spectrum extends to $\approx 10^{21}$ eV, corresponding to centre-of-mass energies around 1000 TeV, energies that will not be in reach of earthbound accelerators in the near future. In these models the hadronic interactions are described by a set of sub-interactions between the participating particles (mainly proton–air or heavy-nucleus–air interactions). These processes are dominated by soft interactions plus occasional semi-hard or even hard interactions, where only the latter can be described using perturbative QCD. These processes are modelled in terms of the formation of a set of colour strings. For the soft interactions semi-empirical phenomenological models must be used. The implementation of such approaches based on extrapolations from accelerator data has been used for simulations of extensive air showers (QGSJET, SIBYLL, DPMJET, VENUS, NEXUS, FLUKA [24–26], see also Appendix 4). Commonly, such models have been integrated into simulation packages such as CORSIKA [27].

15.5.2 Simulation of detector response

The list of particle four-vectors provided by Monte Carlo event-generator packages form a phenomenological basis for a proposed analysis. They may also be used in the context of a *fast simulation* where rough parameterised detector responses are used to smear particle parameters. Finally, they may be fed into full simulations of the response of a specific detector. This last step involves the precise modelling of not only the nature and response of a detector experiment but also the passage of final-state particles through the matter the detector is composed of; this modelling is generally done with Monte Carlo techniques as well with packages such as GEANT [28] and FLUKA [26].

Both Monte Carlo event generation and detector simulation are computationally intensive. Hence, it is common for large detector experiments to employ hundreds or thousands of computers from member institutes to produce Monte Carlo samples. The size of these samples directly impacts the uncertainties associated with signal efficiencies and background-rejection rates in data analysis.

15.5.3 Beyond the detector

Limitations of detector apparatus may often be overcome or at least ameliorated through insight into the nature of the physical processes under study. Such insights are hard to quantify, but in this section some examples of such techniques are presented.

Mass constraint refits

When tracks or jets are known to be from a known particle, their four-vectors may be refit with a *mass constraint*. This technique is common when dealing with decay products of $W^{\pm}$ or Z bosons.

At the Large Electron–Positron Collider (LEP) and, in particular, at the Large Hadron Collider and the Tevatron the event topologies can be quite complicated. In the search for the Higgs boson or supersymmetric particles one has to work out the invariant mass of some anticipated new object very carefully. The better the mass resolution the higher the probability to find a signal associated with low statistics. If it is known for some reason that in the final state a W or a Z has been produced, one can reconstruct these particles from their decay products. For leptonic W

decays this can be rather difficult, since one first has to reconstruct the energy and momentum of the missing neutrino. But also for Z decays into jets the association of low-energy particles to jets may not be unambiguous. Also the jet-energy determination using electromagnetic and hadronic calorimetry may not be very accurate. Therefore the reconstruction efficiency and, in particular, the overall mass resolution would benefit from a kinematic refit to the event assuming the known exact masses for those particles that are known to have been produced.

Extracting efficiency from data

Data can sometimes be used to extract the efficiencies. In many cases one would prefer to obtain the efficiencies and, in general, the characteristics of the various detector components from measurements in a test beam, where the particle type and its momentum are well known. In large experiments with hundreds of detector modules this is difficult to achieve. On top of that, the properties of detector modules may change during the experimental runs, depending on the ambient temperature, radiation levels, pressure and other parameters. Certainly, these parameters are monitored by slow control but, still, one wants to have an on-line calibration to analyse the data with the best set of calibration constants. This can be achieved by using known particles or particle decays. For example, at LEP running at the Z resonance, the decay of the Z into muon pairs presented a very interesting sample of penetrating tracks of known momentum and interaction properties. The *track efficiency* of the tracking device (e.g. a time-projection chamber) could easily be determined from the muon tracks. The efficiency of the muon chambers mounted behind the hadron calorimeter, which also serves as flux return for the magnetic field, could clearly be worked out, since the penetration through the iron for muons of about 46 GeV each (for a Z decay at rest) is guaranteed. Also, detailed properties of the time-projection chamber like magnetic-field inhomogeneities or edge effects due to possible problems of the field cage could be investigated and refitted.

In very much the same way the decay of neutral kaons into charged pions can be used to check on the *track-reconstruction efficiency* or, equivalently, the decay of neutral pions into two photons can be used to investigate the properties of electromagnetic calorimeters in the detector. A particularly clean sample of data is obtained if the photons convert into electron–positron pairs in the gas of a time-projection chamber where the density of the target is well known, and the electron and positron momenta are determined from the track curvature and their energies are then measured in the electromagnetic calorimeter. Such a redundancy gives confidence in the reliability of the calibration parameter.

Finally, it should be mentioned that energetic cosmic-ray muons, which are also available when the accelerator is not running, can be used to check out the detector for efficiency and uniformity of response.

Reconstruction of missing particles

Information from undetected particles may often be recovered when considering a specific physical process. If, for example, an event has been exclusively reconstructed and if, say, a W has decayed leptonically, the total observed event energy does not match the centre-of-mass energy: some energy, i.e. that of the neutrino, is missing. From the knowledge of the centre-of-mass energy and the four-momenta of all detected particles, the energy and momentum of the missing particle can be inferred. This also allows to work out the mass of the parent particle. This *missing-energy* or *missing-momentum technique* works in a clean environment and can be applied to many circumstances, like also in the recovery of neutrino momentum in semileptonic decays of B hadrons or in the search for supersymmetric particles, where the lightest supersymmetric particle is supposed to be stable and normally will escape detection due to its low interaction cross section.

15.5.4 Multivariate techniques

Frequently, a particle physics analysis makes use of a variety of discriminating variables some of which may be partially correlated with each other. Unless clearly motivated by straightforward kinematics or a striking separation between signal and background, the choice of which ones to use and what selection criterion to place on each becomes difficult and often arbitrary.

Multivariate techniques allow selection criteria to be chosen by a prescribed method which frequently reduces many variables to a single discriminant. A wide variety of multivariate techniques exists and these have been used in particle physics analysis. We present a brief summary of some of the most commonly used techniques and then discuss two techniques in more detail below:

- Maximum Likelihood;
- Artificial Neural Networks;
- Genetic Programming [29];
- Genetic Algorithms;

- Support Vector Machines [30, 31] are one of the most innovative recent developments in multivariate data analysis;

- Kernel Probability Density Estimation [32];

- Linear Discriminant Analysis;

- Principal Component Analysis [33].

Maximum-likelihood techniques

In the technique of *maximum likelihood* a likelihood function is introduced that is supposed to characterise the data. The likelihood of a data sample is described by the probability to obtain such a sample under the assumption that the assumed probability distribution describes the data well. The chosen probability distribution normally has a set of parameters that can be adjusted. It is the aim of the maximum-likelihood technique to adjust these parameters in such a way that the likelihood of the sample takes on a maximum value. The actual values of the best fit parameters are called *maximum-likelihood estimates*.

Since this procedure starts with an assumed probability distribution, this method is based on an analytic expression describing the maximisation. It can be applied to any set of data where a smooth function is anticipated to be the best description of the experimental values.

Since the best model assumption is a priori not known, various likelihood functions can be used to test various hypotheses. Within these model assumptions one has the freedom of adjusting the free parameters. The maximum-likelihood estimates are frequently normal distributed so that approximate sample variances and confidence levels can be calculated.

As with many statistical methods the maximum-likelihood technique has to be treated with care for small event samples. The technical problems of required computer time for optimising the model distributions and adjustment parameters which presented problems in the early days of data analysis have been overcome with fast computers [34].

Neural networks

An elegant and effective way to deal with multivariate problems is the use of an *artificial neural network* (NN). NNs are inspired by, and are very crude approximations of, biological cortical neural systems. They can be trained to utilise information available from multiple variables. They take into account correlations between variables and learn to rely upon the given information alone when other variables are not available. Depending on the application, NNs can be trained to identify events of a given topology while reducing the number of background events. At the same time NNs provide additional information like efficiency and purity of an event sample for a given final-state hypothesis.

A discussion of the general NN theory and principles is given elsewhere [17, 35]; here we present a brief overview of the most commonly used NNs in particle physics analyses: *simply connected feedforward backpropagation* NNs. A variety of packages exist for the development and training of NNs for use in physics analyses. These include JETNET [36], SNNS [37], MLPfit [38] and others.

An NN is composed of neurons or *nodes* arranged in *layers*. Two given nodes i and j, which are usually in adjacent layers, are connected to each other via *links* which are assigned a *weight* w_{ij}. Each node is the site of the evaluation of an *activation function* Y which is, dependent upon the values of the activation functions of any neurons it is connected to, multiplied by their weights. For node j, this is given as

$$Y_j = g\left[\left(\sum_i w_{ij}x_i\right) - \theta_j\right] , \qquad (15.12)$$

where x_i is the value of the activation function of node i and θ_j is referred to as the neuron bias. The choice of the activation function is usually a sigmoid function such as $g(x) = \tanh x$ [30].

The first layer is referred to as the *input layer*; each node in acquired values based upon a discriminating variable is derived from final-state physics objects in data or Monte Carlo simulation; one input node is assigned to each discriminating variable. These discriminating variables must be scaled such that the bulk of the distribution lies in the range $[0, 1]$ (or $[-1, 1]$) for effective processing by the NN. There may be multiple *hidden layers*, the nodes of which are fully connected to all of the nodes in previous layers. The final layer is referred to as the *output layer* and provides the discriminating variable which may be used in the selection criteria of the analysis. The choice of overall NN architecture follows no formal rule; instead, a trial-and-error approach or architecture based on previous NN experience is usually chosen.

The NN architecture nomenclature is of the form $X\text{-}Y_1\text{-}\ldots\text{-}Y_N\text{-}Z$ where X denotes the number of nodes in the input layer, Y_i denotes the number of nodes in the ith hidden layer, and Z denotes the number of nodes in the output layer. Figure 15.7 illustrates the architecture of a 6-10-10-1 NN.

Once the NN architecture is specified, the NN is trained, usually using events from Monte Carlo simulation. A *pattern* of input variables selected from a *training sample* is presented to the NN along with information related to the desired output for each pattern (e.g. an output value of 1 for signal patterns and 0 for background patterns). A *learning scheme* adjusts the weights of the links connecting neurons in order to minimise

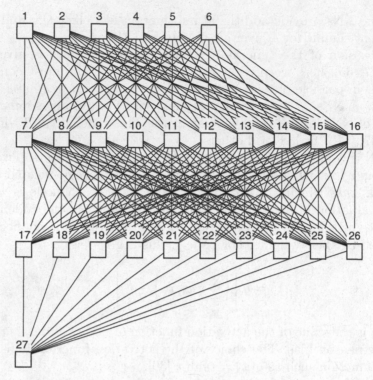

Fig. 15.7. A diagram illustrating the architecture of a 6-10-10-1 neural network. Neurons are depicted as squares while links are the lines joining the neurons. Neurons 1–6 are the six input nodes. Neurons 7–16 comprise the first 10-node hidden layer. Neurons 17–26 comprise the second 10-node hidden layer. Neuron 27 is the output node.

an overall metric denoting how well the NN produces the desired output. As mentioned above, the most commonly used learning scheme is the backpropagation scheme [16–18]. This process of presenting patterns and adjusting weights is repeated many times for a specified number of *training cycles*. Once NN training is finished, NN performance must be evaluated using an independent *testing sample* to avoid bias.

There are many common pitfalls associated with the NN training. These include the correlated issues of NN *overtraining*, testing sample bias and training sample size. Overtraining refers to the use of too many training cycles. In this case, the NN learns the fine-structure details specific to the training sample used rather than providing a more general character of the training sample. This will degrade the performance when evaluated on the testing sample. However, if the number of training cycles is varied and repeatedly evaluated on the testing sample, the testing sample itself becomes biased, and there is a need for another independent *validation sample.* Furthermore, if a large number of input variables is used, a sufficiently large training sample must be used to populate fully the resultant high-dimension domain. If the training sample is too small, the NN will quickly become overtrained or fail to converge in the learning cycle.

15.6 Analysis in action

Perhaps the best way to illustrate some of the methods described above is to consider the case of a sophisticated analysis of collider data. Here the example of an analysis is chosen designed to search for evidence of the production of the *Standard Model Higgs Boson* in electron–positron collisions at centre-of-mass energies near 200 GeV recorded with the LEP detectors in the year 2000 [39].

The Higgs mechanism plays a central rôle for the unification of weak and electromagnetic interactions. Among others it generates the masses for the intermediate vector bosons W and Z. In the electroweak theory the symmetry is broken by the Higgs mechanism. Within this scheme the existence of a single neutral scalar particle, the Higgs boson, is required. The theory, however, gives no clue for the mass of this object.

Measurements at LEP energies below 200 GeV gave no evidence for the production of the Standard Model Higgs boson. In the last year of data taking at LEP large data samples were collected by the four LEP experiments at centre-of-mass energies beyond 200 GeV. The main production process for the Higgs at these energies is supposed to be *Higgsstrahlung* $e^+e^- \to HZ$. Small additional contributions are expected from W and Z boson fusion. The signal processes were simulated extensively by Monte Carlo techniques. For the energy regime of LEP the Higgs boson is expected to decay predominantly into a pair of b quarks, but decays into tau pairs, charm-quark pairs, gluons or W pairs (with one virtual W) are also possible. The main channel of investigation was the decay of the Higgs into b-quark pairs and the decay of the Z into two jets. In addition, the decay channels $H \to b\bar{b}$ and $Z \to \nu\bar{\nu}$ characterised by missing energy, and the channels where the Z decays to lepton pairs, or the Higgs to tau pairs, were investigated.

The search for the Higgs is plagued by background processes which could easily mimic the Higgs production. For example, the WW or ZZ production, which is kinematically possible at centre-of-mass energies exceeding 200 GeV, also produces four-jet final states. Two-photon processes and radiative returns to the Z might also produce signatures that look like the signal.

The main purpose of the analysis is to reduce the background without cutting too much into a possible signal. The identification of b quarks played an important rôle in reducing the background. Due to the high spatial resolution of the vertex detectors displaced vertices from heavy-quark decays could be used for this discrimination. Knowing the signature of the Higgs from Monte Carlo studies, very selective cuts could be applied. For example, the masses of the W and Z could be very close to a possible Higgs mass. Therefore, the reconstruction of their masses was essential to separate a possible signal from more mundane processes. In addition to the *classical cut stream analysis*, also multivariate techniques such as likelihood analysis and, in particular, *neural networks* were extensively used.

It is very important that the different strategies to search for the Higgs using cuts and neural networks have to be established and frozen – after estimating background and optimising selection criteria using Monte Carlo events – *before* the analysis of real events. Modifying the cuts or training the neural networks after the data have been taken and the analysis has been started might introduce a psychological bias, because if one wants to find something, one might be tempted to introduce unconsciously cuts tailored for a signal. Also a blind analysis and a study of real events not in the signal region would help to establish the confidence in a possible discovery.

After extensive and independent analyses of different groups within the same collaboration, ALEPH observed three candidate events consistent with the production of a Standard Model Higgs boson of a mass at around $115\,\text{GeV}$, while OPAL and L3 could explain their candidates with the assumption of background, even though a signal plus background hypothesis was slightly favoured, in contrast to DELPHI which recorded less events compared to the background expectation. A Higgs candidate from ALEPH is shown in Fig. 15.8.

The overall ALEPH evidence for a Higgs at around $115\,\text{GeV}$ can be taken from Fig. 15.9, where the results of the *neural-net analysis* and that of a *cut analysis* are compared. The small excess over background at invariant masses of about $115\,\text{GeV}$ could be an indication of the Higgs production, even though the evidence from these diagrams is not too convincing.

A discovery of the Higgs at a mass of around $115\,\text{GeV}$ by the Tevatron or the Large Hadron Collider at CERN will for sure make the ALEPH collaboration very happy. However, taking the evidence of the four LEP experiments together, a claim for a discovery certainly cannot be made. Instead, combining the results of the four experiments, only a lower bound on the Higgs mass of $114.4\,\text{GeV}$ can be set at 95% confidence level [39].

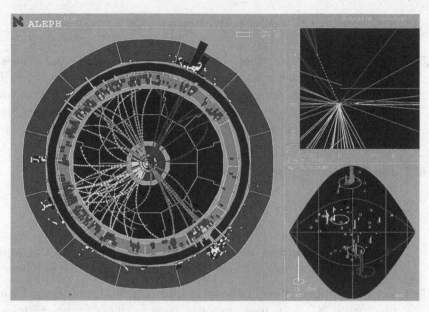

Fig. 15.8. A Higgs candidate assumed to be produced by Higgsstrahlung as observed in the ALEPH experiment at a centre-of-mass energy of 206.7 GeV. Both the Higgs and the Z appear to decay into pairs of b quarks [40].

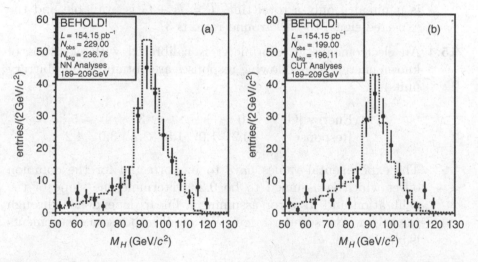

Fig. 15.9. Results of (a) the neural-network analysis and (b) the cut stream analysis in the search for the Higgs boson in ALEPH [40]. Plotted is the distribution of the invariant mass of two jets in a four-jet final state of e^+e^- interactions, see Eqs. (15.10) and (15.11). The dominant background is represented by ZZ, W^+W^- and QCD events (dotted histogram). Reconstructed Zs are the most abundant type of background. The excess events at larger masses ('Higgs candidates') originate from $b\bar{b}$ jets, which are expected to be the dominant decay mode of a Higgs particle in that mass range [41].

15.7   Problems

15.1 In a mixed electron–pion beam a set of electron candidates is
selected by requiring a cut (e.g. on the Cherenkov yield) leading
to N_{acc} accepted events out of a total of N_{tot} particles. Work out
the total number of electron events if no cut were made, given
that the efficiencies for electrons and pions to pass the cut were
ε_e and ε_π. What would happen if $\varepsilon_e = \varepsilon_\pi$?

15.2 The exponential probability density of the variable t $(0 \le t < \infty)$
is given by

$$f(t, \tau) = \frac{1}{\tau} \mathrm{e}^{-t/\tau} \ ,$$

which is characterised by the mean lifetime τ. Work out the expectation value and the variance of the exponential distribution.

15.3 Let the number of events in an experiment with neutrinos from a
reactor be N_1 for a measurement time t_1. This number includes
background from cosmic rays of rate n_μ. With reactor off a number N_2 is obtained during a time t_2. How can the measurement
times t_1 and t_2 be optimised such that the error on the signal rate
is minimal if only a total time $T = t_1 + t_2$ is available and the
expected signal-to-background ratio is 3?

15.4 An electromagnetic calorimeter is calibrated with electrons of
known energy. The following responses are obtained (in arbitrary
units)

Energy [GeV]	0	1	2	3	4	5
Response	0.2	1.0	1.8	2.7	3.0	4.2

The experimental values have to be corrected for the common
offset which is assumed to be 0.2. Determine the slope of the
calibration and its error assuming a linear dependence through
the origin. The standard deviation of all response measurements
is $\sigma = 0.3$.

References

[1]   K. Inami *et al.*, First Observation of the Decay $\tau^- \to \phi K^- \nu_\tau$, *Phys. Lett.*
      B643 (2006) 5–10
[2]   ATLAS Collaboration: http://atlantis.web.cern.ch/atlantis/
[3]   S. Brandt, *Datenanalyse*, 4. Auflage; Spektrum Akademischer Verlag,
      Heidelberg/Berlin (1999); *Data Analysis: Statistical and Computational*

Methods for Scientists and Engineers, 3rd edition, Springer, New York (1998)

[4] W.T. Eadie, D. Drijard, F. James, M. Roos & B. Sadoulet, *Statistical Methods in Experimental Physics*, Elsevier–North Holland, Amsterdam/London (1971)

[5] ALEPH Collaboration, http://aleph.web.cern.ch/aleph/

[6] G. Hanson *et al.*, Evidence for Jet Structure in Hadron Production by e^+e^- Annihilation, *Phys. Rev. Lett.* **35** (1975) 1609–12

[7] X. Artru & G. Mennessier, String Model and Multiproduction, *Nucl. Phys.* **B70** (1974) 93–115; A. Casher, H. Neuberger & S. Nussinov, Chromoelectric-Flux-Tube Model of Particle Production, *Phys. Rev.* **D20** (1979) 179–88; B. Anderson, G. Gustavson, G. Ingelman & T. Sjöstrand, Parton Fragmentation and String Dynamics, *Phys. Rep.* **97** (1983) 31–145

[8] T. Sjöstrand, High-Energy-Physics Event Generation with PYTHIA 5.7 and JETSET 7.4, *Comp. Phys. Comm.* **82** (1994) 74–89

[9] B.R. Webber, A QCD Model for Jet Fragmentation Including Soft Gluon Interference, *Nucl. Phys.* **B238** (1984) 492–528; G.C. Fox & S. Wolfram, A Model for Parton Showers in QCD, *Nucl. Phys.* **B168** (1980) 285–95; R.D. Field & S. Wolfram, A QCD Model for e^+e^- Annihilation, *Nucl. Phys.* **B213** (1983) 65–84; T.D. Gottschalk, A Simple Phenomenological Model for Hadron Production from Low-Mass Clusters, *Nucl. Phys.* **B239** (1984) 325–48

[10] G. Marchesini *et al.*, HERWIG 5.1 – A Monte Carlo Event Generator for Simulating Hadron Emission Reactions with Interfering Gluons, *Comp. Phys. Comm.* **67** (1992) 465–508

[11] R. Barate *et al.* (ALEPH Collaboration), A Measurement of the QCD Colour Factors and a Limit on the Light Gluino, *Z. Phys.* **C76** (1997) 1–14

[12] D. Buskulic *et al.* (ALEPH Collaboration), Quark and Gluon Jet Properties in Symmetric Three-Jet Events, *Phys. Lett.* **B384** (1996) 353–64; R. Barate *et al.* (ALEPH Collaboration), The Topology Dependence of Charged Particle Multiplicities in Three-Jet Events, *Z. Phys.* **C76** (1997) 191–9

[13] W. Bartel *et al.*, Experimental Study of Jets in Electron–Positron Annihilation, *Phys. Lett.* **B101** (1981) 129–34

[14] S. Bethke, Z. Kunszt, D.E. Soper & W.J. Stirling, New Jet Cluster Algorithms: Next-to-leading Order QCD and Hadronization Corrections, *Nucl. Phys.* **B370** (1992) 310–34

[15] S. Catani *et al.*, New Clustering Algorithm for Multijet Cross Sections in e^+e^- Annihilation, *Phys. Lett.* **B269** (1991) 432–8; W.J. Stirling, Hard QCD Working Group – Theory Summary, *J. Phys. G: Nucl. Part. Phys.* **17** (1991) 1567–74

[16] J. Hertz, A. Krogh & R.G. Palmer, *Introduction to the Theory of Neural Computation*, Santa Fe Institute, Addison-Wesley, Redwood City CA (1991)

[17]   R. Rojas, *Theorie der neuronalen Netze*, Springer, Berlin (1993); *Neuronal Networks. A Systematic Introduction*, Springer, New York (1996); *Theorie neuronaler Netze: eine systematische Einführung*, Springer, Berlin (1996)

[18]   D. McAuley, *The BackPropagation Network: Learning by Example* (1997) on-line: www2.psy.uq.edu.au/~brainwav/Manual/BackProp.html

[19]   H.F. Teykal, *Elektron- und Pionidentifikation in einem kombinierten Uran-TMP- und Eisen-Szintillator-Kalorimeter*, RWTH Aachen, PITHA 92/28 (1992)

[20]   B. Rensch, *Produktion der neutralen seltsamen Teilchen K_s und Λ^0 in hadronischen Z-Zerfällen am LEP Speicherring*, Univ. Heidelberg, HD-IHEP 92-09 (1992)

[21]   R. Eckart, *Stan Ulam, John von Neumann, and the Monte Carlo Method*, Los Alamos Science, Special Issue (15) (1987) 131–7

[22]   M. Metropolis & S. Ulam, The Monte Carlo Method, *J. Am. Stat. Ass.* **44** (1949) 335–41

[23]   S. Jadach, *Practical Guide to Monte Carlo*, hep-th/9906207

[24]   J.L. Pinfoldi, Links between Astroparticle Physics and the LHC, *J. Phys. G: Nucl. Part. Phys.* **31** (2005) R1–R74

[25]   H.J. Drescher, M. Bleicher, S. Soff & H. Stöcker, Model Dependence of Lateral Distribution Functions of High Energy Cosmic Ray Air Showers, *Astroparticle Physics* **21** (2004) 87–94

[26]   www.fluka.org/ (2005)

[27]   D. Heck *et al.*, Forschungszentrum Karlsruhe, Report FZKA 6019 (1998); D. Heck *et al.*, Comparison of Hadronic Interaction Models at Auger Energies, *Nucl. Phys. B Proc. Suppl.* **122** (2002) 364–7

[28]   R. Brun, F. Bruyant, M. Maire, A.C. McPherson & P. Zanarini, GEANT3 CERN-DD/EE/84-1 (1987); wwwasdoc.web.cern.ch/wwwasdoc/geant_html3/geantall.html

[29]   K. Cranmer & R. Sean Bowman, *PhysicsGP: A Genetic Programming Approach to Event Selection*, physics/0402030

[30]   V. Vapnik, *The Nature of Statistical Learning Theory*, Springer, New York (1995)

[31]   P. Vannerem *et al.*, *Classifying LEP Data with Support Vector Algorithms*, Proceedings of AIHENP99, Crete, April 1999, hep-ex/9905027

[32]   L. Holmström, S.R. Sain & H.E. Miettinen, A New Multivariate Technique for Top Quark Search, *Comp. Phys. Comm.* **88** (1995) 195–210

[33]   H. Wind, Principal Component Analysis and its Applications to Track Finding, in *Formulae and Methods in Experimental Data Evaluation*, Vol. 3, European Physical Society, CERN/Geneva (1984) pp. d1–k16

[34]   Engineering Statistics, NIST/SEMATECH e-Handbook of Statistical Methods, www.itl.nist.gov/div898/handbook/ (2005)

[35]   C. Peterson & T. Rögnvaldsson, An Introduction to Artificial Neural Networks, in C. Verkerk (ed.), *14th CERN School of Computing – CSC '91*, LU TP 91-23, CERN-92-02 (1991) 113–70; L. Lönnblad, C. Peterson & T. Rögnvaldsson, Pattern Recognition in High Energy Physics with Artificial Neural Networks – JETNET 2.0, *Comp. Phys. Comm.* **70** (1992)

167–82; S.R. Amendolia, Neural Networks, in C.E. Vandoni & C. Verkerk (eds.), *1993 CERN School of Computing* (1993) 1–35

[36]   L. Lönnblad, C. Peterson & T. Rögnvaldsson, *JETNET 3.0: A Versatile Artificial Neural Network Package*, CERN-TH-7135-94 (1994); *Comp. Phys. Comm.* **81** (1994) 185–220

[37]   A. Zell *et al.*, *SNNS – Stuttgart Neural Network Simulator User Manual, Version 4.0*, University of Stuttgart, Institute for Parallel and Distributed High Performance Computing (IPVR), Department of Computer Science, Report 6/95 (1995)

[38]   J. Schwindling & B. Mansoulié, *MLPfit: A Tool for Designing and Using Multi-Layer Perceptrons*, on-line: http: //schwind.home.cern.ch/schwind/MLPfit.html

[39]   G. Abbiendi, The ALEPH Collaboration, the DELPHI Collaboration, the L3 Collaboration, and the OPAL Collaboration, Search for the Standard Model Higgs Boson at LEP, *Phys. Lett.* **B565** (2003) 61–75

[40]   ALEPH Collaboration, http://aleph.web.cern.ch/aleph/alpub/seminar/wds/Welcome.html

[41]   R. Barate *et al.* (ALEPH Collaboration), Observation of an Excess in the Search for the Standard Model Higgs Boson at ALEPH, *Phys. Lett.* **B495** (2000) 1–17

16

Applications of particle detectors outside particle physics

> There are no such things as applied sciences, only applications of science.
>
> *Louis Pasteur*

There is a large number of applications for radiation detectors. They cover the field from medicine to space experiments, high energy physics and archaeology [1–4].

In medicine and, in particular, in nuclear medicine, imaging devices are usually employed where the size and function of the inner organs can be determined, e.g. by registering γ rays from radioactive tracers introduced into the body.

In geophysics it is possible to search for minerals by means of natural and induced γ radioactivity.

In space experiments one is frequently concerned with measuring solar and galactic particles and γ rays. In particular, the scanning of the radiation belts of the Earth (Van Allen belts) is of great importance for manned space missions. Many open questions of astrophysical interest can only be answered by experiments in space.

In the field of nuclear physics, methods of α-, β- and γ-ray spectroscopy with semiconductor detectors and scintillation counters are dominant [5]. High energy and cosmic-ray physics are the main fields of application of particle detectors [6–11]. On the one hand, one explores elementary particles down to dimensions of 10^{-17} cm, and on the other, one tries by the measurement of PeV γ rays (10^{15} eV) to obtain information on the sources of cosmic rays.

In archaeology absorption measurements of muons allow one to investigate otherwise inaccessible structures, like hollow spaces such as chambers in pyramids. In civil and underground engineering, muon absorption measurements allow one to determine the masses of buildings.

In the following, examples of experiments are presented which make use of the described detectors and measurement principles.

16.1   Radiation camera

The imaging of inner organs or bones of the human body by means of X rays or γ radiation is based on the radiation's specific absorption in various organs. If X rays are used, the image obtained is essentially a shadow recorded by an X-ray film or any other X-ray position-sensitive detector. X rays are perfectly suited for the imaging of bones; the images of organs, however, suffer from a lack of contrast. This is related to the nearly identical absorption characteristics of tissue and organs.

In the early days [12] X rays were just imaged with simple X-ray films. Figure 16.1 shows the first picture ever taken with X rays [13]. In comparison, Fig. 16.2 shows a modern X-ray image of the hands [14]. X-ray imaging is still a very important tool in medical diagnostics. The imaging of bones is a standard technique. However, modern X-ray devices also allow the imaging of tissue; e.g., in mammography very small microcalcifications as early indications of breast cancer can be detected with X rays.

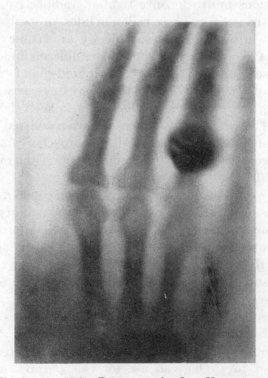

Fig. 16.1.   The hand of Mrs Röntgen: the first X-ray image, 1895 [13].

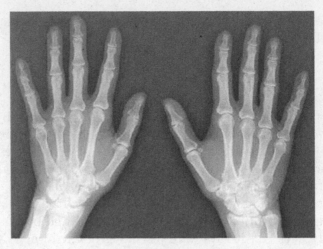

Fig. 16.2. Modern X-ray image of the hands [14].

Even though other techniques of radiation cameras [15], to be described in the following, are powerful tools in medical diagnosis, X-ray imaging is still the choice of the day for many applications.

If organ functions are to be investigated, radioactive tracers can be administered to the patient. These radionuclides, properly integrated into some molecule, will be deposited specifically in certain organs, thereby supplying an image of the organ and its possible malfunctions. Possible tracers for the skeleton are ^{90}Sr, for the thyroid gland ^{131}I or ^{99}Tc, for the kidney again ^{99}Tc and ^{198}Au for the liver. In general, it is advisable to use γ-emitting tracers with short half-lives to keep the radiation load on the patient as low as possible. The γ radiation emitted from the organ under investigation has to be recorded by a special camera so that its image can be reconstructed, e.g. by a scintillation camera introduced by Anger in 1957 [16].

A single small γ-ray detector, e.g. a scintillation counter, has fundamental disadvantages because it can only measure the activity of one picture element (pixel) at a time. In this method, much information remains unused, the time required for a complete picture of the organ is impractically long and the radiation load for the patient is large if many pixels have to be measured – and this is normally necessary for an excellent spatial resolution.

Therefore, a *gamma camera* was developed which allows one to measure the total field of view with a single large-area detector. Such a system, however, also requires the possibility to detect and reconstruct the point of origin of the γ rays. One can use for this purpose a large NaI(Tl) inorganic scintillator, which is viewed by a matrix of photomultipliers

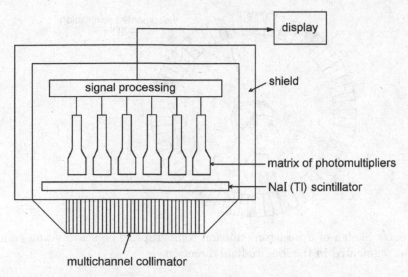

Fig. 16.3.   Sketch of a large-area gamma camera [3, 17].

(Fig. 16.3, [3, 17]). Gamma radiation coming from the human body is collimated by a multichannel collimator to maintain the information about the direction of incidence. The amount of light recorded by a certain photomultiplier is linearly related to the γ activity of the organ part positioned beneath it. The light information of the photomultipliers provides a projected image of the organ based on its specific absorption for the γ-radiating tracer. Organ malfunctions are recognised by a characteristic modification of the γ activity.

The dose for the patient can be reduced if every photon can be used for image reconstruction. This is the aim of *Compton cameras* which can provide excellent image qualities at the expense of requiring complicated reconstruction algorithms. Compton cameras or Compton telescopes are also used in γ-ray astronomy [18].

Positron emission tomography (PET) provides a means to reconstruct three-dimensional images of an organ. This method uses positron emitters for imaging. The positrons emitted from the radionuclides will stop within a very short range ($\approx$ mm) and annihilate with an electron from the tissue into two monoenergetic γ rays,

$$e^+ + e^- \rightarrow \gamma + \gamma \, . \tag{16.1}$$

Both γ rays have 511 keV energy each, since the electron and positron masses are completely converted into radiation energy. Because of momentum conservation the γ rays are emitted back to back. If both γ rays are recorded in a segmented scintillation counter, which completely surrounds

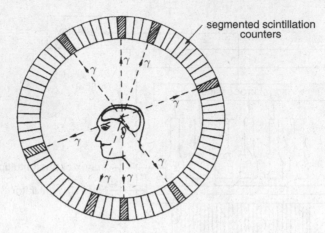

Fig. 16.4.   Sketch of a positron emission tomograph. The scintillation counters are also segmented in the longitudinal direction.

the organ to be investigated, the γ rays must have been emitted from a line connecting the two fired modules. By measuring a large number of γ pairs, the three-dimensional structure of the organ can be reconstructed and its possible malfunctions can be recognised (Fig. 16.4).

PET technology is also an excellent tool to probe, e.g. the structure of the brain, far more powerful than is possible by an electroencephalogram (EEG). In a PET scan, where blood or glucose is given a positron emitter tag and injected into the bloodstream of the patient, the brain functions can be thoroughly investigated. If the patient is observed performing various functions such as seeing, listening to music, speaking or thinking, the particular region of the brain primarily responsible for that activity will be preferentially supplied with the tagged blood or glucose to provide the energy needed for the mental process. The annihilation γ rays emitted from these regions of mental activity allow one to reconstruct detailed pictures of regional brain glucose uptake, highlighting the brain areas associated with various mental tasks [19, 20]. The characteristic γ rays of the tagging radioisotope or the 511 keV radiation from positron annihilation can be measured with a high-resolution scintillation counter (NaI(Tl) or BGO) or semiconductor counter (high-purity germanium detector). The mental activity is directly proportional to the local brain radioactivity. Figure 16.5 shows the different response of a human to language and music [21].

This technique does not only allow to image mental processes but it can also be used to identify malfunctions of the brain because the tagged blood or glucose is differently processed by healthy and diseased tissue.

Commonly used positron emitters integrated into radiopharmaceutical compounds include ^{11}C (half-life 20.4 minutes), ^{15}O (2.03 minutes), ^{18}F

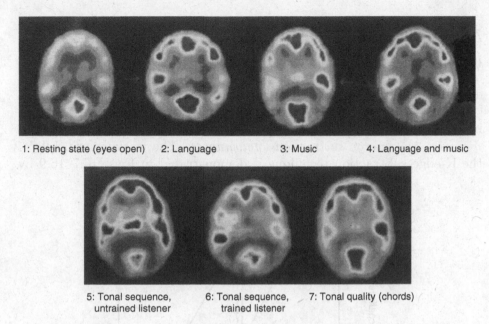

1: Resting state (eyes open)   2: Language   3: Music   4: Language and music

5: Tonal sequence,   6: Tonal sequence,   7: Tonal quality (chords)
untrained listener   trained listener

Fig. 16.5.   Different response of the human brain to language and music [21].

(110 minutes), ^{75}Br (98 minutes), ^{76}Br (16 hours), ^{86}Y (14.7 hours), ^{111}In (2.8 days), ^{123}Xe (2.08 hours), and ^{124}I (4.15 days). Positron sources typically used in high energy physics for calibration, e.g. ^{22}Na (2.6 years), cannot be used for PET technology, because of their long half-life. To get reasonable images with isotopes like ^{22}Na, high-activity sources would have to be used, which would present an unacceptable high radiation dose for the patient. Therefore, one has to find a compromise between activity, radiation dose, half-life and compatibility with metabolic activity. It is important that the radioisotopes are integrated into molecules that in the ideal case are suitable and selective for the human organ under investigation; e.g., ^{11}C is easily integrated into sugar molecules, and sodium fluoride can also be used as pharmaceutical compound [22].

16.2   Imaging of blood vessels

X-ray images of the chest clearly show the spinal column and the ribs, but the heart or the blood vessels are hardly visible. The reason for missing the blood vessels is that physicswise they are not different from the surrounding tissue, so that there is no image contrast.

An injection of iodine into the blood vessels under investigation enhances the contrast significantly, because of the strong absorption of X rays by iodine ($Z = 53$, absorption cross section $\propto Z^5$). The image quality can be much improved exposing the patient to X rays with energy just below the K-absorption edge of iodine and to X rays with energy just above the K edge of iodine (Fig. 16.6). The absorption cross section of tissue varies smoothly across the K edge of iodine, while the iodine attenuates X rays just above the K edge much stronger than below.

The two exposures can be subtracted from each other providing an image of the iodine-containing blood vessel alone (*K-edge subtraction technique* or *dual-energy subtraction angiography*). The working principle of the K-edge subtraction technique is demonstrated in Figs. 16.7–16.9 by

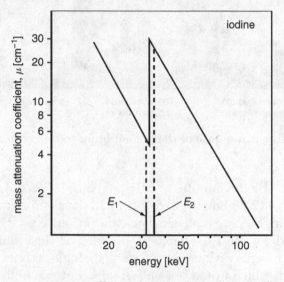

Fig. 16.6.   Mass attenuation coefficient by the photoelectric effect in the vicinity of the K edge in iodine.

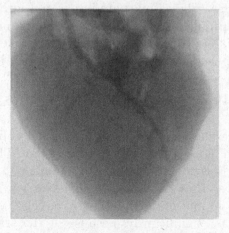

Fig. 16.7.   Imaging the structure of a leaf by the K-edge subtraction technique; image taken below the K edge [23, 24].

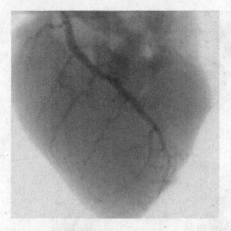

Fig. 16.8.   Imaging the structure of a leaf by the K-edge subtraction technique; image taken above the K edge [23, 24].

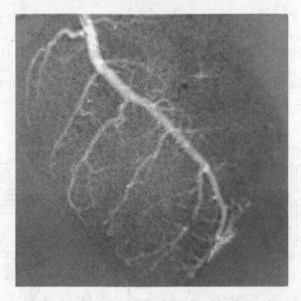

Fig. 16.9.   Imaging the structure of a leaf by the K-edge subtraction technique; difference of the images above and below the K edge [23, 24].

imaging the structure of a leaf [23, 24]. Figure 16.10 shows the aorta and coronary arteries in five successive time frames after the iodine injection using this technique [24, 25].

The required two different energies can be selected from a synchrotron-radiation beam by use of two different monochromators. The fans of monochromatised X rays pass through the chest of the patient and are

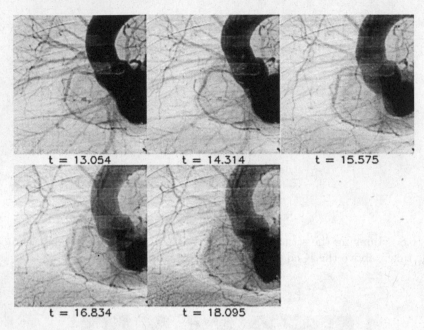

Fig. 16.10.　K-edge subtracted images of the human aorta and close-by coronary arteries in five time frames after an iodine injection [24, 25]. The time is given in seconds. The diameter of the human aorta (darkest part) is around 50 mm. The coronary arteries emerge from the aorta with typical diameters of 3 mm to 5 mm narrowing down to 1 mm and below.

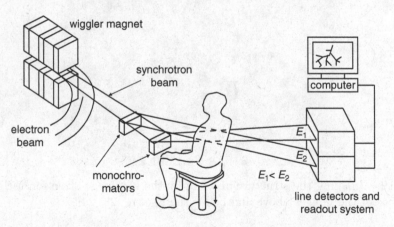

Fig. 16.11.　Preparation of two monochromatic synchrotron beams for the imaging of blood vessels [26–28].

detected in an X-ray detector (Fig. 16.11). In clinical applications of this technique multiwire proportional or drift chambers have been used. Even better resolutions can be obtained with micropattern detectors for the detection of the X rays.

16.3 Tumour therapy with particle beams

It has been known for a long time that tissue, in particular tumour tissue, is sensitive to ionising radiation. Therefore, it is only natural that tumours have been treated with various types of radiation like γ rays and electrons. γ rays are easily available from radioactive sources like ^{60}Co and electrons can be accelerated to MeV energies by relatively inexpensive linear accelerators. The disadvantage of γ rays and electrons is that they deposit most of their energy close to the surface. To reduce the surface dose and to optimise the tumour treatment requires rotating the source or the patient so that the surface dose is distributed over a larger volume. In contrast, protons and heavy ions deposit most of their energy close to the end of the range (Bragg peak, see Fig. 16.12). The increase in energy loss at the Bragg peak amounts to a factor of about 5 compared to the surface dose depending somewhat on the particle's energy. Heavy ions offer in addition the possibility to monitor the destructive power of the beam by observing annihilation radiation by standard PET techniques. The annihilation radiation is emitted by β^+-active nuclear fragments produced by the incident heavy-ion beam itself.

Other techniques of tumour treatment use negative pions which also benefit from the Bragg peak and even additional energy deposits due to star formation. In addition, tumours can also be treated with neutrons. The target for cell-killing is the DNA (deoxyribonucleic acid) in the cell nucleus. The size of the DNA molecule compares favourably well with

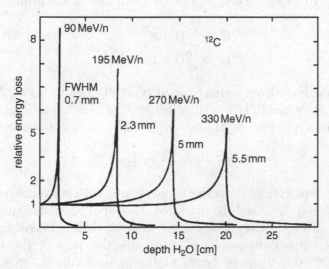

Fig. 16.12. Energy loss of carbon ions ^{12}C in water as a functions of depth [29, 30].

the width of the ionisation track of a heavy ion. The DNA contains two strands containing identical information. A damage of one strand by ionising radiation can easily be repaired by copying the information from the unaffected strand to the damaged one. Therefore, the high ionisation density at the end of a particle's range matches well with the requirement to produce double-strand breaks in the DNA which the cell will not survive.

In *hadron therapy* heavy ions like ^{12}C seem to be optimal for this purpose. Ions heavier than carbon would even be more powerful in destroying tumour tissue, however, their energy loss in the surrounding tissue and in the entrance region already reaches a level where the fraction of irreparable damage is too high, while for lighter ions (like ^{12}C) a mostly reparable damage is produced in the healthy tissue outside the target tumour. The cell-killing rate in the tumour region thus benefits from

- the increased energy loss of protons and ions at the end of the range and

- the increased biological effectiveness due to double-strand breaks at high ionisation density.

The cell-killing rate is eventually related to the equivalent dose H in the tumour region. In addition to the energy loss by ionisation and excitation carbon ions can also fragment leading to the production of lighter carbon ions which are positron emitters. For the ^{12}C case lighter isotopes like ^{11}C and ^{10}C are produced. Both isotopes decay with short half-lives $T_{1/2}(^{11}\text{C}) = 20.38\,\text{min}$, $T_{1/2}(^{10}\text{C}) = 19.3\,\text{s}$ to boron according to

$$^{11}\text{C} \rightarrow {}^{11}\text{B} + e^+ + \nu_e \; , \tag{16.2}$$

$$^{10}\text{C} \rightarrow {}^{10}\text{B} + e^+ + \nu_e \; . \tag{16.3}$$

The positrons have a very short range typically below 1 mm. After coming to rest they annihilate with electrons of the tissue giving off two monochromatic photons of 511 keV which are emitted back to back,

$$e^+ + e^- \rightarrow \gamma + \gamma \; . \tag{16.4}$$

These photons can be detected by positron-emission-tomography techniques and can be used to monitor the spatial distribution of the destructive effect of heavy ions on the tumour tissue. These physical and biological principles are employed in an effective way by the *raster-scan technique* [29–32]. A pencil beam of heavy ions (diameter ≈ 1 mm) is aimed at the tumour. The beam location and spread is monitored by tracking chambers with high spatial resolution. In the treatment planning the

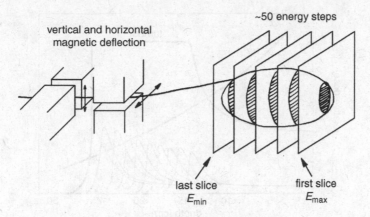

Fig. 16.13. Principle of the raster-scan method [31, 32].

tumour is subdivided into three-dimensional pixels ('voxels'). Then the dose required to destroy the tumour, which is proportional to the beam intensity, is calculated for every voxel. For a fixed depth in tissue an area scan is performed by magnetic deflection sweeping the beam across the area in a similar way as a TV image is produced (Fig. 16.13).

The tumour volume is filled from the back by energy variation (proportional to range variation) of the beam. Typically 50 energy steps are used starting at the rear plane. For a depth profile from 2 cm to 30 cm one has to cover energies from 80 MeV per nucleon to 430 MeV per nucleon. When the beam energy is reduced the required dose for the plane under irradiation is calculated using the damage that the more energetic beam had already produced in its entrance region. This ensures that the lateral (caused by magnetic deflection) and longitudinal scanning (by energy variation) covers the tumour completely. The result of such a scan is shown in Fig. 16.14 in comparison to the effect of ^{60}Co γ rays. The superposition of the individual energy-loss distributions for fixed energies results in a uniform dose distribution over the tumour volume.

As explained, the most effective tumour treatment for deep-seated well-localised tumours takes advantage of heavy ions. On the other hand, protons as charged particles also undergo ionisation energy loss producing a Bragg peak at the end of their range. Protons are easily available at accelerators and they are also frequently used for tumour treatment (*proton therapy*). Figure 16.15 shows the relative dose deposition as a function of depth in water for protons in comparison to γ rays, electrons and neutrons.

In earlier investigations on the possibility of tumour treatment with charged-particle beams, pions, especially negative pions, have also been used for this kind of treatment. Just as protons and heavy ions, pions

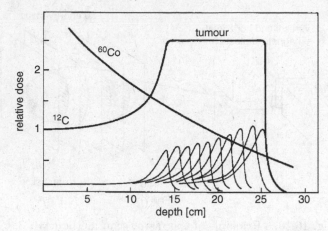

Fig. 16.14.   Superposition of individual energy-loss distributions resulting in a uniform dose profile in the tumour region [29].

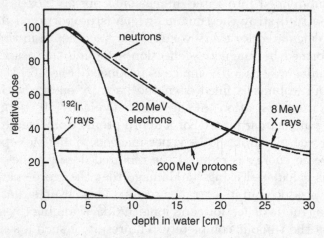

Fig. 16.15.   Comparison of depth–dose curves for neutrons, γ rays (produced by a 8 MV-driven X-ray tube), 200 MeV protons, 20 MeV electrons and ^{192}Ir γ rays (161 keV) [33].

lose their energy in matter by ionisation. Up to the end of their range their energy loss is relatively small. At the end of the range their energy loss increases considerably in very much the same way as for protons and heavy ions. In addition, negative pions are captured by atoms forming pionic atoms. By cascade transitions the pions reach orbitals very close to the nucleus and, finally, they are captured by the nucleus. This process is much faster than the decay of free pions. A large number of light fragments like protons, neutrons, Helium-3, Tritons (= ^{3}H) and α particles can result from pion capture which is called *star formation*. The fragments will deposit their energy locally at the end of the pions' range. In addition the relative biological effectiveness of the fragments is rather high. Because of this effect, the Bragg peak of ionisation is considerably amplified. The depth profile of the energy deposition of negative pions showing the contributions of various mechanisms can be seen from Fig. 16.16.

The relative biological effectiveness of negative pions was measured in vivo and determined to be about a factor of 3. In addition to the much more favourable depth profile compared to γ rays, one therefore gains about a factor of 3 in destructive power for sick tissue.

In addition to *radiotherapy* with charged particles fast neutrons are used for the treatment of brain tumours. The neutron treatment works along the following lines: the tumour is sensitised with a boron compound before neutron treatment is started. Neutrons have a large cross section for the reaction

$$n + {}^{10}\text{B} \rightarrow {}^{7}\text{Li} + \alpha + \gamma \ . \tag{16.5}$$

In this interaction short-range α particles with a high biological effectiveness are produced. In this neutron-induced reaction α particles of 2 MeV with a range of several microns are generated. This ensures that the

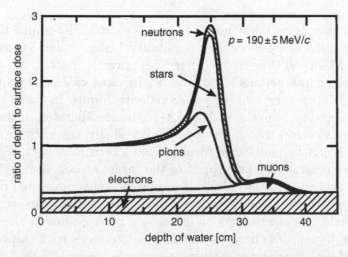

Fig. 16.16. Depth profiles of the energy deposition of a beam of negative pions with a small admixture of muons and electrons. The particle momentum is 190 ± 5 MeV/c. The pion ionisation-loss contribution is marked as 'pions', while 'stars' and 'neutrons' indicate the contributions from nuclear fragments and neutrons produced in pion nuclear interactions, respectively. The small contribution of muons and electrons originates from the contamination of muons and electrons in the negative pion beam [3, 34]. The relative biological effectiveness of muons and electrons was assumed to be unity.

destructive action of the α particles is limited to the local tissue. Clinical tests have shown that best results are obtained with epithermal neutrons (approximately 1 keV). Such neutron beams can be produced by interaction of 5 MeV protons on light target materials, e.g. lithium or beryllium.

A possible direct irradiation with neutrons without sensitising the tumour has the clear disadvantage that neutrons show a similar dose–depth curve like ^{60}Co γ rays thus producing a high amount of biologically very effective damage in the healthy tissue around the tumour.

The ionisation-dose profile of charged particles has been known for a long time from nuclear and particle physics. The instrumentation originally developed for elementary particle physics experiments has made it possible to design and monitor particle beams with great precision which can then be used for tumour therapy. Heavy ions seem to be ideal projectiles for tumour treatment. They are suitable for well-localised tumours. The availability of *treatment facilities* is increasing [31]. Naturally, such a facility requires an expensive and complex accelerator for charged particles. For beam steering and control sophisticated particle detectors and interlock systems are necessary to ensure the safety of patients.

16.4   Surface investigations with slow protons

A large number of non-destructive methods exist to determine the chemical composition of surfaces, one possibility being *proton-induced X-ray emission* (PIXE). If slow charged particles traverse matter, the probability for nuclear interactions is rather low. In most cases the protons lose their kinetic energy by ionising collisions with atoms. In these ionisation processes electrons from the K, L and M shells are liberated. If these shells are filled by electron transitions from higher shells, the excitation energy of the atom can be emitted in form of characteristic X rays. This X-ray fluorescence represents a fingerprint of the target atom. Alternatively, the excitation energy of the atomic shell can be directly transferred radiationless to an electron in one of the outer shells which then can escape from the atom as Auger electron. With increasing atomic number the emission probability ('yield') of characteristic X rays increases with respect to the Auger-electron emission probability. It varies between 15% at $Z = 20$ and reaches nearly 100% for $Z \leq 80$. On the other hand – if the Auger electrons and their energy are measured – their kinetic energy is also a characteristic of the atom and can be used for identification. This Auger-electron spectroscopy (AES) [35] works, however, only for very thin samples because the range of the low-energy electrons is rather short.

The overall photon yield per incident proton depends on the properties of the target, like atomic number, density and thickness of the sample.

The total photon yield can be controlled by the intensity of the primary proton beam.

The measurement of proton-induced characteristic X rays is – quite in contrast to the application of electrons – characterised by a low background of bremsstrahlung. The probability for proton bremsstrahlung is negligible. Only a very low-intensity continuous spectrum will be produced by bremsstrahlung of δ electrons created by the protons. Therefore, the characteristic X rays can be studied in a simple, clear and nearly background-free environment.

The X rays can be recorded in lithium-drifted silicon semiconductor counters, which are characterised by a high energy resolution. An experimental set-up of a typical PIXE system is sketched in Fig. 16.17 [36].

A proton beam of several μA with typical energies of several MeV traverses a thin aluminium scattering foil, which widens the proton beam without a sizeable energy loss. The beam is then collimated and impinges on a selected area of the material to be investigated. A step motor provides a means to move the sample in a well-defined way. This is required to investigate the homogeneity of an alloy over large areas.

The energy of characteristic X rays increases with the atomic number Z according to

$$E_K \propto (Z-1)^2 \qquad (16.6)$$

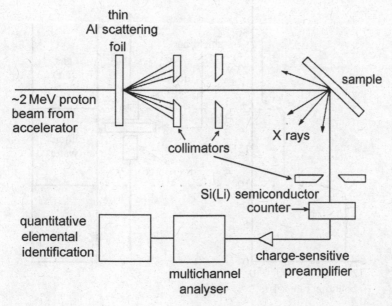

Fig. 16.17. Set-up of a PIXE detector for the investigation of surface structure with slow protons [36].

(*Moseley's law*). The energy resolution of a scintillation counter and certainly that of a silicon lithium-drifted semiconductor counter is sufficient to separate characteristic X rays of elements with Z differing by only 1 unit. Elements from phosphorus ($Z = 15$) up to lead ($Z = 82$) can be identified by this method down to concentrations of less than 1 ppm ($= 10^{-6}$).

The PIXE technique is increasingly applied in the fields of biology, materials science, art and archaeology, and in all cases where a quick, sensitive, non-destructive method of surface investigation is required.

16.5   Gamma- and neutron-backscatter measurements

Measurements to determine the level of some material in a container are based on absorption techniques. Normally, they are performed under well-defined conditions in a laboratory. In applications in geology, for example, in the investigation of boreholes, mainly the chemical composition of the material in the walls of the borehole is of interest. This is particularly true in the search for deposits of certain materials, e.g. oil or rare metals.

Such a search can be done by the *gamma-backscatter method* (Fig. 16.18): a radioactive source like ^{226}Ra emits 186 keV γ rays

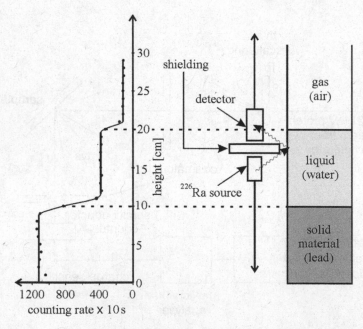

Fig. 16.18.   Gamma-backscatter method for the identification of physicochemical properties of deposits [37].

isotropically. A scintillation counter records γ rays backscattered from the surrounding material. The detector itself is shielded against direct radiation from the source. The cross section for backscattering depends on the density and atomic number of the borehole material. Figure 16.18 demonstrates the working principle of this technique [37]. The counting rate of the scintillation counter as a function of the height reflects the different materials in the layer consisting of lead, water and air. The profile of the backscatter intensities exhibits clear element-specific differences. The experimentally determined backscatter rates therefore allow to infer informations on the density and chemical abundance of the scattering material. Sample measurements on air, water, aluminium, iron and lead show a clear correlation between the backscatter intensity and the product of density ϱ and atomic number Z. The backscatter intensities can be fitted over a wide range by the function $R \propto (\varrho \cdot Z)^{0.2}$ (Fig. 16.19 [37]).

In very much the same way borehole investigations can be done using the *neutron-backscatter technique* [38, 39]. Fast neutrons emitted from an artificial neutron source are scattered in the surrounding material. The scattering cross section associated with high energy transfer is largest for low atomic numbers. Since oil contains hydrogen in form of hydrocarbons, oil is very effective in slowing down the neutrons. If oil is present, the flux of slow neutrons will be high close to the source while in the absence of oil the fast neutrons will hardly be moderated. The intensity ratio of two measurements – one near the source, the other at some distance (60–80 cm) – provides information on the hydrogen concentration near the borehole. The measurement of the backscattered neutrons can be done with BF_3 counters or scintillation counters made from LiI.

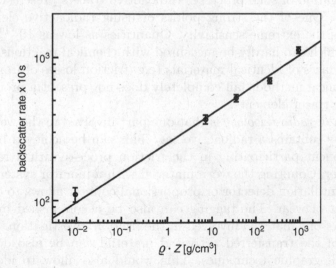

Fig. 16.19. Material dependence of the gamma-backscatter rate [37].

In addition to the neutron-backscatter measurement, gamma rays emitted from nuclei after neutron capture can also be used as an indicator for the chemical configuration of the borehole material. For a clear identification of the scattering material the energies of the gamma rays have to be accurately measured, since these energies are characteristic for the scattering nuclei and can serve as fingerprints for the chemical abundance in the borehole.

Also *aerial surveys* are possible to locate, e.g. uranium deposits [39]. Radiation detectors can be carried on board of planes or helicopters to scan large areas in relatively short periods. As detectors, large-area scintillation counters ($10 \times 10 \times 40 \, \text{cm}^3$, NaI(Tl), CsI(Tl), BGO) can be used. The detection of 1.46 MeV γ rays from ^{40}K decays or daughters from the uranium or thorium decay series indicate the presence of uranium. The existence of ^{40}K along with uranium and thorium in natural ores is often the result of the same geochemical conditions that concentrated the main mineral-bearing ores. The identification of these characteristic γ-ray emitters only requires scintillation detectors with moderate resolution. This is because the low-energy γ rays from ^{230}Th (67.7 keV) and ^{238}U (49.6 keV) are well separated from the ^{40}K emission line (1461 keV).

16.6   Tribology

Tribology deals with the design, friction, wear and lubrication of interacting surfaces in relative motion as, e.g., in bearings or gears. For the investigation of such processes radioactive tracers present distinctive advantages. One of the strong points of using radioactive elements for tribology is the extreme sensitivity. Quantities as low as 10^{-10} g can be detected which can hardly be measured with chemical reactions. Also the wear of surfaces of identical materials (e.g. friction losses of iron on iron) where chemical methods fail completely does not present a problem with radioactive tracer elements.

The idea of *radio-tribology* is that one part involved in the wear or friction process contains a radioactive tag. This can be achieved by coating one component participating in the friction process with a radioactive surface layer. Counting the worn material with a monitor system consisting of a scintillation detector or proportional counter allows to determine the amount of wear. The tagging can also be accomplished by neutron activation of one material involved in the friction investigation. The measurement of the transferred activated material can be also determined by autoradiographic techniques. This would also allow to identify the positions where maximum wear occurs.

In car industries the dependence of the wear on special lubricants is of particular interest. Valve-seat-wear measurements show a distinct dependence on the oil used for lubrication. Figure 16.20 shows the valve moving through a seat that contains an activated zone. The amount of wear can be determined from the activity in the lubricant.

The measurement of the time-dependent wear of the valve seat allows to get information about the long-term behaviour of the oil. The on-line-recorded activity also permits to derive a warning for the due oil exchange.

Similar techniques can also be used for all kinds of gear and bearings. As an example, Fig. 16.21 shows the influence of the motor oil on crankshaft wear. The oil labelled 5 is obviously the best. However, the lubrication deteriorates after 100 h of operation.

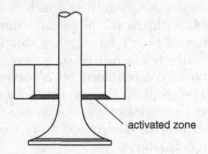

Fig. 16.20. Schematic drawing showing the valve moving through a partially activated zone of the valve seat [40].

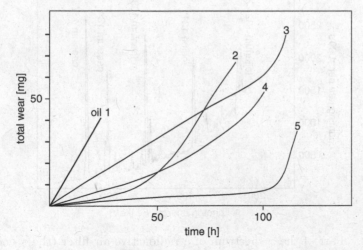

Fig. 16.21.   Influence of different motor-oil products on crankshaft wear [40, 41].

16.7   Identification of isotopes in radioactive fallout

The γ-ray spectrum of a mixture of isotopes can be used to determine
quantitatively the radionuclides it contains. Detectors well suited for this
application are high-resolution germanium semiconductor counters, into
which lithium ions have been drifted, or high-purity germanium crystals.
The atomic number of germanium is sufficiently large so that the γ rays
emitted from the sample are absorbed with high probability via photoelectric effect, thereby producing distinct γ-ray lines. The well-defined
photopeaks or full-absorption peaks are used for the identification of the
radioisotope. Figure 16.22 shows part of the γ-ray spectrum of an air filter
shortly after the reactor accident in Chernobyl [42]. Apart from the γ-ray
lines originating from the natural radioactivity, some *Chernobyl isotopes*
like ^{137}Cs, ^{134}Cs, ^{131}I, ^{132}Te and ^{103}Ru are clearly recognisable by their
characteristic γ energies.

The identification of pure β-ray emitters, which cannot be covered with
this method, is possible with the use of silicon lithium-drifted semiconductor counters. Because of their relatively low atomic number ($Z = 14$),
these detectors are relatively insensitive to γ rays. β-ray emitting isotopes can be quantitatively determined by a successive subtraction of
calibration spectra. The identification of the isotopes is based on the
characteristic maximum energies of the continuous β-ray spectra. The
maximum energies can best be determined from the linearised electron
spectra (Fermi–Kurie plot) [43].

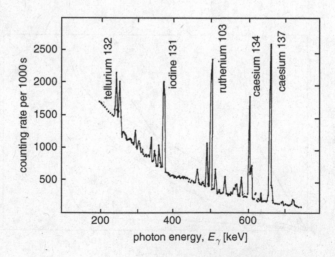

Fig. 16.22.   Part of the γ spectrum of a radioactive air filter (the γ energies of
some 'Chernobyl isotopes' are indicated) [42].

16.8 Search for hidden chambers in pyramids

In the large Cheops pyramid in Egypt several chambers were found: the King's, Queen's, underground chamber and the so-called 'Grand Gallery' (Fig. 16.23). In the neighbouring Chephren pyramid, however, only one chamber, the Belzoni chamber (Fig. 16.24) could be discovered. Archaeologists suspected that there might exist further, undetected chambers in the Chephren pyramid.

It was suggested to 'X ray' the pyramids using muons from cosmic radiation [44]. Cosmic-ray muons can easily penetrate the material of the pyramid. Of course, in this process their intensity is slightly reduced. The intensity reduction is related to the amount of material between the outer wall of the pyramid and the position of the detector. An enhanced

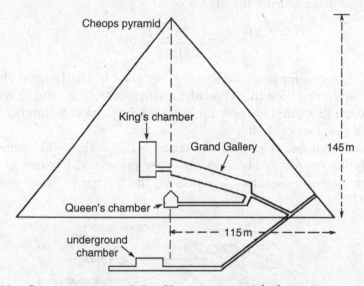

Fig. 16.23. Inner structure of the Cheops pyramid [44], © 1970 by the AAAS.

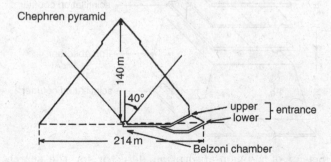

Fig. 16.24. Structure of the Chephren pyramid [44], © 1970 by the AAAS.

relative intensity in a certain direction would indicate the presence of some hollow space which might represent an undiscovered chamber (*muon X-ray technique*).

The intensity of muons as a function of depth $I(h)$ can be approximated by

$$I(h) = k \cdot h^{-\alpha} \qquad \text{with} \quad \alpha \approx 2 \ . \tag{16.7}$$

Differentiating Eq. (16.7) yields

$$\frac{\Delta I}{I} = -\alpha \frac{\Delta h}{h} \ . \tag{16.8}$$

In the case of the Chephren pyramid muons traversed typically about 100 m material before reaching the Belzoni chamber. Consequently, for an anticipated chamber height of $\Delta h = 5$ m, a relative intensity enhancement compared to neighbouring directions of

$$\frac{\Delta I}{I} = -2 \frac{(-5\,\text{m})}{100\,\text{m}} = 10\% \tag{16.9}$$

would be expected for a muon detector installed in the Belzoni chamber.

The detector used for this type of measurement (Fig. 16.25) consisted of a telescope ($2 \times 2\,\text{m}^2$) of three large-area scintillation counters and four wire spark chambers [44, 45].

The spark-chamber telescope was triggered by a three-fold coincidence of scintillation counters. The iron absorber prevented low-energy muons from triggering the detector. Because of their large multiple-scattering

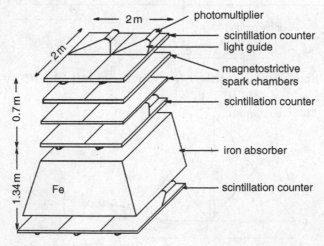

Fig. 16.25.   Set-up of a muon-absorption detector for the search for hidden chambers in the Chephren pyramid [44], © 1970 by the AAAS.

angles, low-energy muons would only produce a fuzzy image of possible chambers. Spark chambers with magnetostrictive readout were used for the track reconstruction of the recorded muons.

The detector was installed approximately at the centre of the base of the Chephren pyramid inside the Belzoni chamber (see Fig. 16.24). It had been suspected that just above the Belzoni chamber there might be additional cavities. Therefore, the range of acceptance of the muon telescope was restricted to zenith angles of about 40° with complete azimuthal coverage. The measured azimuthal variation of the intensity for a fixed zenith angle clearly shows the corners of the pyramid, thus proving the working principle of the method. The section of the pyramid scanned by the detector was subdivided into cells of 3° × 3°. In total, several million muons were recorded. The azimuthal- and zenith-angle variation of the muon flux was compared to a simulated intensity distribution, which took into account the known details of the pyramid structure and the properties of the detector. This allowed one to determine deviations from the expected muon rates. Since the angular distributions of cosmic-ray muons agreed with the simulation within the statistics of measurement, no further chambers in the pyramid could be revealed. The first measurement only covered a fraction of the pyramid volume, but later the total volume was subjected to a *muon X-ray photography*. This measurement also showed that within the resolution of the telescope no further chambers existed in the Chephren pyramid.

A similar muon X-ray technique has also been used to probe the internal structure and composition of a volcano [46].

16.9   Random-number generators using radioactive decays

The need for random numbers satisfying the highest statistical requirements is increasing. Since the advent of publicly available cryptographic software like *Pretty Good Privacy (PGP)*, there has been discussion about how cryptographic keys should be generated. While PGP uses the time between two keystrokes on a keyboard and the value of the key pressed as a source of randomness, this is clearly not enough when it comes to high-security applications. International laws and decrees governing digital signature schemes require that the keys are truly random. Sources of randomness known to physicists are, e.g. radioactive decays or the noise of a diode. The use of radioactivity is superior over thermal noise, since it is virtually independent of the conditions of the environment (pressure, temperature, chemical environment). To the opposite, thermal noise of diodes is temperature-dependent and the consecutive bits are correlated, so a cryptographic treatment of the random numbers is needed in order to obtain useful cryptographic keys. If an adversary can gain access to the device using a diode as the source of randomness, he can alter the temperature and therefore change the output. Radioactive decays are much harder to influence and thus more secure to manipulation.

At the heart of such a device a proportional counter can be used. An incandescent mantle containing thorium-232 can serve as radioactive source. Thorium-232 undergoes α decays with 4.083 MeV energy. The rationale behind the use of the incandescent mantle is that the exemption limit for natural radioactive sources like Th-232 is relatively generous. Therefore, such a low-activity source of natural radioactivity does not require special measures and precautions from the point of view of radiation protection. Of course, also other natural radioactive sources like potassium-40 could be used with the same advantage.

Even though the thin walls of the cathode cylinder of the proportional counter absorb most of the α particles, photons from the γ transitions of thorium-232 or its decay products are recorded. The detection of an ionising particle results in a sudden small decline in the applied high voltage. This pulse is fed through a capacitor to an amplifier and from there to a discriminator, which may, or may not, lift the signal to standard TTL level for a certain time (typically 100 ns). Whenever such a low–high transition occurs, a *toggle flip-flop* is read out. A toggle flip-flop periodically changes its state from logical '0' to logical '1' and vice versa and is typically clocked at high frequencies (e.g. 15 MHz). Since the time difference between two pulses from the proportional counter is not predictable, the sequence of output bits from the toggle flip-flop should be random.

The working principle of such a device is shown in Fig. 16.26. Individual signals from the detector whose appearance we believe to be unpredictable in time are shaped and compared in time with the current state of a freely running flip-flop. If the flip-flop is in a logical '1' state when the random signal arrives, the resulting random bit is set to '1'. If the random signal arrives when the flip-flop is in the low state, the random bit is set to '0'. The resulting bits are stored in a buffer and can be accessed and processed by the CPU. The number of outgoing random bits per time unit by this type of device is therefore directly given by the activity of the random source. If the random sequence is used as a cryptographic key material or as a seed for another pseudo-random-number generator, only a low number of random bits per time unit is required and the activity of the chosen radioactive source can be moderate (e.g. a few hundred Bq) [47].

Random numbers can be created in a fast way by using strong sources. It is, however, not necessary that the rates are comparable to the clock

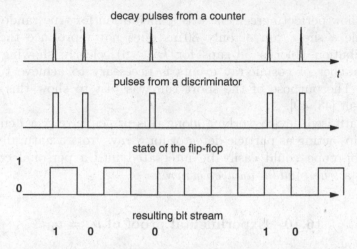

Fig. 16.26.   Working principle of the true random-number generator.

frequency. The principle also works with lower intensities at the expense of longer exposure times to create the desired randomness.

The circuitry with the toggle flip-flop can also be shown to be able to produce random bits in a computer simulation. One can use Poisson-distributed pseudo random numbers to simulate the time difference between two radioactive decays. Also the dead time of the counter and the electronics can be integrated into the simulation. Several statistical and cryptographic tests on the simulated data demonstrate that the output bits can be considered random [48, 49].

To investigate how well the simulation agrees with theory, one can, e.g., compare the results for 4-bit patterns with expectation. If one divides a bit string of length n into substrings of length 4, then the 16-bit patterns $0000, 0001, 0010, 0011, \ldots, 1111$ should occur equally likely.

In very much the same way the particles from a radioactive source can also be replaced by muons from cosmic rays passing through the proportional tube or through a scintillation counter. Consider a bit string generated in this way which consists of substrings of identical bits, either 0s or 1s. Such a substring is called a *run*. A run of 0s is called a *gap*, while a run of 1s is called a *block*. Since the probability for the occurrence of a 0 or 1 in a truly random bit string should be exactly 0.5 and should not depend on the value of the predecessor, one expects to have runs of length 1 with probability 1/2, runs of length 2 with probability 1/4, runs of length 3 with probability 1/8 and so on. In general a run of length k occurs with probability $p_k = 1/2^k$.

The results from different cosmic-ray samples recorded by a plastic scintillator show exactly this behaviour. The exposure times for the different cosmic data sets were on the order of a few seconds. The longer samples show perfect agreement with expectation for true random numbers, while a short run of only 30 ms does not reproduce the tail of the distribution (blocks or gaps for $k \geq 6$), clearly showing that a certain number of cosmic-ray events is necessary to achieve true randomness. The purpose of the short run was just to show this obvious requirement [48, 49].

A miniaturised device working along this principle (e.g. a $1\,\mathrm{cm}^2$ small silicon chip) acting as particle detector for γ rays from a naturally occurring radioisotope could easily be integrated into a personal computer providing a *true random-number generator*.

16.10   Experimental proof of $\nu_e \neq \nu_\mu$

Neutrinos are produced in weak interactions, e.g. in the β decay of the free neutron,

$$n \to p + e^- + \bar{\nu} \ , \tag{16.10}$$

and in the decay of charged pions,

$$\pi^+ \to \mu^+ + \nu \ , $$
$$\pi^- \to \mu^- + \bar{\nu} \ . \tag{16.11}$$

(For reasons of lepton-number conservation one has to distinguish between neutrinos (ν) and antineutrinos ($\bar{\nu}$).) The question now arises whether the antineutrinos produced in the β decay and π^- decay are identical particles or whether there is a difference between the electron- and muon-like neutrinos.

A pioneering experiment at the AGS accelerator (Alternating Gradient Synchrotron) in Brookhaven with optical spark chambers showed that electron and muon neutrinos are in fact distinct particles (*two-neutrino experiment*). The Brookhaven experiment used neutrinos from the decay of pions. The 15 GeV proton beam of the accelerator collided with a beryllium target, producing – among other particles – positive and negative pions (Fig. 16.27, [50]).

Charged pions decay with a lifetime of $\tau_0 = 26\,\mathrm{ns}$ ($c\tau_0 = 7.8\,\mathrm{m}$) into muons and neutrinos. In a decay channel of $\approx 20\,\mathrm{m}$ length practically all pions have decayed. The muons produced in this decay were stopped in an iron absorber so that only neutrinos could emerge from the iron block.

Let us assume for the moment that there is no difference between electron and muon neutrinos. Under this assumption, neutrinos would be expected to be able to initiate the following reactions:

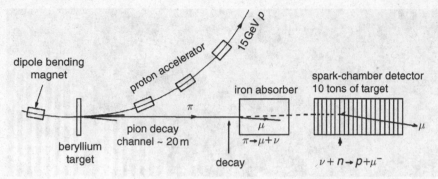

Fig. 16.27. Production of a neutrino beam at the 15 GeV AGS proton synchrotron [50].

$$\nu + n \rightarrow p + e^- \ ,$$
$$\bar{\nu} + p \rightarrow n + e^+ \ ,$$
$$\nu + n \rightarrow p + \mu^- \ ,$$
$$\bar{\nu} + p \rightarrow n + \mu^+ \ . \tag{16.12}$$

If, however, electron and muon neutrinos were distinct particles, neutrinos from the pion decay would only produce muons.

The cross sections for neutrino–nucleon interactions in the GeV range are only on the order of magnitude 10^{-38} cm^2. Therefore, to cause the neutrinos to interact at all in the spark-chamber detector, it had to be quite large and very massive. Ten one-ton modules of optical spark chambers with aluminium absorbers were used for the detection of the neutrinos. To reduce the background of cosmic rays, anti-coincidence counters were installed. The spark-chamber detector can clearly identify muons and electrons. Muons are characterised by a straight track almost without interaction in the detector, while electrons initiate electromagnetic cascades with multiparticle production. The experiment showed that neutrinos from the pion decay only produced muons, thereby proving that electron and muon neutrinos are distinct elementary particles.

Figure 16.28 shows the 'historical' record of a neutrino interaction in the spark-chamber detector [50]. A long-range muon produced in the neutrino interaction is clearly visible. At the primary vertex a small amount of hadronic activity is seen, which means that the interaction of the neutrino was inelastic, possibly

$$\nu_\mu + n \ \rightarrow \ \mu^- + p + \pi^0 \ , \tag{16.13}$$

with subsequent local shower development by the π^0 decay into two photons.

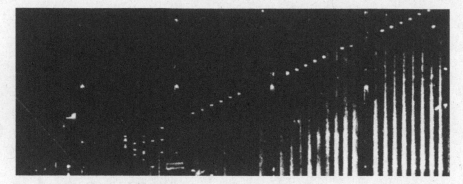

Fig. 16.28.   Muon production in a neutrino–nucleon interaction [50, 51].

Fig. 16.29.   Muon production by muon neutrinos in a multiplate spark chamber in a CERN experiment [52, 53].

Later, the experimental result was confirmed in an experiment at the European Organisation for Nuclear Research (CERN). Figure 16.29 shows a neutrino interaction (ν_μ) in the CERN experiment, in which a high-energy muon is generated via the reaction

$$\nu_\mu + n \to p + \mu^- \, , \tag{16.14}$$

which produces a straight track in the spark-chamber system. The recoil proton can also be clearly identified from its short straight track [52, 53].

16.11   Detector telescope for γ-ray astronomy

In the field of γ-*ray astronomy* the detection of point sources that emit photons in the MeV range and at even higher energies is an interesting topic. The determination of the γ-ray spectra emitted from the source may also provide a clue about the acceleration mechanism for charged particles and the production of energetic γ rays [54, 55]. For energies in excess of several MeV the electron–positron pair production is the dominating photon interaction process. The schematic set-up of a detector for γ-ray astronomy is shown in Fig. 16.30.

The telescope is triggered by a coincidence of elements from the segmented shower counter with an anti-coincidence requirement of the outer veto counter. This selects photons that converted in the tracking device. In the track detector (drift-chamber stack or silicon pixel detector) the produced e^+e^- pair is registered, and the incident direction of the γ ray is reconstructed from the tracks of the electron and positron. The total-absorption scintillator calorimeter can be made from a thick caesium-iodide crystal doped with thallium. Its task is to determine the energy of the γ ray by summing up the energies of the electron–positron pair.

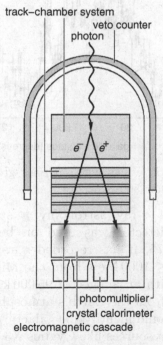

Fig. 16.30.   Schematic set-up of a satellite experiment for the measurement of γ rays in the GeV range [56].

Fig. 16.31.   Photograph of the COS-B detector [58].

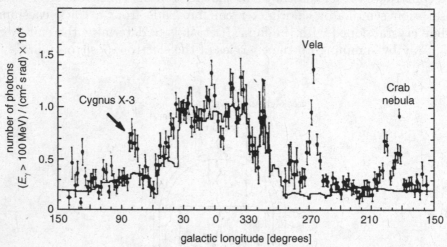

Fig. 16.32.   Distribution of γ rays with energies greater than 100 MeV as a function of galactic longitude [59].

In the early days of γ-ray astronomy a spark-chamber telescope (Fig. 16.31) as track detector was used on board the COS-B satellite [57] launched in 1975. It has recorded γ rays in the energy range between $30\,\text{MeV} \leq E_\gamma \leq 1000\,\text{MeV}$ from the Milky Way. COS-B had a highly eccentric orbit with an apogee of 95 000 km. At this distance the background originating from the Earth's atmosphere is negligible.

The COS-B satellite could identify the galactic centre as a strong γ-ray source. In addition, point sources like Cygnus X3, Vela X1, Geminga and the Crab Nebula could be detected [57].

Figure 16.32 shows the intensity distribution of γ rays with energies greater than 100 MeV as a function of the galactic longitude in a band of

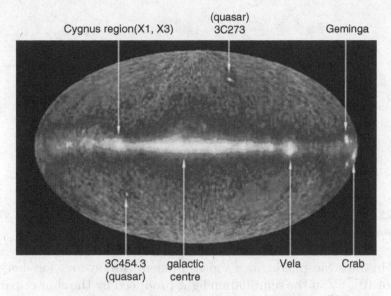

Fig. 16.33.   All-sky survey in the light of gamma rays [60, 61].

±10° galactic latitude. These data were recorded with the SAS-2 satellite [59]. The solid line is the result of a simulation, which assumes that the flux of cosmic γ rays is proportional to the column density of the interstellar gas. In this representation the Vela pulsar appears as the brightest γ-ray source in the energy range greater than 100 MeV.

In an all-sky survey with γ-ray detectors on board the Compton Gamma Ray Observatory (CGRO) a large number of γ-ray sources, including extragalactic ones, could be discovered (Fig. 16.33).

16.12   Measurement of extensive air showers with the Fly's Eye detector

High-energy charged particles and photons produce hadronic and electromagnetic cascades in the atmosphere. In a classical technique for registering these *extensive air showers* (EAS) the shower particles are sampled by a large number of scintillation counters or water Cherenkov counters normally installed at sea level [62], like in [63]. The scintillation counters typically cover 1% of the lateral shower distribution and give information on the number of shower particles at a depth far beyond the shower maximum. Clearly, the energy of the primary particle initiating the cascade can only be inferred with a large measurement error. It would be much better to detect the complete longitudinal development of the

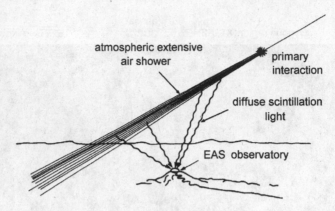

Fig. 16.34. Measurement principle for extensive air showers (EAS) via the scintillation light produced in the atmosphere.

shower in the atmosphere. Such a measurement can be done for energies in excess of 10^{17} eV, if the scintillation light produced by the shower particles in the atmosphere is registered (Fig. 16.34). This can be achieved with the 'Fly's Eye' experiment. The original *Fly's Eye detector* in Utah consisted of 67 mirrors of 1.6 m diameter each [64–67]. Each mirror had in its focal plane 12 to 14 photomultipliers. The individual mirrors had slightly overlapping fields of view. An extensive air shower passing through the atmosphere in the vicinity of the Fly's Eye experiment is only seen by some of the photomultipliers. From the fired phototubes, the longitudinal profile of the air shower can be reconstructed. The total recorded light yield is proportional to the shower energy [68].

Such a Fly's Eye experiment was installed in Utah, USA, for the measurement of high-energy primary cosmic rays (Fig. 16.35). The disadvantage connected with this measurement technique is that the detection of the weak scintillation light can only be done on clear, moonless nights. This detection technique has been further improved with the High Resolution (HiRes) telescope, also in Utah, and the new extended installation, the Telescope Array (TA) which is now under construction at the HiRes site [70]. It is also being used in the larger Auger air-shower array in Argentina [71].

The main scientific aim of these large air-shower arrays is to find the sources of the most energetic cosmic rays and to investigate whether primary protons exceeding a certain threshold energy ($\approx 6 \cdot 10^{19}$ eV) get attenuated by the omnipresent blackbody radiation through the Greisen–Zatsepin–Kuzmin (GZK) cutoff [56] like

$$p + \gamma \;\rightarrow\; \Delta^+ \rightarrow p + \pi^0 \,,$$
$$\rightarrow n + \pi^- \,. \tag{16.15}$$

Fig. 16.35.   Photograph of the 'Fly's Eye' experiment [64, 69].

Recently, it has been shown that the measurement of geosynchrotron emission in the radio band by the relativistic shower electrons in the Earth's magnetic field presents an attractive alternative for the detection of large air showers [72]. The advantage of this method is its 100% duty time compared to the $\approx$ 10% duty time of the optical measurement.

The individual mirrors of air-fluorescence detectors can also be separately operated as *Cherenkov telescopes* (e.g. [73, 74]). With such telescopes the Cherenkov radiation of highly relativistic shower particles in the atmosphere is measured. Cherenkov mirror telescopes provide a means to detect γ-ray point sources, which emit in the energy range in excess of 1 TeV. A high angular resolution of these telescopes allows one to suppress the large background of hadron-induced showers, which is isotropically distributed over the sky, and to identify γ-ray-induced cascades from point sources unambiguously. In this particular case one takes advantage of the fact that γ rays travel along straight lines in the galaxy, while charged primary cosmic rays do not carry any directional information on their origin because they become randomised by irregular galactic magnetic fields.

The *imaging air Cherenkov telescopes* (IACT) have provided valuable information on possible sources which are discussed as candidates for the origin of cosmic rays (e.g. the galaxy M87). Up to date such devices (e.g. [74]) can measure primary γ rays down to 20 GeV, thereby widening the field of view in γ-ray astronomy, because γ rays of this low energy are not absorbed through $\gamma\gamma$ processes in interactions with blackbody or infrared photons.

16.13   Search for proton decay with water Cherenkov counters

In certain theories that attempt to unify the electroweak and strong interactions, the proton is no longer stable. In some models it can decay, violating baryon and lepton conservation number, according to

$$p \rightarrow e^+ + \pi^0 \ . \tag{16.16}$$

The originally predicted proton lifetime on the order of 10^{30} years requires large-volume detectors to be able to see such rare decays. One possibility for the construction of such a detector is provided by *large-volume water Cherenkov counters* (several thousand tons of water). These Cherenkov detectors contain a sufficiently large number of protons to be able to see several proton decays in a measurement time of several years if the theoretical prediction were correct. The proton-decay products are sufficiently fast to emit Cherenkov light.

Large-volume water Cherenkov detectors require ultra-pure water of high transparency to be able to register the Cherenkov light using a large number of photomultipliers. The phototubes can either be installed in the volume or at the inner surfaces of the detector. Directional information and vertex reconstruction of the decay products is made possible by fast timing methods on the phototubes. Short-range charged particles from nucleon decays produce a characteristic ring of Cherenkov light (Fig. 16.36), where the outer radius r_a is used to determine the distance of the decay vertex from the detector wall and the inner radius r_i approximately reflects the range of the charged particle in water until it falls below the Cherenkov threshold. The measured light yield allows one to determine the energy of the particles.

Two such water Cherenkov detectors were installed in the Kamioka zinc mine in Japan (KamiokaNDE = Kamioka Nucleon Decay Experiment) and the Morton-Thiokol salt mine in Ohio, USA (Irvine–Michigan–Brookhaven (IMB) experiment) [75–77].

In spite of running these detectors over several years, no proton decay was detected. From this result, new limits on the lifetime of the proton were determined to be $\tau \geq 10^{33}$ years.

The large-volume water Cherenkov counters have been spectacularly successful, however, in registering neutrinos emitted by the supernova

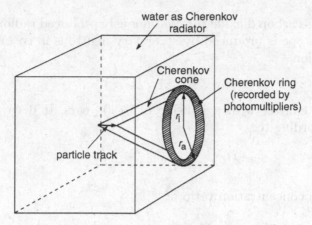

Fig. 16.36.   Principle of Cherenkov-ring production in an experiment searching for proton decay.

1987A. The KamiokaNDE, SuperKamiokande and SNO (Sudbury Neutrino Observatory) experiments were even able to detect solar neutrinos because of their low detection threshold for electron energies [76, 78].

The precise measurement of solar and atmospheric neutrinos with these detectors led to the discovery of neutrino oscillations, a major step for discovering physics beyond the Standard Model of elementary particles. The big *sea-water* and *ice Cherenkov counters* have also opened up a new window for astronomy. Meaningful neutrino telescopes for the detection of energetic galactic and extragalactic neutrinos, however, have to be much larger. Such devices are presently being built in the antarctic (*IceCube*) and prepared in the Mediterranean. For precision results on high-energy neutrino astronomy detectors even larger than IceCube will be required. To instrument larger volumes with photomultipliers on strings is prohibitive for cost reasons. However, a new technique of measuring the sound waves associated with energetic interactions in the antarctic ice using *glaciophones* looks promising. The thermoacoustically generated sound waves can be detected by appropriate glaciophones, e.g. by piezo-electric sensors. The main advantage of this technique is that the attenuation of the acoustic signal is very much weaker compared to the optical signal.

16.14   Radio-carbon dating

The dating of archaeological objects of biological origin can be performed with the *radio-carbon dating* method [79, 80]. The Earth's atmosphere contains in its carbon dioxide the continuously produced radioisotope ^{14}C. This radioisotope is produced by secondary neutrons in cosmic radiation via the reaction

$$n + {}^{14}_{7}N \rightarrow {}^{14}_{6}C + p .$$ (16.17)

^{14}C is a β^- emitter with a half-life of 5730 years. It decays back into nitrogen according to

$$^{14}_{6}C \rightarrow {}^{14}_{7}N + e^- + \bar{\nu}_e .$$ (16.18)

In this way a concentration ratio of

$$r = \frac{N({}^{14}_{6}C)}{N({}^{12}_{6}C)} = 1.2 \cdot 10^{-12}$$ (16.19)

is formed. All plants and, as a consequence of eating vegetable matter, also animals and humans have ^{14}C. Therefore, the isotopic ratio produced in the atmosphere is also formed in the entire biosphere. With the death of a living organism the radio-carbon incorporation comes to an end. The radioactive decay of ^{14}C now reduces the $^{14}C/^{12}C$ ratio. By comparing the ^{14}C activity of an archaeological object and a biological object with the equilibrium value of the present time, the age of the object can be determined.

An experimental problem arises because of the low beta activity of archaeological objects. The maximum energy of the electrons emitted in the ^{14}C decay is only 155 keV. Therefore, a very sensitive detector is required for their detection. If the radioisotope ^{14}C is part of a gas ($^{14}CO_2$), a methane-flow counter can be used (a so-called low-level counter). This detector has to be shielded passively by lead and actively by anti-coincidence counters against background radiation. The methane-flow counter is constructed in such a way that the sample to be investigated – which does not necessarily have to be in the gaseous state – is introduced into the detector volume. This is to prevent energy losses of electrons when entering the counter. A steady methane flow through the detector guarantees a stable gas amplification.

Due to systematic and statistical errors, radio-carbon dating is possible for archaeological objects with ages between 1000 and 75 000 years. In recent times, however, it has to be considered that the concentration ratio r is altered by burning ^{14}C-poor fossil fuels and also by nuclear-weapon tests in the atmosphere. As a consequence r is no longer constant in time. Therefore, a time calibration must first be performed. This can be done by measuring the radio-carbon content of a sample of known age [79].

16.15 Accident dosimetry

Occasionally, the problem arises of determining a radiation dose after radiation accidents if no dosimeter information was available. It is possible to estimate the body dose received after the accident has happened by the *hair-activation method* [81]. Hair contains sulphur with a concentration of 48 mg S per gram hair. By neutron irradiation (e.g. after reactor accidents) the sulphur can be converted to phosphorus according to

$$n + {}^{32}\text{S} \rightarrow {}^{32}\text{P} + p \ . \tag{16.20}$$

In this reaction the radioisotope ${}^{32}\text{P}$ is produced, which has a half-life of 14.3 days. In addition to this particular reaction, the radioactive isotope ${}^{31}\text{Si}$ is formed in the following manner:

$$n + {}^{34}\text{S} \rightarrow {}^{31}\text{Si} + \alpha \ . \tag{16.21}$$

The ${}^{31}\text{Si}$ isotope renders the determination of the phosphorus activity difficult. The half-life of ${}^{31}\text{Si}$, however, is only 2.6 hours. Therefore one waits for a certain amount of time until this activity has decayed before attempting to measure the ${}^{32}\text{P}$ activity. In case of surface contaminations, careful cleaning of the hair has to precede the activity measurement.

${}^{32}\text{P}$ is a pure β-ray emitter. The maximum energy of the electrons in this decay is 1.71 MeV. Because of the normally low expected event rates, a detector with high efficiency and low background is required. An actively and passively shielded end-window counter is a suitable candidate for this kind of measurement. Knowing the activation cross section for the Reaction (16.20), the measured ${}^{32}\text{P}$ activity can be used to infer the radiation dose received.

16.16 Problems

16.1 Particle-detector systems are frequently calibrated with laser beams; e.g., laser beams produce straight tracks in a time-projection chamber to monitor the field uniformity or the drift velocity. Also for air Cherenkov telescopes or air scintillation telescopes laser calibration is employed. What is the force experienced by a detector when it is hit by a 10 mW laser beam of which a fraction $\epsilon = 50\%$ is reflected?

16.2 Carbon-14 dating works well if the age of the samples to be age-determined is on the order of magnitude of its half-life. For geological lifetimes other techniques must be employed. The primordial isotope ^{238}U has a half-life of $4.51 \cdot 10^9$ years. It decays eventually via thorium, actinium, radium, radon and polonium into the stable lead isotope ^{206}Pb. In a rock sample an isotopic ratio of $r = N(^{206}\text{Pb})/N(^{238}\text{U}) = 6\%$ is found. What is the age of the rock if all the ^{206}Pb has been produced by ^{238}U and if all other half-lives in the decay chain can be neglected?

16.3 A geostationary air-watch satellite to measure the scintillation light produced by extensive air showers in the atmosphere is naturally exposed to sunlight for almost all of its orbit. What kind of temperature will the satellite get if its emissivity and absorption – assumed to be the same – are independent of the frequency?

16.4 Neutron detection is frequently done with BF$_3$ counters using the reaction

$$^{10}\text{B} + n \rightarrow {}^{7}\text{Li} + \alpha \ ,$$

where the Q value of the reaction for reaching the ground state of ^{7}Li is $2.8\,\text{MeV}$. What fraction of this energy goes to the α particle?

16.5 The luminosity at e^+e^- colliders is normally determined by small-angle elastic scattering (called Bhabha scattering in this case). For that purpose track-sensitive electromagnetic calorimeters are installed close to the beam line in backward and forward direction. In these forward calorimeters with an acceptance starting at a polar angle of θ_0 (typically $30\,\text{mrad}$) high rates of scattered e^+e^- events (Bhabha events) are measured. The neutral-current Z exchange does not contribute at small scattering angles. In a measurement of the cross section for $\sigma(e^+e^- \rightarrow Z \rightarrow \text{hadrons})$ the accuracy is determined by the precision of the luminosity measurement. If the statistical error of the luminosity determination dominates, and if the accuracy for the Z cross section has to be improved by a factor of 2, by how much would the luminosity calorimeters have to be moved closer to the beam (θ_new)?

16.6 Air Cherenkov telescopes measure the Cherenkov light emitted from electrons and positrons which are produced in the process of the development of the induced electromagnetic cascade. At $100\,\text{GeV}$ primary photon energy the shower does not reach sea level. Therefore, conventional air-shower techniques using ground-level particle sampling have no chance to measure primary photons of this energy. Estimate the number of photons per m^2 produced by a $100\,GeV$ γ ray from some galactic source at sea level.

For details of electromagnetic showers see Sect. 8.1.

References

[1] K. Kleinknecht, *Detektoren für Teilchenstrahlung*, Teubner, Stuttgart (1984, 1987, 1992); *Detectors for Particle Radiation*, Cambridge University Press, Cambridge (1986)

[2] P.B. Cushman, Electromagnetic and Hadronic Calorimeters, in F. Sauli (ed.), *Instrumentation in High Energy Physics*, World Scientific, Singapore (1992)

[3] N.A. Dyson, *Nuclear Physics with Application in Medicine and Biology*, John Wiley & Sons Inc. (Wiley Interscience), New York (1981); *Radiation Physics with Applications in Medicine and Biology*, Ellis Horwood, New York (1993)

[4] F. Sauli, *Applications of Gaseous Detectors in Astrophysics, Medicine and Biology*, CERN-PPE-92-047; *Nucl. Instr. Meth.* **323** (1992) 1–11

[5] K. Siegbahn (ed.), *Alpha, Beta and Gamma-Ray Spectroscopy*, Vol. 1 and Vol. 2, Elsevier–North Holland, Amsterdam (1968)

[6] F. Sauli (ed.), *Instrumentation in High Energy Physics*, World Scientific, Singapore (1992)

[7] T. Ferbel (ed.), *Experimental Techniques in High Energy Nuclear and Particle Physics*, World Scientific, Singapore (1991)

[8] S. Hayakawa, *Cosmic Ray Physics*, John Wiley & Sons Inc. (Wiley Interscience), New York (1969)

[9] K. Kleinknecht & T.D. Lee (eds.), *Particles and Detectors; Festschrift for Jack Steinberger*, Springer Tracts in Modern Physics Vol. 108, Springer, Berlin/Heidelberg (1986)

[10] D.J. Miller, Particle Physics and its Detectors, *Nucl. Instr. Meth.* **310** (1991) 35–46

[11] G. Hall, Modern Charged Particle Detectors, *Contemp. Phys.* **33** (1992) 1–14

[12] W.C. Röntgen, *Eine neue Art von Strahlen*, Sitzungsberichte der Würzburger Physikalisch-medicinischen Gesellschaft, Würzburg (December 1895)

[13] Otto Glasser, *Wilhelm Conrad Röntgen and the Early History of the Röntgen Rays*, National Library of Medicine, London (1933)

[14] RadiologyInfo, Radiological Society of North America, Inc. (RSNA) (2006); www.radiologyinfo.org/en/photocat/photos_more_pc.cfm?pg=bonerad

[15] E. Hell, W. Knüpfer & D. Mattern, The Evolution of Scintillating Medical Detectors, *Nucl. Instr. Meth.* **A454** (2000) 40–8

[16] H.O. Anger, Scintillation Camera, *Rev. Sci. Instr.* **2a** (1958) 27–33

[17] H.O. Anger, Scintillation Camera with Multichannel Collimators, *J. Nucl. Med.* **5** (1964) 515–31

[18] R.W. Todd, J.M. Nightingale & D.B. Everett, A Proposed γ-Camera, *Nature* **251** (1974) 132–4; J.B. Martin *et al.*, A Ring Compton Scatter Camera for Imaging Medium Energy Gamma Rays, *IEEE Trans. Nucl. Sci.* **49(3)**, Part 1 (2002) 817–21; T. Conka-Nurdan *et al.*, Silicon Drift Detector Readout Electronics for a Compton Camera, *Nucl. Instr. Meth.* **A523(3)** (2004) 435–40

[19] G. Montgomery, The Mind in Motion, *Discover*, Charlottesville, VA (1989) 58–61

[20] V.J. Stenger, *Physics and Psychics*, Prometheus, Buffalo, New York (1990)

[21] Solomon H. Snyder, *Frontiers of Science*, National Geographic Society, Library of Congress (1982)

[22] www.triumf.ca/welcome/petscan.html

[23] W. Thomlinson & D. Chapman, private communication (2005)

[24] St. Fiedler, *Synchrotron Radiation Angiography: Dead Ends and Perspectives*; www.lightsource.ca/bioimaging/Saskatoon_2004_sf.pdf

[25] H. Elleaume, S. Fiedler, B. Bertrand, *et al.*, First Human Transvenous Coronary Angiography Studies at the ESRF, *Phys. Med. Biol.* **45** (2000) L39–L43

[26] J.R. Schneider, Hasylab; wissenschaftlicher Jahresbericht DESY (2000) 116–28

[27] T. Dill *et al.*; Intravenous Coronary Angiography with the System NIKOS IV, www-hasylab.desy.de/science/annual_reports/1998/part1/contrib/26/2601

[28] www.physik.uni-siegen.de/walenta/angio/index.html

[29] G. Kraft, *Radiobiology of Heavy Charged Particles*, GSI preprint 96-60 (November 1996)

[30] G. Kraft, *Tumour Therapy with Ion Beams*, invited paper at the SAMBA Symposium at the University of Siegen 1999, *Nucl. Instr. Meth.* **A454** (2000) 1–10

[31] G. Kraft, The Impact of Nuclear Science on Medicine, *Nucl. Phys.* **A654** (1999) 1058C–67C

[32] G. Kraft, Radiotherapy with Heavy Charged Particles, www.gsi.de

[33] Medical Radiation Group, National Accelerator Center South Africa, www.nac.ac.za/public; Brochure for Specialised Radiotherapy, www.nac.ac.za/public/default.htm; U. Amaldi & G. Kraft, Radiotherapy with Beams of Carbon, *Rep. Progr. Phys.* **68** (2005) 1861–82

[34] S.B. Curtis & M.R. Raju, A Calculation of the Physical Characteristics of Negative Pion Beams Energy Loss, Distribution and Bragg Curves, *Radiat. Res.* **34** (1968) 239–47

[35] C.L. Briant & R.P. Messmer, *Auger Electron Spectroscopy (Treatise on Materials Science and Technology)*, Academic Press, Cambridge (1988); M. Prutton & M.M. El Gomati, *Scanning Auger Electron Microscopy*, John Wiley & Sons Inc., New York (2006)

[36] S.E. Hunt, *Nuclear Physics for Engineers and Scientists*, John Wiley & Sons Inc. (Wiley Interscience), New York (1987)

[37] C. Grupen, *Grundkurs Strahlenschutz*, Springer, Berlin (2003)

[38] James E. Martin, *Physics for Radiation Protection*, John Wiley & Sons Inc., New York (2000)

[39] Bicron, Detector Application, Information Note: Geological Exploration, Newbury, Ohio, USA (1997)

[40] S. Sailer, Daimler Benz AG, Central Materials Technology, Stuttgart, Proceedings of the International Atomic Energy Agency (1982)

[41] A. Gerve, The Most Important Wear Methods of Isotope Technology, *Kerntechnik* **14** (1972) 204–12

[42] U. Braun, *Messung der Radioaktivitätskonzentration in biologischen Objekten nach dem Reaktorunfall in Tschnernobyl und ein Versuch einer Interpretation ihrer Folgen*, Diploma Thesis, University of Siegen (1988)

[43] C. Grupen *et al.*, *Nuklid-Analyse von Beta-Strahlern mit Halbleiterspektrometern im Fallout*, Symp. *Strahlenmessung und Dosimetrie*, Regensburg (1966) 670–81

[44] L. Alvarez *et al.*, Search for Hidden Chambers in the Pyramids, *Science* **167** (1970) 832–9

[45] F. El Bedewi *et al.*, Energy Spectrum and Angular Distribution of Cosmic Ray Muons in the Range 50–70 GeV, *J. Phys.* **A5** (1972) 292–301

[46] H. Tanaka *et al.*, Development of a Two-fold Segmented Detection System for Near Horizontally Cosmic-Ray Muons to Probe the Internal Structure of a Volcano, *Nucl. Instr. Meth.* **A507** (2003) 657–69

[47] L. Smolik, private communication (2006); L. Smolik, *True Random Number Generation Using Quantum Mechanical Effects*, Proc. of Security and Protection of Information 2003, pp. 155–60, Brno (Brünn), Czech Republic (2003)

[48] C. Grupen, I. Maurer, S. Schmidt & L. Smolik, *Generating Cryptographic Keys by Radioactive Decays*, 3rd International Symposium on Nuclear and Related Techniques, 22–26 October 2001, NURT, Cuba, ISBN 959-7136-12-0 (2001)

[49] L. Smolik, F. Janke & C. Grupen, Zufallsdaten aus dem radioaktiven Zerfall, *Strahlenschutzpraxis* Heft 2 (Mai 2003) 55–7

[50] G. Danby, J.M. Gaillard, K. Goulianos, L.M. Lederman, N. Mistry, M. Schwartz & J. Steinberger (Columbia U. and Brookhaven) 1962, Observation of High Energy Neutrino Reactions and the Existence of Two Kinds of Neutrinos, *Phys. Rev. Lett.* **9** (1962) 36–44

[51] *Oscillating Neutrinos*, CERN-Courier **20(5)** (1980) 189–90

[52] O.C. Allkofer, W.D. Dau & C. Grupen, *Spark Chambers*, Thiemig, München (1969)

[53] H. Faissner, *The Spark Chamber Neutrino Experiment at CERN*, CERN-Report **63-37** (1963) 43–76

[54] R. Hillier, *Gamma Ray Astronomy*, Clarendon Press, Oxford (1984)

[55] P.V. Ramana Murthy & A.W. Wolfendale, *Gamma Ray Astronomy*, Cambridge University Press, Cambridge (1986)

[56] C. Grupen, *Astroparticle Physics*, Springer, New York (2005)

[57] G.F. Bignami *et al.*, The COS-B Experiment for Gamma-Ray Astronomy, *Space Sci. Instr.* **1** (1975) 245–52; ESRO: The Context and Status of Gamma Ray Astronomy (1974) 307–22

[58] Photo MBB-GmbH, *COS-B, Satellit zur Erforschung der kosmischen Gammastrahlung*, Unternehmensbereich Raumfahrt, München (1975)

[59] C.E. Fichtel *et al.*, SAS-2 Observations of Diffuse Gamma Radiation in the Galactic Latitude Interval $10° < |b| \leq 90°$, *Astrophys. J. Lett. Ed.* **217(1)** (1977) L9–L13; *Proc. 12th ESLAB Symp.*, Frascati (1977) 95–9

[60] CGRO-project, Scientist Dr. Neil Gehrels, Goddard Space Flight Center, http://cossc.gsfc.nasa.gov/docs/cgro/index.html (1999)

[61] http://cossc.gsfc.nasa.gov/docs/cgro/cossc/cossc.html

[62] P. Baillon, *Detection of Atmospheric Cascades at Ground Level*, CERN-PPE-91-012 (1991)

[63] M. Teshima *et al.*, Expanded Array for Giant Air Shower Observation at Akeno, *Nucl. Instr. Meth.* **A247** (1986) 399–411; M. Takeda *et al.*, Extension of the Cosmic-Ray Energy Spectrum beyond the Predicted Greisen-Zatsepin-Kuz'min Cutoff, *Phys. Rev. Lett.* **81** (1998) 1163–6; K. Shinozaki & M. Teshima, AGASA Results, *Nucl. Phys. B Proc. Suppl.* **136** (2004) 18–27

[64] R.M. Baltrusaitis *et al.*, The Utah Fly's Eye Detector, *Nucl. Instr. Meth.* **A240** (1985) 410–28

[65] J. Linsley, The Highest Energy Cosmic Rays, *Scientific American* **239(1)** (1978) 48–65

[66] J. Boone *et al.*, *Observations of the Crab Pulsar near 10^{15}–10^{16} eV*, 18th Int. Cosmic Ray Conf. Bangalore, India, Vol. 9 (1983) 57–60; University of Utah, UU-HEP 84/3 (1984)

[67] G.L. Cassiday *et al.*, Cosmic Rays and Particle Physics, in T.K. Gaisser (ed.), *Am. Inst. Phys.* **49** (1978) 417–41

[68] C. Grupen, Kosmische Strahlung, *Physik in unserer Zeit* **16** (1985) 69–77

[69] G.L. Cassiday, private communication (1985)

[70] High Resolution Fly's Eye: http://hires.physics.utah.edu/; and www.telescopearray.org/

[71] Auger collaboration, www.auger.org/

[72] H. Falcke *et al.* – LOPES Collaboration, Detection and Imaging of Atmospheric Radio Flashes from Cosmic Ray Air Showers, *Nature* **435** (2005) 313–6

[73] HESS collaboration, www.mpi-hd.mpg.de/hfm/HESS/

[74] MAGIC collaboration, wwwmagic.mppmu.mhg.de/collaboration/

[75] K.S. Hirata *et al.*, Observation of a Neutrino Burst from the Supernova SN 1987 A, *Phys. Rev. Lett.* **58** (1987) 1490–3

[76] K.S. Hirata *et al.*, *Observation of ^8B Solar Neutrinos in the Kamiokande II Detector*, Inst. f. Cosmic Ray Research, ICR-Report 188-89-5 (1989)

[77] R. Bionta *et al.*, Observation on a Neutrino Burst in Coincidence with Supernova SN 1987 A in the Large Magellanic Cloud, *Phys. Rev. Lett.* **58(14)** (1987) 1494–6

[78]   www.sno.phy.queensu.ca/

[79]   W. Stolz, *Radioaktivität*, Hanser, München/Wien (1990)

[80]   M.A. Geyh & H. Schleicher, *Absolute Age Determination: Physical and Chemical Dating Methods and Their Application*, Springer, Berlin (1990)

[81]   E. Sauter, *Grundlagen des Strahlenschutzes*, Siemens AG, Berlin/München (1971); *Grundlagen des Strahlenschutzes*, Thiemig, München (1982)

Résumé

Measure what is measurable, and make measurable what is not so.

Galileo Galilei

The scope of detection techniques is very wide and diverse. Depending on the aim of the measurement, different physics effects are used. Basically, each physics phenomenon can be the basis for a particle detector. If complex experimental problems are to be solved, it is desirable to develop a multipurpose detector which allows one to unify a large variety of different measurement techniques. This would include a high (possibly 100%) efficiency, excellent time, spatial and energy resolution with particle identification. For certain energies these requirements can be fulfilled, e.g. with suitably instrumented calorimeters. Calorimetric detectors for the multi-GeV and for the eV range, however, have to be basically different.

The discovery of new physics phenomena allows one to develop new detector concepts and to investigate difficult physics problems. For example, superconductivity provides a means to measure extremely small energy depositions with high resolution. The improvement of such measurement techniques, e.g. for the discovery and detection of Weakly Interacting Massive Particles (WIMPs), predicted by supersymmetry or cosmological neutrinos, would be of large astrophysical and cosmological interest.

In addition to the measurement of low-energy particles, the detection of extremely small changes of length may be of considerable importance. If one searches for gravitational waves, relative changes in length of $\Delta\ell/\ell \approx 10^{-21}$ have to be detected. If antennas with a typical size of $1\,\mathrm{m}$ were chosen, this would correspond to a measurement accuracy of $10^{-21}\,\mathrm{m}$ or one millionth of the diameter of a typical atomic nucleus. This ambitious goal has not yet been reached, but it is expected to be attained in the near future using Michelson interferometers with extremely long lever arms.

510

Since it would be bold to assume that the physical world is completely understood (in the past and also recently [1] this idea has been put forward several times), there will always be new effects and phenomena. Experts in the field of particle detection will pick up these effects and use them as a basis for the development of new particle detectors. For this reason a description of detection techniques can only be a snapshot. 'Old' detectors will 'die out' and new measurement devices will move to the forefront of research. Occasionally an old detector, already believed to be discarded, will experience a renaissance. The holographic readout of vertex bubble chambers for three-dimensional event reconstruction is an excellent example of this. But also in this case it was a new effect, namely the holographic readout technique, that has triggered this development.

Reference

[1] S.W. Hawking, *Is the End in Sight for Theoretical Physics? – An Inaugural Lecture*, Press Syndicate of the University of Cambridge (1980)

17

Glossary

Knowledge is a process of piling up facts; wisdom lies in their
simplification.

Harold Fabing and Ray Marr

The glossary summarises the most important properties of detectors along
with their main fields of application. An abridged description of the
characteristic interactions of particles is also presented.

17.1 Interactions of charged particles and radiation with matter

Charged particles interact mainly with the electrons of matter. The
atomic electrons are either excited to higher energy levels ('excitation')
or liberated from the atomic shell ('ionisation') by the charged particles.
High-energy ionisation electrons which are able themselves to ionise are
called δ rays or 'knock-on electrons'. In addition to the ionisation and
excitation of atomic electrons, bremsstrahlung plays a particular rôle,
especially for primary electrons as charged particles.

Energy loss by ionisation and excitation is described by the
Bethe–Bloch formula. The basic features describing the mean energy loss
per unit length $(\mathrm{d}E/\mathrm{d}x)$ for heavy particles are given by

$$-\left.\frac{\mathrm{d}E}{\mathrm{d}x}\right|_{\mathrm{ion}} \propto z^2 \cdot \frac{Z}{A} \cdot \frac{1}{\beta^2}\left[\ln(a \cdot \gamma^2 \beta^2) - \beta^2 - \frac{\delta}{2}\right] \, , \qquad (17.1)$$

where

z – charge of the incident particle,

Z, A – atomic number and atomic weight of the target,

512

β, γ – velocity and Lorentz factor of the incident particle,

δ – parameter describing the density effect,

a – parameter depending on the electron mass and the ionisation energy of the absorber.

Typical average values of the energy loss by ionisation and excitation are around $2\,\text{MeV}/(\text{g}/\text{cm}^2)$. The energy loss in a given material layer fluctuates, which is not described by a Gaussian, but is characterised, in particular for thin absorber layers, by a high asymmetry (Landau distribution).

Detectors only measure the energy deposited in the sensitive volume. This is not necessarily the same as the energy loss of the particle in the detector, since a fraction of the energy can escape from the detector volume as, e.g., δ rays.

The energy loss of a charged particle in a detector leads to a certain number of free charge carriers n_T given by

$$n_\text{T} = \frac{\Delta E}{W} \;, \tag{17.2}$$

where ΔE is the energy deposited in the detector and W is a characteristic energy which is required for the production of a charge-carrier pair ($W \approx 30\,\text{eV}$ in gases, $3.6\,\text{eV}$ in silicon, $2.8\,\text{eV}$ in germanium).

Another interaction process of charged particles particularly important for electrons is **bremsstrahlung**. The bremsstrahlung energy loss can essentially be parametrised by

$$-\left.\frac{\mathrm{d}E}{\mathrm{d}x}\right|_{\text{brems}} \propto z^2 \cdot \frac{Z^2}{A} \cdot \frac{1}{m_0^2} \cdot E \;, \tag{17.3}$$

where m_0 and E are the projectile mass and energy, respectively. For electrons ($z = 1$) one defines

$$-\left.\frac{\mathrm{d}E}{\mathrm{d}x}\right|_{\text{brems}} = \frac{E}{X_0} \;, \tag{17.4}$$

where X_0 is the **radiation length** characteristic for the absorber material.

The **critical energy** E_c characteristic for the absorber material is defined as the energy at which the energy loss of electrons by ionisation and excitation on the one hand and bremsstrahlung on the other hand are equal:

$$-\left.\frac{\mathrm{d}E}{\mathrm{d}x}(E_\text{c})\right|_{\text{ion}} = -\left.\frac{\mathrm{d}E}{\mathrm{d}x}(E_\text{c})\right|_{\text{brems}} = \frac{E_\text{c}}{X_0} \;. \tag{17.5}$$

Multiple Coulomb scattering of charged particles in matter leads to a deviation from a straight trajectory. It can be described by an rms planar scattering angle

$$\sigma_\theta = \sqrt{\langle \theta^2 \rangle} \approx \frac{13.6\,\mathrm{MeV}/c}{p\beta} \sqrt{\frac{x}{X_0}}\,, \tag{17.6}$$

where

p, β – momentum and velocity of the particle,

x – material traversed in units of radiation lengths X_0.

In addition to the interaction processes mentioned so far, direct electron-pair production and photonuclear interactions come into play at high energies. Energy losses by Cherenkov radiation, transition radiation and synchrotron radiation are of considerable interest for the construction of detectors or applications, but they play only a minor rôle as far the energy loss of charged particles is concerned.

Neutral particles like neutrons or neutrinos first have to produce charged particles in interactions before they can be detected via the interaction processes described above.

Photons of low energy ($< 100\,\mathrm{keV}$) are detected via the photoelectric effect. The cross section for the **photoelectric effect** can be approximated by

$$\sigma^{\mathrm{photo}} \propto \frac{Z^5}{E_\gamma^{7/2}}\,, \tag{17.7}$$

where at high γ energies the dependence flattens to $\propto E_\gamma^{-1}$. In the photoelectric effect one electron (usually from the K shell) is removed from the atom. As a consequence of the rearrangement in the atomic shell, either characteristic X rays or Auger electrons are emitted.

In the region of medium photon energies ($100\,\mathrm{keV}$–$1\,\mathrm{MeV}$) the scattering on quasifree electrons dominates (**Compton scattering**). The cross section for the Compton effect can be approximated by

$$\sigma^{\mathrm{Compton}} \propto Z \cdot \frac{\ln E_\gamma}{E_\gamma}\,. \tag{17.8}$$

At high energies ($\gg 1\,\mathrm{MeV}$) electron-pair production is the dominating interaction process of photons,

$$\sigma^{\mathrm{pair}} \propto Z^2 \cdot \ln E_\gamma\,. \tag{17.9}$$

The above photoprocesses lead to an absorption of X rays or γ radiation which can be described by an absorption law for the photon intensity according to

$$I = I_0\,\mathrm{e}^{-\mu x} . \tag{17.10}$$

μ is a characteristic absorption coefficient which is related to the cross sections for the photoelectric effect, Compton effect and pair production. Compton scattering plays a special rôle since the photon is not completely absorbed after the interaction like in the photoelectric effect or for pair production, but only shifted to a lower energy. This requires the introduction and distinction of attenuation and absorption coefficients.

Charged and also neutral particles can produce further particles in inelastic interaction processes. The strong interactions of hadrons can be described by characteristic nuclear interaction and collision lengths.

The electrons produced by ionisation – e.g. in gaseous detectors – are thermalised by collisions with the gas molecules. Subsequently, they are normally guided by an electric field to the electrodes. The directed motion of electrons in the electric field is called drift. Drift velocities of electrons in typical gases for usual field strengths are on the order of $5\,\mathrm{cm}/\mu\mathrm{s}$. During the drift the charged particles (i.e. electrons and ions) are subject to transverse and longitudinal diffusion caused by collisions with gas molecules.

The presence of inclined magnetic fields causes the electrons to deviate from a drift parallel to the electric field.

Low admixtures of electronegative gases can have a considerable influence on the properties of gas detectors.

17.2 Characteristic properties of detectors

The quality of a detector can be expressed by its measurement resolution for time, track accuracy, energy and other characteristics. Spatial resolutions of 10–$20\,\mu\mathrm{m}$ can be obtained in silicon strip counters and small drift chambers. Time resolutions in the subnanosecond range are achievable with resistive-plate chambers. Energy resolutions in the eV range can be reached with cryogenic calorimeters.

In addition to resolutions, the efficiency, uniformity and time stability of detectors are of great importance. For high-rate applications also random coincidences and dead-time corrections must be considered.

17.3 Units of radiation measurement

The radioactive decay of atomic nuclei (or particles) is described by the decay law

$$N = N_0 \, e^{-t/\tau} \qquad\qquad (17.11)$$

with the lifetime $\tau = 1/\lambda$ (where λ is the decay constant). The half-life $T_{1/2}$ is smaller than the lifetime ($T_{1/2} = \tau \cdot \ln 2$).

The activity $A(t)$ of a radioactive isotope is

$$A(t) = -\frac{\mathrm{d}N}{\mathrm{d}t} = \lambda \cdot N \qquad\qquad (17.12)$$

with the unit **Becquerel** (= 1 decay per second).

The absorbed dose D is defined by the absorbed radiation energy $\mathrm{d}W$ per unit mass,

$$D = \frac{\mathrm{d}W}{\varrho \, \mathrm{d}V} = \frac{\mathrm{d}W}{\mathrm{d}m} \; . \qquad\qquad (17.13)$$

D is measured in **Grays** ($1\,\mathrm{Gy} = 1\,\mathrm{J/kg}$). The old unit of the absorbed dose was rad ($100\,\mathrm{rad} = 1\,\mathrm{Gy}$).

The biological effect of an energy absorption can be different for different particle types. If the physical energy absorption is weighted by the relative biological effectiveness (RBE), one obtains the equivalent dose H, which is measured in **Sieverts** (Sv),

$$H\,\{\mathrm{Sv}\} = RBE \cdot D\,\{\mathrm{Gy}\} \; . \qquad\qquad (17.14)$$

The old unit of the equivalent dose was rem ($1\,\mathrm{Sv} = 100\,\mathrm{rem}$). The **equivalent radiation dose due to natural radioactivity** amounts to about $3\,\mathrm{mSv}$ per year. Persons working in **regions of controlled access** are typically limited to a maximum of $20\,\mathrm{mSv}$ per year. The lethal dose for humans (50% probability of death within 30 days) is around $4000\,\mathrm{mSv}$.

17.4 Accelerators

Accelerators are in use in many different fields, such as particle accelerators in nuclear and elementary particle physics, in nuclear medicine for tumour treatment, in material science, e.g. in the study of elemental composition of alloys, and in food preservation. Present-day particle physics experiments require very high energies. The particles which are accelerated must be charged, such as electrons, protons or heavier ions. In some cases – in particular for colliders – also antiparticles are required.

Such particles like positrons or antiprotons can be produced in interactions of electrons or protons. After identification and momentum selection, they are then transferred into a storage-ring system where they are accelerated to higher energies. Beams of almost any sufficiently long-lived particles can be produced by colliding a proton beam with an external target, and selecting the desired particle species by sophisticated particle identification.

Most accelerators are circular (synchrotrons). For very high-energy electron machines ($\geq 100\,\text{GeV}$), linear accelerators must be used because of the large energy loss due to synchrotron radiation in circular electron accelerators. For future particle physics investigations also neutrino factories are considered.

17.5 Main physical phenomena used for particle detection and basic counter types

The main interaction process for ionisation counters is described by the Bethe–Bloch formula. Depending on a possible gas amplification of the produced electron–ion pairs, one distinguishes ionisation chambers (without gas amplification), proportional counters (gain $\propto \text{d}E/\text{d}x$), Geiger counters, and streamer tubes (saturated gain, no proportionality to the energy loss). Ionisation processes can also be used for liquids and solids (without charge-carrier multiplication).

Solid-state detectors have gained particular importance for high-resolution tracking as strip, pixel and voxel devices and also because of their intrinsically high energy resolution.

The excitation of atoms, also described by the Bethe–Bloch formula, is the basis of scintillation counters which are read out by standard photomultipliers, multianode photomultipliers or silicon photodiodes. For particle-identification purposes Cherenkov and transition-radiation counters play a special rôle.

17.6 Historical track detectors

17.6.1 Cloud chambers

Application: Measurement of rare events in cosmic rays; demonstration experiment; historical importance.

Construction: Gas–vapour mixture close to the saturation vapour pressure. Additional detectors (e.g. scintillation counters) can trigger the expansion to reach the supersaturated state of the vapour.

Measurement principle, readout: The droplets formed along the ionisation track in the supersaturated vapour are photographed stereoscopically.

Advantage: The cloud chamber can be triggered.

Disadvantages: Very long dead and cycle times; tiresome evaluation of cloud-chamber photographs.

Variation: In non-triggerable diffusion cloud chambers a permanent zone of supersaturation can be maintained.

17.6.2   Bubble chambers

Application: Precise optical tracking of charged particles; studies of rare and complex events.

Construction: Liquid gas close to the boiling point; superheating of liquid by synchronisation of the bubble-chamber expansion with the moment of particle incidence into the chamber.

Measurement principle, readout: The bubbles formed along the particle track in the superheated liquid are photographed stereoscopically.

Advantages: High spatial resolution; measurement of rare and complex events; lifetime determination of short-lived particles possible.

Disadvantages: Extremely tedious analysis of photographically recorded events; cannot be triggered but only synchronised; insufficient mass for the absorption of high-energy particles.

Variation: Holographic readout allows three-dimensional event reconstruction with excellent spatial resolution (several μm).

17.6.3   Streamer chambers

Application: Investigation of complex events with bubble-chamber quality in a detector which can be triggered.

Construction: Large-volume detector in a homogeneous strong electric field. A high-voltage signal of very short duration induces streamer discharges along the ionisation track of charged particles.

Measurement principle, readout: The luminous streamers are photographed stereoscopically head-on.

Advantages: High-quality photographs of complex events. Diffusion suppression by addition of oxygen; targets can be mounted inside the sensitive volume of the detector.

Disadvantages: Demanding event analysis; the very short high-voltage signals (100 kV amplitude, 2 ns duration) may interfere with the performance of other detectors.

17.6.4   Neon-flash-tube chambers

Applications: Investigation of rare events in cosmic rays; studies of neutrino interactions; search for nucleon decay.

Construction: Neon- or neon/helium-filled, sealed cylindrical glass tubes or spheres ('Conversi spheres'), or polypropylene tubes with normal gas-flow operation.

Measurement principle, readout: A high-voltage pulse applied to the chamber causes those tubes which have been hit by charged particles to light up in full length. The discharge can be photographed or read out electronically.

Advantages: Extremely simple construction; large volumes can be instrumented at low cost.

Disadvantages: Long dead times; low spatial resolution; no three-dimensional space points but only projections.

17.6.5   Spark chambers

Applications: Somewhat older track detector for the investigation of cosmic-ray events; spectacular demonstration experiment.

Construction: Planar, parallel electrodes mounted in a gas-filled volume. The spark chamber is usually triggered by a coincidence of external detectors (e.g. scintillation counters).

Measurement principle, readout: The high gas amplification causes a plasma channel to develop along the particle track; spark formation occurs. The readout of chambers with continuous electrodes is done photographically. For wire spark chambers a magnetostrictive readout or readout via ferrite cores is possible.

Advantages: Simple construction.

Disadvantages: Low multitrack efficiency, can be improved by current limitation ('glass spark chamber'); tedious analysis of optically recorded events.

17.6.6   Nuclear emulsions

Application: Permanently sensitive detector; mostly used in cosmic rays or as vertex detector with high spatial resolution in accelerator experiments.

Construction: Silver-bromide or silver-chloride microcrystals embedded in gelatine.

Measurement principle, readout: Detection of charged particles similar to light recording in photographic films; development and fixation of tracks. Analysis is done under the microscope or with a CCD camera with subsequent semi-automatic pattern recognition.

Advantages: 100% efficient; permanently sensitive; simple in construction; high spatial resolution.

Disadvantages: Non-triggerable; tedious event analysis.

17.6.7   Plastic detectors

Application: Heavy ion physics and cosmic rays; search for magnetic monopoles; radon-concentration measurement.

Construction: Foils of cellulose nitrate usually in stacks.

Measurement principle, readout: The local damage of the plastic material caused by the ionising particle is etched in sodium hydroxide. This makes the particle track visible. The readout is done as in nuclear emulsions.

Advantages: Extremely simple, robust detector; perfectly suited for satellite and balloon-borne experiments; permanently sensitive; adjustable threshold to suppress the detection of weakly ionising particles.

Disadvantages: Non-triggerable; complicated event analysis.

17.7   Track detectors

17.7.1   Multiwire proportional chamber

Application: Track detector with the possibility of measuring the energy loss. Suitable for high-rate experiments if small sense-wire spacing is used (see also microstrip detectors).

Construction: Planar layers of proportional counters without partition walls.

Measurement principle, readout: Analogous to the proportional counter; with high-speed readout (FADC = Flash ADC) the spatial structure of the ionisation can be resolved.

Advantages: Simple, robust construction; use of standard electronics.

Disadvantages: Electrostatic repulsion of anode wires; limited mechanical wire stability; sag for long anode wires for horizontal construction. Ageing problems in harsh radiation environments.

Variations: (1) Straw chambers (aluminised mylar straws with central anode wire); wire breaking in a stack of many straws affects only the straw with a broken wire.

(2) Segmentation of cathodes possible to obtain spatial coordinates.

17.7.2 Planar drift chamber

Application: Track detector with energy-loss measurement.

Construction: For the improvement of the field quality compared to the multiwire proportional chamber, potential wires are introduced between the anode wires. In general, far fewer wires are used than in a multiwire proportional chamber.

Measurement principle, readout: In addition to the readout as in the multiwire proportional chambers, the drift time of the produced charge carriers is measured. This allows – even at larger wire spacings – a higher spatial resolution.

Advantages: Drastic reduction of the number of anode wires; high track resolution.

Disadvantages: Spatial dependence of the track resolution due to charge-carrier diffusion and primary ionisation statistics; left–right ambiguity of drift-time measurement (curable by double layers or staggered anode wires).

Variations: (1) 'Electrodeless' chambers: field shaping by intentional ion deposition on insulating chamber walls.

(2) Time-expansion chambers: introduction of a grid to separate the drift space from the amplification region allowing adjustable drift velocities.

(3) Induction drift chamber: use of anode and potential wires with very small spacing. Readout of induced signals on the potential wires to solve the left–right ambiguity; high-rate capability.

17.7.3   Cylindrical wire chambers

Cylindrical proportional and drift chambers

Application: Central detectors in storage-ring experiments with high track resolution; large solid-angle coverage around the primary vertex.

Construction: Concentric layers of proportional chambers (or drift chambers). The drift cells are approximately trapezoidal or hexagonal. Electric and magnetic fields (for momentum measurement) are usually perpendicular to each other.

Measurement principle, readout: The same as in planar proportional or drift chambers. The coordinate along the wire can be determined by charge division, by measuring the signal propagation time on the wire, or by stereo wires. Compact multiwire drift modules with high-rate capability can be constructed.

Advantages and disadvantages: High spatial resolution; danger of wire breaking; $\vec{E} \times \vec{B}$ effect complicates track reconstruction.

Jet drift chambers

Application: Central detector in storage-ring experiments with excellent particle-identification properties via multiple measurements of the energy loss.

Construction: Azimuthal segmentation of a cylindrical volume into pie-shaped drift spaces; electric drift field and magnetic field for momentum measurement are orthogonal. Field shaping by potential strips; staggered anode wires to resolve the left–right ambiguity.

Measurement principle, readout: As in common-type drift chambers; particle identification by multiple $\mathrm{d}E/\mathrm{d}x$ measurement.

Advantage: High spatial resolution.

Disadvantages: $\vec{E} \times \vec{B}$ effect complicates track reconstruction. Complicated structure; danger of wire breaking.

Time-projection chamber (TPC)

Application: Practically 'massless' central detector mostly used in storage-ring experiments; accurate three-dimensional track reconstruction; electric drift field and magnetic field (for track bending) are parallel.

Construction, measurement principle, readout: There are neither anode nor potential wires in the sensitive volume of the detector. The produced charge carriers drift to the endcap detectors (in general, multiwire proportional chambers) which supply two track coordinates; the third coordinate is derived from the drift time.

Advantages: Apart from the counting gas there is no material in the sensitive volume (low multiple scattering, high momentum resolution; extremely low photon conversion probability). Availability of three-dimensional coordinates, energy-loss sampling and high spatial resolution.

Disadvantages: Positive ions drifting back into the sensitive volume will distort the electric field (can be avoided by an additional grid ('gating')); because of the long drift times the TPC cannot be operated in a high-rate environment.

Variation: The TPC can also be operated with liquid noble gases as a detector medium and supplies digital three-dimensional 'pictures' of bubble-chamber quality (requires extremely low-noise readout since in liquids usually no gas amplification occurs).

17.7.4 *Micropattern gaseous detectors*

Application: Vertex detectors of high spatial resolution; imaging detectors of high granularity.

Construction: Miniaturised multiwire proportional chamber with 'anode wires' on plastics or ceramic substrates; electrode structures normally produced using industrial microlithographic methods. Possible problems with ion deposition on dielectrics which may distort the field.

Measurement principle, readout: Electron avalanches measured on miniaturised electrode structures.

Advantages: Very high spatial resolution; separation of gas amplification region and readout structure possible.

Disadvantages: Sensitivity to harsh radiation environment, ageing problems, discharges may destroy the electrode structure.

Variation: Many different kinds of miniature structures, micromegas, gas electron multiplier (GEM), and so on.

17.7.5 *Semiconductor track detectors*

Application: Strip, pixel or voxel counters of very high spatial resolution, frequently used as vertex detectors in colliding-beam experiments or light-weight trackers in satellite experiments.

Construction: p–n or p–i–n semiconductor structures mostly of silicon with pitch of $\approx$ 20–50 µm for strip counters and $\approx$ 50 µm × 100 µm for pixel counters.

Measurement principle, readout: Charge carriers (electron–hole pairs) are liberated by ionisation energy loss and collected in an electrical drift field.

Advantages: Extremely high spatial resolution ($\approx$ 10 µm). The intrinsically high energy resolution is related to the low energy required for the production of an electron–hole pair (3.65 eV in Si). It can be taken advantage of for dE/dx measurements.

Disadvantages: Ageing in harsh radiation environments; only specially treated silicon counters are radiation tolerant. A beam loss into a silicon pixel counter produces pin holes and can even disable the whole counter.

Variation: p–n or p–i–n structures can be custom tailored to the intended measurement purpose. The problem of a large number of channels in strip counters can be circumvented with the silicon drift chamber.

17.7.6   Scintillating fibre trackers

Application: Small-diameter scintillating fibres can be individually viewed by multianode photomultipliers allowing high spatial resolutions.

Construction: Bundles of fibres (diameter 50 µm–1 mm) arranged in a regular lattice. The individual fibres are optically separated by a very thin cladding.

Measurement principle, readout: The scintillation light created by the energy loss of charged particles is guided by internal reflection to the photosensitive readout element at the ends of the fibres.

Advantages: Compact arrangements with high spatial resolution. Better radiation tolerance compared to other tracking detectors.

Disadvantages: The readout by photomultipliers is difficult in high magnetic fields and it is space consuming.

17.8   Calorimetry

17.8.1   Electromagnetic calorimeters

Application: Measurement of electron and photon energies in the range above several hundred MeV.

Construction: Total-absorption detectors in which the energy of electrons and photons is deposited via alternating processes of bremsstrahlung and pair production. In sampling calorimeters the energy deposition is usually only sampled in constant longitudinal depths.

Measurement principle, readout: Depending on the type of sampling detector used, the deposited energy is recorded as charge signal (e.g. liquid-argon chambers) or as light signal (scintillators) and correspondingly processed. For the complete absorption of 10 GeV electrons or photons about 20 radiation lengths are required.

Advantages: Compact construction; the relative energy resolution *improves* with increasing energy ($\sigma/E \propto 1/\sqrt{E}$).

Disadvantages: Sampling fluctuations, Landau fluctuations as well as longitudinal and lateral leakage deteriorate or limit the energy resolution.

Variation: By using a segmented readout, calorimeters can also provide excellent spatial resolution. Homogeneous 'liquid' calorimeters with strip readout provide 1 mm coordinate resolution for photons, almost energy-independent. Also 'spaghetti calorimeters' should be mentioned in this respect. Wavelength-shifting techniques allow compact construction of many modules (e.g. tile calorimeters).

17.8.2 Hadron calorimeters

Application: Measurement of hadron energies above 1 GeV; muon identification.

Construction: Total-absorption detector or sampling calorimeter; all materials with short nuclear interaction lengths can be considered as sampling absorbers (e.g. uranium, tungsten; also iron and copper).

Measurement principle, readout: Hadrons with energies > 1 GeV deposit their energy via inelastic nuclear processes in hadronic cascades. This energy is, just as in electron calorimeters, measured via the produced charge or light signals in the active detector volume.

Advantage: Improvement of the relative energy resolution with increasing energy.

Disadvantages: Substantial sampling fluctuations; large fractions of the energy remain 'invisible' due to the break-up of nuclear bonds and due to neutral long-lived particles or muons escaping from the detector volume. Therefore, the energy resolution of hadron calorimeters does not reach that of electron–photon calorimeters.

Variation: By compensation methods, the signal amplitudes of electron- or photon- and hadron-induced cascades for fixed energy can be equalised. This is obtained, e.g. by partially regaining the invisible energy. This compensation is of importance for the correct energy measurement in jets with unknown particle composition.

17.8.3   Calibration and monitoring of calorimeters

Calorimeters have to be calibrated. This is normally done with particles of known identity and momentum. In the low-energy range β and γ rays from radioisotopes can also be used for calibration purposes. To guarantee time stability, the calibration parameters have to be permanently monitored during data taking. This requires special on-line calibration procedures ('slow control').

17.8.4   Cryogenic calorimeters

Application: Detection of low-energy particles or measurement of extremely low energy losses.

Construction: Detectors that experience a detectable change of state even for extremely low energy absorptions.

Measurement principle: Break-up of Cooper pairs by energy depositions; transitions from the superconducting to the normal-conducting state in superheated superconducting granules; detection of phonons in solids.

Readout: With extremely low-noise electronic circuits, e.g. SQUIDs (Superconducting Quantum Interference Devices).

Advantages: Exploitation in cosmology for the detection of 'dark matter' candidates. Also usable for non-ionising particles.

Disadvantages: Extreme cooling required (milli-Kelvin range).

17.9   Particle identification

The aim of particle-identification detectors is to determine the mass m_0 and charge z of particles. Usually, this is achieved by combining information from different detectors. The main inputs to this kind of measurement are

(i) the momentum p determined in magnetic fields: $p = \gamma m_0 \beta c$

   (β – velocity, γ – Lorentz factor of the particle);

(ii) particle's time of flight τ: $\tau = s/(\beta \cdot c)$

   (s – flight path);

(iii) mean energy loss per unit length: $-\dfrac{\mathrm{d}E}{\mathrm{d}x} \propto \dfrac{z^2}{\beta^2} \ln \gamma$;

(iv) kinetic energy in calorimeters: $E_{\mathrm{kin}} = (\gamma - 1) m_0 c^2$;

 (v) Cherenkov light yield: $\propto z^2 \sin^2 \theta_{\mathrm{c}}$;

   ($\theta_{\mathrm{c}} = \arccos(1/n\beta)$, n – index of refraction);

(vi) yield of transition-radiation photons ($\propto \gamma$).

The measurement and identification of neutral particles (neutrons, photons, neutrinos, etc.) is done via conversion into charged particles on suitable targets or inside the detector volume.

17.9.1 Charged-particle identification

Time-of-flight counters

Application: Identification of particles of different mass with known momenta.

Construction, measurement principle, readout: Scintillation counters, resistive-plate chambers or planar spark counters for start–stop measurements; readout with time-to-amplitude converters.

Advantage: Simple construction.

Disadvantages: Only usable for 'low' velocities ($\beta < 0.99$, $\gamma < 10$).

Identification by ionisation losses

Application: Particle identification.

Construction: Multilayer detector for individual $\mathrm{d}E/\mathrm{d}x$ measurements.

Measurement principle, readout: The Landau distributions of the energy loss are interpreted as probability distributions. For a fixed momentum different particles are characterised by different energy-loss distributions. The reconstruction of these distributions with as large a number of measurements as possible allows for particle identification. In a simplified method the truncated mean of the energy-loss distribution can be used for particle identification.

Advantages: The $\mathrm{d}E/\mathrm{d}x$ measurements can be obtained as a by-product in multiwire proportional, jet or time-projection chambers. The measurement principle is simple.

Disadvantages: In certain kinematical ranges the mean energy losses for different charged particles overlap appreciably. The density effect of the energy loss leads to the same dE/dx distribution for all singly charged particles at high energies ($\beta\gamma \approx$ several hundred) even in gases.

Identification using Cherenkov radiation

Application: Mass determination (threshold Cherenkov counters) in momentum-selected beams; velocity determination (differential Cherenkov counter).

Construction: Solid, liquid or gaseous transparent radiators; phase mixtures (aerogels) to cover indices of refraction not available in natural materials.

Measurement principle, readout: Cherenkov-light emission for particles with $v > c/n$ (n – index of refraction) due to asymmetric polarisation of the radiator material. Readout with photomultipliers or multiwire proportional chambers with photosensitive gas. Application in γ-ray astronomy (Imaging Air Cherenkov Telescopes).

Advantages: Simple method of mass determination; variable and adjustable threshold for gas Cherenkov counters via gas pressure; Cherenkov-light emission can also be used for calorimetric detectors; also imaging systems possible (ring-imaging Cherenkov counter (RICH)).

Disadvantages: Low photon yield (compared to scintillation); Cherenkov counters only measure the velocity β (apart from z); this limits the application to not too high energies.

Transition-radiation detectors

Application: Measurement of the Lorentz factor γ for particle identification.

Construction: Arrangement of foils or porous dielectrics with the number of transition layers as large as possible (discontinuity in the dielectric constant).

Measurement principle, readout: Emission of electromagnetic radiation at boundaries of materials with different dielectric constants. Readout by multiwire proportional chambers filled with xenon or krypton for effective photon absorption.

Advantages: The number or, more precisely, the total energy radiated as transition-radiation photons is proportional to the *energy* of the charged particle. The emitted photons are in the X-ray range and therefore are easy to detect.

Disadvantages: Separation of the energy loss from transition radiation and from ionisation is difficult. Effective threshold effect of $\gamma \approx 1000$.

17.9.2 Particle identification with calorimeters

Particle identification with calorimeters is based on the different longitudinal and lateral development of electromagnetic and hadronic cascades.

Muons can be distinguished from electrons, pions, kaons and protons by their high penetration power.

17.9.3 Neutron detection

Applications: Detection of neutrons in various energy ranges for radiation protection, at nuclear reactors, or in elementary particle physics.

Construction: Borontrifluoride counters; coated cellulose-nitrate foils or LiI(Eu)-doped scintillators.

Measurement principle: Neutrons – as electrically neutral particles – are induced to produce charged particles in interactions, which are then registered with standard detection techniques.

Disadvantages: Neutron detectors typically have a low detection efficiency.

17.10 Neutrino detectors

Application: Measurement of neutrinos in astroparticle physics and accelerator experiments.

Construction: Large-volume detectors using water or ice for cosmic-ray, solar, galactic or extragalactic neutrinos. Massive detectors at accelerators at large neutrino fluxes. Massive bubble chambers.

Measurement principle, readout: Conversion of the different neutrino flavours in weak interactions into detectable charged particles, which are measured by standard tracking techniques or via Cherenkov radiation.

Advantages: Substantial gain in physics understanding. Search for point sources in the sky (neutrino astronomy).

Disadvantages: Large-scale experiments require new techniques for deployment in water and ice. Rare event rates. Background from mundane sources requires excellent particle identification.

17.11   Momentum measurement

Applications: Momentum spectrometer for fixed-target experiments at accelerators, for investigations in cosmic rays, and at storage rings.

Construction: A magnet volume is either instrumented with track detectors or the trajectories of incoming and outgoing charged particles are measured with position-sensitive detectors.

Measurement principle, readout: Detectors determine the track of charged particles in a magnetic field; the track bending together with the strength of the magnetic field allows one to calculate the momentum.

Advantages: For momenta in the GeV/c range high momentum resolutions are obtained. The momentum determination is essential for particle identification.

Disadvantages: The momentum resolution is limited by multiple scattering in the magnet and in the detectors, as well as by the limited spatial resolution of the detectors. The momentum resolution *deteriorates* with momentum ($\sigma/p \propto p$). For high momenta the required detector length becomes increasingly large.

17.12   Ageing

- Ageing in wire chambers is caused by the production of molecule fragments in microplasma discharges during avalanche formation. Depositions of carbon, silicates or oxides on anode, potential and cathode wires can be formed.

- Ageing effects can be suppressed by a suitable choice of gases and gas admixtures (e.g. noble gases with additions containing oxygen). In addition, one must be careful to avoid substances which tend to form polymers (e.g. carbon-containing polymers, silicon compounds, halides and sulphur-containing compounds).

- Ageing effects can also be reduced by taking care in chamber set-up and by a careful selection of all components used for chamber construction and the gas-supply system.

- Ageing in scintillators leads to loss of transparency.

- Ageing in semiconductor counters (silicon) leads to the creation of defects and interstitials and type inversion.

17.13 Example of a general-purpose detector

With the idea of general-purpose detectors one usually associates big experiments such as at previous e^+e^- colliders (ALEPH, DELPHI, L3, OPAL), at B factories (Belle, BABAR), at the Large Hadron Collider at CERN (ATLAS, CMS, LHCb, ALICE), or in astroparticle physics experiments (IceCube, Auger experiment, ANTARES, PAMELA). Also large cosmic-ray experiments or spaceborne experiments require sophisticated instrumentation.

One important aspect of such general-purpose detectors is tracking with high spatial resolution with the possibility to identify short-lived particles (e.g. B mesons). Particle identification can be done with Cherenkov detectors, transition radiation, time-of-flight measurements or multiple dE/dx sampling. Momentum measurements and calorimetric techniques for electrons, photons and hadrons are essential to reconstruct event topologies and to identify missing particles like neutrinos or supersymmetric particles, which normally go undetected. These properties are described for the example of the Belle detector operating at the e^+e^- storage ring at KEK. The main objectives of this experiment are to study B physics, CP violation and rare B decays with the aim to determine the angles in the unitarity triangle, which are relevant for the understanding of the elements in the Cabibbo–Kobayashi–Maskawa matrix and electroweak interactions as a whole. In cosmic-ray experiments on the other hand, a large coverage at affordable cost is required, e.g. for neutrino astronomy and/or particle astronomy at $\geq$ EeV energies, while space experiments necessitate compact instruments with excellent spatial resolution and particle identification at limited payloads.

17.14 Electronics

The readout of particle detectors can be considered as an integral part of the detection system. There is a clear tendency to integrate even sophisticated electronics into the front-end part of a detector. The front-end electronics usually consists of preamplifiers, but discriminators can also be integrated. The information contained in analogue signals is normally extracted by analogue-to-digital converters (ADCs). With fast flash ADCs even the time structure of signals can be resolved with high accuracy. Particular care has to be devoted to problems of noise, cross-talk, pickup and grounding. Logic decisions are normally made in places which are also accessible during data taking. These logic devices usually have to handle large numbers of input signals and are consequently configured in different levels. These trigger levels – which can be just coincidences in the most simple case – allow a stepwise decision on whether to accept an event or not. Modern trigger systems also make massive use of microprocessors for the handling of complex event signatures. Events which pass the trigger decision are handed over to the data-acquisition system.

For good data quality, on-line monitoring and slow control are mandatory.

For simpler detection techniques the amount of electronics can be substantially reduced. The operation of visual detectors uses only very few electronic circuits and some detectors, like nuclear emulsions or plastic detectors, require no electronics at all.

17.15   Data analysis

The raw data provided by the detectors consist of a collection of analogue and digital signals and preprocessed results from the on-line data acquisition. The task of the data analysis is to translate this raw information off-line into physics quantities.

The detector data are first used to determine, e.g., the energy, momentum, arrival direction and identity of particles which have been recorded. This then allows one to reconstruct complete events. These can be compared with some expectation which is obtained by combining physics events generators based on a theory with detector simulation. A comparison between recorded and simulated data can be used to fix parameters which are not given by the theory. A possible disagreement requires the modification of the model under test, or it may hint at the discovery of new physics. As an example for the problems encountered in data analysis, the search for the Higgs particle at LEP is discussed.

17.16   Applications

There is a wide range of applications for particle detectors. Mostly, these detectors have been developed for experiments in elementary particle physics, nuclear physics and cosmic rays. However, there are plenty of applications also in the fields of astronomy, cosmology, biophysics, medicine, material science, geophysics and chemistry. Even in domains like arts, civil engineering, environmental science, food preservation, pest control and airport screening, where one would not expect to see particle detectors, one finds interesting applications.

18
Solutions

Chapter 1

1.1 $100 \, \text{keV} \ll m_e c^2 \Rightarrow$ classical (non-relativistic) treatment acceptable

$$E_{\text{kin}} = \frac{1}{2} m_e v^2 \Rightarrow v = \sqrt{\frac{2 E_{\text{kin}}}{m_e}} = 1.9 \cdot 10^8 \, \text{m/s} \; ,$$

$$\text{range } s = \frac{1}{2} a t^2 \; , \quad v = at \; ,$$

$$\Rightarrow t = \frac{2s}{v} = 2.1 \cdot 10^{-12} \, \text{s} = 2.1 \, \text{ps} \; .$$

1.2 $m_\mu c^2 \ll 1 \, \text{TeV}$, therefore in this approximation $m_\mu \approx 0$;

$$R = \int_E^0 \frac{\mathrm{d}E}{\mathrm{d}E/\mathrm{d}x} = \int_0^E \frac{\mathrm{d}E}{a + bE} = \frac{1}{b} \ln\left(1 + \frac{b}{a} E\right) \; ,$$

$$R(1 \, \text{TeV}) = 2.64 \cdot 10^5 \, \text{g/cm}^2$$

$$\hat{=} 881 \, \text{m rock} \; (\varrho_{\text{rock}} = 3 \, \text{g/cm}^3 \text{ assumed}) \; .$$

1.3

$$\frac{\sigma(E)}{E} = \frac{\sqrt{F} \cdot \sqrt{n}}{n} = \frac{\sqrt{F}}{\sqrt{n}} \; ;$$

n is the number of produced electron–hole pairs,

$$n = \frac{E}{W} \; .$$

$W = 3.65\,\mathrm{eV}$ is the average energy required for the production of an electron–hole pair in silicon:

$$\frac{\sigma(E)}{E} = \frac{\sqrt{F \cdot W}}{\sqrt{E}} = 8.5 \cdot 10^{-4} = 0.085\% \ .$$

1.4

$$R = \int_{E_{\mathrm{kin}}}^{0} \frac{\mathrm{d}E_{\mathrm{kin}}}{\mathrm{d}E_{\mathrm{kin}}/\mathrm{d}x} = \int_{0}^{E_{\mathrm{kin}}} \frac{E_{\mathrm{kin}}\,\mathrm{d}E_{\mathrm{kin}}}{az^2 \ln(bE_{\mathrm{kin}})}$$

$$\approx \frac{1}{az^2} \int_{0}^{E_{\mathrm{kin}}} \frac{E_{\mathrm{kin}}\,\mathrm{d}E_{\mathrm{kin}}}{(bE_{\mathrm{kin}})^{1/4}} \approx \frac{1}{a\sqrt[4]{b}\,z^2} \int_{0}^{E_{\mathrm{kin}}} E_{\mathrm{kin}}^{3/4}\,\mathrm{d}E_{\mathrm{kin}}$$

$$= \frac{4}{7a\sqrt[4]{b}\,z^2}\, E_{\mathrm{kin}}^{7/4} \propto E_{\mathrm{kin}}^{1.75} \ ;$$

experimentally, the exponent is found to vary depending on the energy range and the type of particle. For low-energy protons with energies between several MeV and 200 MeV it is found to be 1.8, and for α particles with energies between 4 MeV and 7 MeV, it is around 1.5 [1, 2].

1.5 Longitudinal- and transverse-component momentum conservation requires, see Fig. 18.1:

longitudinal component $\quad h\nu - h\nu' \cos\Theta_\gamma \quad = p\cos\Theta_e,$

transverse component $\quad h\nu' \sin\Theta_\gamma \quad\quad\quad = p\sin\Theta_e,$

$(c = 1$ assumed):

$$\cot\Theta_e = \frac{h\nu - h\nu' \cos\Theta_\gamma}{h\nu' \sin\Theta_\gamma} \ .$$

Because of

$$\frac{h\nu'}{h\nu} = \frac{1}{1 + \varepsilon(1 - \cos\Theta_\gamma)} \ :$$

$$\cot\Theta_e = \frac{1 + \varepsilon(1 - \cos\Theta_\gamma) - \cos\Theta_\gamma}{\sin\Theta_\gamma} = \frac{(1 + \varepsilon)(1 - \cos\Theta_\gamma)}{\sin\Theta_\gamma} \ .$$

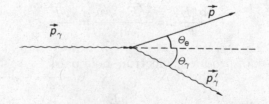

Fig. 18.1.   Kinematics of Compton scattering.

Because of

$$1 - \cos \Theta_\gamma = 2 \sin^2 \frac{\Theta_\gamma}{2}$$

one gets

$$\cot \Theta_e = (1 + \varepsilon) \frac{2 \sin^2 \frac{\Theta_\gamma}{2}}{\sin \Theta_\gamma} \ .$$

With $\sin \Theta_\gamma = 2 \sin(\Theta_\gamma/2) \cdot \cos(\Theta_\gamma/2)$ follows

$$\cot \Theta_e = (1 + \varepsilon) \frac{\sin(\Theta_\gamma/2)}{\cos(\Theta_\gamma/2)} = (1 + \varepsilon) \tan \frac{\Theta_\gamma}{2} \ .$$

This relation shows that the scattering angle of the electron can never exceed 90°.

1.6 $q_\mu + q_e = q'_\mu + q'_e \Rightarrow$

$$\begin{pmatrix} E_\mu \\ \vec{p}_\mu \end{pmatrix} \begin{pmatrix} m_e \\ \vec{0} \end{pmatrix} = \begin{pmatrix} E'_\mu \\ \vec{p}'_\mu \end{pmatrix} \begin{pmatrix} E'_e \\ \vec{p}'_e \end{pmatrix} \ , \quad m_e E_\mu = E'_\mu E'_e - \vec{p}'_\mu \cdot \vec{p}'_e \ .$$

Head-on collision gives maximum energy transfer $\Rightarrow \cos \Theta = 1$:

$$m_e E_\mu = E'_\mu E'_e - \sqrt{E'^2_\mu - m^2_\mu} \sqrt{E'^2_e - m^2_e}$$

$$= E'_\mu E'_e - E'_\mu E'_e \sqrt{1 - \left(\frac{m_\mu}{E'_\mu} \right)^2} \sqrt{1 - \left(\frac{m_e}{E'_e} \right)^2}$$

$$= E'_\mu E'_e \left\{ 1 - \left[1 - \frac{1}{2} \left(\frac{m_\mu}{E'_\mu} \right)^2 + \cdots \right] \left[1 - \frac{1}{2} \left(\frac{m_e}{E'_e} \right)^2 + \cdots \right] \right\}$$

$$= E'_\mu E'_e \left[\frac{1}{2} \left(\frac{m_\mu}{E'_\mu} \right)^2 + \frac{1}{2} \left(\frac{m_e}{E'_e} \right)^2 + \cdots \right] \ ,$$

$$2 m_e E_\mu \approx \frac{E'_e}{E'_\mu} m^2_\mu + \frac{E'_\mu}{E'_e} m^2_e \Rightarrow 2 m_e E_\mu E'_e E'_\mu = E'^2_e m^2_\mu + E'^2_\mu m^2_e \ ,$$

$$m^2_e E'^2_\mu \ll m^2_\mu E'^2_e \Rightarrow 2 m_e E_\mu E'_e E'_\mu \approx E'^2_e m^2_\mu \ ,$$

energy conservation: $E'_\mu + E'_e = E_\mu + m_e, m_e \ll E_\mu$;

$$2m_e E_\mu (E_\mu - E'_e) = m_\mu^2 E'_e = 2m_e E_\mu^2 - 2m_e E_\mu E'_e \ ,$$

$$E'_e = \frac{2m_e E_\mu^2}{m_\mu^2 + 2m_e E_\mu} = \frac{E_\mu^2}{E_\mu + \frac{m_\mu^2}{2m_e}} = \frac{E_\mu^2}{E_\mu + 11\,\text{GeV}} \ ,$$

therefore $E'_e = 90.1\,\text{GeV}$.

1.7 Argon: $Z = 18$, $A = 40$, $\varrho = 1.782 \cdot 10^{-3}\,\text{g/cm}^3$,

$$\phi(E)\,\mathrm{d}E = 1.235 \cdot 10^{-4}\,\text{GeV}\,\frac{\mathrm{d}E}{\beta^2 E^2} = \alpha\,\frac{\mathrm{d}E}{\beta^2 E^2} \ .$$

For a $10\,\text{GeV}$ muon $\beta \approx 1$,

$$P(> E_0) = \int_{E_0}^{E_{\max}} \phi(E)\,\mathrm{d}E = \alpha \int_{E_0}^{E_{\max}} \frac{\mathrm{d}E}{E^2} = \alpha \left(\frac{1}{E_0} - \frac{1}{E_{\max}} \right) \ ,$$

$$E_{\max} = \frac{E_\mu^2}{E_\mu + 11\,\text{GeV}} = 4.76\,\text{GeV} \ ,$$

$$P(> E_0) = 1.235 \cdot 10^{-4} \left(\frac{1}{10} - \frac{1}{4760} \right) = 1.235 \cdot 10^{-5} \approx 0.0012\% \ .$$

1.8 The sea-level muon spectrum can be approximated by

$$N(E)\,\mathrm{d}E \propto E^{-\alpha}\,\mathrm{d}E \ , \quad \text{where } \alpha \approx 2 \ ,$$

$$\frac{\mathrm{d}E}{\mathrm{d}x} = \text{constant}\ (= a) \ \Rightarrow \ E = a \cdot h \ (h - \text{depth}),$$

$$I(h) = \text{const}\ h^{-\alpha} \ ,$$
$$\left| \frac{\Delta I}{I} \right| = \frac{\alpha h^{-\alpha-1} \Delta h}{h^{-\alpha}} = \alpha\,\frac{\Delta h}{h} = 2 \cdot \frac{1}{100} = 2\% \ .$$

Chapter 2

2.1

$$\rho(\text{Al}) = 2.7\,\text{g/cm}^3 \rightarrow \mu = (0.189 \pm 0.027)\,\text{cm}^{-1} \ ,$$
$$I(x) = I_0 \exp(-\mu \cdot x) \rightarrow x = 1/\mu \cdot \ln(I_0/I) \ .$$

Statistical error of the count rates:

$$\sqrt{I_0}/I_0 = 1/\sqrt{I_0} \approx 4.2\% \ , \quad \sqrt{I}/I = 1/\sqrt{I} \approx 5.0\% \ .$$

The fractional error of I_0/I is

$$\sqrt{(4.2\%)^2 + (5.0\%)^2} \approx 6.5\% \ .$$

Hence $I_0/I = 1.440_{\pm 6.5\%}$.

Since $x \propto \ln(I_0/I) = \ln r \rightarrow \mathrm{d}x \propto \mathrm{d}r/r$, so that the absolute error in $\ln r$ is equal to the fractional error in I_0/I.

Therefore, $\ln(I_0/I) = \ln 1.44 \pm 0.065 \approx 0.365 \pm 0.065 \approx 0.37_{\pm 18\%}$.

The fractional error in μ was 14.3%, so the fractional error in x is

$$\sqrt{(18\%)^2 + (14.3\%)^2} \approx 23\% \ .$$

Therefore

$$x = 1/\mu \cdot \ln(I_0/I) = 1.93\,\mathrm{cm}_{\pm 23\%} = (1.93 \pm 0.45)\,\mathrm{cm} \ .$$

2.2

$$P(n, \mu) = \frac{\mu^n \cdot e^{-\mu}}{n!} \ , \quad n = 0, 1, 2, 3, \ldots \ \rightarrow$$

$$P(5, 10) = \frac{10^5 \cdot e^{-10}}{5!} \approx 0.0378 \ ,$$

$$P(2, 1) = \frac{1^2 \cdot e^{-1}}{2!} \approx 0.184 \ , \quad P(0, 10) = \frac{10^0 \cdot e^{-10}}{0!} \approx 4.5 \cdot 10^{-5} \ .$$

2.3 The true dead-time-corrected rate at $d_1 = 10\,\mathrm{cm}$ is

$$R_1^* = \frac{R_1}{1 - \tau\,R_1} \ .$$

Because of the inverse square law $(\propto 1/r^2)$ the true rate at $d_2 = 30\,\mathrm{cm}$ is

$$R_2^* = \left(\frac{d_1}{d_2}\right)^2 R_1^* \ ;$$

and because of $R_2^* = R_2/(1 - \tau\,R_2)$ one gets

$$\left(\frac{d_1}{d_2}\right)^2 \frac{R_1}{1 - \tau\,R_1} = \frac{R_2}{1 - \tau\,R_2} \ .$$

Solving for τ yields

$$\tau = \frac{\left(\frac{d_2}{d_1}\right)^2 R_2 - R_1}{\left[\left(\frac{d_2}{d_1}\right)^2 - 1\right] R_1 R_2} = 10\,\mu\text{s} \; .$$

Chapter 3

3.1

$$\text{dose} = \frac{\text{absorbed energy}}{\text{mass unit}} = \frac{\text{activity} \cdot \text{energy per Bq} \cdot \text{time}}{\text{mass}}$$

$$= \frac{10^9\,\text{Bq} \cdot 1.5 \cdot 10^6\,\text{eV} \cdot 1.602 \cdot 10^{-19}\,\text{J/eV} \cdot 86\,400\,\text{s}}{10\,\text{kg}}$$

$$= 2.08\,\text{J/kg} = 2.08\,\text{Gy} \; .$$

Here, also a common unit for the energy, like eV (electron volt), in addition to Joule, is used: $1\,\text{eV} = 1.602 \cdot 10^{-19}\,\text{J}$.

3.2 The decrease of the activity in the researcher's body has two components. The total decay rate λ_{eff} is

$$\lambda_{\text{eff}} = \lambda_{\text{phys}} + \lambda_{\text{bio}} \; .$$

Because of $\lambda = \frac{1}{\tau} = \frac{\ln 2}{T_{1/2}}$ one gets

$$T_{1/2}^{\text{eff}} = \frac{T_{\text{phys}}\,T_{\text{bio}}}{T_{\text{phys}} + T_{\text{bio}}} = 79.4\,\text{d} \; .$$

Using $\dot{D} = \dot{D}_0\,e^{-\lambda t}$ and $\dot{D}/\dot{D}_0 = 0.1$ one has[†]

$$t = \frac{1}{\lambda} \ln\left(\frac{\dot{D}_0}{\dot{D}}\right) = \frac{T_{1/2}^{\text{eff}}}{\ln 2} \ln\left(\frac{\dot{D}_0}{\dot{D}}\right) = 263.8\,\text{d} \; .$$

[†] The notation $\dot{D}_0$ describes the dose rate at $t = 0$. $\dot{D}_0$ does not represent the time derivative of the constant dose D_0 (which would be zero, of course).

A mathematically more demanding calculation allows to work out the dose that the researcher has received in this time span:

$$D_{\text{total}} = \int_0^{263.8\,\text{d}} \dot{D}_0\, e^{-\lambda t}\, dt = \dot{D}_0 \left(-\frac{1}{\lambda}\right) e^{-\lambda t}\Big|_0^{263.8\,\text{d}}$$

$$= \frac{\dot{D}_0}{\lambda} \left(1 - e^{-\lambda \cdot 263.8\,\text{d}}\right) \ .$$

With

$$\lambda = \frac{1}{\tau} = \frac{\ln 2}{T_{1/2}^{\text{eff}}} = 8.7 \cdot 10^{-3}\,\text{d}^{-1}$$

one obtains $(1\,\mu\text{Sv/h} = 24\,\mu\text{Sv/d})$

$$D_{\text{total}} = \frac{24\,\mu\text{Sv/d}}{\lambda} (1 - 0.1) = 2.47\,\text{mSv} \ .$$

The 50-year dose equivalent commitment $D_{50} = \int_0^{50\,\text{a}} \dot{D}(t)\, dt$ is worked out to be

$$D_{50} = \int_0^{50\,\text{a}} \dot{D}_0\, e^{-\lambda t} dt = \frac{\dot{D}_0}{\lambda} \left(1 - e^{-\lambda \cdot 50\,\text{a}}\right) \approx \frac{\dot{D}_0}{\lambda} = 2.75\,\text{mSv} \ .$$

3.3 The recorded charge ΔQ is related to the voltage drop ΔU by the capacitor equation

$$\Delta Q = C\,\Delta U = 7 \cdot 10^{-12}\,\text{F} \cdot 30\,\text{V} = 210 \cdot 10^{-12}\,\text{C} \ .$$

The mass of the air in the ionisation chamber is

$$m = \varrho_{\text{L}}\, V = 3.225 \cdot 10^{-3}\,\text{g} \ .$$

This leads to an ion dose of

$$I = \frac{\Delta Q}{m} = 6.5 \cdot 10^{-8}\,\text{C/g} = 6.5 \cdot 10^{-5}\,\text{C/kg} \ .$$

Because of $1\,\text{R} = 2.58 \cdot 10^{-4}\,\text{C/kg}$, this corresponds to a dose of 0.25 Röntgen or, respectively, because of $1\,\text{R} = 8.8\,\text{mGy}$,

$$D = 2.2\,\text{mGy} \ .$$

3.4 The total activity is worked out to be

$$A_{\text{total}} = 100\,\text{Bq/m}^3 \cdot 4000\,\text{m}^3 = 4 \cdot 10^5\,\text{Bq} \ .$$

This leads to the original activity concentration in the containment area of

$$A_0 = \frac{4 \cdot 10^5 \, \text{Bq}}{500 \, \text{m}^3} = 800 \, \text{Bq/m}^3 \ .$$

3.5 For the activity one has

$$A = \lambda N = \frac{1}{\tau} N = \frac{\ln 2}{T_{1/2}} N \ ,$$

corresponding to

$$N = \frac{A \, T_{1/2}}{\ln 2} = 1.9 \cdot 10^{12} \ \text{cobalt nuclei}$$

and $m = N \, m_{\text{Co}} = 0.2 \, \text{ng}$. Such a small amount of cobalt can hardly be detected with chemical techniques.

3.6 The radiation power is worked out to be

$$S = 10^{17} \, \text{Bq} \cdot 10 \, \text{MeV} = 10^{24} \, \text{eV/s} = 160 \, \text{kJ/s} \ ;$$

the temperature increase is calculated to be

$$\Delta T = \frac{\text{energy deposit}}{m \, c} = \frac{160 \, \text{kJ/s} \cdot 86\,400 \, \text{s/d} \cdot 1 \, \text{d}}{120\,000 \, \text{kg} \cdot 0.452 \, \text{kJ/(kg K)}} = 255 \, \text{K} \ .$$

This temperature rise of 255 °C eventually leads to a temperature of 275 °C.

3.7 X rays are attenuated according to

$$I = I_0 \, \text{e}^{-\mu x} \ \Rightarrow \ \text{e}^{\mu x} = \frac{I_0}{I} \ .$$

This leads to

$$x = \frac{1}{\mu} \, \ln\left(\frac{I_0}{I}\right) = 30.7 \, \text{g/cm}^2 \ ,$$

and accordingly

$$x^* = \frac{x}{\varrho_{\text{Al}}} = 11.4 \, \text{cm} \ .$$

3.8 With modern X-ray tubes the patient gets an effective whole-body dose on the order of 0.1 mSv. For a holiday spent at an altitude of 3000 m at average geographic latitudes the dose rate by cosmic rays amounts to about 0.1 µSv/h corresponding to 67 µSv in a period of 4 weeks [3]. If, in addition, the radiation load due to terrestrial radiation is also taken into account (about $40\,\mu\text{Sv}$ in 4 weeks), one arrives at a total dose which is very similar to the radiation dose received by an X ray of the chest. It has to be mentioned, however, that older X-ray tubes can lead to higher doses, and that the period over which the dose is applied is much shorter for an X ray, so that the dose rate in this case is much higher compared to the exposure at mountain altitudes.

3.9 The effective half-life for ^{137}Cs in the human body is

$$T_{1/2}^{\text{eff}} = \frac{T_{1/2}^{\text{phys}}\, T_{1/2}^{\text{bio}}}{T_{1/2}^{\text{phys}} + T_{1/2}^{\text{bio}}} = 109.9\,\text{d} \ .$$

The remaining amount of ^{137}Cs after three years can be worked out by two different methods:

a) the period of three years corresponds to $\frac{3\cdot 365}{109.9} = 9.9636$ half-lives:

$$\text{activity}(3\,\text{a}) = 4 \cdot 10^6 \cdot 2^{-9.9636} = 4006\,\text{Bq} \ ;$$

b) on the other hand, one can consider the evolution of the activity,

$$\text{activity}(3\,\text{a}) = 4 \cdot 10^6 \cdot e^{-3\,\text{a}\cdot\ln 2/T_{1/2}^{\text{eff}}} = 4006\,\text{Bq} \ .$$

3.10 The specific dose constants for β and γ radiation of ^{60}Co are

$$\Gamma_\beta = 2.62 \cdot 10^{-11}\,\text{Sv}\,\text{m}^2/\text{Bq}\,\text{h} \ , \quad \Gamma_\gamma = 3.41 \cdot 10^{-13}\,\text{Sv}\,\text{m}^2/\text{Bq}\,\text{h} \ .$$

For the radiation exposure of the hands the β dose dominates. Assuming an average distance of $10\,\text{cm}$ and an actual handling time of the source with the hands of $60\,\text{s}$, this would lead to a partial-body dose of

$$H_\beta = \Gamma_\beta\,\frac{A}{r^2}\,\Delta t = 2.62 \cdot 10^{-11} \cdot \frac{3.7 \cdot 10^{11}}{0.1^2} \cdot \frac{1}{60}\,\text{Sv} = 16.1\,\text{Sv} \ .$$

The whole-body dose, on the other hand, is related to the γ radiation of ^{60}Co. For an average distance of $0.5\,\text{m}$ and an exposure time of 5 minutes the whole-body dose is worked out to be

$$H_\gamma = \Gamma_\gamma\,\frac{A}{r^2}\,\Delta t = 42\,\text{mSv} \ .$$

Actually, a similar accident has happened to an experienced team of technicians in Saintes, France, in 1981. The technicians should have under no circumstances handled the strong source with their hands! Because of the large radiation exposure to the hands and the corresponding substantial radiation hazard, the hands of two technicians had to be amputated. For a third technician the amputation of three fingers was unavoidable.

3.11 After the first decontamination the remaining activity is $N(1 - \varepsilon)$, where N is the original surface contamination. After the third procedure one has $N(1 - \varepsilon)^3$. Therefore, one gets

$$N = \frac{512\,\mathrm{Bq/cm^2}}{(1 - \varepsilon)^3} = 64\,000\,\mathrm{Bq/cm^2} \ .$$

The third decontamination reduced the surface contamination by

$$N(1 - \varepsilon)^2\,\varepsilon = 2048\,\mathrm{Bq/cm^2} \ .$$

The number of decontaminations to reduce the level to $1\,\mathrm{Bq/cm^2}$ can be worked out along very similar lines ($N_n = N/(\mathrm{Bq/cm^2})$):

$$N(1 - \varepsilon)^n = 1\,\mathrm{Bq/cm^2} \ \rightarrow \ (1 - \varepsilon)^n = \frac{1}{N_n} \ \rightarrow$$

$$n \cdot \ln(1 - \varepsilon) = \ln\left(\frac{1}{N_n}\right) = -\ln N_n \ \rightarrow \ n = \frac{-\ln N_n}{\ln(1 - \varepsilon)} = 6.9 \ ,$$

i.e. $\approx$ seven times.

Chapter 4

4.1

$$\begin{aligned}
s = E_{\mathrm{CMS}}^2 &= (q_1 + q_2)^2 \\
&= (E_1 + E_2)^2 - (\vec{p}_1 + \vec{p}_2)^2 \\
&= E_1^2 - p_1^2 + E_2^2 - p_2^2 + 2E_1 E_2 - 2\vec{p}_1 \cdot \vec{p}_2 \\
&= 2m^2 + 2E_1 E_2 (1 - \beta_1 \beta_2 \cos \Theta)
\end{aligned}$$

because $p = \gamma m_0 \beta = E\beta \quad (c = 1 \text{ assumed})$.

In cosmic rays $\beta_1 \approx 1$ and $\beta_2 = 0$, since the target is at rest ($E_2 = m$); also $2E_1 m \gg 2m^2$:

$$s \approx 2mE_1 \ .$$

Under these conditions, one gets

$$E_{\text{lab}} = E_1 = \frac{s}{2m} = \frac{(14\,000\,\text{GeV})^2}{2 \cdot 0.938\,\text{GeV}} = 1.045 \cdot 10^8\,\text{GeV} \approx 10^{17}\,\text{eV} \ .$$

4.2 Centrifugal force $F = \dfrac{mv^2}{R} = evB_{\text{St}}$:

$$B_{\text{St}} = \frac{m}{e} \cdot \frac{v}{R} \ . \tag{18.1}$$

$$(4.13) \ \Rightarrow \ \frac{\mathrm{d}}{\mathrm{d}t}(mv) = e|\vec{E}| = \frac{eR}{2}\frac{\mathrm{d}B}{\mathrm{d}t} \ \Rightarrow \ mv = \frac{eR}{2}\,B \ . \tag{18.2}$$

Compare Eqs. (18.1) and (18.2):

$$B_{\text{St}} = \frac{1}{2}\,B \ ,$$

which is called the *Wideroe condition*.

4.3

$$m(\text{Fe}) = \varrho \cdot 300\,\text{cm} \cdot 0.3\,\text{cm} \cdot 1\,\text{mm} = 68.4\,\text{g} \ ,$$

$$\begin{aligned}
\Delta T &= \frac{\Delta E}{m(\text{Fe}) \cdot c} = \frac{2 \cdot 10^{13} \cdot 7 \cdot 10^3\,\text{GeV} \cdot 1.6 \cdot 10^{-10}\,\text{J/GeV} \cdot 3 \cdot 10^{-3}}{0.56\,\text{J/(g}\cdot\text{K)} \cdot 68.4\,\text{g}} \\
&= 1754\,\text{K}
\end{aligned}$$

$\Rightarrow$ the section hit by the proton beam will melt.

4.4 Effective bending radius

$$\rho = \frac{27\,\text{km} \cdot 2/3}{2\pi} = 2866\,\text{m} \ ,$$

$$\frac{mv^2}{\rho} = evB \Rightarrow p = eB\rho \ ,$$

$$pc = eB\rho c \ ,$$

$$10^9\,pc\,[\text{GeV}] = 3 \cdot 10^8\,B\,[\text{T}] \cdot \rho\,[\text{m}] \ ,$$

$$pc\,[\text{GeV}] = 0.3\,B\,[\text{T}] \cdot \rho\,[\text{m}] \ ,$$

$$pc^{\text{max}}(\text{LEP}) = 116\,\text{GeV} \ ,$$

$$pc^{\text{max}}(\text{LHC}) = 8.598\,\text{TeV} \ .$$

4.5 Magnetic potential $V = -g \cdot x \cdot y$
with g – quadrupole field strength or gradient of the quadrupole;

$$\vec{B} = -\operatorname{grad}V = (gy, gx) \ ;$$

the surface of the magnet must be an equipotential surface $\Rightarrow$

$$V = -g \cdot x \cdot y = \text{const} \;\; \Rightarrow \;\; x \cdot y = \text{const} \;\; \Rightarrow \;\; \text{hyperbolas} \; .$$

Chapter 5

5.1

$$R_{\text{true}} = \frac{R_{\text{measured}}}{1 - \tau_{\text{D}} \cdot R_{\text{measured}}} = 2 \, \text{kHz} \; . \tag{18.3}$$

5.2 For vertical incidence

$$\Delta E = \frac{\mathrm{d}E}{\mathrm{d}x} \cdot d \; , \tag{18.4}$$

for inclined incidence $\Delta E(\Theta) = \Delta E / \cos \Theta$;
measured energy for vertical incidence: $E_1 = E_0 - \Delta E$,
measured energy for inclined incidence: $E_2 = E_0 - \Delta E / \cos \Theta$,

$$E_1 - E_2 = \Delta E \left(\frac{1}{\cos \Theta} - 1 \right) \; ;$$

plot $E_1 - E_2$ versus $\left(\frac{1}{\cos \Theta} - 1 \right) \Rightarrow$ gives a straight line with a slope ΔE. With the known $\mathrm{d}E/\mathrm{d}x$ (from tables) for semiconductors Eq. (18.4) can be solved for d.

5.3

$$q = \begin{pmatrix} E \\ \vec{p} \end{pmatrix} \; , \quad q' = \begin{pmatrix} E' \\ \vec{p}' \end{pmatrix} \; , \quad q_\gamma = \begin{pmatrix} h\nu \\ \vec{p}_\gamma \end{pmatrix}$$

are the four-momentum vectors of the incident particle, the particle after Cherenkov emission, and the emitted Cherenkov photon;

$$q' = q - q_\gamma \; ,$$

$$E'^2 - p'^2 = (q - q_\gamma)^2 = \begin{pmatrix} E - h\nu \\ \vec{p} - \vec{p}_\gamma \end{pmatrix}^2$$

$$= E^2 - 2h\nu E + h^2 \nu^2 - (p^2 + p_\gamma^2 - 2\vec{p} \cdot \vec{p}_\gamma) \; .$$

Since $E^2 = m^2 + p^2$ and $\vec{p}_\gamma = \hbar\vec{k}$:

$$0 = -m^2 + m^2 + p^2 - 2h\nu E + h^2 \nu^2 - p^2 + 2p\hbar k \cos \Theta - \hbar^2 k^2 \; ,$$

$$2p\hbar k \cos\Theta = 2h\nu E - h^2\nu^2 + \hbar^2 k^2,$$

$$\cos\Theta = \frac{2\pi\nu E}{pk} + \frac{\hbar k}{2p} - \frac{2\pi h\nu^2}{2pk} \; ;$$

because of $\frac{c}{n} = \nu \cdot \lambda = \frac{2\pi\nu}{k}$ one has $(c = 1)$

$$\cos\Theta = \frac{E}{np} + \frac{\hbar k}{2p} - \frac{\hbar k}{2pn^2}$$

with $E = \gamma m_0, \gamma = \dfrac{1}{\sqrt{1 - \beta^2}}$, and $p = \gamma m_0 \beta$ one gets

$$\cos\Theta = \frac{1}{n\beta} + \frac{\hbar k}{2p} \left(1 - \frac{1}{n^2}\right) .$$

Normally $\hbar k/2p \ll 1$, so that the usually used expression for the Cherenkov angle is quite justified.

5.4 Let us assume that the light flash with the total amount of light, I_0, occurs at the centre of the sphere. In a first step the light intensity qI_0, where $q = S_\mathrm{p}/S_\mathrm{tot}$ arrives at the photomultiplier. The majority of the light $((1 - q)I_0)$ misses the PM tube and hits the reflecting surface. Then, let us select a small pad S_1 anywhere on the sphere, at a distance r from the photomultiplier and calculate how much light reflected by this pad reaches the photomultiplier after just one reflection (see Fig. 5.46). Denoting the total amount of light reflected from S_1 as ΔJ_0 we find

$$\Delta I_1^{\mathrm{PM}} = \frac{\Delta J_0}{\pi} \cos\chi \, \Delta\Omega = \frac{\Delta J_0}{\pi} \cos\chi \, \frac{S_\mathrm{p} \cos\chi}{(2R \cos\chi)^2} = \Delta J_0 \, q \; .$$

Since ΔI_1^{PM} has no angular dependence, this value can be simply integrated over the sphere which gives the total amount of light collected by the photomultiplier after the first reflection:

$$I_1^{\mathrm{PM}} = I_0 q + I_0 (1 - q)(1 - \mu)q \; .$$

The iteration of this argument leads to an expression for the total amount of light collected by the photomultiplier after an infinite number of reflections:

$$I_{\mathrm{tot}}^{\mathrm{PM}} = I_0 q + I_0 (1 - q)(1 - \mu)q + I_0 (1 - q)^2 (1 - \mu)^2 q + \cdots$$

$$= I_0 q \frac{1}{1 - (1 - \mu)(1 - q)} \; . \tag{18.5}$$

Then the light collection efficiency, $\eta = I_{\mathrm{tot}}^{\mathrm{PM}}/I_0$, is

$$\eta = \frac{q}{\mu + q - \mu q} \approx \frac{q}{\mu + q} \; . \tag{18.6}$$

Similar considerations for non-focussing Cherenkov counters were presented already a long time ago by M. Mando [4].

Chapter 6

6.1 If a small-diameter tube is submerged in a liquid, the liquid level will rise in the tube because the saturation vapour pressure of the concave liquid surface in the tube is smaller than the corresponding pressure over the planar liquid surface (capillary forces). An equilibrium condition is obtained for an elevation h of

$$2\pi r \sigma = \pi \varrho r^2 h g \; , \tag{18.7}$$

where

r – radius of the capillary vessel,

σ – surface tension,

ϱ – density of the liquid,

g – acceleration due to gravity.

For convex droplets the barometric scale formula

$$p_r = p_\infty \exp\left(\frac{Mgh}{RT}\right)$$

with

M – molar mass,

R – gas constant,

T – temperature

can be combined with (18.7) to give

$$\ln(p_r/p_\infty) = \frac{M}{RT}\frac{2\sigma}{\varrho r} \; .$$

With numbers:

$$M \quad = 18\,\text{g/mol for water} \qquad (46\,\text{g/mol for } C_2H_5OH),$$

$$\sigma \quad = 72.8\,\text{dyn/cm for water} \quad (22.3\,\text{dyn/cm for } C_2H_5OH),$$

$$\varrho \quad = 1\,\text{g/cm}^3 \qquad\qquad (0.79\,\text{g/cm}^3 \text{ for } C_2H_5OH),$$

$$R \quad = 8.31\,\text{J/mol K},$$

$$T \quad = 20\,°\text{C},$$

$$p_r/p_\infty = 1.001,$$

$$\to r \quad = 1.08 \cdot 10^{-6}\,\text{m} \qquad (1.07 \cdot 10^{-6}\,\text{m}),$$

i.e., droplets of diameter $\approx 2\,\mu\text{m}$ will form.

If the droplets are electrically charged, the mutual repulsive action will somewhat reduce the surface tension.

6.2 Increase of electron number

$$\mathrm{d}n_e = (\alpha - \beta)n_e\,\mathrm{d}x \ ;$$

$\alpha = $ first Townsend coefficient,

$\beta = $ attachment coefficient,

$$n_e = n_0\,\mathrm{e}^{(\alpha-\beta)d} \ ,$$
$$\mathrm{d}n_{\text{ion}} = \beta n_{e.}\mathrm{d}x \ ,$$
$$\mathrm{d}n_{\text{ion}} = \beta n_0\,\mathrm{e}^{(\alpha-\beta)x}\,\mathrm{d}x \ ,$$
$$n_{\text{ion}} = \beta n_0 \int_0^d \mathrm{e}^{(\alpha-\beta)x}\,\mathrm{d}x$$
$$= \frac{n_0\beta}{\alpha-\beta}\left[\mathrm{e}^{(\alpha-\beta)d} - 1\right] \ ,$$
$$\frac{n_e + n_{\text{ion}}}{n_0} = \frac{n_0\,\mathrm{e}^{(\alpha-\beta)d} + \frac{n_0\beta}{\alpha-\beta}\left[\mathrm{e}^{(\alpha-\beta)d} - 1\right]}{n_0}$$
$$= \frac{1}{\alpha-\beta}\left\{(\alpha-\beta)\,\mathrm{e}^{(\alpha-\beta)d} + \beta\left[\mathrm{e}^{(\alpha-\beta)d} - 1\right]\right\}$$
$$= \frac{1}{\alpha-\beta}\left(\alpha\,\mathrm{e}^{(\alpha-\beta)d} - \beta\right)$$
$$= \frac{1}{18}\left(20\,\mathrm{e}^{18} - 2\right) = 7.3 \cdot 10^7 \ .$$

6.3

$$\sqrt{\langle\theta^2\rangle} = \frac{13.6\,\text{MeV}}{\beta cp}\sqrt{\frac{x}{X_0}}\left[1 + 0.038\ln(x/X_0)\right] \ ,$$
$$\beta cp = 12.86\,\text{MeV} \ .$$

For electrons of this energy $\beta \approx 1 \Rightarrow p = 12.86\,\text{MeV}/c$. More precisely, one has to solve the equation

$$\beta c \gamma m_0 \beta c = 12.86\,\text{MeV}\ ,$$

$$\frac{\beta^2}{\sqrt{1-\beta^2}} = \frac{12.86\,\text{MeV}}{m_0 c^2} = 25.16 = \alpha\ ,$$

$$\beta^2 = \sqrt{1-\beta^2} \cdot \alpha \quad \Rightarrow \quad \beta^4 = \alpha^2 - \alpha^2 \beta^2\ ,$$

$$\beta^4 + \alpha^2 \beta^2 - \alpha^2 = 0\ ,$$

$$\beta^2 = -\frac{\alpha^2}{2} + \sqrt{\frac{\alpha^4}{4} + \alpha^2} = 0.998\,42\ ,$$

$$\gamma = 25.16\ ,$$

$$p = 12.87\,\text{MeV}/c\ .$$

Chapter 7

7.1

$$\Delta t = \frac{T_1 + T_3}{2} - T_2\ ; \tag{18.8}$$

resolution on Δt:

$$\sigma^2(\Delta t) = \left(\frac{\sigma_1}{2}\right)^2 + \left(\frac{\sigma_3}{2}\right)^2 + \sigma_2^2 = \frac{3}{2} \cdot \sigma_t^2\ ; \tag{18.9}$$

for one wire one has

$$\sigma_t = \sqrt{\frac{2}{3}}\,\sigma(\Delta t) = 5\,\text{ns} \;\rightarrow\; \sigma_x = v \cdot \sigma_t = 250\,\mu\text{m}\ . \tag{18.10}$$

Correspondingly, the spatial resolution on the vertex is (Fig. 18.2)

$$\sin\frac{\alpha}{2} = \frac{\sigma_x}{\sigma_z} \;\rightarrow\; \sigma_z = \frac{\sigma_x}{\sin\frac{\alpha}{2}} = 500\,\mu\text{m}\ . \tag{18.11}$$

7.2 What matters is the transverse packing fraction. A simple geometrical argument (Fig. 18.3) leads to the maximum area that can be covered.

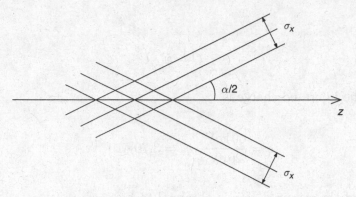

Fig. 18.2. Illustration of the vertex resolution σ_z as derived from the track resolution σ_x.

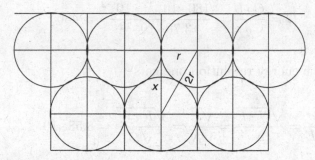

Fig. 18.3. Determination of the maximum packing fraction for a scintillating fibre tracker.

From

$$r^2 + x^2 = (2r)^2 \tag{18.12}$$

one gets $x = \sqrt{3} \cdot r$ and finds the fraction

$$\frac{\pi r^2/2}{r \cdot \sqrt{3}r} = \pi/(2\sqrt{3}) \approx 90.7\% \ , \tag{18.13}$$

$$A_{\text{fibre}} = \pi \cdot 0.5^2 \, \text{cm}^2 = 0.785 \, \text{mm}^2 \ \rightarrow \ N = \frac{A \cdot \pi/(2\sqrt{3})}{A_{\text{fibre}}} = 46\,211 \ . \tag{18.14}$$

7.3

$$\frac{mv^2}{\rho} = evB \ , \tag{18.15}$$

$$\rho = \frac{mv}{eB} = \frac{9.1 \cdot 10^{-31} \, \text{kg} \cdot 0.1 \cdot 10^6 \, \text{m/s}}{1.6 \cdot 10^{-19} \, \text{A s} \cdot B} \le 10^{-5} \, \text{m} \ , \tag{18.16}$$

$$\rightarrow \ B \ge 0.057 \, \text{T} = 570 \, \text{Gauss} \ . \tag{18.17}$$

7.4

$$Q = C \cdot U \ . \tag{18.18}$$

The liberated charge is

$$q = \frac{60 \, \text{keV}}{26 \, \text{eV}} \cdot q_e = 3.70 \cdot 10^{-16} \, \text{A s} \ , \tag{18.19}$$

and the required gain is obtained to be

$$G = \frac{C \cdot U}{q} = \frac{180 \cdot 10^{-12} \cdot 10^{-2}}{3.70 \cdot 10^{-16}} = 4865 \ . \tag{18.20}$$

For the energy resolution one gets

$$\frac{\sigma}{E} = \frac{\sqrt{N \cdot F}}{N} = \frac{\sqrt{F}}{\sqrt{N}} = \frac{\sqrt{F \cdot W}}{\sqrt{E}} = 8.58 \cdot 10^{-3} \ , \tag{18.21}$$

i.e. $(60 \pm 0.5) \, \text{keV}$.

7.5 The horizontal force (tension) F_h does not change along the wire, whereas the vertical one F_v is position-dependent, more precisely, the vertical force at the left boundary is diminished by the weight of the wire to the left of position x:

$$F_v(x) = F_v - \int_{x'=x_1}^{x} \varrho g \sqrt{1 + y'^2(x)} \, dx \ , \quad dm = \varrho \, ds \ ,$$
$$ds = \sqrt{dx^2 + dy^2} = \sqrt{1 + y'^2(x)} \, dx \ .$$

From the above assumptions the slope $y'(x)$ reads

$$y'(x) = -\frac{F_v(x)}{F_h} = -\frac{F_v}{F_h} + \frac{\varrho g}{F_h} \int_{x'=x_1}^{x} \sqrt{1 + y'^2(x')} \, dx' \ ,$$

with $L(x) = \int_{x'=x_1}^{x} \sqrt{1 + y'^2(x')} \, dx'$ being the length of the wire measured from the left boundary. Differentiating this equation leads to a differential equation for y' that can be directly integrated by separation of variables,

$$y''(x) = \frac{\varrho g}{F_\mathrm{h}} \sqrt{1 + y'^2(x)} \; , \qquad \frac{\frac{\mathrm{d}}{\mathrm{d}x} y'(x)}{\sqrt{1 + y'^2(x)}} = \frac{\varrho g}{F_\mathrm{h}} \; .$$

Its solution is

$$\mathrm{arsinh}\, y'(x) = \frac{\varrho g}{F_\mathrm{h}} x + c \; , \quad y'(x) = \sinh\left(\frac{\varrho g}{F_\mathrm{h}} x + c \right) \; ,$$

and a subsequent integration straightforwardly leads to the curve

$$y(x) = \frac{F_\mathrm{h}}{\varrho g} \cosh\left(\frac{\varrho g}{F_\mathrm{h}} x + c \right) + y_0 \; ,$$

where the integration constant c and the horizontal force F_h are to be determined from the geometry and the total length L of the wire. This solution for the form of the wire shows that it is a catenary rather than a parabola. In a symmetric environment and/or for an appropriate choice of the coordinate system the constants can be chosen to be $c = 0$ and $y_0 = -\frac{F_\mathrm{h}}{\varrho g}$. This also guarantees $y(x = 0) = 0$. For the further calculation we set the horizontal tension to $T := F_\mathrm{h}$.

The sag of the wire will be small compared to its length. Therefore, the cosh can be expanded into a series

$$\cosh\left(\frac{\varrho g x}{T} \right) = 1 + \frac{1}{2} \left(\frac{\varrho g x}{T} \right)^2 + \cdots$$

giving

$$y(x) = \mathrm{sag} = -\frac{T}{\varrho g} + \frac{T}{\varrho g} \left[1 + \frac{1}{2} \left(\frac{\varrho g x}{T} \right)^2 + \cdots \right] \; ,$$

$$x = \frac{\ell}{2} \Rightarrow y\left(\frac{\ell}{2} \right) = \frac{1}{2} \frac{\varrho g}{T} \left(\frac{\ell}{2} \right)^2 = \frac{\varrho g \ell^2}{8T} \; ,$$

$$\varrho = \frac{\mathrm{d}m}{\mathrm{d}s} = \pi r_\mathrm{i}^2 \varrho^*$$

mass per unit length, with $\varrho^* = $ density of the wire material,

$$y\left(\frac{\ell}{2} \right) = \frac{1}{8} \pi r_\mathrm{i}^2 \cdot \varrho^* \cdot \frac{g}{T} \ell^2 \; .$$

For a tension of 50 g, corresponding to $T = m_T \cdot g = 0.49\,\mathrm{kg\,m/s^2}$, $\ell = 1\,\mathrm{m}$, $\varrho^*(\mathrm{tungsten}) = 19.3\,\mathrm{g/cm^3} = 19.3 \cdot 10^3\,\mathrm{kg/m^3}$, and $r_\mathrm{i} = 15\,\mathrm{\mu m}$ one gets a sag of $34\,\mathrm{\mu m}$.

Chapter 8

8.1 If ε_1 and ε_2 are the energies of the two photons and ψ the opening angle between them, the two-gamma invariant mass squared is:

$$m_{\gamma\gamma}^2 = (\varepsilon_1 + \varepsilon_2)^2 - (\vec{p_1} + \vec{p_2})^2 = 4\varepsilon_1\varepsilon_2 \sin^2(\psi/2) \ .$$

Using the common error-propagation formula one gets for the relative m^2 uncertainty:

$$\frac{\delta(m^2)}{m^2} = \sqrt{\left[\frac{\delta(\varepsilon_1)}{\varepsilon_1}\right]^2 + \left[\frac{\delta(\varepsilon_2)}{\varepsilon_2}\right]^2 + \cot^2\frac{\psi}{2}\,\delta_\psi^2} \ ,$$

where $\delta(\varepsilon_i)$ and δ_ψ are the energy and angular resolution, respectively. The angular distribution is peaked near ψ_{min} ($\sin(\psi_{min}/2) = m_\eta/E_0$), so that one can take as an estimation a value of $\psi_{min} = 31.8°$. Since

$$\frac{\delta(m^2)}{m^2} = \frac{m_1^2 - m_2^2}{m^2} = \frac{(m_1 + m_2)(m_1 - m_2)}{m^2} = 2\frac{\delta m}{m}$$

or, just by differentiating,

$$\frac{\delta(m^2)}{m^2} = 2m\frac{\delta(m)}{m^2} = 2\frac{\delta m}{m} \ ,$$

one gets

$$\frac{\delta m}{m} = \frac{1}{2}\sqrt{2 \cdot (0.05)^2 + \cot^2(15.9°)(0.05)^2} \approx 9.5\% \ .$$

One can see that in this case the angular accuracy dominates the mass resolution.

8.2 The photon interaction length in matter is $\lambda = (9/7)\,X_0$. Then the probability that the photon passes the aluminium layer without interaction is

$$W_n = \exp\left(-\frac{L}{\lambda}\right) = \exp\left(-\frac{7}{18}\right) = 0.68 \ .$$

In this case the calorimeter response function remains unchanged, namely, it is close to a Gaussian distribution $g(E, E_0)$, where E is the measured energy and E_0 is the incident photon energy.

If the photon produces an e^+e^- pair in the aluminium, at a distance x from the calorimeter, the electron and positron lose part of their energy:

$$\Delta E = 2\varepsilon_{\mathrm{MIP}}x \ ,$$

where $\varepsilon_{\mathrm{MIP}} = (\mathrm{d}E/\mathrm{d}x)_{\mathrm{MIP}}$ is the specific ionisation loss. For Al one has $\varepsilon_{\mathrm{MIP}} = 1.62\,\mathrm{MeV}/(\mathrm{g/cm^2})$ and $X_0 = 24\,\mathrm{g/cm^2}$ resulting in ΔE to vary from $0\,\mathrm{MeV}$ to $39\,\mathrm{MeV}$. As one can see, e.g. for $100\,\mathrm{MeV}$, the measured energy spectrum will consist of a narrow peak ($g(E, E_0)$) comprising 68% of the events, and a wide spectrum ranging from $0.6\,E_0$ to the full energy E_0 containing the other 32%. For a $1\,\mathrm{GeV}$ photon the events with pair production cannot be resolved from the main peak and just result in increasing the width of it.

To estimate the resulting rms one can use a simplified form of the probability density function (PDF):

$$\varphi(E) = pf_1(E) + (1-p)g(E, E_0) \ ,$$

where p is the photon conversion probability in Al and $f_1(E)$ is just an uniform distribution between $E_{\mathrm{min}} = E_0 - \Delta E_{\mathrm{max}}$ and E_0. The modified rms can be calculated as

$$\sigma_{\mathrm{res}}^2 = p\sigma_1^2 + (1-p)\sigma_0^2 + p(1-p)(E_1 - E_0)^2 \ ,$$

where $\sigma_1, E_1, \sigma_0, E_0$ are the rms and average values for $f_1(E)$ and $g(E, E_0)$, respectively. For $f_1(E)$ one has $E_1^{\mathrm{min}} = E_0 - 2\varepsilon_{\mathrm{MIP}}L = E_0 - 39\,\mathrm{MeV}$ and $\sigma_1 = 2 \cdot \varepsilon_{\mathrm{MIP}}L/\sqrt{12} = \varepsilon_{\mathrm{MIP}}L/\sqrt{3} = 11\,\mathrm{MeV}$ (see Chap. 2, Eq. (2.6)). One has to consider that the energy loss ΔE varies uniformly between $0\,\mathrm{MeV}$ and $39\,\mathrm{MeV}$ with an average value of $E_1 = 19.5\,\mathrm{MeV}$, and this value has to be used in the formula for σ_{res}. With these numbers one gets $\sigma_{\mathrm{res}} \approx 11\,\mathrm{MeV}$ for the $100\,\mathrm{MeV}$ photon and $\sigma_{\mathrm{res}} \approx 17\,\mathrm{MeV}$ for the $1\,\mathrm{GeV}$ photon.

8.3 When the pion interacts at depth t, the energy deposited in the calorimeter is a sum of the pion ionisation losses before the interaction (E_{ion}) and the shower energy (E_{sh}) created by the π^0,

$$E_{\mathrm{C}} = E_{\mathrm{ion}} + E_{\mathrm{sh}} \ , \quad E_{\mathrm{ion}} = \frac{\mathrm{d}E}{\mathrm{d}X}tX_0 = E_{\mathrm{cr}}t \ ,$$

$$E_{\mathrm{sh}} = (E_0 - E_{\mathrm{ion}})\int_0^{L-t}\left(\frac{\mathrm{d}E}{\mathrm{d}t}\right)\mathrm{d}t \ ,$$

where E_{cr} is the critical energy. Formula (8.7) describes the electromagnetic shower development. For this estimation one can take

$$\frac{\mathrm{d}E}{\mathrm{d}t} = E_\gamma F(t) \ ,$$

where E_γ is the energy of both photons from the π^0 decay and t is the thickness measured in radiation lengths X_0. Let us assume that the resolution of the calorimeter is $\sigma_E/E = 2\%$ and the condition of the correct particle identification as a pion is

$$\Delta E(t_{\mathrm{c}}) = (E_e - E_{\mathrm{C}}) > 3\sigma_E \ ,$$

where E_e is the energy deposition for an electron in the calorimeter.

For an electron–positron shower of 200–500 MeV in the NaI absorber, the parameter a in Formula (8.7) can be roughly estimated as $a \approx 2$. Then Formula (8.7) simplifies,

$$\frac{1}{E_\gamma}\frac{\mathrm{d}E}{\mathrm{d}t} = \frac{1}{4}\left(\frac{t}{2}\right)^2 \exp(-t/2) \ ,$$

and can be easily integrated for any t. To find t_{c} one has to tabulate the function $\Delta E(t_{\mathrm{c}})$ numerically. The calculated dependencies of E_{ion}, E_{sh} and E_{C} on t are shown in Fig. 18.4. Since $\sigma_E = 2\% \cdot E = 10$ MeV for a 500 MeV shower, and $E_e - E_{\mathrm{C}} > 3\sigma_E$ is required, one has to ask for $E_{\mathrm{C}} < 470$ MeV. Reading this limit from the

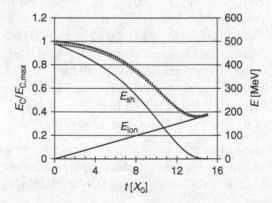

Fig. 18.4.   The calculated dependencies of E_{ion}, E_{sh} and E_{C} (lower line with triangle symbols) on t. The upper line with diamond symbols shows the ratio of the energy deposition in the calorimeter to its maximum value (without leakage). Even when the charge exchange occurs in the very beginning of the calorimeter, some part of the energy leaks through the rear side.

figure leads to $t_c \approx 4$ which corresponds to a thickness of $38\,\text{g/cm}^2$. Working out the interaction probability W from the interaction length $\lambda_{\text{int}} = 151\,\text{g/cm}^2$ and with the knowledge of the charge-exchange probability of 0.5, one obtains the pion misidentification probability P to be

$$P_{\text{M}} = 0.5\,W(t < t_c) = 0.5\,[1 - \exp(-t_c/\lambda_{\text{int}})] \approx 0.12 \ .$$

The probability of misidentification of the electron as pion is much lower.

Chapter 9

9.1 Convert momenta to total energy:

$$E = c\sqrt{p^2 + m_0^2 c^2} = \begin{cases} 3.0032\,\text{GeV} & \text{for} \quad 3\,\text{GeV}/c \\ 4.0024\,\text{GeV} & \text{for} \quad 4\,\text{GeV}/c \\ 5.0019\,\text{GeV} & \text{for} \quad 5\,\text{GeV}/c \end{cases} ,$$

$$m_0 = 139.57\,\text{MeV}/c^2 \ ,$$

$$\gamma = \frac{E}{m_0 c^2} = \begin{cases} 21.518 & \text{for} \quad 3\,\text{GeV}/c \\ 28.677 & \text{for} \quad 4\,\text{GeV}/c \\ 35.838 & \text{for} \quad 5\,\text{GeV}/c \end{cases} ,$$

$$\beta = \sqrt{1 - \frac{1}{\gamma^2}} = \begin{cases} 0.9989195 & \text{for} \quad 3\,\text{GeV}/c \\ 0.9993918 & \text{for} \quad 4\,\text{GeV}/c \\ 0.9996106 & \text{for} \quad 5\,\text{GeV}/c \end{cases} ,$$

$$\cos\theta_c = \frac{1}{n\beta} \quad \Rightarrow \quad \theta_c = \arccos\left(\frac{1}{n\beta}\right) \ ;$$

	$3\,\text{GeV}/c$	$4\,\text{GeV}/c$	$5\,\text{GeV}/c$
Lucite	47.8°	47.8°	47.8°
silica aerogel	12.40°–21.37°	12.52°–21.44°	12.58°–21.47°
Pyrex	47.08°	47.10°	47.11°
lead glass	58.57°	58.59°	58.60°

9.2

$$m_K = 493.677 \, \text{MeV}/c^2 \; , \quad n_{\text{water}} = 1.33 \; ,$$

$$E_K = c\sqrt{p^2 + m_K^2 c^2} = 2.2547 \, \text{GeV} \; ;$$

$$\beta = \sqrt{1 - \frac{1}{\gamma^2}} = 0.9757 \quad \Rightarrow \quad \theta_C = 39.59° \; ,$$

$$\frac{\mathrm{d}E}{\mathrm{d}L} = \frac{\mathrm{d}N}{\mathrm{d}L} \cdot h\nu = \frac{\mathrm{d}N}{\mathrm{d}L} \frac{hc}{\lambda} = 2\pi\alpha z^2 hc \int_{\lambda_1}^{\lambda_2} \left(1 - \frac{1}{\beta^2 n^2}\right) \frac{\mathrm{d}\lambda}{\lambda^3} \; ;$$

assume $n \neq f(\lambda)$, then

$$\frac{\mathrm{d}E}{\mathrm{d}L} = 2\pi\alpha z^2 hc \left(1 - \frac{1}{\beta^2 n^2}\right) \frac{1}{2} \left(\frac{1}{\lambda_1^2} - \frac{1}{\lambda_2^2}\right)$$

$$= \pi\alpha z^2 hc \left(1 - \frac{1}{\beta^2 n^2}\right) \left(\frac{1}{\lambda_1^2} - \frac{1}{\lambda_2^2}\right) \; ;$$

$$h = 2\pi\hbar = 41.36 \cdot 10^{-22} \, \text{MeV s} \; ,$$

$$c = 3 \cdot 10^{17} \, \text{nm/s} \; , \quad n = 1.33 \quad \text{for water} \; ,$$

$$\lambda_1 = 400 \, \text{nm}, \; \lambda_2 = 700 \, \text{nm}$$

$$\rightarrow \frac{\mathrm{d}E}{\mathrm{d}L} = 0.49 \, \text{keV/cm} \; .$$

9.3

$$E_p = c\sqrt{p^2 + m^2 \cdot c^2} = 5.087 \, \text{GeV} \; .$$

If water is considered as Cherenkov medium, one has

$$\beta = \sqrt{1 - \frac{1}{\gamma^2}} = 0.9805 \quad \Rightarrow \quad \theta_C = 40.1° \quad \text{in water} \; ,$$

$$N = 203.2 \quad \text{photons per cm} \; ,$$

$$n = 12 \text{ photoelectrons} = N \cdot x \cdot \eta_{\text{PM}} \cdot \eta_{\text{Geom}} \cdot \eta_{\text{Transfer}} \; ,$$

$$x = \frac{n}{N \cdot \eta_{\text{PM}} \cdot \eta_{\text{Geom}} \cdot \eta_{\text{Transfer}}} = 1.48 \, \text{cm} \; .$$

A counter of $\approx 1.5 \, \text{cm}$ thickness is required for the assumed collection/conversion efficiencies.

9.4

$$n_{\text{Lucite}} = 1.49 \ ,$$

threshold energy for electrons

$$\beta > \frac{1}{n} = 0.67 \ \Rightarrow \ \gamma = \frac{1}{\sqrt{1 - \beta^2}} = 1.35 \ \Rightarrow \ E = 689 \,\text{keV} \ ,$$

$$\frac{\mathrm{d}^2 N}{\mathrm{d}x \, \mathrm{d}T} = \frac{1}{2} K \cdot z^2 \frac{Z}{A} \frac{1}{\beta^2} \frac{1}{T^2} \ ;$$

T – kinetic energy of the δ rays,

$$K = 4\pi N_{\text{A}} \ [\text{mol}^{-1}]/\text{g} \ r_e^2 m_e c^2 = 0.307 \,\text{MeV}/(\text{g/cm}^2) \ ,$$

$$\frac{\mathrm{d}N}{\mathrm{d}T} = \frac{1}{2} \cdot 0.307 \frac{\text{MeV}}{\text{g/cm}^2} \cdot \frac{6}{12} \frac{1}{\beta^2} \frac{1}{T^2} x$$

$$= 0.171 \frac{\text{MeV}}{\text{g/cm}^2} \cdot \frac{1}{T^2} x \ \to \ N = x \cdot \int_T^\infty 0.171 \frac{\text{MeV}}{\text{g/cm}^2} \cdot \frac{1}{T'^2} \, \mathrm{d}T' \ ,$$

$$N = 0.171 \frac{\text{MeV}}{\text{g/cm}^2} \cdot x \frac{1}{T} \ ,$$

$T_{\text{threshold}} = 689 \,\text{keV} - 511 \,\text{keV} = 178 \,\text{keV}$.
This gives $N = 9.6 \ \delta$ rays above threshold. These electrons are distributed according to a $1/T^2$ spectrum. The maximum transferable energy to electrons by $3 \,\text{GeV}/c$ protons is

$$E_{\text{kin}}^{\text{max}} = \frac{E^2}{E + m_p^2/2m_e} = 3.56 \,\text{MeV} \ .$$

However, the $1/T^2$ dependence of the δ rays is strongly modified close to the kinematic limit (the spectrum gets steeper). The 9.6 δ rays should be taken from a $1/T^2$ spectrum by a suitable Monte Carlo. Here we argue that the chance to find a δ ray with more than $1 \,\text{MeV}$ is only

$$P = \left(\frac{178}{1000} \right)^2 \approx 3\% \ .$$

Therefore, we average the energies over the range 178 keV to 1 MeV,

$$\langle T \rangle = \frac{\int_{178\,\text{keV}}^{1\,\text{MeV}} T \cdot \frac{1}{T^2}\,dT}{\int_{178\,\text{keV}}^{1\,\text{MeV}} \frac{1}{T^2}\,dT} = 372\,\text{keV} \ ,$$

$$\beta_{372\,\text{keV}} = \sqrt{1 - \frac{1}{\gamma^2}} = 0.815 \ , \quad \gamma = 1.73 \ ,$$

$$\cos\Theta = \frac{1}{n\beta} = 0.82 \ \Rightarrow \ \Theta = 34.6° \ ,$$

$$N_{\text{Photons}} = 9.6 \cdot 490 \cdot \sin^2\Theta \cdot 0.08 = 121 \text{ photons} \ ,$$

where $x = 0.08$ cm is the range of the δ rays of 372 keV (see Chap. 1).

$$n = N_{\text{Photons}} \cdot \eta_{\text{PM}} \cdot \eta_{\text{Geom}} \cdot \eta_{\text{Transfer}} = 0.97$$

if all efficiencies are assumed to be 20%.

$$\Rightarrow \ \epsilon = 1 - e^{-n} = 62\%$$

is the efficiency for δ rays.

9.5 Imaging Air Cherenkov telescopes measure γ-ray cascades initiated in the atmosphere. Because of the large cross section of photons these cascades are initiated at large altitudes, where the refractive index is smaller than the value given at STP. Does the observed angle of 1° allow to determine the typical altitude where these showers develop?

Density variation in the atmosphere

$$\varrho = \varrho_0 \cdot e^{-h/h_0} \ ,$$

where $h_0 = 7.9$ km for an isothermal atmosphere.

The index of refraction n varies with the permittivity ε like $n = \sqrt{\varepsilon}$. Since $\varepsilon - 1 \propto \varrho$, one has

$$n^2 = \varepsilon - 1 + 1 \propto \varrho + 1 \ \rightarrow \ \frac{\varrho(h)}{\varrho_0} = \frac{n^2(h) - 1}{n_0^2 - 1} \ .$$

Because of $\Theta = 1° \rightarrow n(h) = 1.000\,152$, if $\beta = 1$ is assumed.

$$\Rightarrow \ \frac{\varrho_0}{\varrho(h)} = 1.94 \ \Rightarrow \ h = h_0 \ln\frac{n_0^2 - 1}{n^2(h) - 1} \approx 5235\,\text{m} \ .$$

9.6 Since

$$\frac{\mathrm{d}E}{\mathrm{d}x} = a\frac{mz^2}{E_{\mathrm{kin}}} \cdot \ln\left(b\frac{E_{\mathrm{kin}}}{m}\right)$$

a measurement of

$$\frac{\mathrm{d}E}{\mathrm{d}x} \cdot E_{\mathrm{kin}}$$

identifies $m \cdot z^2$, since the logarithmic term is usually comparable for non-relativistic particles, and the Lorentz factor is always close to unity. Therefore a measurement of $(\mathrm{d}E/\mathrm{d}x)E_{\mathrm{kin}}$ provides a technique for particle identification.

Let us first assume that muons and pions of 10 MeV kinetic energy can be treated non-relativistically, and that we can approximate the Bethe–Bloch formula in the following way:

$$\frac{\mathrm{d}E}{\mathrm{d}x} = K \cdot z^2 \frac{Z}{A}\frac{1}{\beta^2} \cdot \ln\left(\frac{2m_e c^2 \beta^2 \gamma^2}{I}\right)$$

with $K = 0.307\,\mathrm{MeV}/(\mathrm{g/cm^2})$ and $\beta^2 = (2 \cdot E_{\mathrm{kin}})/(m \cdot c^2)$ in the classical approximation. (The correction terms characterising the saturation effect (Fermi plateau) should be rather small in this kinematic domain.)

For singly charged particles one has

$$\frac{\mathrm{d}E}{\mathrm{d}x} = K\frac{Z}{A}\frac{mc^2}{2E_{\mathrm{kin}}} \cdot \ln\left(\frac{2m_e c^2}{I}\frac{2E_{\mathrm{kin}}}{mc^2}\gamma^2\right)$$

$$= 0.076\,75\,\frac{\mathrm{MeV}}{\mathrm{g/cm^2}}\frac{mc^2}{E_{\mathrm{kin}}} \cdot \ln\left(14\,600 \cdot \frac{E_{\mathrm{kin}}}{mc^2}\gamma^2\right)$$

leading to $6.027\,\mathrm{MeV}/(\mathrm{g/cm^2})$ for muons and $7.593\,\mathrm{MeV}/(\mathrm{g/cm^2})$ for pions. Since $\Delta x = 300\,\mathrm{\mu m} \cdot 2.33\,\mathrm{g/cm^3} = 6.99 \cdot 10^{-2}\,\mathrm{g/cm^2}$, one gets $\Delta E(\mathrm{muons}) = 0.421\,\mathrm{MeV}$ and $\Delta E(\mathrm{pions}) = 0.531\,\mathrm{MeV}$.

For muons one would obtain $\Delta E \cdot E_{\mathrm{kin}} = 4.21\,\mathrm{MeV^2}$ and for pions, correspondingly, $\Delta E \cdot E_{\mathrm{kin}} = 5.31\,\mathrm{MeV^2}$.

Neither of these results agrees with the measurement. Redoing the calculation and dropping the assumption that muons and pions can be treated in a non-relativistic fashion, one gets for a consideration of the non-approximated Bethe–Bloch formula and a full relativistic treatment for muons, $\Delta E \cdot E_{\mathrm{kin}} = 4.6\,\mathrm{MeV^2}$, and for pions, correspondingly, $\Delta E \cdot E_{\mathrm{kin}} = 5.7\,\mathrm{MeV^2}$. The difference to the earlier result mainly comes from the correct relativistic treatment. Therefore the above measurement identified a pion.

For the separation of the beryllium isotopes it is justified to use the non-relativistic approach

$$\Delta E = 0.076\,75 \, \frac{\text{MeV}}{\text{g/cm}^2} \, z^2 \frac{mc^2}{E_{\text{kin}}} \cdot \ln\left(14\,600 \cdot \frac{E_{\text{kin}}}{mc^2}\gamma^2\right) \cdot \Delta x \ .$$

This leads to $\Delta E \cdot E_{\text{kin}} = 3056 \, \text{MeV}^2$ for ^{7}Be and $\Delta E \cdot E_{\text{kin}} = 3744 \, \text{MeV}^2$ for ^{9}Be.

The full non-approximated consideration gives results which differ only by about 1%.

Therefore the measured isotope is ^{9}Be, and ^{8}Be did not show up because it is highly unstable and disintegrates immediately into two α particles.

Chapter 10

10.1 The neutrino flux ϕ_ν is given by the number of fusion processes $4p \to \,^4\text{He} + 2e^+ + 2\nu_e$ times 2 neutrinos per reaction chain:

$$\phi_\nu = \frac{\text{solar constant}}{\text{energy gain per reaction chain}} \cdot 2$$

$$\approx \frac{1400 \, \text{W/m}^2}{26.1 \, \text{MeV} \cdot 1.6 \cdot 10^{-13} \, \text{J/MeV}} \cdot 2 \approx 6.7 \cdot 10^{10} \, \text{cm}^{-2}\,\text{s}^{-1} \ .$$

10.2

$$(q_{\nu_\alpha} + q_{e-})^2 = (m_\alpha + m_{\nu_e})^2 \ , \quad \alpha = \mu, \tau \ ;$$

assuming m_{ν_α} to be small ($\ll m_e, m_\mu, m_\tau$) one gets

$$2E_{\nu_\alpha} m_e + m_e^2 = m_\alpha^2 \quad \Rightarrow \quad E_{\nu_\alpha} = \frac{m_\alpha^2 - m_e^2}{2m_e} \quad \Rightarrow$$

$$\alpha = \mu : E_{\nu_\mu} = 10.92 \, \text{GeV} \ , \quad \alpha = \tau : E_{\nu_\tau} = 3.09 \, \text{TeV} \ ;$$

since solar neutrinos cannot convert into such high-energy neutrinos, the proposed reactions cannot be induced.

10.3 The interaction rate is

$$R = \sigma_{\text{N}} N_{\text{A}} [\text{mol}^{-1}]/\text{g} \, d \, A \, \phi_\nu \ ,$$

where σ_{N} is the cross section per nucleon, $N_{\text{A}} = 6.022 \times 10^{23} \, \text{mol}^{-1}$ is the Avogadro constant, d the area density of the target, A the target area and ϕ_ν the solar neutrino flux. With $d \approx 15 \, \text{g}\,\text{cm}^{-2}$,

$A = 180 \times 30\,\mathrm{cm}^2$, $\phi_\nu \approx 7 \cdot 10^{10}\,\mathrm{cm}^{-2}\,\mathrm{s}^{-1}$, and $\sigma_N = 10^{-45}\,\mathrm{cm}^2$ one gets $R = 3.41 \cdot 10^{-6}\,\mathrm{s}^{-1} = 107\,\mathrm{a}^{-1}$. A typical energy of solar neutrinos is $100\,\mathrm{keV}$, i.e., $50\,\mathrm{keV}$ are transferred to the electron. Consequently, the total annual energy transfer to the electrons is

$$\Delta E = 107 \cdot 50\,\mathrm{keV} = 5.35\,\mathrm{MeV} = 0.86 \cdot 10^{-12}\,\mathrm{J}\ .$$

With the numbers used so far the mass of the human is $81\,\mathrm{kg}$. Therefore, the equivalent annual dose comes out to be

$$H_\nu = \frac{\Delta E}{m}\,w_R = 1.06 \cdot 10^{-14}\,\mathrm{Sv}\ ,$$

actually independent of the assumed human mass. The contribution of solar neutrinos to the normal natural dose rate is negligible, since

$$H = \frac{H_\nu}{H_0} = 5.3 \cdot 10^{-12}\ .$$

10.4 Four-momentum conservation yields

$$q_\pi^2 = (q_\mu + q_\nu)^2 = m_\pi^2\ . \tag{18.22}$$

In the rest frame of the pion the muon and neutrino are emitted in opposite directions, $\vec{p}_\mu = -\vec{p}_{\nu_\mu}$,

$$\left(\frac{E_\mu + E_\nu}{\vec{p}_\mu + \vec{p}_{\nu_\mu}}\right)^2 = (E_\mu + E_\nu)^2 = m_\pi^2\ . \tag{18.23}$$

Neglecting a possible non-zero neutrino mass for this consideration, one has

$$E_\nu = p_{\nu_\mu}$$

with the result

$$E_\mu + p_\mu = m_\pi\ .$$

Rearranging this equation and squaring it gives

$$E_\mu^2 + m_\pi^2 - 2E_\mu m_\pi = p_\mu^2\ ,$$
$$2E_\mu m_\pi = m_\pi^2 + m_\mu^2\ ,$$
$$E_\mu = \frac{m_\pi^2 + m_\mu^2}{2m_\pi}\ . \tag{18.24}$$

For $m_\mu = 105.658\,369\,\mathrm{MeV}$ and $m_{\pi\pm} = 139.570\,18\,\mathrm{MeV}$ one gets $E_\mu^{\mathrm{kin}} = E_\mu - m_\mu = 4.09\,\mathrm{MeV}$. For the two-body decay of the kaon,

$K^+ \rightarrow \mu^+ + \nu_\mu$, (18.24) gives $E_\mu^{\text{kin}} = E_\mu - m_\mu = 152.49\,\text{MeV}$ ($m_{K^\pm} = 493.677\,\text{MeV}$).

The neutrino energies are then just given by

$$E_\nu = m_\pi - E_\mu = 29.82\,\text{MeV}$$

for pion decay and

$$E_\nu = m_K - E_\mu = 235.53\,\text{MeV}$$

for kaon decay.

10.5 The expected difference of arrival times Δt of two neutrinos with velocities v_1 and v_2 emitted at the same time from the supernova is

$$\Delta t = \frac{r}{v_1} - \frac{r}{v_2} = \frac{r}{c}\left(\frac{1}{\beta_1} - \frac{1}{\beta_2}\right) = \frac{r}{c}\frac{\beta_2 - \beta_1}{\beta_1\,\beta_2}\ . \tag{18.25}$$

If the recorded electron neutrinos had a rest mass m_0, their energy would be

$$E = mc^2 = \gamma m_0 c^2 = \frac{m_0 c^2}{\sqrt{1 - \beta^2}}\ , \tag{18.26}$$

and their velocity

$$\beta = \left(1 - \frac{m_0^2 c^4}{E^2}\right)^{1/2} \approx 1 - \frac{1}{2}\frac{m_0^2 c^4}{E^2}\ , \tag{18.27}$$

since one can safely assume that $m_0 c^2 \ll E$. This means that the neutrino velocities are very close to the velocity of light. Obviously, the arrival-time difference Δt depends on the velocity difference of the neutrinos. Using (18.25) and (18.27), one gets

$$\Delta t \approx \frac{r}{c}\frac{\frac{1}{2}\frac{m_0^2 c^4}{E_1^2} - \frac{1}{2}\frac{m_0^2 c^4}{E_2^2}}{\beta_1\beta_2} \approx \frac{1}{2}\,m_0^2 c^4\,\frac{r}{c}\frac{E_2^2 - E_1^2}{E_1^2\,E_2^2}\ . \tag{18.28}$$

The experimentally measured arrival-time differences and individual neutrino energies allow in principle to work out the electron-neutrino rest mass

$$m_0 = \left(\frac{2\Delta t}{r\,c^3}\frac{E_1^2\,E_2^2}{E_2^2 - E_1^2}\right)^{1/2}\ . \tag{18.29}$$

10.6 The interaction cross section of high-energy neutrinos was measured at accelerators to be

$$\sigma(\nu_\mu N) = 6.7 \cdot 10^{-39} \, E_\nu \, [\text{GeV}] \, \text{cm}^2/\text{nucleon} \; . \qquad (18.30)$$

For 100 TeV neutrinos one would arrive at a cross section of $6.7 \cdot 10^{-34} \, \text{cm}^2/\text{nucleon}$. For a target thickness of 1 km an interaction probability W per neutrino of

$$W = N_A [\text{mol}^{-1}]/\text{g} \, \sigma \, d \, \varrho = 4 \cdot 10^{-5} \qquad (18.31)$$

is obtained ($d = 1\,\text{km} = 10^5\,\text{cm}$, $\varrho(\text{ice}) \approx 1\,\text{g/cm}^3$).

The total interaction rate R is obtained from the integral neutrino flux Φ_ν, the interaction probability W, the effective collection area $A_{\text{eff}} = 1\,\text{km}^2$ and a measurement time t. This leads to an event rate of

$$R = \Phi_\nu \, W \, A_{\text{eff}} \qquad (18.32)$$

corresponding to 250 events per year. If a target volume of $1\,\text{km}^3$ is fully instrumented, the effective collection area will be even larger.

Chapter 11

11.1

$$\frac{\mathrm{d}E}{\mathrm{d}x} = a + bE$$

is a good approximation for the energy loss, where a represents the ionisation loss and b stands for the losses due to pair production, bremsstrahlung and photonuclear interactions. For 1 TeV muons one finds [5]

$$a \approx 2.5 \, \text{MeV}/(\text{g/cm}^2) \; ,$$
$$b \approx 7.5 \cdot 10^{-6} \, (\text{g/cm}^2)^{-1} \; .$$

For 3 m of iron ($\varrho \cdot x = 2280 \, \text{g/cm}^2$) one gets

$$\Delta E = 3\,\text{m} \, \frac{\mathrm{d}E}{\mathrm{d}x} = 22.8 \, \text{GeV} \; .$$

Because of energy-loss fluctuations, one gets a radiative tail in the momentum distributions of an originally monoenergetic muon beam as sketched in Fig. 18.5 [5].

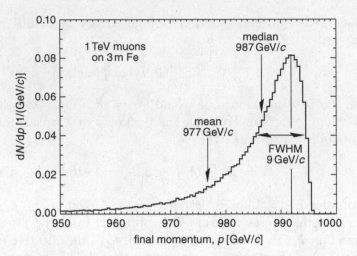

Fig. 18.5. The momentum distribution of $1\,\mathrm{TeV}/c$ muons after traversing $3\,\mathrm{m}$ of iron [6].

11.2 The production probability can be determined along the lines of Eq. (1.25) and the references given in that context. For argon ($Z = 18$, $A = 36$, $\varrho = 1.782 \cdot 10^{-3}\,\mathrm{g/cm^3}$) the column density is

$$d = 0.5346\,\mathrm{g/cm^2} \ .$$

Bending radii from $5\,\mathrm{cm}$ to $20\,\mathrm{cm}$ correspond to momenta

$$p\,[\mathrm{GeV}/c] = 0.3\,B\,[\mathrm{T}] \cdot R\,[\mathrm{m}]$$

of $30\,\mathrm{MeV}/c$ to $120\,\mathrm{MeV}/c$. The δ-electron differential energy spectrum for high-momentum muons can be approximated by

$$\phi(\varepsilon)\,\mathrm{d}\varepsilon = 2C m_e c^2 \frac{\mathrm{d}\varepsilon}{\varepsilon^2} \ ,$$

where ε is the energy of the δ electron and m_e the electron rest mass [7, 8]. With $C = 0.150\,\frac{Z}{A}\,\mathrm{g^{-1}\,cm^2}$ one gets

$$P = \int_{30\,\mathrm{MeV}}^{120\,\mathrm{MeV}} \phi(\varepsilon)\,\mathrm{d}\varepsilon = 0.150 \cdot \frac{Z}{A} \left(\frac{1}{30} - \frac{1}{120} \right) \frac{\mathrm{cm^2}}{\mathrm{g}}$$

$$= 1.875 \cdot 10^{-3}\,\frac{\mathrm{cm^2}}{\mathrm{g}} \ ,$$

$$P \cdot d = 10^{-3} = 0.1\% \quad \text{per track} \ .$$

For 100 tracks one has a 10% probability that one of the charged particles will create a δ electron with the properties in question.

11.3 (a)

$$\Theta_\rho = \Theta_\varphi \quad \text{for} \quad n = \frac{1}{2} \quad \Rightarrow \quad \Theta = \pi\sqrt{2} = 255.6° \ ,$$

$$B(\rho) = B(\rho_0) \left(\frac{\rho_0}{\rho}\right)^{1/2} \ .$$

(b)

$$\frac{\mathrm{d}E}{\mathrm{d}x}(10\,\mathrm{keV}) = 27\,\frac{\mathrm{keV}}{\mathrm{cm}} \cdot \pi\sqrt{2} \cdot \rho_0 \cdot \frac{p}{p_\mathrm{atm}}$$

$$= 27 \cdot 10^3 \cdot \pi\sqrt{2} \cdot 50 \cdot \frac{10^{-3}}{760}\,\mathrm{eV} = 7.9\,\mathrm{eV} \ ,$$

i.e., just about one or perhaps even zero ionisation processes will occur.

11.4

$$B_y \cdot \ell \propto x \ , \quad B_x \cdot \ell \propto y \ ,$$
$$\ell = \mathrm{const} \ \Rightarrow \ B_y = g \cdot x \ , \quad B_x = g \cdot y \ .$$

This leads to a magnetic potential of

$$V = -g \cdot x \cdot y$$

with

$$g = \frac{\partial B_y}{\partial x} = \frac{\partial B_x}{\partial y} \ ,$$

where g is called the gradient of the quadrupole;

$$-\,\mathrm{grad}\,V = -\left(\frac{\partial V}{\partial x}\vec{e}_x + \frac{\partial V}{\partial y}\vec{e}_y\right) = \underbrace{g \cdot y\,\vec{e}_x}_{B_x} + \underbrace{g \cdot x\,\vec{e}_y}_{B_y} \ .$$

Since the surface of the yoke must have constant potential, one has

$$V = V_0 = -g \cdot x \cdot y \ , \quad \text{i.e.} \quad x \propto \frac{1}{y} \ ,$$

which means that the surface of the yoke must be hyperbolic.

Chapter 12

12.1

$$\tau(T^*) = \frac{1}{12}\tau(T) \; , \quad \tau_0 \, e^{E_a/kT^*} = \frac{1}{12}\tau_0 \, e^{E_a/kT} \; ;$$

solve for $\frac{kT^*}{kT} \rightarrow$

$$\frac{kT^*}{kT} = \frac{1}{1 - \frac{kT}{E_a}\ln 12} = 1.18 \; .$$

$\rightarrow$ The temperature has to be increased by 18%.

12.2

$$\frac{\Delta U^{-*}}{\Delta U^-} = \frac{-\frac{N\,e}{C\ln[r_a/(1.1\,r_i)]}\ln[r_0/(1.1\,r_i)]}{-\frac{N\,e}{C\ln(r_a/r_i)}\ln(r_0/r_i)} = \frac{1 - \frac{\ln 1.1}{\ln(r_0/r_i)}}{1 - \frac{\ln 1.1}{\ln(r_a/r_i)}} \approx 0.88 \; .$$

$\rightarrow$ The gain is decreased by 12%.

Chapter 13

13.1 Assume Poisson statistics:

$$\text{efficiency} = 50\% \; \Rightarrow \; e^{-m} = 0.5 \; \Rightarrow \; m = 0.6931 \; ,$$

$$N = \frac{m}{\eta_{\text{PM}} \cdot \eta_{\text{Geom}} \cdot \eta_{\text{Transfer}}} = 43.32 \; ,$$

$$\frac{dN}{dx} = 490 \sin^2 \theta_C \, \text{cm}^{-1} \cdot 150 \, \text{cm} = 43.32 \; ,$$

$$\sin^2 \theta_C = 5.89 \cdot 10^{-4} \; ,$$

$$\theta_C = 1.39° \; ;$$

$$\cos \theta_C = \frac{1}{n\beta} \Rightarrow \beta = \frac{1}{n \, \cos \theta_C} \; .$$

Index of refraction of CO_2 at 3 atm:

$$n = 1.00123 \Rightarrow \beta = 0.99907$$

$$\Rightarrow \gamma = 23.14$$

$$\Rightarrow E_\pi = 3.23 \, \text{GeV} \; .$$

13.2 By grouping the photons into pairs one can work out the invariant mass of the different $\gamma\gamma$ combinations,

$$m^2 = (q_{\gamma_i} + q_{\gamma_j})^2 = 2 \cdot E_{\gamma_i} \cdot E_{\gamma_j} (1 - \cos\theta) \ .$$

One finds for $m(\gamma_1, \gamma_2) = 135\,\text{MeV}$ and for $m(\gamma_3, \gamma_4) = 548\,\text{MeV}$, i.e., the four photons came from a π^0 and an η.

13.3

$$\Delta t = \frac{L \cdot c}{2 \cdot p^2} (m_2^2 - m_1^2) = \frac{L \cdot c}{2 \cdot p^2} (m_2 - m_1)(m_2 + m_1) \ ;$$

if $m_1 \approx m_2 \rightarrow$

$$\Delta t = \frac{L \cdot c}{2 \cdot p^2} \cdot 2m \cdot \Delta m \ ;$$

since

$$p^2 = \gamma^2 \cdot m^2 \cdot \beta^2 \cdot c^2$$

one gets

$$\Delta t = \frac{L \cdot c}{\gamma^2 \cdot \beta^2 \cdot c^2} \cdot \frac{\Delta m}{m} \ ;$$

$\rightarrow$

$$\frac{\Delta m}{m} = \gamma^2 \cdot \frac{c^2 \cdot \beta^2}{L \cdot c} \cdot \Delta t \ .$$

For $\beta \approx 1$ one has

$$\frac{\Delta m}{m} = \gamma^2 \cdot \frac{c}{L} \cdot \Delta t = \gamma^2 \frac{\Delta t}{t} \ . \tag{18.33}$$

For a momentum of $1\,\text{GeV}/c$ the flight-time difference for muons and pions is

$$\Delta t = \frac{L}{c} \cdot \left(\frac{1}{\beta_1} - \frac{1}{\beta_2} \right) \ .$$

From $\gamma\beta mc^2 = 1\,\text{GeV}$ one gets $\gamma_\mu \cdot \beta_\mu = 9.46$, $\gamma_\pi \cdot \beta_\pi = 7.16$ corresponding to $\beta_\mu = 0.989$, $\gamma_\mu = 9.57$ and $\beta_\pi = 0.981$, $\gamma_\pi = 7.30$. With these values the flight-time difference becomes $\Delta t = 27.5\,\text{ps}$. The absolute flight times for pions and muons are not very different (this is the problem!), namely $t_\mu = 3.37\,\text{ns}$ and $t_\pi = 3.40\,\text{ns}$, resulting in

$$\frac{\Delta t}{t} \approx 8.12 \cdot 10^{-3} \ .$$

This excellent value is, however, spoiled by the factor γ^2 in Eq. (18.33), leading to a relatively poor mass resolution.

13.4 The quantity E_{CM}^2 is equal to the kinematical invariant $s = (p_+ + p_-)^2$, where p_+ and p_- are the positron and electron four-momenta, respectively. Then this value can be expressed as

$$s = (p_+ + p_-)^2 = 2m_e^2 + 2(E_+ E_- - \vec{p}_+ \vec{p}_-) \ .$$

Neglecting the electron mass and the angle between the beams, 22 mrad, one gets

$$E_{\text{CM}} = 2\sqrt{E_+ E_-} = 10.58 \, \text{GeV} \ .$$

Considering the finite crossing angle of 22 mrad results in a decrease of the centre-of-mass energy of 200 keV only!

13.5 Since a particle energy loss is recovered at every revolution when it passes the RF cavities, let us calculate first the probability of the emission of a bremsstrahlung photon carrying away more than 1% of the particle's energy. The number of photons which are emitted along the path ΔX in the energy interval $[\varepsilon, \varepsilon + \text{d}\varepsilon]$ to first approximation is (see Rossi's book [7])

$$\text{d}n = \frac{\Delta X}{X_0} \frac{\text{d}\varepsilon}{\varepsilon} \ .$$

Integration of this expression from ε_0 to the beam energy, E_0, gives the required probability

$$w_1 = \frac{\Delta X}{X_0} \ln \frac{E_0}{\varepsilon_0} \ .$$

The density of the residual gas (assuming air, having the density of $1.3 \cdot 10^{-3} \, \text{g/cm}^3$ at $100 \, \text{kPa}$) is $1.3 \cdot 10^{-15} \, \text{g/cm}^3$ which results in $w_1 \approx 0.5 \cdot 10^{-10}$, which means that after an average of $1/w_1 \approx 2 \cdot 10^{10}$ revolutions a bremsstrahlung process with an energy transfer of more than 1% of the beam energy occurs. This corresponds to a beam lifetime of $t_{\text{b}} \approx 2 \cdot 10^5 \, \text{s}$. In a real experiment with intensively colliding beams, the beam lifetime is much shorter and it is determined by other effects, such as beam–beam interactions, the Touschek effect,[‡] nuclear interactions of the electrons with the residual gas, interactions with the ambient blackbody photons of room temperature and so on.

[‡] An effect observed in electron–positron storage rings in which the maximum particle concentration in the counterrotating electron bunches is limited at low energies by the loss of electrons in Møller scattering [9].

13.6 The differential cross section for this process is expressed as (see, for example, [5] (2006), p. 325)

$$\frac{\mathrm{d}\sigma}{\mathrm{d}\varOmega} = \frac{\alpha^2}{4s}(1 + \cos^2\theta) .$$

Integration of this formula over the mentioned solid angle and converting the 'natural units' of the cross-section formula into numerical values by using $\hbar c = 0.1973\,\mathrm{GeV\,fm}$ results in

$$\sigma_{\mathrm{det}} = \frac{\pi\alpha^2}{s}\left(z_0 + \frac{z_0^3}{3}\right) = \frac{65.1\,\mathrm{nb}}{s\,[\mathrm{GeV^2}]}\left(z_0 + \frac{z_0^3}{3}\right) = \frac{70.5\,\mathrm{nb}}{s\,[\mathrm{GeV^2}]} ,$$

where $z_0 = \cos\theta_0$. Thus, one gets $\sigma_{\mathrm{det}} = 0.63\,\mathrm{nb}$ at $E_{\mathrm{CM}} = 10.58\,\mathrm{GeV}$ corresponding to a muon event rate of $6.3\,\mathrm{Hz}$.

Chapter 14

14.1 When the overall resolution of a system is determined by the convolution of multiple Gaussian distributions, the individual resolutions add in quadrature:

$$\Delta t = \sqrt{\Delta t_1^2 + \Delta t_2^2} = \sqrt{100^2 + 50^2}\,\mathrm{ps} = 112\,\mathrm{ps} .$$

14.2 (a)

$$Q_{\mathrm{n}} = \sqrt{Q_{\mathrm{ni}}^2 + Q_{\mathrm{nv}}^2} = \sqrt{120^2 + 160^2}\,\mathrm{eV} = 200\,\mathrm{eV} .$$

(b)

$$Q_{\mathrm{n}} = \sqrt{Q_{\mathrm{ni}}^2 + Q_{\mathrm{nv}}^2} = \sqrt{10^2 + 160^2}\,\mathrm{eV} = 160\,\mathrm{eV} .$$

After cooling, the current noise contribution is not discernible.

14.3 (a) The two Gaussian peaks are adequately resolved at $\sigma_E = \Delta E/3$, so since the spacing between the two peaks is $\Delta E = (72.87 - 70.83)\,\mathrm{keV} = 2.04\,\mathrm{keV}$, the required resolution is $\sigma_E = 0.68\,\mathrm{keV}$ or $1.6\,\mathrm{keV}$ FWHM. Note that in systems dominated by electronic noise it is more useful to specify absolute resolution rather than relative resolution, as the linewidth is essentially independent of energy.

(b) Since the individual resolutions add in quadrature, $\sigma_E^2 = \sigma_{\mathrm{det}}^2 + \sigma_{\mathrm{n}}^2$, the allowable electronic noise is $\sigma_{\mathrm{n}} = 660\,\mathrm{eV}$.

14.4 (a) The noise current sources are the detector bias current, contributing $i_{nd}^2 = 2eI_d$, and the bias resistor with $i_{nb}^2 = 4kT/R_b$. The noise voltage sources are the series resistance and the amplifier, contributing $e_{nR}^2 = 4kTR_s$ and $e_{na}^2 = 10^{-18}\,\mathrm{V^2/Hz}$, respectively. The shape factors for a CR–RC shaper are $F_i = F_v = 0.924$. This results in an equivalent noise charge

$$Q_n^2 = i_n^2 T_s F_i + C_d^2 e_n^2 \frac{F_v}{T_s} \ ,$$

$$Q_n^2 = \left(2eI_d + \frac{4kT}{R_b}\right) \cdot T_s \cdot F_i + C_d^2 \cdot (4kTR_s + e_{na}^2) \cdot \frac{F_v}{T_s} \ ,$$

$$Q_n^2 = (3.2 \cdot 10^{-26} + 1.66 \cdot 10^{-27}) \cdot 10^{-6} \cdot 0.924\,\mathrm{C}^2 + \qquad (18.34)$$

$$+\, 10^{-20} \cdot (1.66 \cdot 10^{-19} + 10^{-18}) \cdot \frac{0.924}{10^{-6}}\,\mathrm{C}^2 \ .$$

The detector bias current contributes $1075\,e$, the bias current $245\,e$ the series resistance $246\,e$ and the amplifier $601\,e$. These add in quadrature to yield the total noise of $Q_n = 1280\,e$ or $4.6\,\mathrm{keV}$ rms ($10.8\,\mathrm{keV}$ FWHM).

(b) As calculated in (a) the current noise contribution is

$$Q_{ni} = \sqrt{1075^2 + 245^2}\,e = 1103\,e$$

and the voltage noise contribution is

$$Q_{nv} = \sqrt{246^2 + 601^2}\,e = 649\,e \ .$$

Minimum noise results when the current and voltage noise contributions are equal. From Eq. (14.18) this condition yields the optimum shaping time

$$T_{s,\mathrm{opt}} = C_i \frac{e_n}{i_n} \sqrt{\frac{F_v}{F_i}} \ .$$

This yields $T_{s,\mathrm{opt}} = 589\,\mathrm{ns}$ and $Q_{n,\mathrm{min}} = 1196\,e$.

(c) Without the bias resistor, the noise is $1181\,e$. For the resistor to add 1% to the total, its noise may be 2% of $1181\,e$ or $24\,e$, so $R_b > 34\,\mathrm{M\Omega}$.

14.5 (a) Equation (14.26) yields the timing jitter

$$\sigma_t = \frac{\sigma_n}{(dV/dt)_{V_T}} \ .$$

The noise level is $\sigma_n = 10\,\mu\text{V}$ and the rate of change is

$$\frac{dV}{dt} \approx \frac{\Delta V}{t_r} = \frac{10 \cdot 10^{-3}\,\text{V}}{10 \cdot 10^{-9}\,\text{s}} = 10^6\,\text{V/s} \ ,$$

yielding the timing jitter

$$\sigma_t = \frac{10 \cdot 10^{-6}}{10^6}\,\text{s} = 10\,\text{ps} \ .$$

(b) For the $10\,\text{mV}$ signal the threshold of $5\,\text{mV}$ is at 50% of the rise time, so the comparator fires at $(5+1)\,\text{ns}$, whereas for the $50\,\text{mV}$ signal the threshold is at 10% of the rise time, so the comparator fires at $(1+1)\,\text{ns}$. The time shift is $4\,\text{ns}$. Note that the time t_0 drops out, so it can be disregarded.

Chapter 15

15.1

$$N_{\text{acc}} = \varepsilon_e N_e + \varepsilon_\pi N_\pi = \varepsilon_e N_e + \varepsilon_\pi (N_{\text{tot}} - N_e)$$

Solving for N_e gives

$$N_e = \frac{N_{\text{acc}} - \varepsilon_\pi N_{\text{tot}}}{\varepsilon_e - \varepsilon_\pi} \ .$$

In case of $\varepsilon_e = \varepsilon_\pi$ there would obviously be no chance to determine N_e.

15.2

$$E[t] = \frac{1}{\tau} \int_0^\infty t\,e^{-t/\tau}\,dt = \tau \ ,$$

$$\sigma^2[t] = \frac{1}{\tau} \int_0^\infty (t-\tau)^2\,e^{-t/\tau}\,dt$$

$$= \frac{1}{\tau} \left[\int_0^\infty t^2\,e^{-t/\tau}\,dt - \int_0^\infty 2t\tau\,e^{-t/\tau}\,dt + \tau^2 \int_0^\infty e^{-t/\tau}\,dt \right]$$

$$= \frac{1}{\tau} \left(2\tau^3 - 2\tau^3 + \tau^3 \right) = \tau^2 \ .$$

15.3 Source rate

$$n_\nu = \frac{N_1}{t_1} - \frac{N_2}{t_2} = (n_\nu + n_\mu) - n_\mu \ .$$

Standard deviation from error propagation:

$$\sigma_{n_\nu} = \left[\left(\frac{\sigma_{N_1}}{t_1}\right)^2 + \left(\frac{\sigma_{N_2}}{t_2}\right)^2\right]^{1/2} = \left(\frac{N_1}{t_1^2} + \frac{N_2}{t_2^2}\right)^{1/2}$$

$$= \left(\frac{n_\nu + n_\mu}{t_1} + \frac{n_\mu}{t_2}\right)^{1/2} .$$

$t_1 + t_2 = T$ is fixed. Therefore, $dT = dt_1 + dt_2 = 0$. Squaring and differentiating σ_{n_ν} with respect to the measurement times gives

$$2\sigma_{n_\nu}\, d\sigma_{n_\nu} = -\frac{n_\nu + n_\mu}{t_1^2}\, dt_1 - \frac{n_\mu}{t_2^2}\, dt_2 .$$

Setting

$$d\sigma_{n_\nu} = 0$$

yields the optimum condition ($dt_1 = -dt_2$):

$$\frac{n_\nu + n_\mu}{t_1^2}\, dt_2 - \frac{n_\mu}{t_2^2}\, dt_2 = 0 \quad \Rightarrow \quad \frac{t_1}{t_2} = \sqrt{\frac{n_\nu + n_\mu}{n_\mu}} = \sqrt{\frac{n_\nu}{n_\mu} + 1} = 2 .$$

15.4 $y = mE$, m – slope. The fitted linear relation is obtained from $y + Am = 0$ (C_y – error matrix) with [10]

$$A = -\begin{pmatrix} 0 \\ 1 \\ 2 \\ 3 \\ 4 \\ 5 \end{pmatrix} ,$$

$$C_y = \begin{pmatrix} 0.3^2 & & & & & \\ & 0.3^2 & & & 0 & \\ & & 0.3^2 & & & \\ & & & 0.3^2 & & \\ & 0 & & & 0.3^2 & \\ & & & & & 0.3^2 \end{pmatrix} = 0.09\, I ,$$

$$m = -(A^{\mathrm{T}}A)^{-1}A^{\mathrm{T}}y = \left\{ \begin{pmatrix} 0 & 1 & 2 & 3 & 4 & 5 \end{pmatrix} \begin{pmatrix} 0 \\ 1 \\ 2 \\ 3 \\ 4 \\ 5 \end{pmatrix} \right\}^{-1}.$$

$$\begin{pmatrix} 0 & 1 & 2 & 3 & 4 & 5 \end{pmatrix} \begin{pmatrix} 0 \\ 0.8 \\ 1.6 \\ 2.5 \\ 2.8 \\ 4.0 \end{pmatrix} = \frac{1}{55} \cdot 42.7 \approx 0.7764 ,$$

$$(\Delta m)^2 = (A^{\mathrm{T}}C_y^{-1}A)^{-1}$$

$$= \left\{ \begin{pmatrix} 0 & 1 & 2 & 3 & 4 & 5 \end{pmatrix} 0.09^{-1} \begin{pmatrix} 0 \\ 1 \\ 2 \\ 3 \\ 4 \\ 5 \end{pmatrix} \right\}^{-1}$$

$$= 0.09 \cdot \frac{1}{55} \approx 0.00164 ,$$

$$\rightarrow m = 0.7764 \pm 0.0405 .$$

The data points – corrected for the offset – along with the best fit are shown in Fig. 18.6.

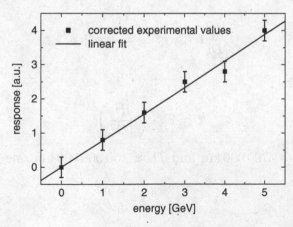

Fig. 18.6. Calibration data, corrected for the offset, along with the best fit calibration function.

Chapter 16

16.1 $P = 10\,\mathrm{mW}$ laser power at the frequency ν; rate of photons $n = P/h\nu$, h – Planck's constant; momentum of the photon (after de Broglie) $p = h/\lambda = h\nu/c$; change of momentum upon reflection $2p = 2h\nu/c$; the force has two components: (a) reflected photons $F_1 = n \cdot 2p \cdot \epsilon = (P/h\nu)2(h\nu/c)\epsilon = 2(P/c)\epsilon$; (b) absorbed photons $F_2 = (P/h\nu)(h\nu/c)(1-\epsilon) = (P/c)(1-\epsilon)$; $F = F_1 + F_2 = \frac{P}{c} \cdot (\epsilon + 1) = 5 \cdot 10^{-11}\,\mathrm{N}$.

16.2 Number of ^{238}U nuclei: $N = N_0 \cdot \mathrm{e}^{-\lambda t}$, where $\lambda = \ln 2/T_{1/2}$; number of lead nuclei: $N_0(1 - \mathrm{e}^{-\lambda t})$. $r = N_0(1 - \mathrm{e}^{-\lambda t})/N_0\,\mathrm{e}^{-\lambda t} = \mathrm{e}^{\lambda t} - 1 = 0.06$, $t = 3.8 \cdot 10^8$ years.

16.3 Total power radiated by the Sun: $P = 4\pi R^2 \sigma T_S^4$, where σ – Boltzmann's constant, T_S – Sun's surface temperature ($\approx 6000\,\mathrm{K}$), R – Sun's radius. The satellite will absorb the power

$$P_1 = \frac{4\pi R^2 \sigma T_S^4}{4\pi D^2} \cdot \pi r^2 \varepsilon = \frac{R^2}{D^2}\sigma T_S^4 \pi r^2 \varepsilon \ ,$$

where D – distance Sun–satellite, r – radius of the satellite, ε – absorption coefficient. Since the emissivity is equal to the absorption, one gets

$$P_2 = 4\pi r^2 \sigma T^4 \cdot \varepsilon$$

for the radiated power by the satellite. At equilibrium one has $P_1 = P_2$, and therefore

$$\frac{R^2}{D^2}\sigma T_S^4 \pi r^2 \cdot \varepsilon = 4\pi r^2 \sigma T^4 \cdot \varepsilon$$

yielding

$$T = T_S \cdot \left(\frac{R^2}{4D^2}\right)^{1/4} \ ;$$

with $R \approx 700\,000\,\mathrm{km}$ and $D \approx 150\,000\,000\,\mathrm{km}$ one obtains $T = 290\,\mathrm{K}$.

16.4

$$E_{\mathrm{Li}} + E_\alpha = 2.8\,\mathrm{MeV} \ , \quad E = \frac{p^2}{2m} \rightarrow \sqrt{2m_{\mathrm{Li}}E_{\mathrm{Li}}} = \sqrt{2m_\alpha E_\alpha}$$

because the lithium nucleus and the α particle are emitted back to back;

$$E_\alpha = \frac{m_{\mathrm{Li}}}{m_\alpha} \cdot (Q - E_\alpha) \rightarrow E_\alpha = \frac{m_{\mathrm{Li}}}{m_{\mathrm{Li}} + m_\alpha} \cdot Q = 1.78\,\mathrm{MeV} \ .$$

16.5 $d\sigma/d\Omega \propto 1/\sin^4 \theta/2 \propto 1/\theta^4$ for Bhabha scattering. The count rate is determined by the lower acceptance boundary,

$$\sigma_{\mathrm{Bhabha}}(\theta_0) = \int_{\theta_0} (d\sigma/d\Omega)2\pi\,d\theta \propto 1/\theta_0^3 \ .$$

Doubling the accuracy of $\sigma(e^+e^- \rightarrow Z)$ by a factor of 2 means

$$\sigma_{\mathrm{Bhabha}}(\theta_{\mathrm{new}}) = 4 \cdot \sigma_{\mathrm{Bhabha}}(\theta_0) \ , \quad 1/\theta_{\mathrm{new}}^3 = 4 \cdot 1/\theta_0^3 \ .$$

This leads to

$$\theta_{\mathrm{new}} = \theta_0 \cdot \sqrt[3]{1/4} = 0.63\,\theta_0 \approx 19\,\mathrm{mrad} \ .$$

16.6 A $100\,\mathrm{GeV}$ γ-induced shower starts at an altitude of $d \approx 20\,\mathrm{km}$ and has just about 100 energetic secondaries which emit Cherenkov light over a distance of $\approx 20\,X_0(= 6000\,\mathrm{m})$. The photon yield in air is ≈ 20 photons/m, leading to a total number of Cherenkov photons of

$$N_\gamma \lesssim 100 \cdot 20 \cdot 6000 = 1.2 \cdot 10^7 \ .$$

These photons will be distributed at sea level over a circular area

$$A = \pi \cdot (d \cdot \tan\theta)^2 \ ,$$

where θ is the Cherenkov angle of relativistic electrons in air at $20\,\mathrm{km}$ altitude ($\approx 1.2°$):

$$A = 550\,000\,\mathrm{m}^2 \ .$$

With an absorption coefficient in air of $\epsilon \approx 30\%$ one gets

$$n = N_\gamma/A \cdot (1 - \epsilon) \lesssim 15/\mathrm{m}^2 \ .$$

References

[1] E. Fenyves & O. Haimann, *The Physical Principles of Nuclear Radiation Measurements*, Akadémiai Kiadó, Budapest (1969)
[2] G. Hertz, *Lehrbuch der Kernphysik*, Bd. 1, Teubner, Leipzig (1966)
[3] C. Grupen, *Grundkurs Strahlenschutz*, Springer, Berlin (2003)

[4]   M. Mandò, Non-focused Cherenkov Effect Counters, *Nuovo Cim.* **12** (1954)
      5–27; in J.V. Jelley, *Cherenkov Radiation and Its Applications*, Pergamon
      Press, London, New York (1958)

[5]   Particle Data Group, Review of Particle Physics, S. Eidelman *et al*, *Phys.
      Lett.* **B592 Vol. 1–4** (2004) 1–1109; W.-M. Yao *et al*, *J. Phys.* **G33** (2006)
      1–1232; http://pdg.lbl.gov

[6]   Donald E. Groom, Nikolai V. Mokhov & Sergei I. Striganov, Muon Stopping
      Power and Range Tables 10-MeV to 100-TeV, *Atom. Data Nucl. Data Tabl.*
      78 (2001) 183–356; A. van Ginneken, Energy Loss and Angular Character-
      istics of High-Energy Electromagnetic Processes, *Nucl. Instr. Meth.* **A251**
      (1986) 21–39

[7]   B. Rossi, *High Energy Particles*, Prentice-Hall, Englewood Cliffs (1952)

[8]   C. Grupen, Electromagnetic Interactions of High Energy Cosmic Ray
      Muons, *Fortschr. der Physik* **23** (1976) 127–209

[9]   C. Bernardini, B. Touschek *et al.*, Lifetime and Beam Size in a Storage
      Ring, *Phys. Rev. Lett.* **10** (1963) 407–9

[10] S. Brandt, *Datenanalyse*, 4. Auflage; Spektrum Akademischer Verlag,
      Heidel-berg/Berlin (1999); *Data Analysis: Statistical and Computational
      Methods for Scientists and Engineers*, 3rd edition, Springer, New York
      (1998)

Appendix 1

Table of fundamental
physical constants

[From Particle Data Group; *Phys. Lett.* **B592** (2004) 1–1109; *J. Phys.*
G33 (2006) 1–1232; P.J. Mohr & B.N. Taylor, CODATA Recommended
Values of the Fundamental Constants: 2002, *Rev. Mod. Phys.* **77** (2005)
1–107; B.N. Taylor & E.R. Cohen, *J. Res. Nat. Inst. Standards and Tech-
nology* **95** (1990) 497–523; R.C. Weast & M.J. Astle (eds.), *Handbook of
Chemistry and Physics*, CRC Press, Boca Raton, Florida (1973).]

Speed of light*	c	$299\,792\,458\,\mathrm{m/s}$
Planck's constant	h	$6.626\,069\,3 \cdot 10^{-34}\,\mathrm{J\,s}$
		$\pm 0.000\,001\,1 \cdot 10^{-34}\,\mathrm{J\,s}$
Planck's constant, reduced	$\hbar = \dfrac{h}{2\pi}$	$1.054\,571\,68 \cdot 10^{-34}\,\mathrm{J\,s}$
		$\pm 0.000\,000\,18 \cdot 10^{-34}\,\mathrm{J\,s}$
		$= 6.582\,119\,15 \cdot 10^{-22}\,\mathrm{MeV\,s}$
		$\pm 0.000\,000\,56 \cdot 10^{-22}\,\mathrm{MeV\,s}$
Electron charge[†]	e	$1.602\,176\,53 \cdot 10^{-19}\,\mathrm{C}$
		$\pm 0.000\,000\,14 \cdot 10^{-19}\,\mathrm{C}$
		$= 4.803\,204\,41 \cdot 10^{-10}\,\mathrm{esu}$
		$\pm 0.000\,000\,41 \cdot 10^{-10}\,\mathrm{esu}$
Gravitational constant	G	$6.674\,2 \cdot 10^{-11}\,\mathrm{m^3/(kg\,s^2)}$
		$\pm 0.001\,0 \cdot 10^{-11}\,\mathrm{m^3/(kg\,s^2)}$

---

* The value of the velocity of light forms the basis for the definition of the length unit, the
metre. 1 m is now defined to be the distance travelled by light in $1/299\,792\,458\,\mathrm{s}$. The quoted
value for the speed of light is therefore exact and without error.
† esu = electrostatic charge unit.

Avogadro number	N_A	$6.022\,141\,5 \cdot 10^{23}\,\text{mol}^{-1}$ $\pm 0.000\,001\,0 \cdot 10^{23}\,\text{mol}^{-1}$
Boltzmann constant	k	$1.380\,650\,5 \cdot 10^{-23}\,\text{J/K}$ $\pm 0.000\,002\,4 \cdot 10^{-23}\,\text{J/K}$
Molar gas constant	$R(= kN_A)$	$8.314\,473\,\text{J/(K mol)}$ $\pm 0.000\,014\,\text{J/(K mol)}$
Molar volume, ideal gas at STP‡	V_{mol}	$22.413\,996 \cdot 10^{-3}\,\text{m}^3/\text{mol}$ $\pm 0.000\,039 \cdot 10^{-3}\,\text{m}^3/\text{mol}$
Permittivity of free space§	$\varepsilon_0 = 1/\mu_0 c^2$	$8.854\,187\,817\ldots \cdot 10^{-12}\,\text{F/m}$
Permeability of free space	μ_0	$4\pi \cdot 10^{-7}\,\text{N/A}^2$ $= 12.566\,370\,614\ldots \cdot 10^{-7}\,\text{N/A}^2$
Stefan–Boltzmann constant	$\sigma = \dfrac{\pi^2 k^4}{60\,\hbar^3 c^2}$	$5.670\,400 \cdot 10^{-8}\,\text{W/(m}^2\,\text{K}^4)$ $\pm 0.000\,040 \cdot 10^{-8}\,\text{W/(m}^2\,\text{K}^4)$
Electron mass	m_e	$0.510\,998\,918\,\text{MeV}/c^2$ $\pm 0.000\,000\,044\,\text{MeV}/c^2$ $= 9.109\,382\,6 \cdot 10^{-31}\,\text{kg}$ $\pm 0.000\,001\,6 \cdot 10^{-31}\,\text{kg}$
Proton mass	m_p	$938.272\,029\,\text{MeV}/c^2$ $\pm 0.000\,080\,\text{MeV}/c^2$ $= 1.672\,621\,71 \cdot 10^{-27}\,\text{kg}$ $\pm 0.000\,000\,29 \cdot 10^{-27}\,\text{kg}$
Unified atomic mass unit (u)	$(1\,\text{g}/N_A)$	$931.494\,043\,\text{MeV}/c^2$ $\pm 0.000\,080\,\text{MeV}/c^2$ $= 1.660\,538\,86 \cdot 10^{-27}\,\text{kg}$ $\pm 0.000\,000\,28 \cdot 10^{-27}\,\text{kg}$
Charge-to-mass ratio of the electron	e/m_e	$1.758\,820\,11 \cdot 10^{11}\,\text{C/kg}$ $\pm 0.000\,000\,20 \cdot 10^{11}\,\text{C/kg}$

‡ Standard temperature and pressure ($0\,^\circ\text{C} \cong 273.15\,\text{K}$ and $1\,\text{atm} = 101\,325\,\text{Pa}$).
§ Because of the fact that the velocity of light c is without error by definition, and because μ_0 is defined to be $\mu_0 = 4\pi \cdot 10^{-7}\,\text{N/A}^2$, ε_0 is also exact.

Fine-structure constant[¶] α	$\alpha^{-1} = \left(\dfrac{e^2}{4\pi\varepsilon_0\hbar c}\right)^{-1}$	137.035 999 11 $\pm 0.000\,000\,46$
Classical electron radius	$r_e = \dfrac{e^2}{4\pi\varepsilon_0 m_e c^2}$	$2.817\,940\,325 \cdot 10^{-15}\,\mathrm{m}$ $\pm 0.000\,000\,028 \cdot 10^{-15}\,\mathrm{m}$
Electron Compton wavelength	$\dfrac{\lambda_e}{2\pi} = \dfrac{\hbar}{m_e c} = \dfrac{r_e}{\alpha}$	$3.861\,592\,678 \cdot 10^{-13}\,\mathrm{m}$ $\pm 0.000\,000\,026 \cdot 10^{-13}\,\mathrm{m}$
Bohr radius	$r_0 = \dfrac{4\pi\varepsilon_0\hbar^2}{m_e e^2} = \dfrac{r_e}{\alpha^2}$	$0.529\,177\,210\,8 \cdot 10^{-10}\,\mathrm{m}$ $\pm 0.000\,000\,001\,8 \cdot 10^{-10}\,\mathrm{m}$
Rydberg energy	$E_{\mathrm{Ry}} = m_e c^2 \alpha^2 / 2$	$13.605\,692\,3\,\mathrm{eV}$ $\pm 0.000\,001\,2\,\mathrm{eV}$
Bohr magneton	$\mu_{\mathrm{B}} = e\hbar/2m_e$	$5.788\,381\,804 \cdot 10^{-11}\,\mathrm{MeV/T}$ $\pm 0.000\,000\,039 \cdot 10^{-11}\,\mathrm{MeV/T}$
Gravitational acceleration, sea level[‖]	g	$9.806\,65\,\mathrm{m/s}^2$
Mass of Earth	$M_{\oplus}$	$5.792\,3 \cdot 10^{24}\,\mathrm{kg}$ $\pm 0.000\,9 \cdot 10^{24}\,\mathrm{kg}$
Solar mass	$M_{\odot}$	$1.988\,44 \cdot 10^{30}\,\mathrm{kg}$ $\pm 0.000\,30 \cdot 10^{30}\,\mathrm{kg}$

[¶] At a four-momentum transfer squared $q^2 = -m_e^2$. At $q^2 = -m_W^2$ the value is approximately $1/128$, where $m_W = 80.40\,\mathrm{GeV}/c^2$ is the mass of the W boson.

[‖] Exact by definition. Actually g varies for different locations on Earth. At the equator $g \approx 9.75\,\mathrm{m/s}^2$, at the poles $g \approx 9.85\,\mathrm{m/s}^2$.

Appendix 2

Definition and conversion of physical units

Physical quantity	Name of unit and symbol
Activity A	1 Becquerel (Bq) = 1 decay per second (s^{-1}) 1 Curie (Ci) = $3.7 \cdot 10^{10}$ Bq
Work, energy W	1 Joule (J) = 1 W s = 1 N m 1 erg = 10^{-7} J 1 eV = $1.602\,177 \cdot 10^{-19}$ J 1 cal = 4.185 5 J kT at 300 K = 25.85 MeV = 1/38.68 eV
Density ϱ	1 kg/m^3 = 10^{-3} g/cm^3
Pressure* p	1 Pascal (Pa) = 1 N/m^2 1 bar = 10^5 Pa 1 atm = $1.013\,25 \cdot 10^5$ Pa 1 Torr (mm Hg) = $1.333\,224 \cdot 10^2$ Pa 1 kp/m^2 = 9.806 65 Pa
Unit of absorbed dose D	1 Gray (Gy) = 1 J/kg 1 rad = 0.01 Gy
Unit of equivalent dose H	1 Sievert (Sv) = 1 J/kg (H {Sv} = $RBE \cdot D$ {Gy} ; RBE = relative biological effectiveness) 1 rem = 0.01 Sv
Unit of ion dose I	1 I = 1 C/kg 1 Röntgen (R) = $2.58 \cdot 10^{-4}$ C/kg = $8.77 \cdot 10^{-3}$ Gy (for absorption in air)

---

* kp stands for kilopond; it is the weight of 1 kg on Earth, i.e. 1 kp = 1 kg $\cdot$ g, where g is the acceleration due to gravity, $g = 9.806\,65$ m s^{-2}.

Entropy S	$1\,\text{J/K}$
Electric field strength E	$1\,\text{V/m}$
Magnetic field strength H	$1\,\text{A/m}$
	$1\,\text{Oersted (Oe)} = 79.58\,\text{A/m}$
Magnetic induction B	$1\,\text{Tesla (T)} = 1\,\text{V\,s/m}^2 = 1\,\text{Wb/m}^2$
	$1\,\text{Gauss (G)} = 10^{-4}\,\text{T}$
Magnetic flux Φ_{m}	$1\,\text{Weber (Wb)} = 1\,\text{V\,s}$
Inductance L	$1\,\text{Henry (H)} = 1\,\text{V\,s/A} \doteq 1\,\text{Wb/A}$
Capacitance C	$1\,\text{Farad (F)} = 1\,\text{C/V}$
Force F	$1\,\text{Newton (N)} = 10^5\,\text{dyn}$
Length l	$1\,\text{inch} = 0.0254\,\text{m}$
	$1\,\text{m} = 10^{10}\,\text{Ångström (Å)}$
	$1\,\text{fermi (fm)} = 10^{-15}\,\text{m}$
	$\qquad (= 1\,\text{femtometre})$
	$1\,\text{astronomical unit (AU)}\,^{\dagger}$
	$\qquad = 149\,597\,870\,\text{km}$
	$1\,\text{parsec (pc)} = 3.085\,68 \cdot 10^{16}\,\text{m}$
	$\qquad = 3.26\,\text{light-years}$
	$\qquad = 1\,\text{AU/1\,arcsec}$
	$1\,\text{light-year (ly)} = 0.3066\,\text{pc}$
Power P	$1\,\text{Watt (W)} = 1\,\text{N\,m/s} = 1\,\text{J/s}$
Mass m	$1\,\text{kg} = 10^3\,\text{g}$
Electric potential U	$1\,\text{Volt (V)}$
Electric current I	$1\,\text{Ampère (A)} = 1\,\text{C/s}$
Charge Q	$1\,\text{Coulomb (C)}$
	$1\,\text{C} = 2.997\,924\,58 \cdot 10^9\,\text{electrostatic}$
	$\qquad\qquad\qquad\text{charge units (esu)}$
Temperature T	$1\,\text{Kelvin (K)}$
	$\text{Celsius (}^\circ\text{C)};\ T\,\{^\circ\text{C}\} = T\,\{\text{K}\} - 273.15\,\text{K}$
Electric resistance R	$1\,\text{Ohm }(\Omega) = 1\,\text{V/A}$
Specific resistivity ϱ	$1\,\Omega\,\text{cm}$
Time t	$1\,\text{s}$
Cross section σ	$1\,\text{barn} = 10^{-24}\,\text{cm}^2$

---

† Fixed by the International Astronomical Union 1996.

Appendix 3

Properties of pure and composite materials

[From Particle Data Group; *Phys. Lett.* **B592** (2004) 1–1109; *J. Phys.* **G33** (2006) 1–1232.]

Properties of pure materials[*]

Material	Z	A	Nuclear inter- action length [g/cm^2]	$\left.\frac{dE}{dx}\right|_{min}$ $\left[\frac{MeV}{g/cm^2}\right]$	Radiation length [g/cm^2]	Density [g/cm^3]	Refractive index at STP[†]
H$_2$ gas	1	1.008	50.8	4.1	61.3	$0.089\,9 \cdot 10^{-3}$	1.000 139 2
He gas	2	4.003	65.1	1.937	94.3	$0.178\,6 \cdot 10^{-3}$	1.000 034 9
Be	4	9.012	75.2	1.594	65.19	1.848	
C	6	12.011	86.3	1.745	42.7	2.265	
N$_2$ gas	7	14.007	87.8	1.825	37.99	$1.25 \cdot 10^{-3}$	1.000 298
O$_2$ gas	8	15.999	91.0	1.801	34.24	$1.43 \cdot 10^{-3}$	1.000 296
Al	13	26.981	106.4	1.615	24.01	2.70	
Si	14	28.086	106.0	1.664	21.82	2.33	3.95
Ar gas	18	39.948	117.2	1.519	19.55	$1.78 \cdot 10^{-3}$	1.000 283
Fe	26	55.845	131.9	1.451	13.84	7.87	
Cu	29	63.546	134.9	1.403	12.86	8.96	
Ge	32	72.610	140.5	1.371	12.25	5.323	
Xe gas	54	131.29	169	1.255	8.48	$5.86 \cdot 10^{-3}$	1.000 701
W	74	183.84	185	1.145	6.76	19.3	
Pb	82	207.2	194	1.123	6.37	11.35	
U	92	238.03	199	1.082	6.00	18.95	

[*] The nuclear interaction length λ_I in g/cm^2 is related to the inelastic cross section by $\lambda_I = A/(N_A \cdot \sigma_{inel})$, where A is given in g/mol, N_A in mol^{-1}, and σ_{inel} in cm^2. There is no unequivocal name for λ_I in the literature. Frequently, λ_I is also called nuclear absorption length λ_a.

[†] Standard temperature and pressure ($0\,°C \cong 273.15\,K$ and $1\,atm = 101\,325\,Pa$). Refractive indices are evaluated at the sodium D line.

582

Properties of composite materials [‡]

Material	Nuclear inter-action length [g/cm^2]	$\dfrac{\mathrm{d}E}{\mathrm{d}x}\Big|_{\min}$ $\left[\frac{\mathrm{MeV}}{\mathrm{g/cm}^2}\right]$	Radiation length [g/cm^2]	Density [g/cm^3]	Refractive index at STP
Air (STP)	90.0	1.815	36.66	$1.29 \cdot 10^{-3}$	1.000 293
H_2O	83.6	1.991	36.08	1.00	1.33
CO_2 gas	89.7	1.819	36.20	$1.977 \cdot 10^{-3}$	1.000 410
Shielding concrete	99.9	1.711	26.70	2.5	
CH_4 gas	73.4	2.417	46.22	$0.717 \cdot 10^{-3}$	1.000 444
C_2H_6 gas	75.7	2.304	45.47	$1.356 \cdot 10^{-3}$	1.001 038
C_3H_8 gas	76.5	2.262	45.20	$1.879 \cdot 10^{-3}$	1.001 029
Isobutane	77.0	2.239	45.07	$2.67 \cdot 10^{-3}$	1.001 900
Polyethylene	78.4	2.076	44.64	≈ 0.93	
Plexiglas	83.0	1.929	40.49	≈ 1.18	≈ 1.49
Polystyrene scintillator	81.9	1.936	43.72	1.032	1.581
BaF_2	145	1.303	9.91	4.89	1.56
BGO	157	1.251	7.97	7.1	2.15
CsI	167	1.243	8.39	4.53	1.80
NaI	151	1.305	9.49	3.67	1.775
Silica aerogel	96.9	1.740	27.25	0.04–0.6	$1.0 + 0.21 \cdot \varrho$
G10	90.2	1.87	33.0	1.7	
Kapton	85.8	1.82	40.56	1.42	
Pyrex Corning (borosilicate)	97.6	1.695	28.3	2.23	1.474
Lead glass (SF-5)	132.4	1.41	10.38	4.07	1.673

[‡] The nuclear interaction length λ_I in g/cm^2 is related to the inelastic cross section by $\lambda_I = A/(N_A \cdot \sigma_{\mathrm{inel}})$, where A is given in g/mol, N_A in mol^{-1}, and σ_{inel} in cm^2. There is no unequivocal name for λ_I in the literature. Frequently, λ_I is also called nuclear absorption length λ_a.

Appendix 4
Monte Carlo event generators*

General-purpose Monte Carlo event generators are designed for generating a wide variety of physics processes. There are literally hundreds of different Monte Carlo event generators; some of these are

- **ARIADNE** [1] is a programme for simulation of QCD cascades implementing the colour dipole model.

- **HERWIG** [2] (Hadron Emission Reactions With Interfering Gluons) is a package based on matrix elements providing parton showers including colour coherence and using a cluster model for hadronisation.

- **ISAJET** [3] is a programme for simulating pp, $p\bar{p}$ and e^+e^- interactions; it is based on perturbative QCD and phenomenological models for parton and beam jet fragmentation including the Fox–Wolfram final-state shower QCD radiation and Field–Feynman hadronisation.

- **JETSET** [4] is a programme for implementing the Lund string model for hadronisation of parton systems. Since 1998, JETSET has been combined with PYTHIA in a single package.

- **PYTHIA** [5] is a general-purpose programme with an emphasis on QCD cascades and hadronisation; it includes several extensions for modelling new physics (e.g. Technicolour).

There are also Monte Carlo event generators which are specifically designed to generate a number of interesting physics processes. They can be interfaced to one or more of the general-purpose event generators above or with other specialised generators.

---

* For a recent review see also Z. Nagy & D.E. Soper, *QCD and Monte Carlo Event Generators*, XIV Workshop on Deep Inelastic Scattering, hep-ph/0607046 (July 2006).

- **AcerMC** [6] models Standard Model background processes in *pp* collisions at the LHC and works with either PYTHIA or HERWIG; it provides a library of massive matrix elements for selected processes and is designed to have an efficient phase-space sampling via self-optimising approaches.

- **CASCADE** [7] models full hadron-level processes for *ep* and *pp* scattering at small $x = 2p/\sqrt{s}$ according to the CCFM [8] evolution equation.

- **EXCALIBUR** [9] computes all four-fermion processes in e^+e^- annihilation which includes QED initial-state corrections and QCD contributions.

- **HIJING** [10] (Heavy Ion Jet INteraction Generator) models mini-jets in *pp*, *pA* and *AA* reactions.

- **HZHA** [11, 12] provides a wide coverage of the production and decay channels of Standard Model and Minimal Super Symmetric Model (MSSM) Higgs bosons in e^+e^- collisions and was heavily used in LEP2 Higgs-boson searches.

- **ISAWIG** [13] works with the ISAJET SUGRA package and general MSSM programs to describe SUSY particles which can be read in by the HERWIG event generator.

- **KK** [14] models two-fermion final-state processes in e^+e^- collisions including multiphoton initial-state radiation and a treatment of spin effects in τ decays.

- **KORALB** [15] provides a simulation of the τ-lepton production in e^+e^- collisions with centre-of-mass energies below 30 GeV including treatment of QED, Z exchange and spin effects; it makes use of the TAUOLA package.

- **KORALZ** [16] provides a simulation of the production and decay processes of τ leptons including spin effects and radiative corrections in e^+e^- collisions with centre-of-mass energies ranging from 20 GeV to 150 GeV.

- **KORALW** [17, 18] provides a simulation of all four-fermion final states in e^+e^- collisions and includes all non-double-resonant corrections to all double-resonant four-fermion processes; it uses the YFSWW package (see below) to include electroweak corrections to W-pair production.

- **LEPTO** [19] models deep inelastic lepton–nucleon scattering.

- **MC@NLO** [20] is a parton-shower package implementing schemes of next-to-leading order matrix-element calculations of rates for QCD processes and makes use of the HERWIG package; it includes the hadroproduction of single vector and Higgs bosons, vector-boson pairs, heavy-quark pairs and lepton pairs.

- **MUSTRAAL** [21] simulates radiative corrections to muon and quark-pair production in e^+e^- collisions near centre-of-mass energies of 91.2 GeV.

- **PANDORA** [22] is a general-purpose parton-level generator for linear collider physics which includes beamstrahlung, initial-state radiation and full treatment of polarisation effects including processes from the Standard Model and beyond; it is interfaced with PYTHIA and TAUOLA in the PANDORA–PYTHIA package.

- **PHOJET** [23] models hadronic multiparticle production for hadron–hadron, photon–hadron and photon–photon interactions using the Dual Parton Model (DPM).

- **PHOTOS** [24] simulates QED single-photon (bremsstrahlung) radiative corrections in decays; it is intended to be interfaced with another package generating decays.

- **RESBOS** (RESummed BOSon Production and Decay) [25] models hadronically produced lepton pairs via electroweak vector-boson production and decay by resumming large perturbative contributions from multiple soft-gluon emissions.

- **RacoonWW** [26] models four-fermion production at e^+e^- colliders including radiative corrections to four-fermion decays from W-pair production; it includes anomalous triple gauge-boson couplings as well as anomalous quartic gauge-boson couplings where applicable.

- **SUSYGEN** [12, 27] models the production and decay of MSSM sparticles (supersymmetric partners of particles) in e^+e^- collisions.

- **TAUOLA** [28] is a library of programs modelling the leptonic and semi-leptonic decays of τ leptons including full-final-state topologies with a complete treatment of spin structure; it can be used with any other package which produces τ leptons.

- **VECBOS** [29] models the leading-order inclusive production of electroweak vector bosons plus multiple jets.

- **YFSWW** [30] provides high-precision modelling of the $W^\pm$ mass and width using the YFS exponentiation technique.

Finally, several packages exist which aid in the evaluation of Feynman diagrams and are able to provide source code for inclusion in a Monte Carlo event generator. Such packages include CompHEP [31], FeynArts/FeynCalc [32], GRACE [33], HELAS (HELicity Amplitude Subroutine for Feynman diagram evaluation) [34] and MADGRAPH [35].

In the field of cosmic rays and astroparticle physics the following Monte Carlo event generators are frequently used. A recent overview including the relevant references is published in [36]. For further references see Sect. 15.5.1.

- **VENUS** (Very Energetic NUclear Scattering) is designed for ultra-relativistic heavy-ion collisions including a detailed simulation of creation, interaction and fragmentation of colour strings. Diffractive and non-diffractive collisions are also treated. It covers cosmic-ray energies up to $2 \cdot 10^7$ GeV.

- **QGSJET** (Quark Gluon String model with Jets) is based on the Gribov–Regge model of strong interactions. It treats nucleus–nucleus interactions and semihard processes. At high energies the collision is described as a superposition of a number of elementary processes based on Pomeron exchange.

- **DPMJET** (Dual Parton Model with JET production) simulates particle production in hadron–nucleus and nucleus–nucleus interactions at high energies. The soft component is described by a supercritical Pomeron. For hard collisions also hard Pomerons are introduced.

- **HDPM** is a phenomenological generator inspired by the Dual Parton Model and adjusted to experimental data.

- **NEXUS** combines VENUS and QGSJET in the framework of a parton-based Gribov–Regge theory with unified soft and hard interactions. The shower development is based on cascade equations.

- **SIBYLL** is a minijet model using a critical Pomeron describing soft processes and strings originating from hard collisions with minijet production of high transverse momenta.

Extensive air showers are frequently generated with the CORSIKA programme [37] where different event generators can be built in. CORSIKA includes also packages for the modelling of the geometry of a special detector, like GEANT [38] and the description of low-energy interactions, e.g. with FLUKA [39].

References

[1] L. Lönnblad, ARIADNE Version 4: A Program for Simulation of QCD Cascades Implementing the Color Dipole Model, *Comp. Phys. Comm.* **71** (1992) 15–31

[2] G. Marchesini *et al.*, HERWIG: A Monte Carlo Event Generator for Simulating Hadron Emission Reactions with Interfering Gluons. Version 5.1, *Comp. Phys. Comm.* **67** (1992) 465–508

[3] F.E. Paige, H. Baer, S.D. Protopopescu & X. Tata, *ISAJET 7.51 – A Monte Carlo Event Generator for pp, pp̄, and e⁺e⁻ Reactions*, www-cdf.fnal.gov/cdfsim/generators/isajet.html, ftp://ftp.phy.bnl.gov/pub/isajet/

[4] T. Sjöstrand, High-Energy Physics Event Generation with PYTHIA 5.7 and JETSET 7.4, *Comp. Phys. Comm.* **82** (1994) 74–90; *Lund University report* LU TP 95–20

[5] T. Sjöstrand, QCD Generators, in G. Altarelli, R.H.P. Kleiss & C. Verzegnassi (eds.), *Z physics at LEP 1 – Event Generators and Software*, CERN-89-08-V-3 (1989) 143–340

[6] B.P. Kersevan & E. Richter-Was, *The Monte Carlo Event Generator AcerMC 1.0 with Interfaces to PYTHIA 6.2 and HERWIG 6.3*, hep-ph/0201302

[7] H. Jung (Lund U.), G.P. Salam (CERN & Paris U., VI-VII), Hadronic Final State Predictions from CCFM: The Hadron Level Monte Carlo Generator CASCADE, *Eur. Phys. J.* **C19** (2001) 351–60

[8] L. Lönnblad & H. Jung (Lund U.), *Hadronic Final State Predictions from CCFM Generators*, 9th International Workshop on Deep Inelastic Scattering (DIS 2001), Bologna, Italy, 27 April–1 May 2001, Published in 'Bologna 2001, Deep inelastic scattering', 467–70

[9] F.A. Berends, R. Pittau & R. Kleiss, *EXCALIBUR – A Monte Carlo Program to Evaluate All Four Fermion Processes at LEP 200 and Beyond*, INLO-PUB-12/94 (1994) and hep-ph/9409326

[10] X.-N. Wang & M. Gyulassy, HIJING: A Monte Carlo Model for Multiple Jet Production pp, pA, and AA Collisions, *Phys. Rev.* **D44** (1991) 3501–16

[11] P. Janot, HZHA, (in part M.L. Mangano, G. Ridolfi (conveners), Event Generators for Discovery Physics), in G. Altarelli, T. Sjöstrand, F. Zwirner (eds.), *Physics at Lep2*, CERN-96-01-V-2 (1996) 309–11

[12] E. Accomando *et al.*, *Event Generators for Discovery Physics*, hep-ph/9602203

[13] H. Baer, F.E. Paige, S.D. Protopopescu & X. Tata, ISAJET 7.48: A Monte Carlo Event Generator for pp, p̄p, and e⁺e⁻ Reactions, preprint BNL-HET-99-43, FSU-HEP-991218, UH-511-952-00 (1999), hep-ph/0001086

[14] S. Jadach, Z. Was & B.F.L. Ward, *The Precision Monte Carlo Generator KK for Two-Fermion Final States in e⁺e⁻ Collisions*, hep-ph/9912214

[15] S. Jadach & Z. Was, Monte Carlo Simulation of the Process $e^+e^- \to \tau^+\tau^-$, Including Radiative $O(\alpha^3)$ QED Corrections, Mass and Spin Effects, *Comp. Phys. Comm.* **36** (1985) 191–211, KORALB version 2.1. An Upgrade with the TAUOLA Library of τ Decays, *Comp. Phys. Comm.* **64** (1991) 267–74

[16] S. Jadach, B.F.L. Ward & Z. Was, The Monte Carlo Program KORALZ Version 4.0 for Lepton or Quark Pair Production at LEP/SLC Energies, *Comp. Phys. Comm.* **79** (1994) 503–22

[17] M. Skrzypek, S. Jadach, W. Placzek & Z. Was, Monte Carlo Program KORALW 1.02 for W-pair Production at LEP2/NLC Energies with Yennie-Frautschi-Suura Exponentiation, *Comp. Phys. Comm.* **94** (1996) 216–48; S. Jadach, W. Placzek, M. Skrzypek, B.F.L. Ward & Z. Was, Monte Carlo Program KORALW 1.42 for All Four-Fermion Final States in e^+e^- Collisions, *Comp. Phys. Comm.* **119** (1999) 272–311

[18] S. Jadach, W. Placzek, M. Skrzypek & B.F.L. Ward, The Monte Carlo Program KoralW Version 1.51 and the Concurrent Monte Carlo KoralW&YFSWW3 with All Background Graphs and First-Order Corrections to W-pair Production, *Comp. Phys. Comm.* **140** (2001) 475–512

[19] G. Ingelman, A. Edin & J. Rathsman, LEPTO 6.5 – A Monte Carlo Generator for Deep Inelastic Lepton-Nucleon Scattering, *Comp. Phys. Comm.* **101** (1997) 108–34

[20] S. Frixione & B.R. Webber, *Matching NLO QCD Computations and Parton Shower Simulations*, JHEP **6** (2002) 29, 1–64; S. Frixione & B.R. Webber, *The MC@NLO 2.3 Event Generator*, Cavendish-HEP-04/09 (GEF-TH-2/2004)

[21] F.A. Berends, R. Kleiss & S. Jadach, Radiative Corrections to Muon Pair and Quark Pair Production in Electron-Positron Collisions in the Z0 Region, *Nucl. Phys.* **B202** (1982) 63–88

[22] M. Iwasaki & M.E. Peskin, *Pandora and Pandora-Pythia: Event Generation for Linear Collider Physics*, on-line: ftp://ftp.slac.stanford.edu/groups/lcd/Generators/PANDORA/ppythia.pdf

[23] R. Engel & J. Ranft, Hadronic Photon-Photon Collisions at High Energies, *Phys. Rev.* **D54** (1996) 4244–62

[24] E. Barberio, B. van Eijk & Z. Was, Photos – A Universal Monte Carlo for QED Radiative Corrections in Decays, *Comp. Phys. Comm.* **66** (1991) 115–28; E. Barberio & Z. Was, PHOTOS – A Universal Monte Carlo for QED Radiative Corrections: Version 2.0, *Comp. Phys. Comm.* **79** (1994) 291–308

[25] C. Balazs & C.P. Yuan, Soft Gluon Effects on Lepton Pairs at Hadron Colliders, *Phys. Rev.* **D56** (1997) 5558–83

[26] A. Denner, S. Dittmaier, M. Roth & D. Wackeroth, RacoonWW1.3: A Monte Carlo Program for Four-Fermion Production at e^+e^- Colliders, *Comp. Phys. Comm.* **153** (2003) 462–507

[27] St. Katsanevas & P. Morawitz, SUSYGEN 2.2 – A Monte Carlo Event Generator for MSSM Particle Production at e^+e^- Colliders, *Comp. Phys. Comm.* **112** (1998) 227–69, on-line: lpscwww.in2p3.fr/d0/generateurs/

[28] S. Jadach, J.H. Kühn & Z. Was, TAUOLA – A Library of Monte Carlo Programs to Simulate Decays of Polarised tau Leptons, *Comp. Phys. Comm.* **64** (1991) 275–99; S. Jadach, M. Jeżabek, J.H. Kühn & Z. Wąs, The τ Decay Library TAUOLA, update with exact $\mathcal{O}(\alpha)$ QED corrections in $\tau \to \mu(e)\nu\bar{\nu}$ decay modes, *Comp. Phys. Comm.* **70** (1992) 69–76; S. Jadach,

Z. Was, R. Decker & J.H. Kühn, The tau Decay Library TAUOLA: Version 2.4, *Comp. Phys. Comm.* **76** (1993) 361–80

[29]   F.A. Berends, H. Kuijf, B. Tausk & W.T. Giele, On the Production of a W and Jets at Hadron Colliders, *Nucl. Phys.* **B357** (1991) 32–64

[30]   S. Jadach, W. Placzek, M. Skrzypek, B.F.L. Ward & Z. Was, The Monte Carlo Event Generator YFSWW3 VERSION 1.16 for W Pair Production and Decay at LEP-2/LC Energies, *Comp. Phys. Comm.* **140** (2001) 432–74

[31]   A. Pukhov *et al.*, *CompHEP – A Package for Evaluation of Feynman Diagrams and Integration over Multi-particle Phase Space. User's Manual for Version 33* hep-ph/9908288

[32]   J. Küblbeck, M. Böhm & A. Denner, Feyn Arts – Computer-Algebraic Generation of Feynman Graphs and Amplitudes, *Comp. Phys. Comm.* **60** (1990) 165–80

[33]   Tl. Tanaka, T. Kaneko & Y. Shimizu, Numerical Calculation of Feynman Amplitudes for Electroweak Theories and an Application to $e^+e^- \rightarrow W^+W^-\gamma$ *Comp. Phys. Comm.* **64** (1991) 149–66

[34]   H. Murayama, I. Wantanabe & K. Hagiwara, *HELAS: HELicity Amplitude Subroutines for Feynman Diagram Evaluations*, KEK Report 91-11 (1992)

[35]   T. Stelzer & W.F. Long, Automatic Generation of Tree Level Helicity Amplitudes, *Comp. Phys. Comm.* **81** (1998) 357–71

[36]   S. Ostapchenko, *Hadronic Interactions at Cosmic Ray Energies*, hep-ph/0612175 (December 2006)

[37]   D. Heck *et al.*, Forschungszentrum Karlsruhe, Report FZKA 6019 (1998); D. Heck *et al.*, Comparison of Hadronic Interaction Models at Auger Energies, *Nucl. Phys. B Proc. Suppl.* **122** (2002) 364–7

[38]   R. Brun, F. Bruyant, M. Maire, A.C. McPherson & P. Zanarini, GEANT3 CERN-DD/EE/84-1 (1987); wwwasdoc.web.cern.ch/wwwasdoc/geant_html3/geantall.html

[39]   www.fluka.org/ (2005)

Appendix 5

Decay-level schemes

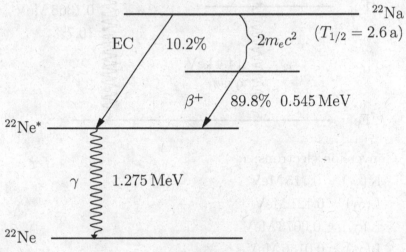

Fig. A5.1.   Decay-level scheme of ^{22}Na.

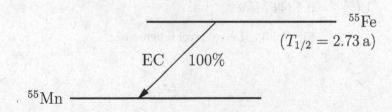

Characteristic X rays from ^{55}Mn:

$K_\alpha = 5.9 \, \text{keV}$

$K_\beta = 6.5 \, \text{keV}$

Fig. A5.2.   Decay-level scheme of ^{55}Fe.

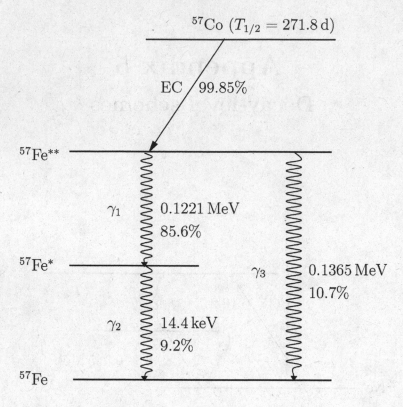

Conversion electrons:

$K(\gamma_1) = 0.115\,\text{MeV}$

$L(\gamma_1) = 0.121\,\text{MeV}$

$K(\gamma_2) = 0.0073\,\text{MeV}$

$L(\gamma_2) = 0.0136\,\text{MeV}$

$K(\gamma_3) = 0.1294\,\text{MeV}$

$L(\gamma_3) = 0.1341\,\text{MeV}$

Fig. A5.3.   Decay-level scheme of ^{57}Co.

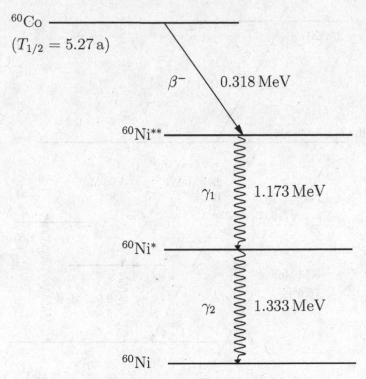

Fig. A5.4.   Decay-level scheme of ^{60}Co.

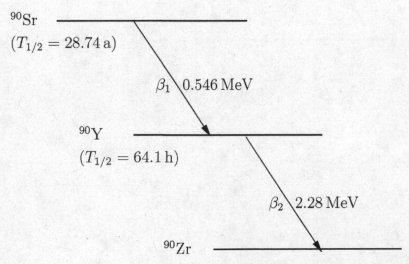

Fig. A5.5.   Decay-level scheme of ^{90}Sr.

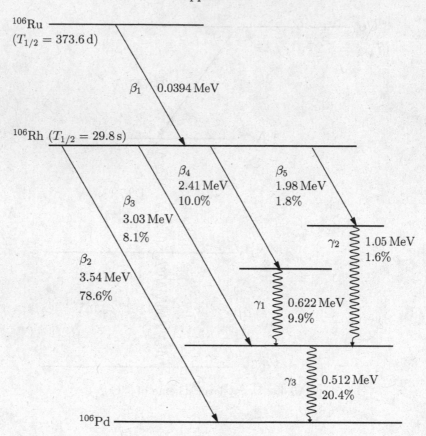

Fig. A5.6.   Decay-level scheme of ^{106}Ru.

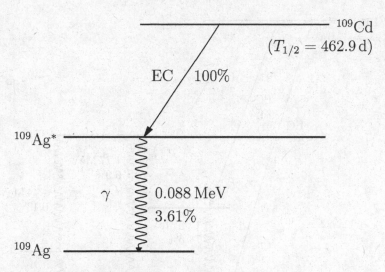

^{109}Cd

$(T_{1/2} = 462.9 \, \text{d})$

EC / 100%

^{109}Ag*

γ 0.088 MeV
3.61%

^{109}Ag

Conversion electrons:

K (γ) = 0.0625 MeV
L (γ) = 0.0842 MeV
M (γ) = 0.0873 MeV

K$_\alpha$ X rays: 0.022 MeV
K$_\beta$ X rays: 0.025 MeV

Fig. A5.7.   Decay-level scheme of ^{109}Cd.

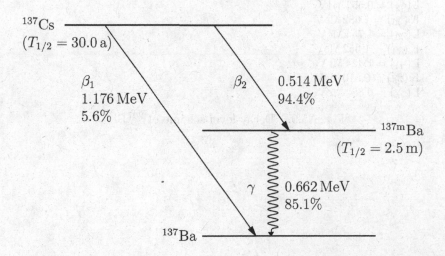

^{137}Cs
$(T_{1/2} = 30.0 \, \text{a})$

β_1
1.176 MeV
5.6%

β_2 0.514 MeV
94.4%

$^{137\text{m}}$Ba

$(T_{1/2} = 2.5 \, \text{m})$

γ 0.662 MeV
85.1%

^{137}Ba

Conversion electrons:
K (γ) = 0.624 MeV
L (γ) = 0.656 MeV

Fig. A5.8.   Decay-level scheme of ^{137}Cs.

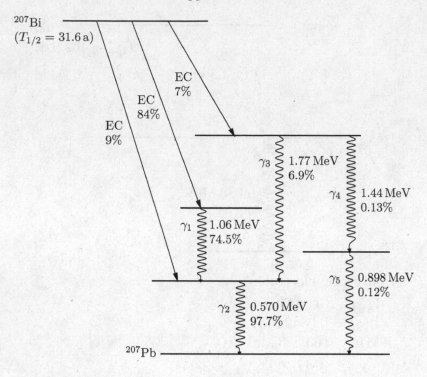

Conversion electrons:
K(γ_1) = 0.976 MeV
L(γ_1) = 1.048 MeV
K(γ_2) = 0.482 MeV
L(γ_2) = 0.554 MeV
K(γ_3) = 1.682 MeV
L(γ_3) = 1.754 MeV
K(γ_4) = 1.352 MeV
L(γ_4) = 1.424 MeV
K(γ_5) = 0.810 MeV
L(γ_5) = 0.882 MeV

Fig. A5.9.   Decay-level scheme of ^{207}Bi.

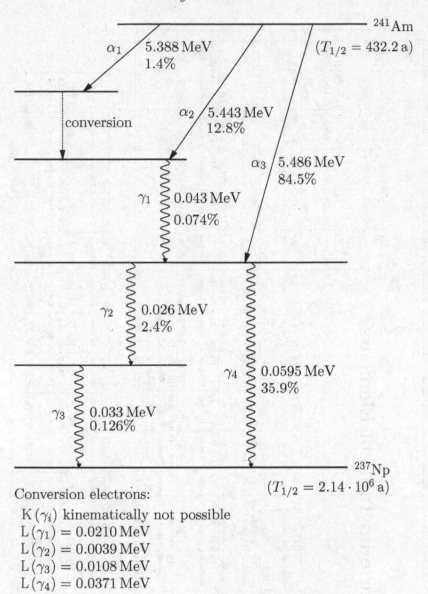

Conversion electrons:

K (γ_i) kinematically not possible
L $(\gamma_1) = 0.0210$ MeV
L $(\gamma_2) = 0.0039$ MeV
L $(\gamma_3) = 0.0108$ MeV
L $(\gamma_4) = 0.0371$ MeV

Fig. A5.10.   Decay-level scheme of ^{241}Am.

Periodic Table of Elements

Group

Ia	IIa	IIIb	IVb	Vb	VIb	VIIb	VIIIb	VIIIb	VIIIb	Ib	IIb	IIIa	IVa	Va	VIa	VIIa	VIIIa
1 H Hydrogen 1.01																	2 He Helium 4.00
3 Li Lithium 6.94	4 Be Beryllium 9.01											5 B Boron 10.81	6 C Carbon 12.01	7 N Nitrogen 14.01	8 O Oxygen 16.00	9 F Fluorine 19.00	10 Ne Neon 20.18
11 Na Sodium 22.99	12 Mg Magnesium 24.31											13 Al Aluminium 26.98	14 Si Silicon 28.09	15 P Phosphorus 30.97	16 S Sulfur 32.07	17 Cl Chlorine 35.45	18 Ar Argon 39.95
19 K Potassium 39.10	20 Ca Calcium 40.08	21 Sc Scandium 44.96	22 Ti Titanium 47.87	23 V Vanadium 50.94	24 Cr Chromium 52.00	25 Mn Manganese 54.94	26 Fe Iron 55.85	27 Co Cobalt 58.93	28 Ni Nickel 58.69	29 Cu Copper 63.55	30 Zn Zinc 65.39	31 Ga Gallium 69.72	32 Ge Germanium 72.64	33 As Arsenic 74.92	34 Se Selenium 78.96	35 Br Bromine 79.90	36 Kr Krypton 83.80
37 Rb Rubidium 85.47	38 Sr Strontium 87.62	39 Y Yttrium 88.91	40 Zr Zirconium 91.22	41 Nb Niobium 92.91	42 Mo Molybdenum 95.94	43 Tc Technetium 97.91	44 Ru Ruthenium 101.07	45 Rh Rhodium 102.91	46 Pd Palladium 106.42	47 Ag Silver 107.87	48 Cd Cadmium 112.41	49 In Indium 114.82	50 Sn Tin 118.71	51 Sb Antimony 121.76	52 Te Tellurium 127.60	53 I Iodine 126.90	54 Xe Xenon 131.29
55 Cs Cesium 132.91	56 Ba Barium 137.33	57-71 Lanthanides 138.91	72 Hf Hafnium 178.49	73 Ta Tantalum 180.95	74 W Tungsten 183.84	75 Re Rhenium 186.21	76 Os Osmium 190.23	77 Ir Iridium 192.22	78 Pt Platinum 195.08	79 Au Gold 196.97	80 Hg Mercury 200.59	81 Tl Thallium 204.38	82 Pb Lead 207.20	83 Bi Bismuth 208.98	84 Po Polonium 208.98	85 At Astatine 209.99	86 Rn Radon 222.02
87 Fr Francium 223.02	88 Ra Radium 226.03	89-103 Actinides 227.03	104 Rf Rutherfordium 261.11	105 Db Dubnium 262.11	106 Sg Seaborgium 263.12	107 Bh Bohrium 262.12	108 Hs Hassium 277.15	109 Mt Meitnerium 268.14	110 Ds Darmstadtium 271.15	111 Rg Roentgenium 272.15							

Lanthanide series	57 La Lanthanum 138.91	58 Ce Cerium 140.12	59 Pr Praseodymium 140.91	60 Nd Neodymium 144.24	61 Pm Promethium 144.91	62 Sm Samarium 150.36	63 Eu Europium 151.96	64 Gd Gadolinium 157.25	65 Tb Terbium 158.93	66 Dy Dysprosium 162.50	67 Ho Holmium 164.93	68 Er Erbium 167.26	69 Tm Thulium 168.93	70 Yb Ytterbium 173.04	71 Lu Lutetium 174.97
Actinide series	89 Ac Actinium 227.03	90 Th Thorium 232.04	91 Pa Protactinium 231.04	92 U Uranium 238.03	93 Np Neptunium 237.05	94 Pu Plutonium 244.06	95 Am Americium 243.06	96 Cm Curium 247.07	97 Bk Berkelium 247.07	98 Cf Californium 251.08	99 Es Einsteinium 252.08	100 Fm Fermium 257.09	101 Md Mendelevium 258.10	102 No Nobelium 259.10	103 Lr Lawrencium 262.11

For each element the atomic number (top left) and atomic mass (bottom) is given. The atomic mass is weighted by the isotopic abundance in the Earth's crust.

Fig. A5.11.   Periodic table of elements.

Index

Underlined page numbers identify a page citation that continues on the pages that follow.

599

Printed in the United States
by Baker & Taylor Publisher Services